Peatland Biogeochemistry
and Watershed Hydrology
at the
Marcell Experimental Forest

Peatland Biogeochemistry and Watershed Hydrology at the Marcell Experimental Forest

Edited by

Randall K. Kolka
USDA Forest Service, MN, USA

Stephen D. Sebestyen
USDA Forest Service, MN, USA

Elon S. Verry
USDA Forest Service, Northern Research Station

Kenneth N. Brooks
University of Minnesota, MN, USA

CRC Press
Taylor & Francis Group
Boca Raton London New York

CRC Press is an imprint of the
Taylor & Francis Group, an **informa** business

CRC Press
Taylor & Francis Group
6000 Broken Sound Parkway NW, Suite 300
Boca Raton, FL 33487-2742

First issued in paperback 2017

Library of Congress Cataloging-in-Publication Data

Peatland biogeochemistry and watershed hydrology at the Marcell Experimental Forest / editor, Randall Kolka ... [et al.].
    p. cm.
Includes bibliographical references and index.
ISBN 978-1-4398-1424-6 (hardcover : alk. paper)
    1. Peatland ecology--Minnesota--Marcell Experimental Forest. 2. Biogeochemistry--Minnesota--Marcell Experimental Forest. 3. Watershed hydrology--Minnesota--Marcell Experimental Forest. 4. Marcell Experimental Forest (Minn.) I. Kolka, Randall.

QH105.M55P43 2011
577.68'7--dc23   2011016706

Visit the Taylor & Francis Web site at
http://www.taylorandfrancis.com

and the CRC Press Web site at
http://www.crcpress.com

Contents

Foreword from the Research and Development Office of the Forest Service

Shortly after the USDA Forest Service was established in 1905, agency leaders began to establish experimental sites to address large-scale problems of forest, range, and watershed management. Concerns about water supplies and water quality led to watershed studies at more than two dozen locations. These projects were the beginnings of today's network of experimental forests and ranges.

Today almost all of the 81 experimental forests are located on national forests. They represent regional landscapes over a very broad range of environmental conditions with sites from Puerto Rico to Alaska. They range from boreal to tropical forests to wetlands to semiarid chaparral to dry desert. This broad spread includes nearly 50° of latitude and reflects a great range of temperature and precipitation.

The subject of this book, the Marcell Experimental Forest (MEF) in Minnesota, was established to study the ecology and hydrology of upland/peatland watersheds. Thanks to 50 years of research and data collection on this site, Forest Service researchers and our partners have developed a fundamental understanding of peatland hydrology, acid rain impacts, nutrient and carbon cycling, trace gas emissions, and controls on mercury transport in boreal watersheds. This fundamental understanding has led to applied research being used to address important policy implications for forestry best management practices, mercury and sulfur emissions, and the impacts of climate change.

The experimental forests continue to serve as living laboratories where Forest Service scientists learn and share results with cooperators and those who care about our forests and wetlands. The network has generated a huge database of information, with some sites boasting data collection for nearly 100 years. Those data sets are invaluable in looking at environmental changes over the last century and in answering many of today's pressing questions at landscape and global scales. The MEF is an exemplary example of leveraging our long-term data into new relevant studies addressing important issues facing us today. Studies at the MEF are using the long-term observations of climate and atmospheric chemistry, carbon and nutrient dynamics, mercury cycling, and ecohydrology to develop new research addressing the effects of climate, land use, and atmospheric deposition changes on forest ecosystems.

The Forest Service's ability to conduct long-term studies on a watershed or landscape scale, often in cooperation with universities and other partners, is one of the greatest strengths of its century-long research and development program. The value of this asset has never been more apparent as scientists around the world work to understand the influence of human activity on local and global environments. Research at the MEF contributes significantly to these efforts. We look forward to another 50 years of successful research on the MEF.

Ann M. Bartuska
Deputy Chief for Research
USDA Forest Service

Michael T. Rains
Director of the Northern Research Station
USDA Forest Service

Foreword from a Scientific Research Perspective

When I was approached to write a foreword for *Peatland Biogeochemistry and Watershed Hydrology of the Marcell Experimental Forest: Results from 50 Years of Research* I had several questions. A quick search of the web answered one of the questions, "What is a foreword?" Second, "Why me?" I have never been to the Marcell Experimental Forest (MEF) and I missed the conference in 2009 that commemorated the first 50 years of the MEF! However, a quick look at the authors and materials in this book made me realize my connections to the many scientists who have worked at the MEF—we have a lot of common interests and the work of the MEF scientists has had a profound influence on my views on peatland hydrology, ecology, and biogeochemistry. I am sure that the table of contents will have the same effect on many peatland scientists—it represents a who's who of peatland research.

The progression of the science at the MEF as described in this book is not unlike that of my own research career. The MEF began as an experimental forest to examine the impacts of forestry in a landscape with a large coverage of peatland ecosystems—mostly bogs and fens that originated in ice-block depressions left among the deposits of the last North American continental ice sheets. This immediately set the MEF apart from many other experimental forests—yes, it had upland forests but it also had much, much more. At the cost of raising the ire of many who study upland forests, the MEF had a much more interesting set of complex issues—the linkages between uplands and peatlands. In the early years, a focus on physical processes at the MEF produced some of the fundamental advances in our understanding of the hydrology of watersheds where peatlands predominate. In fact, much of the early understanding of organic soils (peat) comes from the MEF.

Early in my PhD studies, I went to a conference at the University of Guelph that had as its theme the lowlands and peatlands of the Hudson Bay region. A keynote speaker named Eville Gorham discussed with me the work some of his friends were doing on peatland hydrology at the MEF. The work of Eville's friends became a foundation to peatland hydrology and the publications of Roger Bay and Elon S. Verry are still cited regularly. It was soon after that when my PhD supervisor, Dr. Ming-ko Woo, corrected me, as he often had occasion to do, by giving me one of Roger Bay's papers and saying in his characteristic manner: "Peatlands do not reduce runoff, they have very little storage—read this!"

The breadth of research at the MEF quickly expanded beyond that of its original hydrological and forestry mandate. As described in several chapters, this transition happened because Forest Service scientists saw the MEF as a place for university and other agency scientists to do their research. Beyond hydrology and forestry research, I was reminded through reading this book of the many "firsts" that occurred at the MEF in the areas of peatland biogeochemistry. It was the place where Bob Harriss and his collaborators did some of the first, if not the first, measurements of methane emissions from North American peatlands. This was followed by Patrick Crill's pioneering work on why the methane emissions occurred; Nancy Dise's work on what happened during winter; and Narasinha Shurpali, Shashi Verma, and Robert Clement's study using the eddy covariance technique to continuously measure the flux of methane from peatlands. This was in turn followed by the first eddy covariance measurement of carbon dioxide fluxes for North American peatlands at the Bog Lake fen of the MEF by the same team of scientists. With enough time, some things come full circle. This was apparent to me when I read about some of the newest research at the MEF. I discovered that two of the MEF mercury researchers included a former PhD student of mine (Brian Branfireun) and an academic grandchild (Carl Mitchell)!

One cannot read this book without really appreciating and respecting just how important the research at the MEF has been as a foundation for much of peatland science, particularly from a watershed perspective. Only in Finland was the same effort being made at the same time. I have been told, and it is obvious from some of the references in this book, that there was considerable intellectual exchange between Juhani Paivanen, one of the great Finnish peatland scientists, and Roger Bay. I think it would be fair to divide the thinking on peatlands' hydrological processes and biogeochemistry to before and after MEF—which I did not really appreciate until I read this book. Not only does it include the results of hundreds of papers that I have read and I give to my students to read, but the book obtains a level of synthesis beyond the individual papers. Each chapter is a beautiful synthesis of a set of ideas that pull much of the work together and a few of the chapters recount some very nice hydrological and biogeochemical stories. One also soon realizes that research is still very much alive at the MEF. New research is being put forward and new directions are being forged.

Who should read this book? Anyone who has the slightest interest in the structure and function, the hydrology, and the biogeochemistry of peatlands. My bias toward the book is influenced by my interest in peatland ecosystems, but I am sure that there are those from forestry and silviculture who will read this book and be equally moved by its contribution. Anyone interested in understanding how upland forestry influences both upland and peatland hydrology will find plenty to think about here. While the focus is on the research at the MEF, it has played such a formative role in our thinking about these systems, and although it provides an excellent introduction to the field, there is enough meat to satisfy even the most experienced veterans of peatland research. This book will stand alongside Charman's *Peatlands and Environmental Change* and Rydin, Jeglum, and Hooijer's *The*

Biology of Peatlands as "must" readings for undergraduate summer assistants and incoming graduate students new to peatland research.

I thank, on behalf of the peatland community, the editors and authors for this wonderful book. I will come back to it over and over and have already thought of some new research ideas based on my reading.

Nigel T. Roulet
James McGill Professor of Biogeosciences
McGill University, Montreal

Foreword from a Watershed Management Perspective

The national forests of this country are sustainably managed for multiple uses—water, fisheries, and wildlife, recreation, wood, and range—that provide for people's needs while promoting the health and diversity of forests. A number of complex laws provide the basis for this management. They also require public involvement in the development of project- and forest-level management plans. This ensures that national forests are managed to protect or restore the environment, to maintain or enhance outdoor recreation, and to produce commodities.

In the Lake States there is an abundance of water resources in the form of streams, lakes, wetlands, and groundwater. The Chequamegon–Nicolet National Forest in Wisconsin has over 2,000 mi of streams, 608 lakes greater than 10 ac, 325,000 ac of wetlands, and an abundance of groundwater that interacts with many of these surface waters. Managing a forest for multiple uses results in the manipulation of vegetation or harvest of trees to achieve ecological goals and produce wood products; the construction and maintenance of roads and trails for access and recreation; the construction and maintenance of impoundments for recreation, fisheries, and wildlife; and the suppression or management of fire.

The preceding facts highlight three relevant and important considerations with regard to this book: (1) our national forests are specifically managed to provide an abundant, clean source of water for people and the environment; (2) they are also managed for "other" resources that will very likely affect or be affected by water; and (3) to be successful, such management requires a thorough scientific understanding of Lake States' hydrology.

National forest hydrologists working in the Lake States have a wide range of interesting and challenging tasks. These include serving as local experts about the water resources of their forest; designing and implementing inventory projects to better understand the water resources of the forest; identifying and implementing projects to improve watershed conditions; providing support for other resource management programs such as timber, fisheries, wildlife, recreation, special uses, and engineering; and monitoring to ensure individual projects and Forest Plan implementation have the desired effect of protecting or restoring water resources. While these tasks are many and varied, they can ultimately be distilled into one primary job: integrating the sciences of watershed management and hydrology into management of our national forests.

The Marcell Experimental Forest (MEF) is extremely important to field hydrologists in the Lake States because it provides a scientific foundation for much of their work. Since its inception in 1959, the MEF has been a significant source of fundamental research of water, chemical, and nutrient cycling in forest–peatland watersheds. It has provided a majority of the research concerning the effects of forest management on hydrology and water quality in the Lake States. And, as research at the MEF evolved, it provided important contributions regarding acid deposition, carbon fluxes, and mercury cycling.

The studies on peatland hydrology and physical properties of organic soils provide valuable insights regarding water flow in wetlands, surface–groundwater interactions, differences in water chemistry in relation to water source, and hydrologic differences between peat bogs and fens. The studies of lake and groundwater interactions and water impoundments enhanced the understanding of water movement in the Lake States and provided practical applications for land managers. The outstanding research regarding timber harvesting effects on hydrology and water quality serve as the primary basis for national forests in the Lake States to evaluate the potential impacts of timber harvest on water. These treatments utilizing paired watersheds include the classic upland clearcut harvest of aspen, upland aspen conversion to conifer, and a black spruce bog strip cut.

Without the MEF, the extent of hydrology and watershed management science that would be available for integration with National Forest management in the Lake States would be diminished. That would make my job more difficult and certainly less effective. Over the years, I have used much of the MEF research in the role as local water resource "expert." For example, by making simple conductivity measurements one can generally determine if a stream, lake, or wetland is predominantly surface- or groundwater fed. This tends to be extremely helpful when developing advice for the particular issue at hand. That same concept has been used along with basic water chemistry data and topographic maps to predict the approximate depth to groundwater and likely direction of flow. In a few more specific examples on the forest, we have used the MEF impoundment research to select sites for development and more recently as a basis for removing old impoundments with limited ecological value; we have used the peatland research to protect and restore wetlands by providing for adequate drainage in the upper profile; we have identified lakes susceptible to acidification because they are perched above the groundwater table; and we have used the connection between water chemistry and water source (i.e., groundwater vs. surface water) as one of the foundations for the classification of ecological stream segments on the forest.

The paired watershed study of the effects of upland aspen clearcutting on hydrology and water quality has been particularly useful to national forests throughout the Lake States. Resource professionals and segments of the public remain vigilant about the potential effects of timber harvest on water resources. This study quantified the magnitude and duration of increased water yield and storm runoff associated with such harvest; it showed how peak snowmelt runoff can increase or decrease depending on the proportion of watershed harvested, and it demonstrated that such timber harvesting has limited effects on water quality. These and other results from the MEF have been used for both project- and forest-level analyses to evaluate the potential environmental effects of timber harvest in the Lake States. They have also been used to develop Forest Plan standards or guidelines and project-level design features that will minimize potential adverse impacts to watersheds.

This book provides a concise summary of the rich research history of the MEF. For field hydrologists, it will serve as an invaluable reference and will provide them with a solid scientific foundation in their professional endeavors to protect and manage forest and peatland watersheds of the Lake States.

Dale Higgins
Hydrologist
Chequamgeon-Nicolet National Forest, Wisconsin

Preface

The 50th Anniversary Symposium for the Marcell Experimental Forest (MEF) attracted more than 80 participants from locations across the United States, Canada, the United Kingdom, Israel, and the Netherlands. Although some notable personalities in the history of the MEF were unable to attend, we are happy to report that generations of MEF researchers were there, including Roger Bay (emeritus, USDA Forest Service) who led the scoping effort in 1959 that shortly thereafter led to the establishment of the MEF. If Roger is one of the first MEF persons-of-merit, he is not the last, as exampled by the generation of current undergraduate and graduate students who took a respite from field work at the MEF to attend the symposium.

The symposium event marked the initial presentation of the materials synthesized in this book. The planning had actually started decades earlier when Elon S. Verry, Eville Gorham (emeritus, University of Minnesota), Noel Urban (Michigan Tech University), Steve Eisenreich (Joint Research Centre European Commission), and others started discussions of a book based on the NASA and NSF work at the MEF in the 1980s. A looming 50th anniversary was the final impetus for us to reconsider a book. We have now spent more than two years actively working to scope the book, write chapters, convince others to write to contribute their expertise, coordinate an extensive and impressive group of reviewers, and revise and edit the volume that is now in front of you. By providing financial support for the 50th Anniversary Symposium, the USDA Forest Service, the Blandin Foundation, and the Itasca County Land Department contributed to the development and publication of this book.

The 15 chapters in this book synthesize five decades of research, hundreds of research publications, dozens of graduate theses, and even some previously unpublished studies. The depth and breadth of long-term studies on hydrology, biogeochemistry, ecology, and forest management on peatland watersheds at the MEF are remarkable. Research at the MEF has been at the forefront of many of the disciplines and we are proud to present a synthesis of research for the MEF. The contents of this book are largely focused on landscapes that include peatlands, lakes, and uplands. The topics span pioneering research on hydrology during the 1960s to the innovative studies of atmospheric deposition (1970s); nutrient cycles including carbon and nitrogen (early 1980s); methane emissions (mid-1980s); mercury deposition (1990s); controls on methylmercury production (2000s); and landscape-level carbon storage and cycling (2000s). With the continuation of our existing long-term measurements, initiation of new monitoring programs, and plans for unprecedented studies on climate change advancing, the prospects for future research at the MEF are bright.

As with any large project that spans decades and hundreds of participants, we unfortunately did not have space or time to include every reference or research finding in this book. We apologize to anyone whose work and ideas may have been inadvertently overlooked. We appreciate the contributions of every scientist, graduate student, technician, student worker, volunteer, neighbor, and visitor who has contributed to the success of the MEF! We thank Ann Bartuska, Michael Rains, Nigel Roulet, and Dale Higgins for writing forewords. We gratefully acknowledge the numerous peer-reviewers who provided thoughtful and informed reviews—the chapters have greatly improved. We also wish to acknowledge our departed colleagues, some of whom have recently passed away, who may not be reflected in the author list of the book despite their tremendous contributions to the MEF research program. We thank Clara Schreiber for her invaluable typing and editorial assistance, Marty Jones for his editorial review, Susan Wright (USDA Forest Service) for publication support, and Tim Eaton and Kari Finkler of Eaton and Associates Design Company for graphical support.

Finally, we wish to acknowledge the Forest Service technicians who are largely responsible for the long-term data collection. They have also contributed to every research project that has been completed at the MEF. Their collective efforts exceed the sum of their individual contributions; our research and the MEF are the beneficiaries of their commitment and diligence.

Randall K. Kolka
USDA Forest Service, Grand Rapids, Minnesota

Stephen D. Sebestyen
USDA Forest Service, Grand Rapids, Minnesota

Elon S. Verry
USDA Forest Service, Grand Rapids, Minnesota

Kenneth N. Brooks
University of Minnesota, St. Paul, Minnesota

Editors

Dr. Randall K. Kolka received his BS in soil science from the University of Wisconsin-Stevens Point, and his MS and PhD in soil science with minors in water resources and forest resources from the University of Minnesota. Following the completion of his PhD in 1996, he worked as a research soil scientist with the USDA Forest Service's Southern Research Station on the Savannah River Site in South Carolina, where he conducted research on a multidisciplinary riparian/wetland restoration project. In 1998, he became an assistant professor of forest hydrology and watershed management in the Department of Forestry at the University of Kentucky. In 2002, he became team leader and research soil scientist with the USDA Forest Service's Northern Research Station in Grand Rapids, Minnesota. Currently, he leads a team of scientists addressing the ecology and management of riparian, wetland, and aquatic ecosystems. Dr. Kolka's primary research interests are in both terrestrial and aquatic systems, where he studies water, nutrient, carbon, and mercury storage, cycling, and transport and the effects that forest management, land use change, and climate change have on these processes. He is an adjunct faculty member at five universities, was formerly an associate editor for the journal *Wetlands*, and has published over 100 scientific articles in his career.

Dr. Stephen D. Sebestyen is a catchment scientist whose research focuses on how hydrological and biogeochemical processes interact in ecosystems. He received his BS from Susquehanna University, his MS from Cornell University, and his PhD from the State University of New York College of Environmental Science and Forestry. He started as a research hydrologist with the Northern Research Station of the USDA Forest Service in Grand Rapids, Minnesota, in 2007, and became the third hydrologist to follow Roger Bay and Elon S. Verry on the Marcell Experimental Forest project. He researches the effects of nitrogen pollution on ecosystem functions, carbon cycling, sources of Hg in the landscape, understanding how climate variability affects hydrological processes in forests, and quantifying the effects of landscape disturbance on yields of water and solutes from ecosystems across a range of settings in the United States and abroad. He is adjunct faculty at the University of Minnesota and Bemidji State University.

Dr. Elon S. Verry spent 37 years as a research hydrologist and project leader with the North Central Forest Experiment Station in Grand Rapids, Minnesota, focusing on peatland hydrology and ecology, watershed management, watershed processes (physical and chemical), atmospheric chemistry, riparian forest ecology and management, and stream geomorphology.

He received his BS in forest management and his MS in water resources from the University of Illinois, and his PhD in earth resources from Colorado State University. He has published over 100 scientific articles, book chapters, and a previous book on riparian management. Since "retirement" in 2004, he provides stream restoration services and workshops for Ellen River Partners.

Dr. Kenneth N. Brooks is a professor at the Department of Forest Resources, University of Minnesota, St. Paul. He received his BS in range science and watershed management from Utah State University and his MS and PhD in watershed management from the University of Arizona. Dr. Brooks teaches and conducts research in watershed management as well as forest and wetland hydrology and has authored/coauthored over 100 publications and five books, including three editions of the textbook *Hydrology and the Management of Watersheds*. He is certified as a professional hydrologist by the American Institute of Hydrology.

Contributors

James E. Almendinger
St. Croix Watershed Research
    Station
Science Museum of Minnesota
Marine on St. Croix, Minnesota

Timothy J. Arkebauer
Department of Agronomy and
    Horticulture
University of Nebraska
Lincoln, Nebraska

Paul K. Barten
Department of Natural Resources
    Conservation
University of Massachusetts
Amherst, Massachusetts

Karen B. Bartlett
Institute for the Study of Earth,
    Oceans, and Space
University of New Hampshire
Durham, New Hampshire

Peter C. Bates
Department of Geosciences and
    Natural Resources
Western Carolina University
Cullowhee, North Carolina

Roger R. Bay
Pacific Southwest Research Station
USDA Forest Service
Albany, California

David P. Billesbach
Department of Biological Systems
    Engineering
University of Nebraska
Lincoln, Nebraska

Don H. Boelter
Northern Research Station
USDA Forest Service
Grand Rapids, Minnesota

John B. Bradford
Northern Research Station
USDA Forest Service
Grand Rapids, Minnesota

Brian A. Branfireun
Department of Biology
University of Western Ontario
London, Ontario, Canada

Patrick L. Brezonik
Department of Civil Engineering
University of Minnesota
Minneapolis, Minnesota

Scott D. Bridgham
Center for Ecology and Evolutionary
    Biology
University of Oregon
Eugene, Oregon

Kenneth N. Brooks
Department of Forest Resources
University of Minnesota
St. Paul, Minnesota

Robert J. Clement
School of Geosciences
The University of Edinburgh
Edinburgh, United Kingdom

Jill K. Coleman-Wasik
Water Resources Center
University of Minnesota
St. Paul, Minnesota

James B. Cotner
Department of Ecology, Evolution,
    and Behavior
University of Minnesota
St. Paul, Minnesota

Patrick M. Crill
Department of Geology and
    Geochemistry
Stockholm University
Stockholm, Sweden

Nancy B. Dise
Department of Environmental and
    Geographical Sciences
Manchester Metropolitan
    University
Manchester, United Kingdom

Carrie Dorrance
Northern Research Station
USDA Forest Service
Grand Rapids, Minnesota

Steven Eisenreich
Joint Research Centre European
    Commission
Brussels, Belgium

Art E. Elling
Northern Research Station
USDA Forest Service
Grand Rapids, Minnesota

Daniel R. Engstrom
St. Croix Watershed Research
    Station
Science Museum of Minnesota
Marine on St. Croix, Minnesota

Dawn R. Ferris
Soil Science Society of America
Lexington, Ohio

Peter F. Ffolliott
School of Natural Resources
University of Arizona
Tucson, Arizona

Patricia A. Flebbe
Pacific Southwest Region
USDA Forest Service
Vallejo, California

Jacob A. Fleck
Department of Soil, Water, and
    Climate
University of Minnesota
St. Paul, Minnesota

Avi Gafni
Keren Kayemeth LeIsrael
Jewish National Fund
Makkabim, Israel

Eville Gorham
Department of Ecology, Evolution
    and Behavior
University of Minnesota
St. Paul, Minnesota

David F. Grigal
Department of Soil, Water, and
    Climate
University of Minnesota
St. Paul, Minnesota

D. Phillip Guertin
School of Natural Resources
University of Arizona
Tucson, Arizona

Robert C. Harriss
Houston Advanced Research Center
The Woodlands, Texas

Neal A. Hines
Barr Engineering
Minneapolis, Minnesota

Joannes Janssens
Lambda-Max Ecological Research
Minneapolis, Minnesota

Jeffrey D. Jeremiason
Gustavus Adolphus University
St. Peter, Minnesota

Brian Johnson
Metropolitan Council
St. Paul, Minnesota

Cheryl A. Kelley
Department of Geological Sciences
University of Missouri
Columbia, Missouri

Joon Kim
Global Environment Laboratory
Yonsei University
Seoul, Korea

Jennifer Y. King
Department of Geography
University of California
Santa Barbara, California

Randall K. Kolka
Northern Research Station
USDA Forest Service
Grand Rapids, Minnesota

Richard Kyllander
Northern Research Station
USDA Forest Service
Grand Rapids, Minnesota

Shiang-Yue Lu
Taiwan Forestry Research Institute
Taipei, Taiwan

Tom Malterer
Natural Resources Research
    Institute
University of Minnesota
Duluth, Minnesota

Christopher S. Martens
Department of Marine Sciences
The University of North Carolina at
    Chapel Hill
Chapel Hill, North Carolina

Carl P.J. Mitchell
Department of Physical and
    Environmental Sciences
University of Toronto Scarborough
Toronto, Ontario, Canada

Bruce A. Monson
Minnesota Pollution Control
    Agency
St. Paul, Minnesota

Edward A. Nater
Department of Soil, Water, and
    Climate
University of Minnesota
St. Paul, Minnesota

John L. Neiber
Department of Bioproducts and
    Biosystems Engineering
University of Minnesota
St. Paul, Minnesota

Dale S. Nichols
Northern Research Station
USDA Forest Service
Grand Rapids, Minnesota

Donna M. Olson
Northern Research Station
USDA Forest Service
Grand Rapids, Minnesota

Juhani Päivänen
Department of Forest Ecology
University of Helsinki
Helsinki, Finland

Donald A. Perala
Northern Research Station
USDA Forest Service
Grand Rapids, Minnesota

Steve R. Predmore
Missouri Basin River Forecast
    Center
National Oceanic and Atmospheric
    Administration
Omaha, Nebraska

Daniel I. Sebacher
Atmospheric Sciences Division
NASA Langley Research Center
Hampton, Virginia

Stephen D. Sebestyen
Northern Research Station
USDA Forest Service
Grand Rapids, Minnesota

Narasinha J. Shurpali
Department of Environmental
    Science
University of Eastern Finland
Kuopio, Finland

Kurt A. Smemo
The Holden Arboretum
Kirtland, Ohio

Kelly Smith
Carlton County Soil and Water
    Conservation District
Carlton, Minnesota

Edward B. Swain
Minnesota Pollution Control
    Agency
St. Paul, Minnesota

Noel Urban
Department of Civil and
    Environmental Engineering
Michigan Technological University
Houghton, Michigan

Shashi B. Verma
School of Natural Resources
University of Nebraska
Lincoln, Nebraska

Elon S. Verry
Northern Research Station
USDA Forest Service
Grand Rapids, Minnesota

Peter Weishampel
Department of Natural Resources
Northland College
Ashland, Wisconsin

R. Kelman Wieder
Department of Biology
Villanova University
Villanova, Pennsylvania

Christopher J. Williams
Department of Earth and
    Environment
Franklin and Marshall College
Lancaster, Pennsylvania

Joseph B. Yavitt
Department of Natural Resources
Cornell University
Ithaca, New York

1

Establishing the Marcell Experimental Forest: Threads in Time

Elon S. Verry, Roger R. Bay, and Don H. Boelter

CONTENTS

Introduction

The Marcell Experimental Forest (MEF), carved out of northern Minnesota's aspen, pine, and swamp forests in 1959, turned 50 years old in 2009. Established to investigate the role of peatlands in the northern Lake States region, its instrumented watersheds (Figure 1.1) include nutrient-poor bogs and nutrient-rich fens (organic-soil wetlands) along with glacial-till moraines and deep sandy drift. Unlike most experimental watersheds in the eastern and western mountains of the United States, the MEF is flat, wet, and tied to the regional groundwater system.

The establishment of the MEF and watershed research by the USDA Forest Service in Minnesota parallels the first scientific assessments of forests and water and their interaction. These threads of understanding of this resource date to the mid-1800s when logging in Maine waned and logging in the Lake States supported the westward sweep of settlers into the Midwest and West following the Civil War.

By 1840, loggers were leaving the cutover lands of New England and Maine en route to the Lake States; Yankee farmers homesteaded the cleared areas. In 1864, George Perkins Marsh, President Lincoln's ambassador to Italy and designer of the Washington Monument, also chronicled how man's use of the land changed its water, soil, capacity to produce vegetation, wildlife, and fish. Before becoming a lawyer and diplomat, the versatile

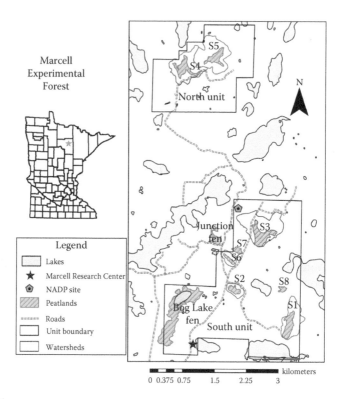

FIGURE 1.1
The MEF in north central Minnesota. The North and South Units are in the Marcell Hills Moraine in central Itasca County, Minnesota. The map shows the six research watersheds (S1–S6), other long-term research sites, and the Marcell Research Center.

Vermonter had many occupations, including logger, lumber dealer, and woolen manufacturer.

Marsh was the first to chronicle the destructive impact of human activities on the environment. Until he published *Man and Nature: Physical Geography as Modified by Human Action*, scientists held that the physical aspects of the Earth were entirely the result of natural phenomena. Marsh suggested that humans were the primary agents of change in the environment. His work, revised in 1874 as *The Earth as Modified by Human Action*, sparked the establishment of forest reserves (1891) that became the national forests (1905).

In the same year that Marsh published *The Earth as Modified by Human Action*, a man who would be instrumental to watershed research in the Lake States and the nation was born in Simbirsk (now Ulyanovsk), Russia. Raphael Zon (Figure 1.2) immigrated to the United States in 1896, studied forestry at Cornell University, and, in 1901, joined Gifford Pinchot, the first chief of the Forest Service, as a forester in the Bureau of Forestry in Washington. While in this post, Zon became interested in the relationship between forests and water. From 1907 to 1920, he was chief of Silvics and Forest Investigation.

FIGURE 1.2
Raphael Zon (1944) and Carlos Bates (1949). Raphael Zon (left) was the first director of the Lake States Forest Experiment Station in St. Paul, Minnesota from 1923 to 1944. In 1927, Bates (right) joined with Zon in St. Paul.

From 1907 to 1912, Zon compiled 1200 references concerning the relationship between forests and water. These references, dating to 1801 (with observations from 1789), represented work in North America, Europe, Russia, India, China, and Africa. Zon's conclusions mirrored those of Marsh. One conclusion was that man and his impact on forest cover and soils changes the quantity and timing of streamflow. In 1923, Zon left Washington to become the first director of the Forest Service's Lake States Forest Experiment Station (LSFES) in St. Paul, Minnesota.

St. Paul was the home of Adolf Meyer, another pioneer in hydrology. A consulting engineer and professor of hydraulic engineering at the University of Minnesota, Meyer published *The Elements of Hydrology* in 1917. A revised edition included precipitation and streamflow data from early U.S. Geological Survey (USGS) streamflow gaging sites in Minnesota and the Dakotas. Meyer concluded that cultural conditions (forest cover or agricultural land) had little impact on retarding the flow of floodwaters; he preferred streamworks of dams and levees to control floods. The grand debate between those who favored forest influences (watershed management), and supporters of hydraulic engineering (dams) began in the early 1900s and persisted into the twenty-first century.

While in Washington, Zon proposed a system of experimental forests and, in 1908, established the Fort Valley Experimental Forest in northern Arizona. A year later, the Fremont Experimental Forest was established near, Colorado Springs, Colorado. Carlos Bates, a graduate forester (University of Nebraska), spent 1907 through early 1909 in Zon's Silvics and Forest Investigations section (Figure 1.2). Like Zon, Bates excelled in both silviculture and forest influences. Zon sent Bates to Colorado in 1909 to open the Fremont Experimental Forest and establish a satellite watershed experiment 480 km away at Wagon Wheel Gap. Bates conducted the world's second controlled (paired watershed) experiment at Wagon Wheel Gap. The first was in Czechoslovakia in 1867 at the upper reaches of the River Becva in Moravia (Nemec and Zeleny 1967).

Following 10 years of paired-watershed calibration at Wagon Wheel Gap, conifers were cut from one of the watershed pairs. The results of that experiment, published in 1927, only fueled the watershed management versus dam debate, which Sartz (1977) discussed in detail. Bates developed the first theories on the movement of water through the soil of mountain watersheds to maintain streamflow during droughts. In 1927, he joined Zon in St. Paul. Beginning in 1928, the Midwest and Great Plains experienced 10 years of consecutive drought as annual precipitation was 7.6–22.9 cm below average. Minnesota had experienced a similar drought from 1882 to 1891 with annual precipitation about 10.2 cm below average. During this period, much of the original white pine in Minnesota and Wisconsin was harvested and sent downstream via the many splash dams that captured the below-average snowmelt. Massive floods in 1892 ended the dry spell, smashed many splash dams, and left the floodplains full of stamped white pine logs released in the rogue log drives. A long summer of log retrieval followed.

During a period of massive soil erosion and dust bowls (1929–1934), Bates focused on the cause and prevention of erosion in southeastern Minnesota and southwestern Wisconsin in an area known as the Driftless Area. In this unglaciated landscape, early farming on steep slopes led to intense gully erosion. In 1934, Bates turned his attention to the dust bowls of the Great Plains as head of the Great Lakes Shelterbelt Project in St. Paul. This was 2 years before multiple, massive floods wiped out the town of Beaver in the Whitewater River basin in southeastern Minnesota. In that flooding, the floodplain was covered with 2 m of fine sand that eroded from Driftless Area gullies. Landscape change, flooding, erosion, and deposition were the norm in 1936 and 1937 even though annual precipitation at St. Paul was 15.2 cm below average.

The Coon Creek basin in southwestern Wisconsin experienced the same erosion and deposition that occurred in the Whitewater River basin. These and other basins that drain into the Mississippi River sent massive amounts of sand into the nation's central transport system; long islands formed in the Mississippi behind dams and locks built for barge transport. It was Coon Creek where Aldo Leopold, considered the father of wildlife management and the U.S. wilderness system, began to apply lessons for healing the land that were developed on his "Sand County" farm near Baraboo, Wisconsin. Coon Creek also was where Hugh Hamond Bennett, known widely as the father of soil conservation, began to incorporate field terraces, contour strip-cropping, and riparian revegetation into farm-management practices in the Midwest.

During the 1920s and 1930s, the public concern grew after Bennett and other environmentalists pointed out the profound relationship between land abuse and the loss of soil, recurrence of floods, and adverse effects on

FIGURE 1.3
In 1956, Zig Zasada (left) was enlisted by Station Director Dick Dickerman (center) to head a station-wide effort to establish experimental forests for the purpose of watershed research (the MEF in Minnesota, Coulee in Wisconsin, and the Udell in Michigan). Sid Weitzman (right) became the first Chief of the Division of Watershed Management in 1956 in St. Paul.

the quality of the nation's rivers and streams. The landmark *Soil Erosion: A National Menace*, authored by Bennett and Ridgely Chapline and published in 1928, was instrumental in the creation of the U.S. Soil Conservation Service in 1935.

In 1933, Zigmond "Zig" Zasada (Figure 1.3) arrived in the tiny town of Bena in north-central Minnesota as foreman of the St. Louis (Missouri) crew in one of the many newly established Civilian Conservation Corps camps. A recent forestry graduate from Syracuse University, Zasada and his crew laid out roads, planted red pine, tended and released pine, and, as the timber sale official on the Chippewa National Forest, conducted the first thinnings in a large area of jack pine and red pine near Cutfoot Sioux Lake and the Winnibigoshish Reservoir on the Mississippi River. Following the removal of jack pine, numerous thinnings, and accelerated growth of red pine, this area was designated as the scenic "Avenue of Pines" on State Highway 46 northwest of Deer River, Minnesota. In 1945, Zasada left the Chippewa and became a forest utilization specialist with the LSFES at St. Paul, where he shared an office with Carlos Bates.

Ralphael Zon retired as director of the LSFES in 1944 and was succeeded by Elwood Demmon. At that time, the Station was located on the third floor of Green Hall at the University of Minnesota. For the next 2 years, silviculture, utilization, soil erosion, and forest influences merged again at Green Hall. A contemporary of Bates was Joe Stoeckeler, a soil scientist who worked with Bates on erosion research in Wisconsin and later was one the first members of the watershed project at Grand Rapids.

Demmon campaigned strongly for research centers outside of St. Paul and established centers in Grand Rapids, Minnesota; Marquette and East Lansing, Michigan; and Rhinelander, Wisconsin. In 1946, Zasada succeeded

Larry Neft as head of the center at Grand Rapids. Zasada directed research on forest economics, entomology, utilization, and silviculture, leaving the center for 2 years to direct the center at Marquette. He returned to Grand Rapids in 1953 and remained there as the center's leader until 1961. It was during this period that Zasada set about adding watershed-management projects at the LSFES.

Murlyn "Dick" B. Dickerman (Figure 1.3), the new station director in St. Paul, added a host of disciplines, including pathology, genetics, engineering, wildlife habitat, recreation, and watershed management, along with office buildings at outlying research centers to house the new staff. The 1957 launching of Sputnik shocked the nation and resulted in an increase in funding for a variety of Federal programs, including basic research in the physical sciences. This favorable development benefited Dickerman, who, since 1954, had backed the development of the Udell Experimental Forest in Cadillac, Michigan; the Coulee Experimental Forest in Lacrosse, Wisconsin; and the MEF near Grand Rapids for watershed research. Zasada was chosen to lead this effort.

Zasada organized a local committee in Grand Rapids that included Wes Libbey of the Izaak Walton League; Charlie Godfrey, the Itasca County land commissioner; H.M. Sword, county commissioner and chairman of the Itasca County Farm Bureau; Buck Hedman, leader of the local realtors association; and George Rossman, publisher of the *Grand Rapids Herald Review*. Local support also was obtained from C.K. Andrews, president of Blandin Paper Co. as well as from Minnesota and Ontario Paper, Sartell Paper Mill, and Northwest Paper Co.

Not everyone saw new knowledge as a need in forestry, but members of the Grand Rapids committee learned how advances in silviculture and watershed management meant real stewardship for Lake States forests and subsequently championed the three watershed research projects. Their efforts paid off when Representative John Blatnik and Senators Edward Thye and Hubert Humphrey (Box 1.1) appropriated $70,000 to hire project leaders for each watershed site (Roger Bay at the MEF, Dave Striffler at the Udell, and Richard Sartz at the Coulee). The funding also enabled the LSFES to establish the position of division director for Watershed Management at St. Paul. In 1956, Sid Weitzman (Figure 1.3), who had been with the Fernow Experimental Forest near Parsons, West Virginia, became the first director of that division. Watershed-management research had been conducted on the Fernow since 1948, and so Weitzman proved an effective mentor to Bay (Figure 1.4), Striffler, Starz, and Stoeckeler, accompanying them on a tour of USDA Forest Service watersheds in Arizona, Colorado, and Utah in 1958.

In 1956, Bay and Weitzman joined the LSFES at St. Paul only days apart. Previously with the National Forest system in Montana, Bay had met Dickerman while earning a Master of Science degree in Forestry from the University of Minnesota. Following discussions with Weitzman, Bay accepted a watershed-research position at Grand Rapids, Minnesota.

BOX 1.1

Working with the local community was one of Zig Zasada's many talents. In 1958, he secured funding for a Forestry Sciences Laboratory in Grand Rapids. The building was dedicated in 1960 with speeches by Richard McArdle, Chief of the USDA Forest Service, and Minnesota Senator Hubert H. Humphrey. While Zasada was driving them to the airport following the ceremony, Humphrey casually inquired if there was anything he could do for the USDA Forest Service when he returned to Washington. McArdle told Humphrey his support would be needed to enact legislation on multiple-use sustained yield that currently was before Congress. A week later, the bill became law, directing that the National Forests be managed for a variety of uses and values, including watershed protection as well as timber, wildlife, rangeland grazing, and recreation. Zasada left Grand Rapids in 1966 to become a staff assistant to the deputy chief of the USDA Forest Service. During the next 3 years, he was instrumental in the establishment of 17 USDA Forest Service research laboratories following numerous, wide-ranging discussions with congressional delegations and university and local officials.

At St. Paul, Stoeckeler had developed a program to study snow accumulation and melt and soil freezing for several forest types and density levels. He also conducted road surveys in northern Minnesota to determine the most effective drainage across roads in swamp areas. In addition to suggesting study areas, Stoeckeler helped Bay set up a small laboratory in Grand Rapids for the collection of soil-moisture data. Study plots were established on the Cutfoot and Pike Bay Experimental Forests on the Chippewa National Forest and on the Big Falls Experimental Forest (State of Minnesota) in large, black spruce peatlands north of Grand Rapids. Bay collected and analyzed the data from snow and soil-freezing measurements and reported results in several publications. Stoeckeler remained in St. Paul working on special projects.

In 1957, Clarence Hawkinson (Figure 1.4) left the Chippewa National Forest to join the staff at Grand Rapids as a forestry technician. He later became the first "Master of the Marcell" as principal field technician. In 1958, Don Boelter, a soil scientist, joined the staff to research the physical properties of organic soils (Figure 1.4).

The Search for Potential Study Watersheds

In 1959, Boelter and Bay began hunting for a site on which to locate an experimental forest for watershed studies. In the relatively flat or rolling landscapes

FIGURE 1.4
Early work at the MEF (clockwise from upper left): Don Boelter preparing organic soil cores for determination of water retention curves; Roger Bay with a local Rotary group; and Clarence Hawkinson changing the chart on an H-type flume at the outlet of S1 watershed and changing the chart on the recording upland well in S3. (Upper left: Roger Bay photographer; upper right: unknown; lower left: Sandy Verry photographer; lower right: Roger Bay photographer.)

of northern Minnesota, it was difficult to determine watershed boundaries. To locate watercourses, Boelter and Bay studied National Forest and county maps and also consulted USGS topographical maps. Upon selecting a potential site, they visited county land offices and studied aerial photographs. On the photos, it was possible to distinguish black spruce stands in swamplands from aspen and other upland forests and thus potential sites on State, county, and National Forest lands. The final test was hiking into areas with small watersheds that contained both a significant amount of peatland and streams suitable for the construction of weirs or flumes for measuring water yield.

Because it was difficult to delineate watershed boundaries and access the large peatlands north of the Chippewa National Forest and within the glacial Lake Agassiz area, Boelter and Bay redirected their efforts north of Grand Rapids in the glacial moraine and outwash area known as the Marcell Hills. This area also had the advantage of short travel times from Grand Rapids and was mostly National Forest land to the south and mostly Itasca County land to the north. In 1959, they settled on a 925 ha area that would formally be established as the North and South Units of the MEF in 1962 (Bay 1962).

In 1960, the first permanent streamflow gage at the MEF was constructed on the outlet stream of the S2 watershed. Unlike other USDA Forest Service experimental watersheds, which used the designation WS, watersheds on the MEF were identified as S1–S8 (S stands for swamp, the common term for wetland forest in the Lake States). Almost immediately, attention was directed toward defining what the Scandinavian terms "bog" and "fen" meant at the MEF with respect to wetland vegetation and water chemistry and to calibrating watersheds for future studies of forest harvesting.

Building a Watershed Research Program in Northern Minnesota

After Dickerman obtained funding to expand the watershed program, the LSFES contracted with Robert Dils, a professor at Michigan State University, to produce a problem analysis for watershed-management research in the Lake States. This was a cooperative effort among the LSFES, Michigan State University, and the Michigan Agricultural Experiment Station. He completed the problem analysis in April 1956. At that time, Stoeckeler reviewed the recommended areas for research and proposed the initiation of various watershed projects.

Several years earlier, the Marcell Ranger District on the Chippewa had constructed an all-weather access road for potential timber sales. The short road led to the future S1 watershed, providing access to most of the South Unit. Access to the North Unit posed a different problem. A rough road had been developed by landowners to access the area during deer-hunting season. Four-wheel drive and luck were needed to drive through mud and snow in the early years. The owners and Itasca County signed a cooperative agreement that allowed use by the experimental forest so long as access during deer hunting season was limited.

Beginning Work on MEF Watersheds

One of the first projects was generating a contour map of the MEF. A survey line was run from a USGS benchmark several miles from the area to establish local benchmarks for the mapping process. Mark Hurd, a private contractor, took aerial photos of the forest and produced a 1.2 m contour map of the area. In the first several years, considerable time was spent constructing flumes and weirs for streamflow measurements and drilling groundwater wells in the bogs and uplands. In the latter, a portable, gas-powered drill was used to auger through the soil and drive a pipe and well point. For the wet peat soil in the bogs, a length of steel pipe (3.2 cm in diameter) was used. A steel circle (with a hole of the pipe diameter), cut on the radius and twisted to form an auger screw, was attached to the end of the well pipe. The well was simply screwed down 6.1–10.7 m by Bay and Boelter.

Streamflow measurements were made with a concrete cutoff V-notch weir that could be placed in solid mineral soil; at other outlets, metal flumes that could be transported on poor forest roads were used. It was believed that the flumes would not pond like the V-notch weir and thus increase the accuracy of measurements. At Lacrosse, Wisconsin, Richard Sartz had been working with a local sheet metal shop to fabricate flumes for use on the Coulee Experimental Forest. The flumes were manufactured in sections and shipped to Grand Rapids. Clarence Hawkinson (Figure 1.4) supervised the manual assembly of the flumes, caulking each section, installing cutoff walls of steel, and placing the flumes in the outlets. Over time, the flumes proved less accurate than V-notch weirs with concrete cutoff walls. In the flat and cold terrain of northern Minnesota, the flumes, supported on large screw rods set in concrete footings, twisted and turned, requiring weekly adjustments to ensure that the flumes and their sheet metal approaches remained level.

Don Boelter developed a series of field and laboratory instruments for measuring the physical properties of organic soils. These included caissons to isolate organic soil in the field where deformation could be avoided in extracting peat samples, and special adaptations of tension plates and pressure chambers in the laboratory to measure water retention in organic soils that are sensitive to changes in hydraulic head. Both auger-hole and piezometer pumping methods adapted to organic soils were used to measure hydraulic conductivity.

Initial research at the MEF emphasized peatland hydrology and ecology and the physical properties of organic soils. It then was expanded to include traditional paired-watershed experiments, with studies of the impact of forest harvest or forest-type conversion on streamflow. Every aspect of the water cycle also has been investigated with respect to water quantity and quality. Table 1.1 provides a timeline of major events on the MEF and Figure 1.4 pictures of early laboratory and field work. In 2004, the MEF increased by an additional 202 ha (total of 1141 ha) following approval of the forest plan for the Chippewa National Forest by the chief of the USDA Forest Service.

Of the three original watershed projects administered by the LSFES, only the MEF maintains an active, wide-reaching research program with fully instrumented watersheds including two control watersheds and four treatment watersheds modeled after the paired-watershed treatments at Wagon Wheel Gap and the Fernow Experimental Forest.

From the beginning, the MEF has fostered cooperative research programs with various universities. Many of the USDA Forest Service scientists

TABLE 1.1

Timeline of Major Events at the MEF

Year	Event
1959	Site selection of the MEF
1961	First routine monitoring of hydrology and climate begins
1961	Field laboratory built
1961	Studies begin to understand peat soil properties, peatland hydrology, and ecosystem energy budgets
1962	Formal establishment of the MEF
1966	First water chemistry samples collected
1967	Ditching experiment in S7 bog initiated
1968	Studies begin to develop the first electric analog model of groundwater flow at S3 fen (Sander 1971)
1969	First stripcutting the S1 bog
1969	Establishment of first upland runoff collectors
1969	Research begins assessing the influence of various canopies on interception
1972	Upland harvest in the S4 watershed
1972	Harvest of the S3 fen
1974	Strips left from first harvest cut in S1 bog
1977	Establishment of atmospheric chemistry sampling that became part of the NADP program in 1978
1978	PHIM (Peatland Hydrologic Impact Model) development begins
1980	Harvest of the S6 upland
1980–1982	Trials of cattle grazing and herbicide applications to convert S6 upland from a deciduous to conifer system
1981	Major research on sulfur, DOC, and nitrogen cycles begins
1983	Gas flux measurements of methane begin
1989	Soil temperature monitoring begins
1990	Establishment of pilot Long-Term Soil Productivity (LTSP) study
1991	Eddy covariance measurement of both carbon dioxide and methane
1993	Mercury studies begin
1993	Establishment of the Mercury Deposition Network (MDN) site
2001	Sulfate deposition experiment on the S6 bog begins
2006	Reestablishment of eddy covariance measurements for carbon dioxide at bog Lake fen
2006	Opening of the Marcell Research Center
2006	Addition of 200 ha to the South Unit of the MEF
2009	Reestablishment of eddy covariance measurements for methane at bog Lake fen
2009	Establishment of study assessing upland forest harvesting effects on mercury cycling
2009	Symposium on 50 years of Research at the MEF
2010	Book produced: *Peatland Biogeochemistry and Watershed Hydrology at the Marcell Experimental Forest: Results from 50 Years of Scientific Research*

at the MEF have been appointed as adjunct professors. The University of Minnesota leads the list in the number of cooperative studies. There are also cooperative programs with the Science Museum of Minnesota's St. Croix Watershed Laboratory, Michigan State University, University of Michigan, Michigan Technological University, University of Wisconsin, North Dakota State University, University of Nebraska, University of New Hampshire, University of Toronto, and Lakehead University, and universities and institutions in Finland, Germany, the Netherlands, Norway, and Russia.

Seeing the potential of new research directions has kept the MEF on the forefront of water research. In 1969, upland runoff plots were established as part of watershed research. During this time, studies were initiated on interception for amount and water chemistry and microclimate. In 1975, work began in cooperation with the Minnesota Agricultural Experiment Station to develop what would become the National Atmospheric Deposition Program. Researchers found an unequivocal link between emissions from coal power plants and acid precipitation and explored the effects of acid rain on lakes in the Lake States region. In 1980, Sandy Verry started worked with researchers at the National Oceanic and Atmospheric Administration to measure greenhouse gases and atmospheric deposition. In 1991, as an MDN test site, the MEF investigated the cycling of total and methylmercury through forest and wetland ecosystems. That same year, it was the first site for studies of soil compaction in the Lake States. During the mid-1980s, the MEF initiated research on carbon pools and cycling that included the major anions and cations, heavy metals, and organic toxins as part of a long-term NSF research program. The MEF also has been the site of one of the most intensive studies of the interaction of mercury and sulfate deposition in North America and has been called the "climate change" experimental forest.

Each year, students prowl the mosquito-infested swamps and forests of the MEF. They now number in the hundreds and hail from numerous states as well as Canada, Finland, Germany, Israel, Korea, the Netherlands, and Taiwan. Housing for students at the MEF, once rudimentary, is critical to study progression, as is the new Marcell Research Center with its modern facilities and living quarters.

The MEF serves as an anchor for water and climate research in the Lake States, a valued partner for research comparing watershed streamflow results in the eastern United States (Hornbeck et al. 1993) as well as for studies of atmospheric chemistry, carbon cycling, mercury in the environment, and climate change. Research at the MEF has settled a significant portion of the debate on forest influences versus engineering: land-use change can affect flows of up to at least the 30 year flood. More importantly, the cycle of erosion and deposition in stream channels is accelerated as they adjust to changes in land use.

Raphael Zon would be pleased that his pioneering efforts to understand the interactions of forests and water continue at the MEF thanks in large part to Zig Zasada, who spearheaded the effort to establish watershed projects in the Lake States. Zig and Marie Zasada spread Zon's ashes in the Cutfoot Experimental Forest in 1956, 3 years before the MEF was established.

References

Bay, R.R. 1962. *Establishment Report, Marcell Experimental Forest (Chippewa National Forest and Adjacent Private, Itasca County, and State of Minnesota Lands)*. Grand Rapids, MN: USDA Forest Service.

Hornbeck J.W., M.B. Adams, E.S. Verry and J.A. Lynch. 1993. Long-term impacts of forest treatments on water yield: a summary for northeastern USA. *Journal of Hydrology* 150:323–344.

Nemec, J.P. and Zeleny, V. 1967. Forest hydrology research in Czechoslovakia. In *International Symposium on Forest Hydrology*, W.E. Sopper and H.W. Lull (eds.). New York: Pergamon press, pp. 31–33.

Sander, J.E. 1971. Bog-watershed relationships utilizing electric analog modeling. PhD dissertation. Michigan State University, East Lancing, MI.

Sartz, R.S. 1977. Carlos Bates: Maverick Forest Service scientist. *Journal of Forest History* 21:31–39.

2

Long-Term Monitoring Sites and Trends at the Marcell Experimental Forest

Stephen D. Sebestyen, Carrie Dorrance, Donna M. Olson,
Elon S. Verry, Randall K. Kolka, Art E. Elling, and Richard Kyllander

CONTENTS

Introduction

Scoping for the Marcell Experimental Forest (MEF) site began in the 1959 (Chapter 1) and the MEF was formally established in 1962 to fill a geographic and ecological void in the experimental forest network of the U.S. Department of Agriculture (USDA) Forest Service (Bay 1962). Six research watersheds were designated and instrumented during the 1960s to study the ecology and hydrology of lowland watersheds with uplands that drain to peatlands. The hydrological and meteorological measurements at these six research watersheds form a core of long-term data on air temperature, precipitation, streamflow, groundwater levels, and water chemistry. New instruments and measurements have been added incrementally to expand baseline monitoring and broaden research themes beyond the original scope of hydrological research. The MEF is one of few long-term research programs on the hydrology and ecology of undrained peatlands in boreal forests. No other site in the Experimental Forest and Range Network of the Forest Service and few sites around the globe have studied the hydrology and biogeochemistry of peatland watersheds with the intensity or longevity as on the MEF. In this chapter, we describe the research sites, report long-term data collected at the MEF, and discuss emerging trends.

Climate

The climate at the MEF is strongly continental with moist warm summers. Winters are dry, cold, and sunny. Air temperatures range from −45°C to +38°C (Figure 2.1). Mean annual temperature since 1961 is 3.4±13.0°C (±1 standard deviation). The monthly mean temperature is 18.9±3.3°C during July and −15.1±8.2°C during January. Lakes begin to freeze in November and are usually ice free by early May.

Mean annual air temperature has increased by 0.4°C per decade (linear regression, $p=0.0005$) since 1961. Most of the warming occurred during winter months (Figure 2.1). Mean air temperature has increased by 0.7°C per decade during winter months from January to March ($p=0.0053$) and by 0.3°C per decade during summer months from June to August ($p=0.0088$).

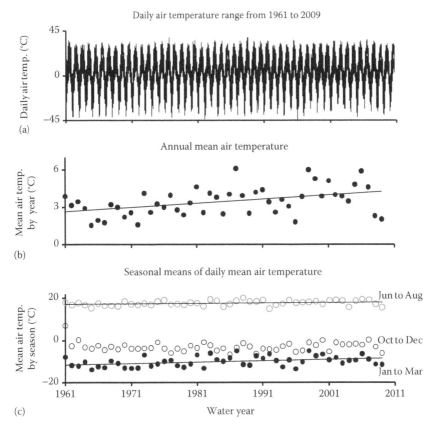

FIGURE 2.1

Daily air temperatures from 1961 to 2009 (a). The mean annual air temperature has increased significantly since 1961 (linear regression, $p \ll 0.05$) (b). The mean air temperature during the January to March and June to August periods has increased significantly since 1961 while no change occurs for the October to December period (c).

Mean air temperature has not significantly changed during autumn months from October through December.

Annual precipitation averages 78.0 cm (± 1 standard deviation of 11.0 cm), ranges from 46.2 to 98.7 cm, and has not significantly changed since 1961 (Figure 2.2a). More precipitation occurs during summer months than winter. For example, mean precipitation is 8.9 cm during August and 2.1 cm during February. About one-third of the precipitation occurs as snowfall. Snow typically begins to accumulate in November or December and melts in late March or April. A mid-winter thaw of several days is common. The maximum annual snow water equivalent under aspen cover has significantly decreased since 1962 ($p = 0.041$) with no change under black spruce cover or in open (treeless) areas (Figure 2.2b). The rate of decline under aspen cover is 1.2 cm per decade.

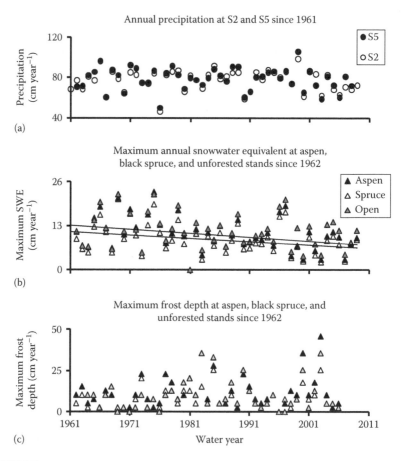

FIGURE 2.2
Annual precipitation during water years has not changed since 1961 (S2) or 1962 (S5) (linear regression, $p > 0.05$) (a). Maximum annual snow water equivalents have decreased under aspen cover since 1961 ($p = 0.041$) with no change under the black spruce canopy in bogs or in the treeless (open grass) site (b). Maximum frost depth under aspen cover, under black spruce cover, and in the grass area has not significantly changed since 1962 (c).

Geographic, Lithological, Geological, and Hydrological Setting

The MEF, which includes uplands, lakes, and wetlands, is spread across two land areas, the North and South Units, situated in the Marcell Hills of north central Minnesota (Chapter 1, Figure 1.1). The MEF was 925 ha at the time of establishment during 1962 (Bay 1962). The size increased to 1141 ha in July

2004 to include land around a new Marcell Research Center headquarters, the Bog Lake fen study area, and other forest plots.

The lakes and peatlands on the MEF formed in ice-block depressions among low-elevation hills that were deposited as glacial moraines and outwash (Wright 1972; Chapter 4). Shallow post-glacial lakes and ice-block depressions slowly filled with organic soils. The peatlands at the MEF include fens, poor fens, and bogs (Gorham 1956, 1990). The organic soils in peatlands are typically less than 3 m deep in glacial lake beds but may exceed 10 m in ice-block depressions. Glacial drifts are 45–55 m thick and form a regional groundwater aquifer above pre-Cambrian Ely greenstone and Canadian Shield granite and gneiss bedrock. The layer directly above the bedrock is 8 m of dense basal till which is overlain by sandy outwash that is up to 35 m thick.

Soils of the MEF were surveyed during autumn 1967 to classify and map soil types (Paulson 1968). Sandy outwash is exposed over one-third of the MEF. These soils include Menahga sands (mixed frigid, Typic Udipsamment), Graycalm loamy sands (isotic, frigid, lamellic Udipsamments), Cutaway loamy sands (fine-loamy, mixed, superactive, frigid oxyaquic Hapludalfs), and Sandwick loamy sands (loamy, mixed, superactive, frigid Arenic Glossaqualf). On the remaining two-thirds of the MEF, the sandy outwash is covered by a clay loam till that contains fragments of limestone and shale and a 10 cm layer of loess. These soils are Warba sandy clay loam (fine-loamy, mixed, superactive, frigid haplic Glossudalfs), Nashwauk sandy loam (fine-loamy, mixed, superactive, frigid oxyaquic Glossudalfs), and Keewatin fine sandy loam (fine-loamy, mixed, superactive, frigid aquic Glossudalfs). Invasive nonnative earthworms have been observed in upland soils throughout the MEF since the 1970s, likely affecting the thickness of the O and A horizons (Hale et al. 2005).

The deep glacial deposits of northern Minnesota form a large regional aquifer. Fens intersect this aquifer and groundwater discharges or laterally flows through the fens. Other peatlands are perched above the aquifer and do not have groundwater inputs from the regional aquifer (Bay 1968, 1969; Verry and Boelter 1975). Clay loams along with a thin layer of glacial flour (silt, very fine sand, and clay) line the peat-filled, ice-block depressions and restrict the vertical flow of water into the underlying sands. Water in such perched peatlands originates solely from precipitation inputs to the perched watershed (Chapters 4 and 7). These wetlands are bogs with surrounding laggs. Hillslopes rise a maximum of 20 m from the nearly flat peatland surfaces to upland summits. However, the bogs are slightly domed to heights about 10–20 cm above the laggs. Water flows to the laggs from both the uplands and raised bogs. On mineral soil hillslopes, the depth to the clay loam soil usually is less than 1 m deep. These clay layers have low hydraulic conductivity and water flows preferentially along lateral pathways in the overlying sandy loams to the peatlands (Timmons et al. 1977; Verry and Timmons 1982; Tracy 1997). See Chapters 4 and 7 for a discussion of flowpath connections among uplands, peatlands, and underlying aquifers.

Atmospheric Deposition

Long-term data collected at the National Atmospheric Deposition Program (NADP) site at the MEF show declines in sulfate wet deposition since 1979 ($p < 0.0001$, Figure 2.3). Total nitrogen in wet deposition at the MEF is in

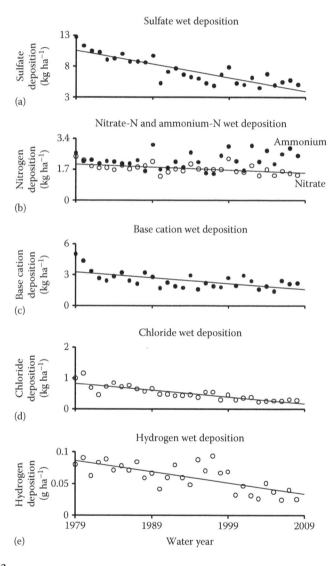

FIGURE 2.3
Wet deposition of sulfate (a), nitrate (b), base cations (c), chloride (d), and hydrogen (e) have decreased (linear regressions, $p < 0.002$) during the decades since 1978.

the same range as that in many forests in the northeastern United States where the effects of nitrogen deposition on ecosystem processes are widely acknowledged (Aber et al. 2003; Driscoll et al. 2003). Ammonium in wet deposition at the MEF has not changed since 1979. Nitrate deposition, which is three- to fourfold larger than ammonium deposition, has decreased since 1979 (Chapter 8). Wet deposition of chloride, base cations (the sum of calcium, magnesium, potassium, and sodium), and hydrogen has decreased since 1979 ($p < 0.0001$). Monson (2009) detected a regional decrease in mercury concentrations in wet deposition during the 1980s and increases from the 1990s to the present. Data from the MEF Mercury Deposition Network (MDN) site were included in that study.

Vegetation and Forest Management

Before European settlement in the late 1800s, white (*Pinus strobus*), red (*Pinus resinosa*), and jack (*Pinus banksiana*) pine were dominant in the overstory canopy, which included mixed northern hardwood species. Today, most forest stands in northern Minnesota are primary successional mixed hardwoods with trembling aspen (*Populus tremuloides*), bigtooth aspen (*Populus grandidentata*), balsam poplar (*Populus balsamifera*), paper birch (*Betula papyrifera*), and balsam fir (*Abies balsamea*). Stands dominated by aspen are prolific on upland soils throughout the region. No remnant virgin pine stands remain on the MEF and most pine stands were planted or regenerated naturally starting during the 1930s (Chapter 12). Red and jack pine grow on sandy upland soils along with mixed stands of aspen, paper birch, balsam fir, white spruce (*Picea glauca*), and red maple (*Acer rubrum*). Basswood (*Tilia americana*), sugar maple (*Acer saccharum*), aspen, and paper birch grow on richer sandy loam tills. A 1968 vegetation survey on the MEF revealed many upland aspen stands that were 50–60 years old (Verry 1969). These stand ages correspond to establishment after large fires during 1915 and 1917 that burned the original logging slash left after clearcutting (Chapters 4 and 12).

Prior to widespread logging of upland forests in the late 1800s and early 1900s, the primary natural disturbances that affected vegetation in northern Minnesota were wildfires, windstorms including tornados, and native insects and diseases. Presettlement fire history of the MEF has not been reconstructed. However, fire was important prior to European settlement as documented by other regional studies (Clark 1990a,b; Heinselman 1999). The effects of past natural disturbance regimes still are evident as many black spruce stands on peatlands are even-aged. This stand structure is suggestive of regeneration after catastrophic disturbance. Blowdown effects are limited to small forest patches. Logging has had a much greater contemporary influence over forest composition and structure than natural disturbance.

At the MEF, upland understory species include mountain maple (*Acer spicatum*), serviceberry (*Amelanchier* spp.), red columbine (*Aquilegia canadensis*), sasparilla (*Aralia nudicaulis*), big-leaved aster (*Aster macrophyllus*), dogwoods (*Cornus stolonifera, Cornus rugosa*), hazels (*Corylus cornuta, Corylus americana*), veiny pea (*Lathyrus venosus*), bracken fern (*Pteridium aquilinum*), dwarf raspberry (*Rubus pubescens*), largeflower bellwort (*Uvularia grandiflora*), lowbush blueberry (*Vaccinium angustifolium*), and arrowwood (*Viburnum dentatum*).

Peatlands may be forested or open (grass or shrub covered with no forest overstory species). When forested, black spruce (*Picea mariana*), eastern tamarack (*Larix laricina*), and northern white cedar (*Thuja occidentalis*) are common overstory species. In the lagg zones that surround raised-dome bogs, speckled alder (*Alnus incana*) and willow (*Salix* spp.) may grow. The peatland understory has a variety of *Sphagnum* species (including *Sphagnum angustifolium, Sphagnum centrale, Sphagnum fuscum*, and *Sphagnum magellanicum*), and bog shrubs including bog birch (*Betula pumila*), leatherleaf (*Chamaedaphne calyculata*), Labrador tea (*Ledum groenlandicum*), cotton grass (*Eriophorum angustifolium*), and bog rosemary (*Andromeda glaucophylla*). Other understory species include grass pink orchid (*Calopogon tuberosus*), pink lady's slipper (*Cypripedium acaule*), round-leaved sundew (*Drosera rotundifolia*), horsetail (*Equisetum arvense*), creeping snowberry (*Gaultheria hispidula*), bog laurel (*Kalmia polifolia*), buckbean (*Menyanthes trifoliata*), Schreber's feathermoss (*Pleurozium schreberi*), haircap moss (*Polytrichum* spp.), pitcher plant (*Sarracenia purpurea*), bog cranberry (*Vaccinium oxycoccos*), blueberry (*Vaccinium angustifolium*), blue flag iris (*Iris versicolor*), and three-leaved false Solomon's seal (*Smilacina trifolia*). Flora in open fens and understory species of forested fens include speckled alder (*Alnus rugosa*), lady fern (*Athyrium filix-femina*), marsh marigold (*Caltha palustris*), sedges (*Carix* spp.), bunchberry (*Cornus canadensis*), jewelweed (*Impatiens capensis*), seeded hop sedge (*Carex oligosperma*), rannock rush (*Scheuchzeria palustris*), and cattail (*Typha latifolia*), as well as many bog species.

Aboveground tree and shrub basal area and stem density of upland and wetland species were estimated during a 1968 vegetation survey of the research watersheds (Verry 1969). Simultaneous measurement of both upland and peatland vegetation on all watersheds has not occurred since 1968 and remeasurement is now a priority. Upland vegetation at the S4 watershed was resurveyed during 1971 after harvest, every 2 years from 1972 to 2002, and during 2008. During 2008, vegetation survey plot centers were monumented for the first time. During all previous surveys, data were collected from the same general locations but not from permanent plots. Black spruce seedlings on the S1 bog were counted during August 1971 to assess reproduction after black spruce stripcutting during 1969 (Chapter 12) and recounted during November 1976 after the leave strips were clearcut during 1974 (Verry and Elling 1978). During September 2009, all trees less than 5 m tall with diameters at breast height greater than 2.4 cm were measured on 163 plots measuring 4.9 m² along a 20 × 20 m grid that spanned the entire S1 bog. Aspen biomass on the uplands of the S6 watershed was measured during 1980 and prior to cattle grazing of the watershed during 1981 and 1982. The basal area of red pine and white spruce on the S6 uplands was calculated during 1983 at the time of planting and heights were remeasured during 1987, 1996, and 1999 on 40 plots measuring 40.5 m². Beyond the research watershed boundaries, 600 points were surveyed along 20 transects across the entire North and South Units to measure aboveground vegetation during 1992 and 1993 (Chapter 9). These data along with measurements of soil and woody debris were used to estimate carbon storage in the landscape. Aboveground biomass of woody, herbaceous, and *Sphagnum* species was measured between 2001 and 2005 on 16 grid points in a 1 km² area of the South Unit that included the S2 watershed (Weishampel et al. 2009).

Instrumentation and Data Collection

Six research watersheds were instrumented in the 1960s (Bay 1962, 1967). Studies at these watersheds were designed to determine effects of forest management practices such as clearcutting of upland, bog, and fen forests as well as the conversion of an upland aspen forest to conifer cover on stream water yields (Chapter 13). Other areas at the MEF are used for short-term, intensive studies and new long-term monitoring sites have been added. At sites such as the S7 peatland ditching experiment (Boelter 1972b, 1972), studies were completed in the 1970s, after which monitoring was discontinued.

Stripchart, Manual, and Electronic Data Recording

Data are recorded from manual field measurements, on paper stripcharts by clock-driven stripchart recorders, and as computer-readable files by electronic dataloggers with time recorded in Central Standard Time. Data are collected and stripcharts are retrieved weekly at soil- and air-temperature stations, precipitation and stream gages, runoff plots, and recording-well sites. Stripchart recorders are cleaned and lubricated and batteries are changed yearly or as needed. Elevations of flume and weir structures as well as water table wells are resurveyed periodically to correct erroneous data that result from frost heaving or settling.

At least 40,000 stripcharts are archived in fireproof safes at the Forestry Sciences Laboratory in Grand Rapids, Minnesota. Stripcharts have not been scanned and the subdaily data have not been digitized except for stream stage and some upland runoff data. The nondigitized hydrological and meteorological data recorded on the stripcharts represent a vast archive of subdaily data that are available for research. As the MEF research budget permits, electronic sensors and data recorders will be obtained to cross-calibrate and eventually replace stripchart recorders.

Meteorological Measurements

A network of three Belfort Universal Recording weighing-type precipitation gages, 17 National Weather Service (NWS) standard 8 in. (20 cm) precipitation gages, and snow courses was setup during the 1960s to quantify water inputs to the research watersheds. Precipitation gages in larger clearings have Alter shields (Alter 1937) and the tops are positioned 1.5 m above the ground. The vegetation around each gage is trimmed as needed to maintain an open forest canopy in any direction extending upward at 45° from the gage orifice. During snow-free months, weekly water levels are measured as the calibrated depth to the closest 0.03 cm of rainfall and the gages are emptied. The gages are charged with antifreeze to melt snow and a thin layer of vegetable oil is added to prevent evaporation during autumn when freezing is first expected, refilled as needed throughout winter, and emptied during spring when temperatures remain above 0°C. When the gage is filled with antifreeze, precipitation is measured weekly with a temperature-compensated spring scale.

Universal weighing bucket gages continuously trace precipitation amounts on stripcharts that are retrieved weekly. Daily precipitation amounts are read from stripcharts in increments of 0.13 cm. At least two standard gages (Brakensiek et al. 1979) are measured in or near each research watershed. These gages are used to weigh precipitation amounts by area for each watershed by the Thiesen polygon method (Brakensiek et al. 1979).

Air temperature is measured at meteorological stations in the uplands and bog at S2 and the uplands at S5. Each station has an NWS louvered shelter (McGuiness et al. 1979) with a Belfort hygrothermograph stripchart recorder, minimum thermometer, maximum thermometer, and Onset HOBO recording digital thermometer. Minimum and maximum temperatures are recorded weekly to verify the continuously recorded air temperature on the hygrothermograph stripcharts. Daily minimum and maximum temperatures are read to the nearest degree Fahrenheit from the stripcharts. Temperatures are converted to degrees Celsius and mean daily air temperatures are calculated by averaging the daily minimum and maximum temperatures. Air temperature, rainfall, and other measurements have been made at automated micrometeorological stations at the Bog Lake fen, the NADP site, and under the aspen canopy in S2 since 2006 or 2007 (Table 2.1).

Snow depth and water equivalent are measured with Mount Rose tubes on snow courses (Figure 2.4) located in each of the dominant forest-cover types (open areas, aspen, aspen-birch, red and jack pine, black spruce, and mixed hardwoods). During the 1960s, measurements were made on as many as 21 snow courses that were distributed among the research watersheds. Measurements were discontinued on many of the courses during the 1970s and two courses were added during 1985. Snow courses under aspen and mixed aspen and birch covers include north, east, south, and west aspect exposures. Each snow course has 10 points with posts for hanging a calibrated weighing

TABLE 2.1

Meteorological Sensors at the Automated Campbell Scientific Meteorological Stations

Measurement	Instrument	Note
Wind speed and direction	RM Young 05103 anemometer	About 2 m above the ground
Photosynthetically active radiation	LI-COR LI190SB PAR sensor	About 2 m above the ground
Air temperature and relative humidity	Vaisala HMP45AC temperature and relative humidity probe	About 1.5 m above the ground
Soil moisture	Campbell Scientific CS616 water content reflectometer	Angled to a 15 cm depth in surficial soil
Surficial soil temperature	Campbell Scientific type-E TCAV soil temperature probe	Vertical in top 5 cm of soil
Rainfall	Texas Instruments TE525WS tipping bucket rain gage	About 1 m above the ground
Snow depth	Campbell Scientific SR50 sonic ranging sensor	S2 site only, about 2 m above the ground
Datalogging	Campbell Scientific CR1000 datalogger	

Note: The stations are located on the South Unit at the Bog Lake fen, the NADP site, and under the forest canopy in the S2 uplands.

FIGURE 2.4
Hydrological technician Art Elling using a Mount Rose snow tube to measure snow depth and water equivalent along the S1 upland snowcourse. The hanging scale is used to weigh the snow tube to determine snow water equivalent. (Photo courtesy of unknown photographer, USDA Forest Service, Grand Rapids, MN.)

scale. Depths in increments of 1.3 cm are measured around each location starting in late February and every 2 weeks thereafter until the snow is gone.

Frost depth is measured at about the time it is maximum, usually in late February. Most sites are measured once during the year except the S2 bog and Junction Fen snow courses where frost depth is measured until peatland ice melts. The frost-depth probe is a Lakes States penetrometer (Stoeckeler and Thames 1958) that was manufactured at a local machine shop from a 1.3-cm-diameter stainless-steel rod with a sharpened and slightly flared point. The bottom 45 cm is marked in intervals of 2.5 cm. The rod is placed on the soil surface and driven through the frozen soil by raising a weight 45 cm and pounding the weight against a smash plate. When the flared point breaks through the frost layer, the rod moves through the soil with less resistance, and frost depth is estimated to the nearest 2.5 cm. If the rod can be pushed into the soil by hand and no frost is encountered deeper, frost depth is recorded as zero. On the rare occasion that frost depth exceeds 45 cm, the depth is recorded as 45 cm.

Maximum snow and frost data are summarized for open areas (i.e., no overstory forest) in uplands and peatlands as well as for upland deciduous, upland conifer (red and jack pine), and bog conifer (black spruce) cover types by averaging data within these five categories. The data are not specific to any one watershed and are representative of the cover types that span the experimental forest.

Streamflow

Streamflow has been measured at as many as seven stream gages. Streamflow currently is measured at five 120° V-notch weirs that replaced earlier H-type flumes or weirs. The V-notch and flume bottoms were set to the elevations of the stream channels that drained bogs. Pools were excavated behind the stream gages and channels were contoured downstream to create a hydraulic drop. Stream gages at the outlets of the S2 and S4 watersheds are downstream from the peatlands. By contrast, stream gages and weir pools at the outlets of the S5 and S6 watersheds are adjacent to the bogs and pooled water backs into the bogs as water levels rise.

Water levels are recorded with a Stevens A35 recorder at the S2 gage and Belfort FW-1 stripchart recorders at the other stream gages. An A35 has a precision of 0.3 cm and an FW-1 has a precision of 0.6 cm. Stage measurements are verified with weekly point-gage measurements (Brakensiek et al. 1979) or a tape measure when spring ice prevents point-gage measurement. Data on stripcharts are stage and time corrected. Archived stripcharts from the S2, S4, S5, and S6 stream gages have been digitized with breakpoint-stage data stored in computer files. Stream discharge is calculated from stage–discharge relationships and daily streamflow is calculated by integrating the hydrograph area.

Insulated wooden shelters with propane heaters are placed over the weirs before the first freeze. If streamflow stops and water levels drop below the V-notch, a heater is turned off to conserve fuel until late spring when the propane lamp is relit to prevent ice formation in the V-notch. Flumes were also enclosed and heated during most winter.

Water Table Elevations

Water table elevations in uplands and peatlands are measured in each of the research watersheds and at Bog Lake fen. Water tables in bogs and fens are recorded at sheltered pools near the center of the peatland (Figure 2.5). Each recording peatland well has a float-driven, Belfort FW-1 stripchart recorder which is placed on a platform that is affixed to three or four galvanized steel posts (Chapter 1). The posts are anchored through peat into mineral soil. Stripcharts are changed weekly. Daily maximum water levels are read from stripcharts, and water levels are reported to the closest 0.3 cm. Recording-well stripcharts have not been digitized. Propane lamp heaters were added to the shelters incrementally between 1990 and 2005 to maintain an unfrozen pool for year-round operation. Before heating, ice was chipped from nearby satellite wells when the recording peatland wells were frozen and daily water levels were extrapolated from water table recession curves.

Non-recording wells made of 3.2-cm-diameter galvanized metal pipe were installed in the uplands and peatlands of most watersheds and monitored for at least several years when ice free. Many of the wells no longer are measured and remain in place. The non-recording wells in the peatlands are anchored in mineral soils beneath the peats (Chapter 1). Upland wells were drilled using mobile drill rigs. Most upland wells penetrate through confining till layers to measure water level in the regional aquifer. The wells terminate in 1.2–1.8 m long sand points. Rudimentary well drilling logs list the drill-hole lithology and well depths. Water table elevations are measured during the first week of every month and are reported relative to mean sea level at a resolution of 0.3 cm. The depth to water table was usually measured with a chalked tape measure. More recently, measurements have been made with an electronic well-depth sounder.

FIGURE 2.5
Hydrological technician Deacon Kyllander changing stripcharts on the FW-1 recorder at the water table well in the S2 bog. The heated shelter covers a small pool of open water and the metal enclosure contains the FW-1 stripchart recorder that is driven by a float, tape, and counterweight system. (South facing photo by A.E. Elling, USDA Forest Service, Grand Rapids, MN.)

Well 305 in the uplands of S3 is a 10-cm-diameter corrugated pipe in which a float and pulley system is used to record water levels. Well 305 is fully penetrating to bedrock and is screened throughout. Water levels are recorded continuously to a resolution of 0.3 cm with a Stevens A71 recorder on Type F stripcharts that are changed every 4 weeks. Daily maximum water levels are reported to the closest 0.3 cm.

Surface and Subsurface Runoff

Surface and subsurface runoff collectors at the S2 and S6 watersheds are based on a USDA Agricultural Research Service (ARS) design (Mutchler 1963). Runoff collectors were measured by the Forest Service for a study initially led by ARS scientists (Timmons et al. 1977). The surface runoff collectors (1.83 m wide and 18–23 m long) are bounded by galvanized sheet metal that is embedded vertically into the surface soil. At the downslope end, a galvanized sheet-metal trough funnels water into a polyvinyl chloride (PVC) pipe that is routed to a holding tank. A 1.82-m-long stainless-steel well point is buried perpendicular to the slope in a trench parallel to the base of the A horizon to capture subsurface runoff above a Bt horizon. Flow from a surface or subsurface collector is routed through a PVC pipe into a 700–800 L polyethylene tank that is enclosed inside a shelter (Figure 2.6). A bottom drain is removed manually to empty the tanks after flow stops, before tanks overflow, or when tanks are overflowing. A tank volume is read from a calibrated staff and recorded when the tanks are drained. Prior to 1982 at S6 and 1981 at S2, the tanks were galvanized metal that was coated with nonreactive paint (Latterell et al. 1974). Water levels in the metal tanks were recorded and total runoff volumes were calculated from stage–volume relationships. During the 1980s, heaters were installed inside the shelters

FIGURE 2.6

The photo of the nearly full tank of the S2S subsurface runoff collector shows the sample collection bucket, floats, counterweights, and water level recorders on March 24, 2009. The surface runoff tank is to the right of the subsurface tank and both are enclosed inside a shelter. The FW-1 stripchart is at the top center and the shaft encoder is near the center. (Photo courtesy of S.D. Sebestyen, USDA Forest Service, Grand Rapids, MN.)

to prevent freezing. Belfort FW-1 stripchart recorders with floats and counterweights have recorded water levels in the S2S and S2N tanks since 1986. Water levels in the S2S tanks have been measured with incremental shaft encoders and recorded every 5 min since April 2008, and since July 2010 at S2N. Each shaft encoder has a resolution of 0.3 cm. Water temperature and specific conductance are measured as runoff water flows through the pipe into the holding tanks.

Verry and Timmons (1982) used recession analysis of the S2 runoff-plot data along with S2 stream water yield to develop equations to separate daily S2 streamflow into bog and upland components. These relationships have been applied to the other research watersheds and estimated daily runoff data are reported on Forest Service (http://nrs.fs.fed.us/ef/marcell/data/) and HydroDB/ClimDB (http://www.fsl.orst.edu/climhy/) websites.

Soil Temperature and Soil Moisture

Temperatures in mineral and organic soils have been measured weekly since 1989 at depths of 5, 10, 20, 30, 40, 50, 100, and 200 cm at a resolution of 0.1°C (Nichols 1998). Temperatures at each depth are read weekly, typically between 10:00 and 13:30 CST, from an Omega handheld model HH-25TC digital type-T thermocouple thermometer. Thermocouples made from insulated type-T (copper/constantan) wire were inserted into holes along a 1.6-cm-diameter wood rod. Rods were inserted into augered holes in mineral soils or holes that were created by inserting and removing an equally wide metal rod in organic soils. Surficial soil (0–5 cm depth) temperatures are measured at the automated meteorological stations at the NADP site, the Bog Lake fen eddy-covariance site, and near the S2 south runoff plots.

Soil water content has been measured in upland mineral soils seasonally since September 1966 by the neutron probe technique (Brakensiek et al. 1979). A Troxler Model 105 Depth Moisture Gage was used until 1990 and a Series 4300 Gage since then. Soil moisture is measured in May before deciduous leaf emergence, in September during autumn leaf fall, and in November when soil water has recharged and soils have not yet frozen. Soil moisture is measured incrementally from 15.2 cm to a maximum of 305.0 cm at intervals of 30.5 cm. The neutron probe access tubes are 3.8-cm-diameter aluminum pipe. The neutron probe is not suitable for measuring soil moisture in the top 15 cm interval. Instead, a soil core is collected and soil moisture is measured gravimetrically. Soil bulk density and soil moisture at 15 bar were measured on soil samples that were collected when the access tubes were installed.

Beginning in 2008 and 2009, volumetric soil water content has been measured along two hillslope transects in S2 using Campbell Scientific CS616 Water Content Reflectometers. Data are recorded every 10 min. The reflectometers and co-located soil-temperature sensors were inserted parallel to the surface slope at the base of a mixed O/A horizon (depth of about 10 cm) and the interface above a Bt horizon (depth of 30–50 cm).

Water Chemistry

Water chemistry samples were first collected during 1964 to measure pH, specific conductance, and calcium concentrations of bog and stream waters at the research watersheds. Samples were collected occasionally during the 1960s and 1970s on a study-by-study basis (Bay 1970; Verry 1975; Timmons et al. 1977) and the number of analytes was expanded to include potassium, magnesium, sodium, phosphorus species, nitrogen species, chloride, sulfate, iron, aluminum, other trace metals, and alkalinity (Verry 1972). By 1969, the chemistry of rain, snow, throughfall, stemflow, upland runoff, soil, ground, stream, and peatland waters had been measured (Verry and Timmons 1977). Routine biweekly sampling of stream or peatland water from fixed locations in some research watersheds began during 1975. Water temperature is measured in the field for each sample and water chemistry is measured on unfiltered water samples. When water is present, peatland water samples are collected by dipping water with a clean ladle from depressions that are lined with 0.1 mm stainless-steel screen. Due to severe droughts and little streamflow at most of the watersheds during 1976 and 1988, few or no samples were collected during those years. At the upland runoff plots, surface and subsurface waters are collected after rainfall or snowmelt events. The runoff samples were dipped from the collection tanks before the 1990s. Water samples are now poured from 5 L stainless-steel pails that are suspended inside the larger polyethylene tanks.

Most samples were analyzed at the Regional Water Quality Lab of the Forest Service at Ely, Minnesota, before 1975; the University of Minnesota Soil Science Laboratory at St. Paul from 1975 to 1995; and the chemistry laboratory of the Forest Service at Grand Rapids from 1995 to the present. Samples from the S2 runoff plots, peatland, and stream were measured at the laboratory of the USDA ARS at Morris, Minnesota, from 1969 to 1973 (Timmons et al. 1977). Water chemistry has been measured with standard methods (APHA 1998). Instruments have been upgraded or replaced and analytical methods have been updated periodically. For example, total nitrogen is now measured colorimetrically by inline digestion and flow injection analysis which supplanted total Kjeldahl nitrogen measurements during 1997. The water chemistry data are currently being compiled from paper records and computer spreadsheets to prepare a comprehensive database of MEF solute concentrations.

Aerial Photographs and LiDAR

Low-altitude oblique aerial photographs of each research watershed have been taken occasionally to document forest cover, forest harvests, and recovery from experimental manipulations. Stereopair vertical aerial photographs from 1959 and 1966 are stored at the Forestry Sciences Laboratory archive at Grand Rapids. An airborne light detection and ranging (LiDAR) survey of ground topography and vegetation height on the South Unit was flown on August 16, 2005 by the Airbone1 Corporation (El Segundo, California). The return data were processed using proprietary techniques and provided to the Forest Service as tiled x, y, z computer files. Mean point density was 0.5 points per square meter.

The MEF Research Sites

The research watersheds were designated as S1, S2, S3, S4, S5, and S6 before differences in peatland types were widely recognized. The "S" before the watershed number corresponds to the "swamp" number. All watershed outlets are accessible by road. Trails through the uplands lead from roads to boardwalks that are used to access stations in the peatlands and minimize disturbance to sensitive wetlands soils and vegetation. When multiple identical instruments are distributed spatially within or among watersheds, the first digit in the site name typically indicates the watershed and the following numbers indicate the order in which the devices were installed. For example, well 201 was the first well installed in watershed 2 and gage 3-1 refers to precipitation gage 1 in watershed 3. In other cases, the letters N, S, E, or W may be added to the name to indicate the cardinal direction of the instrument in relation to the central peatland. For example, S2-E is a soil moisture site to the east of the S2 bog.

Each research watershed has a mineral soil upland and one or more peatlands that range in size from 2.0 to 12.1 ha and cover 12%–33% of the watershed area (Table 2.2). All peatlands have hummock and hollow microtopography (Verry 1984). Water levels are more variable at the S1 bog than the other peatlands. At S1, the water table fluctuates from several centimeters above the peat surface in hollows to a maximum depth of 1.4 m (Figure 2.7). The maximum depth to the water table in the other bog watersheds ranges from 0.7 to 1.0 m. Water tables vary to depths of 0.7 m at the Bog Lake fen and 0.6 m at the S3 fen. Water table elevations in the bogs are more variable than

TABLE 2.2

Areas and Elevations of the Research Watersheds

Watershed	Total Area (ha)	Area of Central Peatland (ha)	Area of Central Peatland (%)	Percent Upland	Outlet Elevation (m)	Maximum Elevation (m)
S1	33.2	8.1	24	76	412	430
S2	9.7	3.2	33	67	420	430
S3	72.0	18.6	23	77	412	429
S4	34.0	8.1	24	76	428	438
S5	52.6	6.1	12	88	422	438
S6	8.9	2.0	23	77	423	435

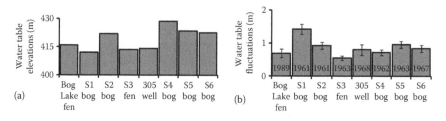

FIGURE 2.7
Mean peatland water table elevations at Bog Lake fen, the research watersheds, and the 305 uplands well (a). The bars show the total fluctuation ranges of the water tables and the error bars show ±1 standard deviation of the mean water elevation (b). The numbers inside the bars indicate the first year of measurements.

in the fens, particularly during droughts and in years that are wetter than normal (Figure 2.8). At the 305 well in the uplands of the S3 watershed, the water table has varied over an interval of 0.8 m since 1967 (Figure 2.7).

Annual precipitation is calculated for water years that begin on November 1 and end on October 31. This period reflects the accumulation of snow that begins in November and does not contribute to streamflow until snowmelt the following spring. Annual water yield via stream runoff is calculated for water years that begin on March 1 and end on the last day of February. The water year reflects the annual hydrologic cycle of Minnesota bog watersheds where intermittent flow begins during spring snowmelt and always stops during winter (Figure 2.9). The stream has flowed at the S2 watershed during every April and May whereas 90% of the Februaries since 1961 have had no streamflow (Figure 2.9b). Streamflow is least likely during February due to freezing air temperatures and the lack of water recharge from the snowpack. Monthly streamflows are highest during April and May in response to melting of the snowpack (Figure 2.9a). Streamflow often stops during summer when rates of evapotranspiration are highest despite the largest monthly inputs of precipitation (Figure 2.9b). About 40% of annual precipitation inputs to the perched watersheds may recharge the regional groundwater aquifer via deep seepage through the clay aquitard (Nichols and Verry 2001). Although annual water yield and streamflow amount during snowmelt have not changed since 1961 (Figure 2.10), snowmelt occurs about 3 weeks earlier in the year. With this change, snowmelt now occurs on average during the third week of April rather than the second week of May (Figure 2.10c).

Daily air temperature and precipitation amount are reported for the South Unit (S2 uplands) and North Unit (S5 uplands) on Forest Service and the HydroDB/ClimbDB websites. Other data reported currently on the websites include daily streamflow, daily peatland water levels, monthly upland water levels, and seasonally-available soil water of upland soils for each of the research watersheds; daily groundwater levels at one site; and maximum annual frost depth, snow depth, and snow water equivalent for hardwood, conifer, and open cover types.

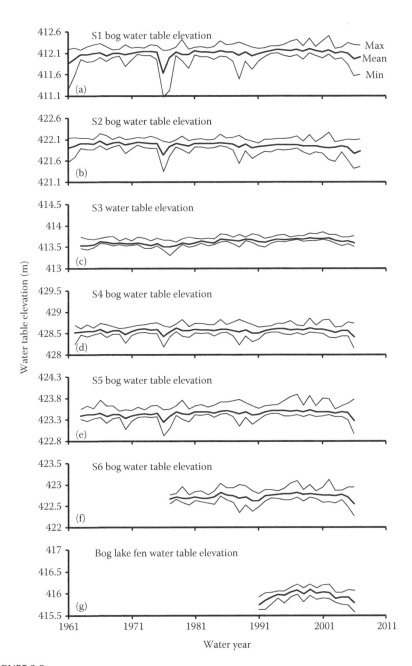

FIGURE 2.8
The mean annual, minimum, and maximum peatland water table elevations at the research watersheds and Bog Lake fen: (a) S1 bog water table elevation; (b) S2 bog water table elevation; (c) S3 bog water table elevation; (d) S4 bog water table elevation; (e) S5 bog water table elevation; (f) S6 bog water table elevation; and (g) Bog Lake fen water table elevation.

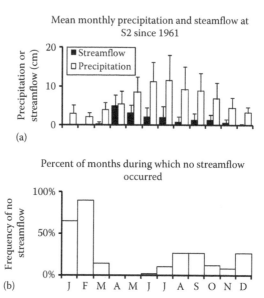

FIGURE 2.9

Mean monthly precipitation and streamflow amounts at the S2 watershed since 1961 (a). The frequency of months during which no streamflow occurs (b). The error bars show ±1 standard deviation.

Soil types and physical properties of mineral (Paulson 1968) and organic (Boelter 1962, 1964a,b, 1965, 1966, 1968, 1969) soils have been described for each research watershed. The chemical properties of peats including carbon, nitrogen, phosphorus, sulfur, calcium, magnesium, sodium, potassium, aluminum, iron, zinc, lead, and mercury content have been measured for various peatlands on the MEF.

S1 Watershed, South Unit

The S1 bog originally was called Cutaway Bog. The instruments, record lengths, and other details for measurements are listed in Table 2.3. The S1 watershed drains to the Prairie River via Cutaway Lake and eventually to the Gulf of Mexico via the Mississippi River. The 8.1 ha peatland is 610 m long, 152 m wide, and surrounded by 25.1 ha of upland hardwood forest (Figure 2.11). The peat fills two adjoining depressions such that the peat is 2–3 m deep near the middle of the bog with deeper pockets to the north and south. The peat is deepest (11 m) near the outlet. The S1 outlet is 412 m above mean sea level and the watershed has a maximum elevation of 430 m above mean sea level. A sand berm separates the S1 bog from an adjacent downgradient bog. Bog water flows through the berm via a stream and lateral subsurface seepage.

The peatland in S1 was used to study the effects of strip clearcutting on black spruce regeneration (Chapter 12), peatland hydrology (Chapters 6 and 13), and harvest effects on micrometeorology (Brown 1972a,b, 1976). During 1969, eight

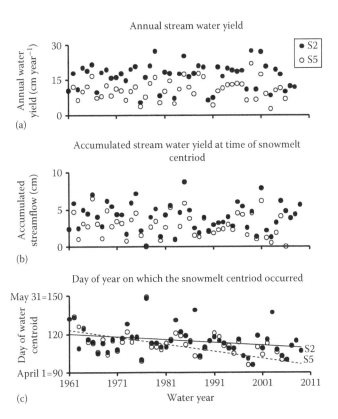

FIGURE 2.10
Annual stream water yield since 1961 (S2) or 1962 (S5) has not changed (linear regression, $p > 0.05$) (a). The accumulated water yield at the centroid of snowmelt has not changed ($p > 0.05$) (b). The day of the snowmelt centroid is about 20 days earlier in the 2000s versus the 1960s at S2 ($p = 0.047$) and S5 ($p = 0.011$) (c). The snowmelt centroid is the day when the accumulated streamflow exceeds 50% of the total flow between the start of snowmelt in March or April and June 1 of any given year, calculated similarly to Hodgkins and Dudley (2006).

33.5 m wide strips were cut and the remaining strips of black spruce were clearcut during 1974. The upland forest had two age classes of mature aspen, 44 and 52 years, with a mean basal area of $22 \, m^2 \, ha^{-1}$ during the 1968 vegetation survey (Verry 1969). Black spruce on the bog also had two age classes, 62 and 73 years. Mean height of black spruce was 12 m, crown closure was 75%, and basal area was $7.0 \, m^2 \, ha^{-1}$ (Verry 1969; Brown 1973). The upland cover has not changed substantially except for aging during the 40 years since the 1968 survey.

Streamflow was measured occasionally at a temporary wooden flume during 1960 and an H-type flume with wood cutoff walls was installed later that year. The cutoff walls heaved during the winter of 1970 and streamflow measurements were curtailed until May 1970 when the walls were again driven into a cemented "ortstein" sand layer. The wood cutoff walls were replaced with interlocking steel piling during October 1974. Ice damage

TABLE 2.3

S1 Instrumentation

Measurement	ID	Instrument	Record Period	Note
Stream stage	S1 flume	0.61 m H-type flume with float driven FW-1 recorder	Daily 9/28/1960 to 12/2/1980	Wood cutoff walls replaced with sheet metal piling during October 1974
Precipitation	1-1	8 in. standard gage	Weekly 1/1/1961 to present	Clearing in black spruce forest
	1-2		Weekly 1/1/1961 to present	Clearing in black spruce forest
	1-3		Weekly 1/1/1961 to present	Clearing in aspen-birch forest
Snow depth, water equivalent, and frost depth	1-1	Mt. Rose snow tube and frost penetrometer	Annually 1962 to 1977	Aspen forest cover
	1-2		Annually 1962 to present	Aspen forest cover
	1-4		Annually 1968 and 1969	Black spruce forest
	1-5		Annually 1962 to 1977	Black spruce forest
	1-7		1972	Black spruce forest
	1-8		1972	Black spruce forest
Bog water table elevation	S1 Bog well a	Float-driven FW-1 recorder	Daily 8/11/1960 to present	In 1969 cut strip, heated since 2004
	S1 Bog well b	Chalked tape measure	Daily 7/25/1969 to 11/15/1972	In uncut strip of black spruce
	Non-recording wells		Occasionally 1961 to 1973	
Upland water table elevations	101	Chalked tape measure or electronic well probe	Monthly 8/21/1962 to present	3.4 m depth
	102		Monthly 8/28/1962 to present	8.3 m depth
	103		Monthly 8/7/1963 to present	18.0 m depth
	104		Monthly 9/17/1963 to 1/15/1973	15.5 m depth
	105		Monthly 11/16/1964 to present	3.9 m depth
	106		Monthly 11/16/1964 to present	5.7 m depth
	107		Monthly 11/16/1964 to 3/7/1967	Depth unknown
Upland soil moisture	S1-N	Neutron probe	Three times a year 9/5/1967 to present	Measured to 3.2 m
	S1-S		Three times a year 9/14/1966 to present	Measured to 3.2 m

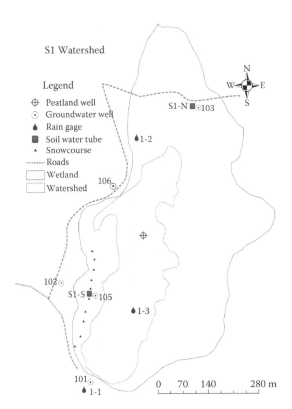

FIGURE 2.11
Map of the S1 watershed showing locations of monitoring equipment.

continued to be a problem for upkeep of the flume until streamflow measurements were discontinued and the flume was removed during December 1980. Mean stream water yield during preharvest water years from 1962 to 1968 was 8.2 ± 4.1 cm (mean $\pm$ standard deviation) and the mean ratio of annual stream runoff to annual precipitation was 0.11 ± 0.03. Biweekly sampling of stream water chemistry began during 1989. Samples are collected about 3 m upstream of the defunct flume cutoff walls.

Precipitation amount has been measured with standard gages at one upland and two bog sites since 1961. Peak annual snow depth and snow water equivalent were measured at two upland snow courses under aspen cover and one peatland snow course under black spruce cover from 1962 to 1977. Measurements still are made on one upland course on the western edge of the bog. Additional courses were measured during 1968, 1969, and 1972 to determine how stripcutting affected snow accumulation and melt. Precipitation, throughfall, and stemflow amounts were measured during 1969 and for several years during the early 1970s (Verry 1976; Verry and Timmons 1977) (Figure 2.12).

Water table elevations in the bog have been measured since 1960. After the 1969 stripcut, the original recording well, then in a clearcut area, was

FIGURE 2.12
Research hydrologist Sandy Verry collecting data at the black spruce stemflow plots in the S1 bog during June 1969. (Photo courtesy of R.R. Bay, USDA Forest Service, Grand Rapids, MN.)

augmented from 1969 to 1972 with a second recording well in an adjoining uncut strip. Water levels were recorded several times a year at 10 other non-recording bog wells from 1961 to 1973. Seven upland groundwater wells were installed from 1962 to 1964. Wells 101, 102, 103, 105, and 106 are measured regularly. Soil moisture is measured three times a year by the neutron probe technique near well 103 at site S1-N and near well 105 at site S1-S.

S2 Watershed, South Unit

The many studies of hydrological and biogeochemical processes at the S2 bog make this site one of the most studied peatlands on the planet. The instruments, record lengths, and other details for measurements at the S2 watershed are listed in Table 2.4. The oval-shaped S2 watershed is 9.7 ha (Figure 2.13). The 3.2 ha peatland is 305 m long and 107 m wide. The S2 outlet is 420 m above mean sea level and the watershed has a maximum elevation of 430 m above mean sea level. Mean stream water yield during water years from 1961 to 2009 was 16.7 ± 5.5 cm and the mean ratio of precipitation to stream runoff was 0.21 ± 0.05. The paleoecology and developmental history of the S2 peatland is described in detail in Chapter 4.

The S2 watershed is a control site that has been compared to forest vegetation and atmospheric deposition experiments on the S1, S3, and S6 watersheds. When inventoried during 1968, mean basal area of stands dominated by aspen was $23.2 \, m^2 \, ha^{-1}$ and the stand age was 50 years (Verry 1969).

TABLE 2.4

S2 Instrumentation

Measurement	ID	Instrument	Record Period	Note
Stream stage	S2 weir	120° V-notch weir with float-driven Stevens A35 recorder	Daily 10/17/1960 to present	Concrete cutoff walls, weir replaced and operational on 7/18/1983
Stream-specific conductance and temperature		Campbell Scientific CS547 sensor	Every 5 min 11/20/2007 to present	Behind V-notch weir plate
Precipitation	South upland/2-1RRG	Belfort weighing bucket gage with stripchart recorder	Daily 1/1/1961 to 9/13/2010	S2 open canopy meteorological station
		ETI Instrument Systems NOAH IV Total Precipitation Gage	Daily 9/13/2010 to present	S2 open canopy meteorological station
	2-1	8 in. standard gage	Weekly 1/1/1961 to present	
	2-2		Weekly 1/1/1961 to present	Clearing in aspen forest
Snow depth, water equivalent, and frost depth	2-1	Mt. Rose snow tube and frost penetrometer	Annually 1962 to present	Aspen-birch cover
	2-2		Annually 1962 to present	Aspen-birch cover
	2-3		Annually 1962 to present	Black spruce cover
Air temperature	South/S2 upland	Belfort hygrothermograph	Daily 1/1/1961 to present	Inside an NWS shelter at the S2 open canopy meteorological station
		Min/max thermometers	Weekly 1/1/1961 to present	
		HOBO Pro series temp sensor	Every 15 min 7/8/1997 to present	
	S2 bog	Belfort hygrothermograph	Daily 6/22/1989 to present	Under black spruce canopy inside an NWS shelter
		Min/max thermometers	Weekly 6/22/1989 to present	
		HOBO Pro series temp sensor	Every 15 min 8/27/1999 to present	
Meteorological station	S2 Forest Met	Campbell Scientific meteorological station, see Table 2.1	Every 30 min 11/6/2007 to present	North-facing slope near S2S runoff collectors and soil moisture sites
Evaporation		Class A evaporation pan	Weekly 1963 and 1964	Beneath the black spruce canopy

(continued)

TABLE 2.4 (continued)

S2 Instrumentation

Measurement	ID	Instrument	Record Period	Note
Evapotranspiration in bog	Plots 3 and 4	Bottomless evapotranspirometers 3.0 m in diameter	Occasionally 1964 to 1966	Measured during growing season, beneath the black spruce canopy
	Plots 2-1, 2-2, and 2-3	Bottomless evapotranspirometers 0.6 m in diameter	Occasionally 1966	
Net solar radiation		Kipp and Zonen pyranometers	Occasionally 1968 and 1969	Above the black spruce canopy
Bog water table elevation	S2 bog well	Float-driven FW-1 recorder	Daily 7/20/1960 to present	Heated since 1992
	Non-recording wells	Chalked tape measure	Occasionally 1961 to 1973	
Peat soil temperature		Type-T thermocouple	Weekly 7/18/1989 to present	Depths of 5, 10, 20, 30, 40, 50, 100, and 200 cm
Upland water table elevations	201	Chalked tape measure or electronic well probe	Monthly 8/7/1962 to 1/15/1973	11.8 m depth
	202		Monthly 8/14/1962 to present	14.1 m depth
	203		Monthly 8/7/1962 to 1/15/1973	15.1 m depth
	204		Monthly 8/21/1962 to 1/15/1973	14.8 m depth
Upland soil moisture	S2-E	Neutron probe	Three times a year	Measured to 2.4 m depth
	S2-S	Campbell Scientific CS616 water content reflectometers	10/31/1967 to present	Measured to 3.0 m depth
	S2-W		Three times a year	Measured to 3.2 m depth
	S2S		4/30/1968 to present	Two depths at down-, mid-, and up-hillslope positions
	S2N		Three times a year 9/14/1966 to 5/6/1974	

Measurement	Site	Method	Frequency/Period	Notes
Upland soil temperature	S2S S2S S2N	Decagon EC-5 sensors Campbell Scientific CS-107 soil temperature sensors	Every 10 min 11/5/2008 to present Every 10 min 11/12/2009 to present Every 10 min 8/12/2008 to present Every 10 min 11/5/2008 to present Every 10 min 11/12/2009 to present	Two depths at down-, mid-, and up-hillslope Co-located with the water content reflectometers
Upland runoff	S2N and S2S S2N and S2S S2N and S2S S2S S2N	Total volumes at surface and subsurface runoff tanks Total volumes at surface and subsurface runoff tanks Tank water depths with float-driven FW-1 recorder at surface and subsurface runoff tanks Tank water depths at surface and subsurface runoff tanks with datalogged shaft encoder Tank water depths at surface and subsurface run-off tanks with datalogged shaft encoder	Event-based 1969 to 1973 Event-based 2/19/1981 to present 3/18/1986 to present Every 5 min 4/25/2008 to present Every 5 min 7/4/2010 to present	Metal tanks Polyethylene tanks Most stripcharts have not been digitized
Upland runoff specific conductance and temperature	S2S	Campbell Scientific CS547 sensor	Every 5 min 11/22/2008 to present	
Soil respiration		LI-COR LI-8100 soil respiration chamber	Every 30 min 7/14/2009 to present	

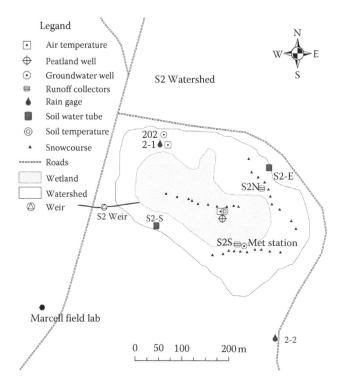

FIGURE 2.13
Map of the S2 watershed showing locations of monitoring equipment.

Mean age of black spruce in the S2 peatland during 1968 was 99 years and basal area was 13.3 m² ha⁻¹. The upland and bog forest cover types have not changed substantially except for aging since the 1968 survey.

A 120° V-notch weir was constructed and stripchart recording was implemented during October 1960. The original S2 weir was removed and the current V-notch weir and concrete cutoff walls were completed on July 18, 1983. In addition to streamflow measurements, specific conductance and stream water temperature have been measured since November 2007 with a probe that is suspended behind the V-notch and recorded every 5 min.

A small area of about 0.06 ha within the S2 uplands was clearcut for the South meteorological station where measurements began during 1961. This station includes an NWS shelter with minimum and maximum thermometers and a Belfort hygrothermograph to record air temperature and relative humidity. A temperature sensor has recorded air temperature every 15 min since 1997. Precipitation amount has been measured with a recording weighing bucket gage and a standard gage (site 2-1) since January 1961 at the South station. Weekly precipitation also is measured with a second standard gage (site 2-2) that is southeast of the S2 watershed. Snow depth, snow water equivalent, and frost depth are measured along a snow course transect that

FIGURE 2.14
Hydrological technician Clarence Hawkinson measuring water levels in the S2 evaporation pan during August 1963. (Photo courtesy of D.H. Boelter, USDA Forest Service, Grand Rapids, MN.)

extends from east to west across most of the bog. Similar measurements are made at upland snow courses on north- and south-facing hillslopes.

Air temperature, relative humidity, snow depth, volumetric soil-water content, rainfall, wind speed, wind direction, and forest-floor soil temperature have been measured every 30 min since November 2007 at a north-facing slope under the aspen canopy. Air temperature also has been measured in the S2 bog under the black spruce canopy with minimum and maximum thermometers since 1989, a Belfort hygrothermograph since 1989, and a HOBO temperature logger since 1999 inside an NWS shelter. Peat temperatures have been measured weekly in organic soils since 1989.

During 1963, two cylinders of galvanized sheet metal (0.9 m tall and 3.0 m in diameter) were inserted into the peat to form bottomless evapotranspirometers (Bay 1966). The top edges were set flush with hummock tops. Water levels were measured weekly inside an observation well to monitor changes in water table elevation from 1963 to 1966. Three additional evapotranspirometers (0.6 m in diameter) were installed and monitored during 1966. An NWS Class A evaporation pan was monitored several times a week during 1963 and 1964 (Figure 2.14). Total and reflected solar radiation were measured with Kipp and Zonen pyranometers that were suspended on a 39.2 m long cableway above the black spruce canopy in S2 (Berglund and Mace 1972). Measurements were made occasionally between August 1968 and August 1969 as instruments cycled back and forth along the cableway that was suspended 13.1 m above the bog surface from two galvanized steel towers.

Water table elevations are reported since July 1960 from a recording bog well site near the high point of the bog dome. Water table elevations were measured several times a year at eight other non-recording bog wells from 1961 to 1973. Four upland groundwater wells were installed during 1962. Water levels have been measured monthly at well 202 on the north side at the South meteorological station since 1962. Other wells to the west (201 near the weir), east (203), and south (204) of the bog were monitored from 1962 to 1972. Paired surface and subsurface runoff collectors were installed on the north and south hillslopes of the S2 bog during 1969 (Timmons et al. 1977). Site S2N is the south-facing runoff plot to the north of the bog and S2S is the north-facing runoff plot south of the bog.

Soil moisture in mineral soils is measured three times a year by the neutron probe technique north (S2-E since 1967) and south (S2-S since 1968) of the bog along the watershed boundaries. Another soil-moisture site, S2-W near the weir, was measured from 1966 to 1975. Long-term measurement of volumetric soil moisture and temperature at down-, mid-, and upslope positions along a north-facing transect began during August 2008. Measurements are recorded every 10 min. Identical measurements at a south-facing transect began during November 2009. These monitoring sites are near each of the upland runoff collectors.

Precipitation, throughfall, and stemflow amount and chemistry were measured during the early 1970s and 1990s at S2 (Verry and Timmons 1977; Kolka et al. 1999). Regular biweekly sampling of stream water began during 1975. Samples are collected at the weir or from a depression in the lagg where the stream coalesces into the stream that flows to the weir. Stormflow samples have been collected at the V-notch with an ISCO automated water sampler since April 2008. Automated sample collection is triggered by threshold changes of stage. A monthly water sample has been collected from well 202 since 1987. Surface and subsurface event water samples from the S2N and S2S runoff collectors were analyzed during 1969, from 1971 to 1973, and since 1981. Higher-frequency runoff samples have been collected with ISCO automated water samplers at S2S since March 2008. Sample collection is triggered as water accumulates in the holding tank. These samples may be collected minutes, hours, or days apart depending on runoff rates. Event sampling of soil water from zero-tension lysimeters began during August 2009 after a year-long equilibration period of the samplers. The soil lysimeters were installed at about 10 cm depths in three soil pits along with soil moisture sensors located near the S2S runoff plots. Lysimeters were installed near the S2N runoff plots during November 2009 and event soil water sampling began during 2010 after equilibration.

Organic soil cores were collected at S2 to study peat physical properties (Boelter 1968; Nichols and Boelter 1984; Grigal et al. 1989), evaporation (Nichols and Brown 1980), paleoecology (Chapter 4), peat accumulation rates (Wieder et al. 1994; Gorham et al. 2003), trace metal accumulation rates (Urban et al. 1987, 1990), cation content (Urban et al. 1995), mineralization rates (Bridgham et al. 1998), phosphorus release rates (Knighton and Stiegler 1981), litter decomposition (Farrish and Grigal 1985), peat decomposition

(Farrish and Grigal 1988), atmospheric deposition of DDT (Rapaport et al. 1985), plant uptake of potassium (Buttleman and Grigal 1985), peatland sulfur transformations (Urban et al. 1989; Novák and Wieder 1992), methane production (Yavitt et al. 1997, 2000, 2005), and anaerobic methane oxidation (Smemo and Yavitt 2007). Continuing a history of chamber-based flux measurements of carbon trace gases at S2 (Harriss et al. 1985; Sebacher et al. 1986; Crill et al. 1988; Dise 1992, 1993), an automated soil respiration chamber was deployed at S2 during 2009 to measure carbon dioxide fluxes from a south-facing upland site every 30 min.

S3 Watershed, South Unit

The S3 watershed is 72.0 ha with an 18.6 ha fen. The fen is 610 m long and is 305 m wide where a lobe juts toward the east (Figure 2.15). Some publications incorrectly refer to the fen as the S3 bog. The instruments, record lengths, and other details for measurements at the S3 watershed are listed in Table 2.5. The S3 outlet elevation is 412 m above mean sea level and the maximum elevation in the watershed is 429 m. The stream draining the S3 fen flows perennially unlike streams in the other MEF research watersheds. Streamflow from S3 drains to Wilson Lake via a series of downgradient peatlands and eventually flows to the Mississippi River.

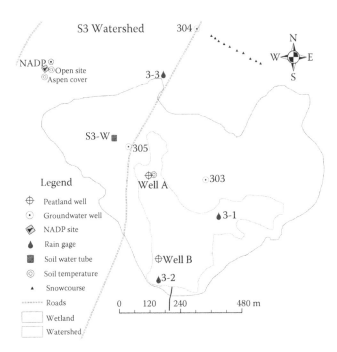

FIGURE 2.15
Map of the S3 watershed showing locations of monitoring equipment.

TABLE 2.5

S3 Instrumentation

Measurement	ID	Instrument	Record Period	Note
Stream stage		Stage measurement with float-driven FW-1 recorder	Daily 8/17/1967 to 11/26/1975	In an open channel during ice-free periods
Precipitation	3-1 3-2 3-3	8 in. standard gage	Weekly 1/1/1961 to present Weekly 1/1/1961 to present Weekly 1/1/1961 to present	Clearing in alder and brush Clearing in black spruce forest Clearing in aspen forest
Snow depth, water equivalent, and frost depth	3-1 3-2 3-3	Mt. Rose snow tube and frost penetrometer	Annually 1962 to 1972 Annually 1962 to 1972 Annually 1962 to 1972	Aspen/jack pine-birch/ balsam fir cover Aspen-jack pine/ jack pine-birch/ birch cover Black spruce cover
Fen water table elevation	S3 fen well A S3 fen well B Non-recording wells	Float-driven FW-1 recorder Chalked tape measure	Daily 9/1/1960 to present Daily 10/26/1966 to 11/28/1967 Occasionally 1961 to 1973	Heated since 1993
Peat soil temperature		Type-T thermocouple	Weekly 12/6/1989 to present	Depths of 5, 10, 20, 30, 40, 50, 100, and 200 cm
Upland water table elevations	301 302 303 304 306 307 308 309 310 311 312 313 305 recording well	Chalked tape measure or electronic well probe Float-driven Stevens A71 recorder	Monthly 8/7/1962 to 1/15/1973 Monthly 8/7/1962 to 1/15/1973 Monthly 10/23/1962 to present Monthly 8/21/1962 to present Monthly 9/1/1967 to 1/21/1971 Monthly 9/22/1967 to 1/21/1971 Monthly 9/22/1967 to 1/21/1971 Monthly 7/14/1968 to 1/21/1971 Monthly 5/27/1969 to 1/21/1971	5.4 m depth 7.3 m depth 16.7 m depth 11.8 m depth 19.8 m depth 5.3 m depth 18.3 m depth 6.2 m depth 5.7 m depth 9.1 m depth 7.1 m depth 13.7 m depth

TABLE 2.5 (continued)

S3 Instrumentation

Measurement	ID	Instrument	Record Period	Note
			Monthly 5/27/1969 to 1/21/1971	
			Monthly 7/11/1969 to 10/2/1970	
			Monthly 8/8/1969 to 10/2/1970	
			Daily 8/23/1967 to present	
Upland soil moisture	S3-E S3-W	Neutron probe	Three times a year 9/14/1966 to present	Measured to 3.2 m Measured to 2.9 m
			Three times a year 9/14/1966 to present	

The upland forest on the north side was logged during the early 1960s as part of a Chippewa National Forest timber sale that was in progress while the MEF was established. The clearcut area was control-burned on October 4, 1963 and subsequently planted with red pine and white spruce. The remainder of the S3 uplands had aspen and two stands of different-aged jack pine during the 1968 vegetation survey when mean basal area in uplands was 37.6 m^2 ha^{-1} (Verry 1969). Prior to clearcutting on the fen, overstory vegetation in the fen included black spruce with balsam fir, tamarack, bog birch, and northern white cedar. The black spruce trees were 70–100 years old and 15–18 m tall with a basal area of 14.0 m^2 ha^{-1}. The fen was clearcut during the winter of 1972 to 1973 to study black spruce and white cedar regeneration (Chapter 12). Logging slash was scattered and burned on July 13, 1973 on 12 ha (86% of the surface) of the fen. The remaining 6.6 ha were not burned. During March 1974, 400,000 black spruce seeds per hectare were dispersed across the entire fen and 130,000 cedar seeds per hectare were sown in five 6.1-m-wide strips in the burned areas. Today, black spruce, alder, and willow are the most abundant overstory species. Recruitment of white cedar was poor as mortality was nearly 97% between 1974 and 1987 when black spruce and white cedar stocking were inventoried.

Hydrological monitoring began at S3 during 1961 when 10 peatland wells were installed. Peatland well A, the long-term site for peatland water table measurements, was instrumented with an FW-1 stripchart recorder. The other nine fen wells were non-recording and measured occasionally until 1973. Water levels at a second site, well B, were recorded on stripcharts from October 1966 to November 1967. Peat temperatures have been measured weekly near the recording well since December 1989.

Upland well number 302 was drilled in S3 as a part of a demonstration of a Hossfeld Prospecting drill in 1962. Upland wells (301, 303, and 304) were installed along the west and north sides of the fen during 1962 and other wells were drilled from 1967 to 1969. Wells 303 and 304 have been monitored monthly until the present. Other wells were drilled as part of seismic refraction study (Sander 1978; Chapter 4) and monitored for several years to calibrate an electric analog model of hydrology at the S3 fen (Sander 1976). Water levels at well 305 have been recorded on stripcharts since August 1967 and maximum daily water levels are reported. Soil moisture has been measured three times a year to the west of the fen (site S3-W) since 1966. Soil moisture was measured east of the fen (site S3-E) from 1966 to 1970.

Vegetation was cleared along a 30-m-long path downgradient from the S3 outlet. This 3–5-m-wide strip was ditched using dynamite during the summer of 1967 to create a channel in which to measure surface water flow. Plywood cutoff walls were erected to form a stable control area within the ditch. Water velocity and level were measured occasionally to develop a predictive equation for stream water yield using the regression between water stage and discharge. Water level in the ditch was recorded on stripcharts during the ice-free period from 1967 to 1975. Since this calibration period, the stream has not been gaged because beaver repeatedly built dams downstream of S3. The dams elevated water levels in the S3 drainage ditch. A regression between water level at the fen well in the S3 peatland and streamflow has been used to calculate monthly and annual water yields from S3 from 1963 to the present. Mean annual stream runoff at S3 was 6777 ± 3994 cm from 1964 to 2006. This value is the specific discharge relative to the surface delineation of the watershed and is large because groundwater inputs are from an aquifer that extends beyond the delineation of the watershed based on surface topography.

Precipitation inputs have been measured at three rain standard gages since 1961. Gage 3-1 is in the fen in the southeast lobe, 3-2 is in the fen near the watershed outlet, and 3-3 is off Wilderness Lake Road beyond the northern watershed boundary. Snow depth, snow water equivalent, and frost depth were measured from 1962 to 1972 along a snow course that extended across the fen and two upland courses alongside the fen. Snow depth, snow water equivalent, and frost depth have been measured since 1985 in a red pine stand that is north of the S3 watershed.

Routine stream water samples were collected biweekly from 1975 to 1987 and from 1994 to the present. Samples are dipped with a ladle from the slowly flowing water where several stream channels coalesce into the S3 outlet stream.

S4 Watershed, North Unit

The S4 watershed is 34.0 ha and has a maximum elevation of 438 m above mean sea level. An 8.1 ha diamond-shaped peatland (Figure 2.16) has a black

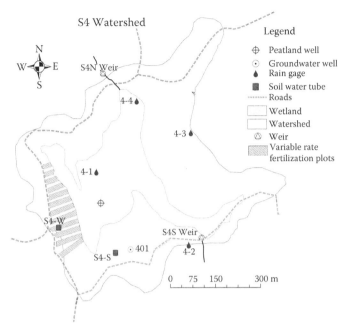

FIGURE 2.16
Map of the S4 watershed showing locations of monitoring equipment.

spruce forest that surrounds a 0.4 ha open bog and a small perennial pond (0.02 ha). The pond greatly expands in size during snowmelt. The instruments, record lengths, and other details for measurements are listed in Table 2.6. The S4 watershed has more sections of road within the boundaries than other MEF watersheds. The watershed sits atop the continental divide of the Mississippi and Hudson Bay drainages. Surface water flows from two outlets at 428 m above mean sea level and streamflow is measured at both outlets. About 70% of the stream water flows from the north outlet (stream gage S4N) to the Hudson Bay drainage with the rest flowing through the south outlet (stream gage S4S) to the Mississippi River (Verry 1972). Streamflow from both weirs is added to calculate the stream water yield from the entire S4 watershed. Mean stream water yield was 19.4 ± 5.1 cm year^{-1} and the runoff ratio was $46\% \pm 9\%$ prior to upland clearcutting which began during 1970.

The 25.9 ha uplands of the S4 watershed were harvested progressively between 1970 and 1972. The upland forest was predominantly a 51-year-old mature aspen stand at the time that clearcutting began. About 2% of the upland forest was a mixed cover of 57-year-old paper birch and aspen. Basal area was 21.8 m^2 ha^{-1} and the S4 upland forest was understocked relative to the other research watersheds at the MEF during the 1968 vegetation survey (Verry 1969). Merchantable timber was removed from the S4 uplands during 1970 and 1972 (Chapter 13). All remaining non-merchantable trees on the

TABLE 2.6

S4 Instrumentation

Measurement	ID	Instrument	Record Period	Note
Stream stage	S4N flume	0.61 m H-type flume with float-driven FW-1 recorder	Daily 3/29/1962 to 5/8/1984	Metal cutoff walls; measurements stopped 5/1984 when water was leaking and resumed when the flume was replaced with the weir
	S4N weir	120° V-notch weir with float driven FW-1 recorder	Daily 11/9/1984 to present	Concrete cutoff walls
	S4S weir		Daily 4/11/1962 to 4/16/65	Metal cutoff walls
	S4S flume	120° V-notch weir with float driven FW-1 recorder	Daily 4/16/1965 to 11/1/1980	Concrete cutoff walls
	S4S weir	0.45 m H-type flume with float driven FW-1 recorder	Daily 11/3/1980 to present	
		120° V-notch weir with float-driven FW-1 recorder		
Precipitation	4-1	8 in. standard gage	Weekly 1/1/1962 to present	Clearing in black spruce forest
	4-2		Weekly 1/1/1962 to present	Clearing in aspen forest
	4-3		Weekly 1/1/1962 to present	Clearing in aspen forest
	4-4		Weekly 1/1/1962 to present	Clearing in black spruce forest
Snow depth, water equivalent, and frost depth	4-1	Mt. Rose snow tube and frost penetrometer	Annually 1962 to 1972	Aspen cover
	4-2		Annually 1962 to 1972	Aspen cover
	4-3		Annually 1962 to 1972	Black spruce cover/open (treeless)
Bog water table elevation	S4 bog well	Float-driven FW-1 recorder	Daily 11/21/1961 to present	Heated since 2002
	Non-recording wells	Chalked tape measure	Occasionally 1962 to 1973	
Upland water table elevations	401	Chalked tape measure or electronic well probe	Monthly 9/17/1963 to present	16.8 m depth
Upland soil moisture	S4-E	Neutron probe	Three times a year 5/1/1968 to 11/8/1971	Measured to 2.6 m
	S4-S		Three times a year 9/14/1966 to present	Measured to 2.9 m
	S4-W		Three times a year 5/1/1968 to present	Measured to 2.3 m

uplands larger than 8.9 cm in diameter at breast height were cut during the summer of 1972. The black spruce forest on the central peatland of S4 was not harvested. Clearcutting northeast and west of the peatland began during December 1970 and continued through snowmelt in 1971. Half of the upland area, or 34% of the total watershed area, was clearcut before snowmelt in 1971. Cutting resumed the following autumn. When completed during January of 1972, 71% of the total watershed area was clearcut. The regrowing aspen forest was fertilized during 1978 with an ammonium-nitrate fertilizer that was applied by helicopter. Fertilizer was applied at 340 kg ha^{-1} across the uplands except in an area of nearly 3 ha that was reserved to study variable application rates. This study of fertilization at variable rates had 12 blocks in which fertilizer was applied in replicated amounts of 0, 168, 336, and 504 kg ha^{-1} (Berguson and Perala 1988). The fertilized plots were 15.2 m wide and extended west from the peatland upslope to Wilderness Trail Road. The plots were separated by buffers that were 10.0 m wide.

Streamflow at S4N was measured with an H-type flume with cutoff walls made of metal sheet piling from December 1961 to November 1984. The flume was removed and a V-notch weir with concrete cutoff walls was constructed during November 1984. Biweekly sampling of stream water at S4N began during 1975. No samples were collected from 1983 to 1988. Samples are collected upstream of the S4N stream gage and a road crossing to avoid road effects. The stream gage at S4S was built as a V-notch weir with metal cutoff walls. The gage was converted to an H-type flume during 1965. During 1980, the S4S flume was replaced with a V-notch weir with concrete cutoff walls. Routine biweekly water sampling at the S4S weir began in November 2008.

Weekly precipitation inputs have been measured at four standard gages since 1962. One gage is near the bog center, one is along the northeast edge of the peatland, one is on the divide between S4 and S5, and one is west of the S4S weir. Snow depth, snow water equivalent, and frost depth were measured along one snow course in the S4 bog and two courses in the uplands from 1962 to 1972.

Bog water table elevations have been recorded on stripcharts since November 1961. Water table elevations were measured several times a year at four other non-recording bog wells from 1962 to 1973 when ice free. An upland groundwater well, 401, was drilled during May 1963. Monthly measurements began during September 1963. Soil moisture has been measured three times a year at upland sites to the south of the bog (S4-S since 1966) and to the west (S4-W) since 1968. Measurements to the east of the bog (S4-E) were made from 1968 to 1971.

S5 Watershed, North Unit

The S5 watershed drains to the Mississippi River, is 52.6 ha, and is square-shaped. The instruments, record lengths, and other details for measurements are listed in Table 2.7. The outlet elevation is 422 m above mean sea

TABLE 2.7

S5 Instrumentation

Measurement	ID	Instrument	Record Period	Note
Stream stage	S5 flume S5 weir S5 weir	0.76 m wide H-type flume with float-driven FW-1 recorder 120° V-notch weir with float-driven FW-1 recorder 120° V-notch weir with float driven shaft encoder	Daily 3/29/1962 to 5/8/1984 Daily 11/9/1984 to present Every 15 min 4/9/2002 to present	Metal cutoff walls Metal cutoff walls Concrete cutoff walls
Precipitation	North/5-1RRG 5-1 5-2 5-3 4-3	Belfort weighing bucket gage with stripchart recorder 8 in. standard gage	Daily 1/1/1962 to present Weekly 1/1/1962 to present Weekly 1/1/1962 to present Weekly 1/1/1962 to present Weekly 1/1/1962 to present	S5 open-canopy meteorological station S5 open-canopy meteorological station Clearing in aspen forest Clearing in aspen forest Clearing in aspen forest on border with S4
Air temperature	North/S5	Belfort hygrothermograph Min/max thermometers HOBO Pro series temp sensor	Daily 1/1/1962 to present Weekly 1/1/1962 to present Every 15 min 7/8/1997 to present	Inside an NWS shelter at the S5 open-canopy meteorological station
Snow depth, water equivalent, and frost depth	5-1 5-2 5-3 5-4 5-5	Mt. Rose snow tube and frost penetrometer	Annually 1962 to present Annually 1962 to present Annually 1962 to 1974 Annually 1962 to present Annually 1962 to present	Birch-aspen/ birch-balsam fir cover Aspen cover Balsam fir/open/ cedar/mixed hardwood cover Black spruce cover S5 open-canopy meteorological station
Bog water table elevation	S5 bog well Non-recording wells	Float-driven FW-1 recorder Chalked tape measure	Daily 11/21/1961 to present Occasionally 1962 to 1973	Heated since 2002

TABLE 2.7 (continued)

S5 Instrumentation

Measurement	ID	Instrument	Record Period	Note
Upland water table elevations	501 502	Chalked tape measure or electronic well probe	Monthly 2/13/1963 to present Monthly 2/13/1963 to 1/15/1973	10.3 m depth 13.1 m depth
Upland soil moisture	S5	Neutron probe	Three times a year 9/14/1966 to present	Measured to 2.3 m

level and the watershed has a maximum elevation of 438 m. Upland soils and five small satellite wetlands drain into a 6.1 ha central peatland. The central peatland stretches east to west across most of the watershed and has a lobe that extends toward the southern boundary of the watershed (Figure 2.17). Black ash grows on one of the satellite wetlands and black spruce on the others.

The S5 watershed is similar in size to the S1 and S4 watersheds and is a control site for comparison of experimental manipulations on those two watersheds. During 1968, the S5 uplands had stands of aspen (two age classes), white cedar, white spruce, balsam fir, red and white pine, and mixed hardwoods. Mean basal area was 23.6 m² ha⁻¹ and mean age was 100 years. Mean stand age was older than other uplands due to a patch of old-growth white cedar (Verry 1969). The stand age of black spruce on the central peatland was 100 years and basal area was 13.3 m² ha⁻¹. The upland and bog cover types have not changed substantially except for aging since the 1968 survey.

The North meteorological station is a 0.15 ha clearcut area in the uplands north of the S5 bog. Measurements began during 1961. This station includes an NWS shelter with minimum and maximum thermometers to measure weekly extremes of air temperature and a Belfort hygrothermograph to continuously record air temperature and relative humidity. Air temperature has been recorded every 15 min with a temperature sensor since 1997. Precipitation amount is measured with a weighing bucket gage (site 5-1RG) and a standard gage (site 5-1). Weekly precipitation amount has been measured at two other standard gages since 1962. Gage 5-2 is on the S4 watershed boundary and 5-3 is along the southeast edge of the watershed. Snow depth, snow water equivalents, and frost depth have been measured in the forest opening at the S5 meteorological station since 1962. Snow and frost measurements also have been made along one bog and two other upland snow courses since 1962.

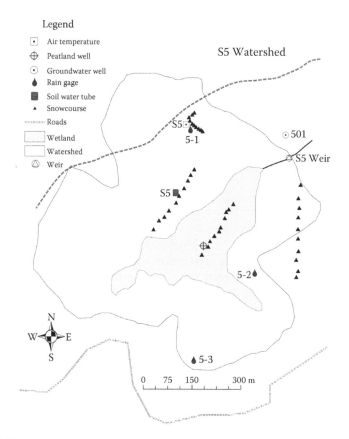

FIGURE 2.17
Map of the S5 watershed showing locations of monitoring equipment.

An H-type flume with metal cutoff walls was installed at the S5 outlet during 1961. The flume was removed and replaced with a V-notch weir with concrete cutoff walls during September 1982. Mean stream water yield from 1962 to 2006 was 11.0 ± 4.1 cm year^{-1}, mean runoff ratio was $14\% \pm 4\%$, and mean annual precipitation was 78.6 ± 11.2 cm year^{-1}. Biweekly sampling of stream water at S5 began during 1975. No samples were collected from 1983 to 1988. Samples are collected upstream of the weir pool.

Bog water table elevations have been recorded on stripcharts since November 1961. Water table elevations were measured several times a year at three other non-recording bog wells from 1961 to 1973 when ice free. Upland well 501 is beyond the eastern border of the watershed near the weir and has been measured monthly since July 1963. Well 502, near precipitation gage 5-2, was measured monthly from 1963 to 1972. Upland soil moisture has been measured three times a year since 1966 by the neutron probe technique at a site north of the S5 bog well.

S6 Watershed, South Unit

The S6 watershed is 8.9 ha with a narrow peatland that is 2.0 ha (Figure 2.18). The instruments, record lengths, and other details for measurements are listed in Table 2.8. The S6 watershed drains via peatlands, Scrapper Lake, and the Prairie River to the Mississippi River. The outlet is 423 m above mean sea level and the watershed has a maximum elevation of 435 m.

During 1980, the upland forest at the S6 watershed was clearcut to study the effects of harvesting and conversion of aspen to conifers (Chapter 13). The watershed was fenced and grazed with cattle from 1980 to 1982 to suppress aspen regeneration, and replanted with white spruce and red pine seedlings during 1983 to convert the uplands cover to conifers (Chapter 12). The herbicide Garlon 4 was sprayed on upland

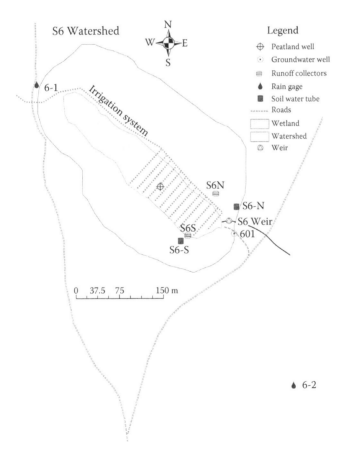

FIGURE 2.18
Map of the S6 watershed showing locations of monitoring equipment.

TABLE 2.8

S6 Instrumentation

Measurement	ID	Instrument	Record Period	Note
Stream stage	S6 flume S6 weir	0.45 m H-type flume with float-driven FW-1 recorder 120° V-notch weir with float-driven FW-1 recorder	Daily 4/9/1964 to 6/5/1974 Daily 3/1/1976 to present	Concrete cutoff walls
Precipitation	6-1 6-2	8 in. standard gage	Weekly 1/1/1964 to present Weekly 1/1/1964 to present	
Snow depth, water equivalent, and frost depth	6-1	Mt. Rose snow tube and frost penetrometer	Annually 1965 to 1969	Black spruce cover
Bog water table elevation	S6 bog well Non-recording wells	Float-driven FW-1 recorder Chalked tape measure	Daily 9/17/1964 to 6/25/1974 and 4/19/1977 to present Occasionally 1968 to 1973	Heated since 2001
Upland water table elevations	601	Chalked tape measure or electronic well probe	Monthly 7/6/1982 to present	
Upland soil moisture	S6-N S6-S	Neutron probe	Three times a year 11/2/1967 to present Three times a year 9/18/1985 to present	Measured to 2.3 m Measured to 3.2 m
Upland runoff	S6N and S6S S6N and S6S	Total volumes at surface and subsurface runoff tanks Total volumes at surface and subsurface runoff tanks	Event-based 4/13/1978 to 9/13/1984 Event-based 10/16/1984 to present	Metal tanks Polyethylene tanks

vegetation on August 3 and 4, 1987 to kill willow, paper birch, and hazel that invaded the site and shaded the shorter conifers. From 2001 to 2008, sulfate was added to the downstream half of the peatland to simulate levels of wet sulfate deposition that occurred during the 1970s prior to sulfur-emission controls (Chapter 11). The study was designed to assess effects of sulfate deposition on methylmercury production in peatlands (Jeremiason et al. 2006).

A metal H-type flume with concrete cutoff walls was constructed during 1963. A crack in the cement wall was found and repaired during autumn 1973. Winter frost heaved the concrete structure during February 1974, creating a crack that was deemed irreparable. The flume was replaced with a V-notch weir and streamflow measurements resumed on March 1, 1976. The weir blade was set 7.6 cm too high, causing water to occasionally back up into the bog until the weir blade height was corrected on July 1, 1977. Water levels were low during the drought of 1976 and the mistake went unnoticed until June 1977, when water levels rose after the drought. Even after correct placement to the depth of the original stream contour, the change from a rectangular to a V shape caused higher water levels that backed into the peatland during high flows. The change in stream gaging structure makes the comparison of post-1976 to pre-1974 data impossible because evaporation and transpiration in the bog are sensitive to water level (Nichols and Brown 1980; Chapter 6). As a result, only 4 years of data from 1976 to 1979 are used to calibrate the watershed. Mean annual runoff was 16 ± 7 cm year^{-1} and mean annual runoff ratio was $20\% \pm 7\%$.

Weekly precipitation inputs have been measured with two standard gages since January 1964. One gage, 6-1, is west of the watershed next to Wilderness Trail Road. The other gage, 6-2, is about 300 m south and east of the S6 weir. Snow depth, snow water equivalent, and frost depth were measured along one snow course under the black spruce and tamarack canopy in the bog from 1962 to 1972.

Bog water table elevations have been recorded on stripcharts since September 1964. Water levels are not reported from July 1974 to April 1977 due to the leaking flume wall and the temporarily incorrect height of the replacement weir notch. The change in stream gaging structure also makes the comparison of post-1977 to pre-1974 water table elevation data impossible. Water table elevations were measured several times a year from 1968 to 1973 at four other non-recording bog wells when ice free.

Paired surface and subsurface runoff collectors were installed on hillslopes to the north (site S6N) and south (site S6S) of the S6 bog during 1978 to measure runoff during snowmelt and storm events. Trees on the runoff plots were hand-felled during the clearcutting to prevent mechanical disturbance of soils on the plots. Well 601 near the eastern border of the watershed has been measured monthly since July 1982. Soil moisture has been measured three times a year by the neutron probe technique at S6-N since 1967 and at S6-S since 1985.

Biweekly sampling of stream water at S6 began during 1978. Samples are collected upstream of the weir pool. Stormflow samples have been collected occasionally with an ISCO automated water sampler since April 2008. Samples from the S6N and S6S runoff collectors have been analyzed since 1979.

Other Study Sites

During other short-duration studies, measurements have been made at research sites beyond the six core research watersheds.

Bog Lake Fen Study Area, South Unit

Bog Lake fen is a poor fen that is about 20 ha (Figure 2.19). The poor fen surface is slightly concave in cross section. The instruments, record lengths, and other details for measurements are listed in Table 2.9. Except for a small area near a pond at the north end with black spruce, Bog Lake fen is an open peatland with sparse tamarack among *Sphagnum* species, sedges, pitcher plant, leather leaf, and bog cranberry. Organic soils are up to 3 m deep where the long-term measurements are made. The primary research activity at the Bog Lake fen is measurement of trace gases. The site also has been used to measure controls on methylmercury production in a series of 40 mesocosms

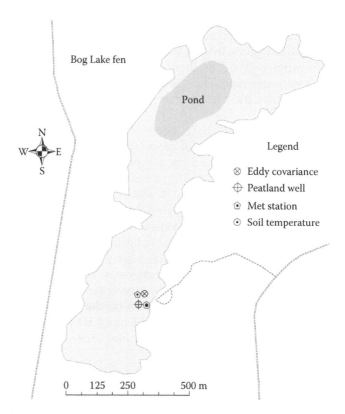

FIGURE 2.19
Map of the Bog Lake fen showing locations of monitoring equipment.

TABLE 2.9

Bog Lake Fen Instrumentation

Measurement	Instrument	Record Period	Note
Water table elevation	Float-driven FW-1 recorder	7/19/90 to present	Heated since 1990
Wind speed and direction	Campbell Scientific CSAT3 3-dimensional sonic anemometer	6/2/06 to present	10 Hz data
CO_2 and H_2O vapor	LI-COR LI-7500 infrared gas analyzer	6/2/06 to present	10 Hz data
Methane	Campbell Scientific trace gas analyzer TGA 100A	4/23/09 to present	10 Hz data
Wind speed and direction, relative humidity, rainfall, PAR, peat temperature, peat soil moisture	Meteorological station, see Table 2.1	6/29/06 to present	Every 30 min
Net radiation	Kipp and Zonen CNR1 net radiometer	6/2/06 to present	
	Radiation and Energy Balance Systems Q7.1 net radiometer	7/10/07 to present	
Peat temperature	Type-T thermocouple	7/18/89 to present	Depths of 5, 10, 20, 30, 40, 50, 100, and 200 cm
Peat heat flux	Hukesflux HFP01SC Heat flux plates	6/2/06 to present	
Soil respiration	LI-COR LI-8100 single long-term soil respiration chamber	2006 and 2007	1 week a month
	LI-COR LI-8100 long-term soil respiration system with four chambers	6/29/06 to present	

(Mitchell et al. 2008; Figure 2.20) and study anaerobic methane oxidation (Smemo and Yavitt 2007).

Bog Lake fen is where the eddy covariance technique was first used to measure peatland methane fluxes during 1990. The system included a tunable diode laser absorption spectrometer and other instruments that are listed in Table 2.10 (Kim and Verma 1992; Verma et al. 1992a,b; Clement et al. 1995). Eddy covariance and chamber measurements of methane fluxes were compared from January 1991 to October 1992 (Verma et al. 1992b; Clement et al. 1995). Vegetation, water level, nitrogen additions, and sulfur additions were evaluated for their impact on methane emissions (Smith 1993). Carbon dioxide fluxes also were measured with chambers on hummocks and hollows to investigate microtopographic effects on gas efflux (Kim and Verma 1992).

FIGURE 2.20

Hydrological technician Donna Olson maintaining the eddy covariance and meteorological systems at Bog Lake fen during November 2009. Mesocosms for several mercury studies are the square boxes to the left of the eddy covariance and meteorological station. Several soil respiration chambers are to the right of Donna. (North-facing photo by Carrie Dorrance, USDA Forest Service, Grand Rapids, MN.)

TABLE 2.10

Bog Lake Fen Instrumentation to Measure Methane Emissions Using Eddy Covariance, July 1990 and January 1991–October 1992

Measurement	Instrument
Eddy correlation	Campbell Scientific 1-dimensional sonic anemometer and krypton hygrometer, A.I.R. Lyman-alpha hygrometers, Kaijo Denki 3-dimensional sonic anemometers, fine-wire thermocouples
Methane	Fast-response, closed-cell, tunable diode laser absorption spectrometer
Wind speed and direction	Cayuga Development three-cup anemometers
Relative humidity	Vaisala HMP35C aspirated ceramic wick psychrometers
Total solar radiation	Eppley Laboratories PSP pyranometer
Photosynthetically active radiation	LI-COR LI-190SA quantum sensors
Net radiation	Radiation and Energy Balance Systems Q6 net radiometer
Leaf area index	LI-COR LAI-2000 area meter
Peat temperature	Platinum resistance thermometers
	Copper-constantan thermocouples
Peat heat flux	Radiation and Energy Balance Systems HFT-1 heat-flow transducers

Source: Kim, J. and Verma, S.B., *Biogeochemistry*, 18(1), 37, 1992; Verma, S.B. et al., *Boundary-Layer Meteorol.*, 58(3), 289, 1992a; Verma, S.B. et al., *Suo*, 43(4–5), 285, 1992b; Clement, R.J. et al., *J Geophys. Res. D, Atmospheres*, 100(D10), 21047, 1995.

Long-term eddy covariance measurement of water vapor and carbon dioxide fluxes was established at Bog Lake fen during June 2006. Measurements now are taken year round. The system includes an open-path infrared gas analyzer, sonic anemometer, and other meteorological instruments (Table 2.10, Figure 2.20). A trace gas analyzer (Campbell Scientific TGA 100A) was added during April 2009 to quantify methane flux.

Soil respiration from peat at Bog Lake fen was measured every 30 min for 1 week each month during 2006 and 2007 using a LI-COR LI-8100 automated soil respiration chamber. Carbon dioxide flux from the chamber was measured on a *Sphagnum* lawn next to the boardwalk. Since July 2009, soil respiration from a hollow, a hummock, and two areas of *Sphagnum* lawn has been measured with a LI-COR LI-8100 four-chamber automated system.

Peat temperatures have been measured weekly in organic soils since 1989 and every 30 min in organic soils south of the eddy covariance system since September 2006. Daily water table elevations have been recorded on stripcharts at a peatland well since July 1990. Biweekly sampling of surface water began during 1991. Samples are dipped with a ladle from a small pool of open water along the edge of the peatland.

S7 Watershed, South Unit

The S7 peatland was the site of a large-scale drainage experiment that was monitored from 1966 to 1970 (Boelter 1972b). The S7 watershed is 7.0 ha and has an oval-shaped bog that is 2.1 ha. The maximum elevation of 434.5 m is on the divide between S6 and S7. The peat is up to 7 m deep in the bog center (Boelter 1972b). A stream drains S7 on the north border where a sandy berm creates a narrow topographic high that separates the perched S7 bog from a downgradient fen. The outlet was 415.9 m above mean sea level before installation of an H-type flume and control structure with stop gates to manipulate water levels. The wooden structure was not maintained after the S7 study was discontinued during 1970. The structure has decayed, water leaks through the wood cutoff walls, and the elevation of the drainage stream now is below that of the experiment.

The black spruce stand in the S7 bog was 75–80 years old when the S7 experiment began. Black spruce trees were removed from a 135-m-long strip through the center of the bog and a drainage ditch that was 2 m wide and 1.25 m deep was excavated with dynamite. Four transects of wells and piezometers extend perpendicular to the ditch and were measured from 1966 to 1970 (Boelter 1972b). The wells were positioned at 1.25, 2.5, 5.0, 10.0, 20.0, and 50.0 m from the ditch. Nests with piezometers screened at 0.5, 1.0, 1.5, 2.0, and 2.5 m were 5.0 and 10.0 m from the ditch.

The upland deciduous forest was clearcut along the S6 and S7 border during 1980 and planted in red pine and white spruce during 1983. The remaining uplands forest is mixed hardwoods including aspen, birch, and sugar maple. During June 2009, three subsurface runoff collectors were installed on the north-facing hillslope. Each runoff collector is 10.0 m wide and drains to a tipping bucket gage to measure outflow. Trees on the hillslopes above two of the three runoff collectors will be harvested during 2012 to study the effects of harvesting on methylmercury production and transport from hillslopes to peatlands.

S8 Bog, South Unit

The S8 bog has scattered black spruce and tamarack trees near a 1.0 ha study area where water levels were manipulated inside experimental chambers (Boelter 1972a). *Sphagnum* spp., Labrador tea, leatherleaf, cotton grass, and sedges grew inside the tanks. A 1.5-m-deep ditch was excavated through the center of the bog during July 1968 to control drainage from six bottomless 1-m-deep tanks (1.5 × 1.8 m). Each water-level treatment (0.0, 0.3, and 0.6 m below the lowest hollow in each tank) was replicated. Water levels were monitored inside the tanks from 1968 to 1970. The study site has been inactive since 1970.

Junction Fen, South Unit

Junction Fen is an open poor fen in the South Unit. This fen has incorrectly been called a bog in other publications. The peatland has no inlet or outlet streams. Vegetation includes few seeded hop sedge, rannock rush, bog cranberry, and *Sphagnum* spp. (*Sphagnum angustifolium*, *Sphagnum capillifolium*, and *Sphagnum fuscum*). Junction Fen is the site of several short-term studies, for example the measurement of methane emissions (Dise 1992, 1993; Dise et al. 1993; Dise and Verry 2001). Long-term data from Junction Fen include peat temperature since 1989, snow depth and snow water equivalents since 1985, and frost depth since 1985.

National Atmospheric Deposition Program/National Trends Network/Mercury Deposition Network, South Unit

The MEF was designated the first operational site of the NADP on July 7, 1978. The site is a 0.23 ha forest clearing on the South Unit and is closest to the S3 watershed (Figure 2.15). Samples of wet deposition are collected weekly from an Aerochem wet–dry collector. Precipitation amount was measured with a recording weighing bucket gage from 1978 to 2010. Stripcharts were collected weekly and forwarded to the NADP central laboratory in Illinois. A digital ETI Instrument Systems NOAH IV Total Precipitation Gage replaced the weighting bucket gage during March 2010. Total daily precipitation is reported in publications and on the website of the NADP (http://nadp.sws. uiuc.edu/). Weekly precipitation has been measured with a standard precipitation gage since January 1980.

The MEF was a pilot site of the MDN from October 1993 to December 1994, during which collectors and protocols for mercury atmospheric deposition

FIGURE 2.21
Hydrological technician Carrie Dorrance downloading data from the ozone sensor at the NADP site on July 21, 2009. (North facing photo by Josh Prososki, USDA Forest Service, Grand Rapids, MN.)

were evaluated at the NADP site (Vermette et al. 1995). Along with 14 other sites, the MEF was part of a transitional mercury-monitoring network during 1995 that became the MDN in 1996.

Air temperature, relative humidity, snow depth, volumetric soil water content, rainfall, wind speed, wind direction, and soil temperature have been measured every 30 min since 2006 (Table 2.1). Temperatures in mineral soils have been measured weekly since July 1989 under the adjacent aspen canopy and in the open under grass cover. Ozone concentration has been measured every 30 min since June 2009 with a 2B Technologies Model 202 Ozone Monitor (Figure 2.21). Atmospheric deposition of DDT (Rapaport et al. 1985) and lead (Eisenreich et al. 1986) has been measured during short-term studies at the NADP site.

Long-Term Soil Productivity Study Area, South Unit

The MEF was also a pilot site for the Long-Term Soil Productivity (LTSP) project. The LTSP is a study of the effects of soil compaction and forest-floor removal on forest productivity and soil physical and chemical properties (Powers et al. 2005). A contiguous area of 4.9 ha was clearcut during the winter of 1991 and individual treatments were superimposed within the clearcut area (Chapter 12). Two control plots and seven treatment plots form an arc to the east of the NADP site. The harvest area extends partially into the western edge of the S3 uplands. Seven 30 × 40 m plots were prepared during the winter of 1991 to simulate mild and severe effects of harvesting practices on site productivity. Pretreatment measurements were made during the fall of 1990. Posttreatment measurements were made 1, 5, 10, and 15 years after harvesting. Overstory and understory biomass were measured for trees and shrubs on each treatment plot. Soil bulk density and soil strength (with a cone penetrometer) were measured on all plots.

Marcell Field Laboratory, South Unit

The Marcell Field Laboratory was the primary support facility at the MEF from 1963 to 2006 when the Marcell Research Center opened. The one-story building (Figure 2.22) had two bunkrooms, a kitchen, a bathroom, laboratory benches, and storage space. The building was razed during July 2010. The stand-alone garage will be retained for storage.

An NWS Class A evaporation pan (McGuiness et al. 1979) was monitored weekly from 1967 to 1972 along with air temperature at the Marcell Field Laboratory. Measurements were made in an open area to the north between the driveway and the aspen forest. Air temperature was measured inside a louvered weather shelter with a hygrothermograph, minimum thermometer, and maximum thermometer.

An open fen northwest of the Marcell Field Laboratory was the site of several studies. During 1963, two cylinders of galvanized sheet metal were inserted to form bottomless evapotranspirometers that were 3.0 m in diameter (Bay 1966). Water levels were measured weekly inside observation wells from 1963 to 1966. Additional evapotranspirometers were installed and monitored during 1965 (0.6 m in diameter) and 1966 (1.5 m in diameter). Air temperature was measured inside a louvered weather shelter with a hygrothermograph, minimum thermometer, and maximum thermometer. Actual

FIGURE 2.22
The Marcell Field Lab in the early 1960s. (West-facing photo by an unknown photographer, USDA Forest Service, Grand Rapids, MN.)

BOX 2.1   TECHNICAL SUPPORT FOR
LONG-TERM MONITORING AT THE MEF

The long-term data collection requires diligence and devotion. Forest Service technicians are responsible for the collection of long-term data, data processing, and instrument maintenance. Hydrological technicians also plow snow and sometimes repair roads. Once each week on Tuesday, two hydrological technicians retrieve stripcharts, make measurements, and collect NADP/MDN samples. Clarence Hawkinson, the first hydrological technician on the MEF project, worked from 1960 until his death in 1974. Art Elling began working at the MEF in September 1969 and retired in January 2005. Richard "Deacon" Kyllander has worked on various projects at the MEF since 1981. With the opening of the Marcell Research Center during 2006, Kyllander became the facility manager and he was the first employee to be stationed at the MEF rather than the Forestry Sciences Laboratory in Grand Rapids. Carrie Dorrance was hired as a hydrological technician during 2005 and he manages the MEF databases. Donna Olson was hired during 2005 to install and maintain the Bog Lake eddy covariance system as well as analyze data.

Other Forest Service employees have played important roles. Bob Barse was a forestry technician for the S1 and S6 harvests. John Elioff has installed, maintained, and measured the Lake States LTSP sites, including the MEF pilot site, since the 1990s. Doris Nelson and Dwight Streblow frequently contributed to data collection during the 1980s and 1990s. Nelson also has assisted chemists Don Nagle, Will Pettit, and John Larson during the analysis of water, soil, and plant-tissue chemistry samples. The facilities managers, Lester Weller, Ramon Sanders, and Eric Troumbly, have maintained the buildings at the MEF, helped with the installation of field instruments, and assisted with weekly measurements as needed. Student workers have been hired during many summers to assist with fieldwork, data collection, and site maintenance.

evaporation was measured with a Class A evaporation pan and rainfall was measured with a standard gage. Solar irradiance was measured with a Belford recording pyrheliometer from July 18 to October 19, 1966. Total and reflected solar radiation were measured with Kipp and Zonen pyranometers that were suspended on a 33.5-m-long cableway 2 m above ericaceous shrubs, sedges, and various *Sphagnum* spp. between August 1968 and August 1969 (Berglund and Mace 1972). The open fen research area was last used for a study of trace gas emissions in which metal chambers were sunk in the peat and enclosed beneath transparent plastic greenhouses to cause warming. The study was not completed.

FIGURE 2.23
The Marcell Research Center on August 7, 2008. (Northeast-facing photo by Stephen Sebestyen, USDA Forest Service, Grand Rapids, MN.)

Marcell Research Center, South Unit

The Marcell Research Center (Figure 2.23), opened in 2006, has a conference room, kitchen and dining area, office for the site manager, and wet chemistry laboratory with deionized water, fume hood, and laboratory benches. An adjacent building has a two-bay garage and heated workshop. Four bunkrooms are available for students and researchers upon request.

References

Aber, J.D., C.L. Goodale, S.V. Ollinger, M.-L. Smith, A.H. Magill, M.E. Martin, R.A. Hallett, and J.L. Stoddard. 2003. Is nitrogen deposition altering the nitrogen status of northeastern forests? *BioScience* 53(4):375–389.

Alter, J.C. 1937. Shielded storage precipitation. *Monthly Weather Review* 66(7):262–265.

APHA 1998. Standard methods of the examination of water and wastewater, 20th edition. American Public Health Association/American Water Works Association Water Environment Federation, Washington, DC.

Bay, R.R. 1962. *Establishment Report, Marcell Experimental Forest (Chippewa National Forest and Adjacent Private, Itasca county, and State of Minnesota Lands)*. Grand Rapids, MN: USDA Forest Service.

Bay, R.R. 1966. Evaluation of an evapotranspirometer for peat bogs. *Water Resources Research* 2(3):437–441.

Bay, R.R. 1967. Techniques of hydrologic research in forested peatlands, USA. In *Proceedings of the XIV IUFRO Congress*. Munich, Germany, September 1967. Section 11, edited, pp. 400–415, IUFRO.

Bay, R.R. 1968. The hydrology of several peat deposits in northern Minnesota, USA. In *Proceedings of the Third International Peat Congress*. Quebec, Canada: National Research Council of Canada, pp. 121–218.

Bay, R.R. 1969. Runoff from small peatland watersheds. *Journal of Hydrology* 9(1):90–102.

Bay, R.R. 1970. Water table relationships on experimental basins containing peat bogs. In *Symposium on the Results of Research on Representative and Experimental Basins. Proceedings of the Wellington Symposium*, December 1970, UNESCO/International Association of Hydrological Sciences, Paris, France, Vol. 96, pp. 360–368.

Berglund, E.R. and A.C. Mace. 1972. Seasonal albedo variation of black spruce and sphagnum-sedge bog cover types. *Journal of Applied Meteorology* 11(5):806–812.

Berguson, W.E. and D.A. Perala. 1988. Aspen fertilization and thinning research results and future potential. In *Minnesota's Timber Supply: Perspectives and Analysis*, eds. A.E. Ek and H.M. Hoganson. Grand Rapids, MN: University of Minnesota, College of Natural Resources and Agricultural Experiment Station, pp. 176–183.

Boelter, D.H. 1962. A study of some physical properties of several peat materials and their relation to field water conditions in the peat bog. PhD dissertation. St. Paul, MN: University of Minnesota.

Boelter, D.H. 1964a. Laboratory techniques for measuring water storage properties of organic soils. *Soil Science Society of America Proceedings* 28(6):823–824.

Boelter, D.H. 1964b. Water storage characteristics of several peats *in situ*. *Soil Science Society of America Proceedings* 33:433–435.

Boelter, D.H. 1965. Hydraulic conductivity of peats. *Soil Science* 100:227–231.

Boelter, D.H. 1966. Hydrologic characteristics of organic soils in Lakes States Watersheds. *Journal of Soil and Water Conservation* 21(2):50–53.

Boelter, D.H. 1968. Important physical properties of peat materials. In *Proceedings of the Third International Peat Congress*. Quebec, Canada: National Research Council of Canada, pp. 150–154.

Boelter, D.H. 1969. Physical properties of peats as related to degree of decomposition. *Soil Science Society of America Proceedings* 33:606–609.

Boelter, D.H. 1972a. Preliminary results of water level control on small plots in a peat bog. In *The Use of Peatland for Agriculture, Horticulture, and Forestry. Fourth International Peat Congress Proceedings*. Ontanienmi, Finland: International Peat Congress, Vol. 3, pp. 347–354.

Boelter, D.H. 1972b. Water table drawdown around an open ditch in organic soils. *Journal of Hydrology* 15(4):329–340.

Brakensiek, D.L., H.B. Osborn, and W.J. Rawls. 1979. Field manual for research in agricultural hydrology. *Agriculture Handbook 224*. Washington, DC: U.S. Department of Agriculture.

Bridgham, S.D., K. Updegraff, and J. Pastor. 1998. Carbon, nitrogen, and phosphorus mineralization in northern wetlands. *Ecology* 79(5):1545–1561.

Brown, J.M. 1972a. Effect of clearcutting a black spruce bog on net radiation. *Forest Science* 18(4):273–277.

Brown, J.M. 1972b. The effect of overstory removal upon surface wind in a black spruce bog. Research note NC-137. St. Paul, MN: USDA Forest Service.

Brown, J.M. 1973. Effect on overstory removal on production of shrubs and sedge in a northern Minnesota bog. *Journal of the Minnesota Academy of Science* 38(2–3):96–97.

Brown, J.M. 1976. Peat temperature regime of a Minnesota bog and the effect of canopy removal. *The Journal of Applied Ecology* 13(1):189–194.

Buttleman, C.G. and D.F. Grigal. 1985. Use of the Rb/K ratio to evaluate potassium nutrition of peatlands. *Oikos* 44(2):253–256.

Clark, J.S. 1990a. Fire and climate change during the last 750 year in northwestern Minnesota. *Ecological Monographs* 60(2):135–159.

Clark, J.S. 1990b. Twentieth-century climate change, fire suppression, and forest production and decomposition in northwestern Minnesota. *Canadian Journal of Forest Research* 20(2):219–232.

Clement, R.J., S.B. Verma, and E.S. Verry. 1995. Relating chamber measurements to eddy correlation measurements of methane flux. *Journal of Geophysical Research, D, Atmospheres* 100(D10):21047–21056.

Crill, P.M., K.B. Bartlett, R.C. Harriss, E. Gorham, E.S. Verry, D.I. Sebacher, L. Madzar, and W. Sanner. 1988. Methane flux from Minnesota peatlands. *Global Biogeochemical Cycles* 2(4):371–384.

Dise, N.B. 1992. Winter fluxes of methane from Minnesota peatlands. *Biogeochemistry* 17(2):71–83.

Dise, N.B. 1993. Methane emission from Minnesota peatlands: spatial and seasonal variability. *Global Biogeochemical Cycles* 7(1):123–142.

Dise, N.B., E. Gorham, and E.S. Verry. 1993. Environmental factors controlling methane emissions from peatlands in northern Minnesota. *Journal of Geophysical Research, D, Atmospheres* 98(D6):10583–10594.

Dise, N.B. and E.S. Verry. 2001. Suppression of peatland methane emission by cumulative sulfate deposition in simulated acid rain. *Biogeochemistry* 53(2):143–160.

Driscoll, C.T., D.R. Whitall, J.D. Aber, E.W. Boyer, M.S. Castro, C.S. Cronan, C.L. Goodale, P.M. Groffman, C.S. Hopkinson, K.F. Lambert, G.B. Lawrence, and S.V. Ollinger. 2003. Nitrogen pollution in the northeastern United States: Sources, effects, and management options. *BioScience* 53(4):357–374.

Eisenreich, S.J., N.A. Metzer, N.R. Urban, and J.A. Robbins. 1986. Response of atmospheric lead to decreased use of lead in gasoline. *Environmental Science and Technology* 20(2):171–174.

Farrish, K.W. and D.F. Grigal. 1985. Mass loss in a forested bog: Relation to hummock and hollow microrelief. *Canadian Journal of Soil Science* 66(2):375–378.

Farrish, K.W. and D.F. Grigal. 1988. Decomposition in an ombrotrophic bog and a minerotrophic fen in Minnesota. *Soil Science* 145(5):353–358.

Gorham, E. 1956. The ionic composition of some bog and fen waters in the English lake district. *Journal of Ecology* 44(1):142–152.

Gorham, E. 1990. Biotic impoverishment in northern peatlands. In *The Earth in Transition: Patterns and Process of Biotic Impoverishment*, ed. G.M. Woodwell. New York: Cambridge University Press, pp. 65–98.

Gorham, E., J.A. Janssens, and P.H. Glaser. 2003. Rates of peat accumulation during the postglacial period in 32 sites from Alaska to Newfoundland, with special emphasis on northern Minnesota. *Canadian Journal of Botany* 81(5):429–438.

Grigal, D.F., S.L. Brovold, W.S. Nord, and L.F. Ohmann. 1989. Bulk density of surface soils and peat in the north central United States. *Canadian Journal of Soil Science* 69(4):895–900.

Hale, C.M., L.E. Frelich, P.B. Reich, and J. Pastor. 2005. Effects of European earthworm invasion on soil characteristics in northern hardwood forests of Minnesota, USA. *Ecosystems* 8(8):911–927.

Harriss, R.C., E. Gorham, D.I. Sebacher, K.B. Bartlett, and P.A. Flebbe. 1985. Methane flux from northern peatlands. *Nature* 315(6021):652–654.

Heinselman, M.L. 1999. *Boundary Waters: Wilderness Ecosystem*. St. Paul, MN: University of Minnesota Press.

Hodgkins, G.A. and R.W. Dudley. 2006. Changes in the timing of winter-spring streamflows in eastern North America, 1913–2002, *Geophys. Res. Lett.*, doi:10.1029/2005GL025593.

Jeremiason, J.D., D.R. Engstrom, E.B. Swain, E.A. Nater, B.M. Johnson, J.E. Almendinger, B.A. Monson, and R.K. Kolka. 2006. Sulfate addition increases methylmercury production in an experimental wetland. *Environmental Science and Technology* 40(12):3800–3806.

Kim, J. and S.B. Verma. 1992. Soil surface CO_2 flux in a Minnesota peatland. *Biogeochemistry* 18(1):37–51.

Knighton, M.D. and J.H. Stiegler. 1981. Phosphorus release following clearcutting of a black spruce fen and a black spruce bog. In *Proceedings of the Sixth International Peat Congress*, August 17–23, 1980, Duluth, MN. Eveleth, MN: W.A. Fisher Company, pp. 577–583.

Kolka, R.K., E.A. Nater, D.F. Grigal, and E.S. Verry. 1999. Atmospheric inputs of mercury and organic carbon into a forested upland/bog watershed. *Water, Air, and Soil Pollution* 113(1):273–294.

Latterell, J.J., D.R. Timmons, R.F. Holt, and E.M. Sherstad. 1974. Sarption of orthophosphate on the surface of water sample containers. *Water Resour. Res.*, 10(4), 865–869.

McGuiness, J.L., W.C. Mills, and P.R. Nixon. 1979. Climate. In *Field Manual for Research in Agricultural Hydrology*, eds. D.L. Brakensiek, H.B. Osborn, and W.J. Rawls. Washington, DC: U.S. Department of Agriculture, pp. 215–237.

Mitchell, C.P.J., B.A. Branfireun, and R.K. Kolka. 2008. Assessing sulfate and carbon controls on net methylmercury production in peatlands: An in situ mesocosm approach. *Applied Geochemistry* 23(3):503–518.

Monson, B.A. 2009. Trend reversal of mercury concentrations in piscivorous fish from Minnesota Lakes. *Environmental Science and Technology* 43(6):1750–1755.

Mutchler, C.K. 1963. *Runoff Plot Design and Installation for Soil Erosion Studies.* Washington, DC: USDA Agricultural Research Service.

Nichols, D.S. 1998. Temperature of upland and peatland soils in a north central Minnesota forest. *Canadian Journal of Soil Science* 78(3):493–509.

Nichols, D.S. and D.H. Boelter. 1984. Fiber size distribution, bulk density, and ash content of peats in Minnesota, Wisconsin, and Michigan. *Soil Science Society of America Journal* 48(6):1320–1328.

Nichols, D.S. and J.M. Brown. 1980. Evaporation from a sphagnum moss surface. *Journal of Hydrology* 48(3–4):289–302.

Nichols, D.S. and E.S. Verry. 2001. Stream flow and ground water recharge from small forested watersheds in north central Minnesota. *Journal of Hydrology* 245(1–4):89–103.

Novák, M. and R.K. Wieder. 1992. Inorganic and organic sulfur profiles in nine Sphagnum peat bogs in the United States and Czechoslovakia. *Water, Air, and Soil Pollution* 65(3):353–369.

Paulson, R.O. 1968. *A Soil Survey of Marcell Experimental Forest.* Grand Rapids, MN: USDA Soil Conservation Service.

Powers, R.F., S.D. Andrew, F.G. Sanchez, R.A. Voldseth, D.S. Page-Dumroese, J.D. Elioff, and D.M. Stone. 2005. The North American long-term soil productivity experiment: Findings from the first decade of research. *Forest Ecology and Management* 220(1–3):31–50.

Rapaport, R.A., N.R. Urban, P.D. Capel, J.E. Baker, B.B. Looney, S.J. Eisenreich, and E. Gorham. 1985. "New" DDT inputs to North America: Atmospheric deposition. *Chemosphere* 14(9):1167–1173.

Sander, J.E. 1976. An electric analog approach to bog hydrology. *Ground Water* 14(1):30–35.

Sander, J.E. 1978. The blind zone in seismic ground-water exploration. *Ground Water* 16(6):394–397.

Sebacher, D.I., R.C. Harriss, K.B. Bartett, S.M. Sebacher, and S.S. Grice. 1986. Atmospheric methane sources: Alaska tundra bogs, an alpine fen, and a subarctic boreal marsh. *Tellus* 38B:1–10.

Smemo, K.A. and J.B. Yavitt. 2007. Evidence for anaerobic CH_4 oxidation in freshwater peatlands. *Geomicrobiology Journal* 24(7):583–597.

Smith, K. 1993. Methane flux of a Minnesota peatland: Spatial and temporal variables and flux prediction from peat temperature and water table elevations. MS thesis. St. Paul, MN: University of Minnesota.

Stoeckeler, J.H. and J.L. Thames. 1958. The lake states penetrometer for measuring depth of soil freezing. *Soil Science* 85(1):47–50.

Timmons, D.R., E.S. Verry, R.E. Burwell, and R.F. Holt. 1977. Nutrient transport in surface runoff and interflow from an aspen-birch forest. *Journal of Environmental Quality* 6(2):188–192.

Tracy, D.R. 1997. Hydrologic linkages between uplands and peatlands. PhD dissertation. St. Paul, MN: University of Minnesota.

Urban, N.R., S.J. Eisenreich, and E. Gorham. 1987. Aluminum, iron, zinc, and lead in bog waters of northeastern North America. *Canadian Journal of Fisheries and Aquatic Sciences* 44:1165–1172.

Urban, N.R., S.J. Eisenreich, and D.F. Grigal. 1989. Sulfur cycling in a forested sphagnum bog in northern Minnesota. *Biogeochemistry* 7(2):81–109.

Urban, N.R., S.J. Eisenreich, D.F. Grigal, and K.T. Schurr. 1990. Mobility and diagenesis of Pb and 210Pb in peat. *Geochimica et Cosmochimica Acta* 54(12):3329–3346.

Urban, N.R., E.S. Verry, and S.J. Eisenreich. 1995. Retention and mobility of cations in a small peatland: Trends and mechanisms. *Water, Air, and Soil Pollution* 79(1–4):201–224.

Verma, S.B., F.G. Ullman, D.P. Billesbach, R.J. Clement, J. Kim, and E.S. Verry. 1992a. Eddy correlation measurements of methane flux in a northern peatland ecosystem. *Boundary-Layer Meteorology* 58(3):289–304.

Verma, S.B., F.G. Ullman, N.J. Shurpali, R.J. Clement, J. Kim, and D.P. Billesbach. 1992b. Micrometeorological measurements of methane and energy fluxes in a Minnesota peatland. *Suo* 43(4–5):285–288.

Vermette, S.J., S.E. Lindberg, and N. Bloom. 1995. Field tests for a regional mercury deposition network—Sampling design and preliminary test results. *Atmospheric Environment* 29(11):1247–1251.

Verry, E.S. 1969. 1968 Vegetation survey of the Marcell Experimental Watersheds. Report GR-W2–61. Grand Rapids, MN: USDA Forest Service.

Verry, E.S. 1972. Effect of an aspen clearcutting on water yield and quality in northern Minnesota. In *Proceedings of a Symposium on Watersheds in Transition*, eds. S.C. Csallany, T.G. McLaughlin, and W.D. Striffler, Fort Collins, CO, June 19–22, 1972. Urbana, IL: American Water Resources Association, pp. 276–284.

Verry, E.S. 1975. Streamflow chemistry and nutrient yields from upland-peatland watersheds in Minnesota. *Ecology* 56(5):1149–1157.

Verry, E.S. 1976. Estimating water yield differences between hardwood and pine forests. Research Paper NC-128. St. Paul, MN: USDA Forest Service.

Verry, E.S. 1984. Microtropography and water table fluctuation in a *Sphagnum* mire. In *Proceedings of the 7th International Peat Congress*. Dublin, Ireland: The Irish National Peat Committee/The International Peat Society, Vol. 2, pp. 21–31.

Verry, E.S. and D.H. Boelter. 1975. The influence of bogs on the distribution of streamflow from small bog-upland watersheds. In *International Symposium on the Hydrology of Marsh-ridden Areas. Proceedings of the Minsk Symposium*, June 1972. Minsk, Byelorussia: UNESCO/International Association of Hydrological Sciences, Vol. 105, pp. 1–11.

Verry, E.S. and A.E. Elling. 1978. Two years necessary for successful natural seeding in nonbrushy black spruce bogs. Research Note NC-229. St. Paul, MN: USDA Forest Service.

Verry, E.S. and D.R. Timmons. 1977. Precipitation nutrients in the open and under two forests in Minnesota. *Canadian Journal of Forest Research* 7(1):112–119.

Verry, E.S. and D.R. Timmons. 1982. Waterborne nutrient flow through an upland-peatland watershed in Minnesota. *Ecology* 63(5):1456–1467.

Weishampel, P., R.K. Kolka, and J.Y. King. 2009. Carbon pools and productivity in a 1-km^2 heterogeneous forest and peatland mosaic in Minnesota, USA. *Forest Ecology and Management* 257(2):747–754.

Wieder, R.K., M. Novák, W.R. Schell, and T. Rhodes. 1994. Rates of peat accumulation over the past 200 years in five *Sphagnum*-dominated peatlands in the United States. *Journal of Paleolimnology* 12(1):35–47.

Wright, H.E. 1972. Quaternary history of Minnesota. In *Geology of Minnesota: A Centennial Volume*, eds. P.K. Sims and G.B. Morey. St. Paul, MN: Minnesota Geological Survey, pp. 515–592.

Yavitt, J.B., C.J. Williams, and R.K. Wieder. 1997. Production of methane and carbon dioxide in peatland ecosystems across North America: Effects of temperature, aeration and organic chemistry of peat. *Geomicrobiology Journal* 14:299–316.

Yavitt, J.B., C.J. Williams, and R.K. Wieder. 2000. Controls on microbial production of methane and carbon dioxide in three *Sphagnum* dominated peatland ecosystems as revealed by a reciprocal field peat transplant experiment. *Geomicrobiology Journal* 17(1):61–88.

Yavitt, J.B., C.J. Williams, and R.K. Wieder. 2005. Soil chemistry versus environmental controls on production of CH_4 and CO_2 in northern peatlands. *European Journal of Soil Science* 56(2):169–178.

3

An Evolving Research Agenda at the Marcell Experimental Forest

Randall K. Kolka, Stephen D. Sebestyen, and John B. Bradford

CONTENTS

Introduction

The Marcell Experimental Forest (MEF) was established to study the effects of forest management on watershed dynamics in upland–peatland landscapes. We are not aware of any other research location that has the depth and breadth of data from these landscapes (Chapter 2). Findings from 50 years of research constitute a foundational knowledge of the hydrology and biogeochemistry of peatland watersheds (Table 3.1). In this chapter, we discuss what historically and currently drives our research program, the importance of networks in our research portfolio, summarize our past and current research, assess what expert panels anticipate are important future research topics, and align our expectations of future research at the MEF.

Drivers of Research at the MEF

The evolution of research at the MEF is similar to that for other watershed studies throughout the USDA Forest Service's network of experimental forests and ranges. Experimental forests with gaged watersheds were established to assess the impact of forest management on water (Holscher 1967).

TABLE 3.1

Time Series of Research Themes on the MEF

1960s
Increasing lowland forest productivity
Understand peatland hydrology, water and energy budgets, peat soil properties
1970s
Determining harvesting effects on water quality and quantity
Hydrologic modeling of peatland watersheds
1980s
Effects of atmospheric deposition and ecosystem acidification
Methane production from peatlands
Biogeochemistry of carbon and nutrients
1990s
Net ecosystem exchange of carbon in peatlands
Hydrologic cycling of mercury and organic carbon
2000s
Hotspots of biogeochemical activity in peatland landscapes
Controls on mercury methylation
Hydrologic source areas of water and solutes

Initial studies at the MEF measured hydrological response to clearcut harvests of upland and peatland forests (Chapter 13). Funding was mainly congressionally appropriated to the Forest Service and there was little pursuit of extramural funding for research. In addition to personnel, instrumentation, and study costs, Forest Service funding was used to support collaborative research agreements with academic and other agency scientists. Beginning in the 1970s with the Clean Water and Clean Air Acts, the focus of many Forest Service watershed experimental forests shifted to research on biogeochemical cycling and the effect of acid rain on terrestrial and aquatic systems. The Clean Air Act also was the impetus for establishing the National Atmospheric Deposition Program (NADP) in the late 1970s and the Mercury Deposition Network (MDN) in the early 1990s. During the 1970s and 1980s, appropriated funds were sufficient to conduct directed research.

Since the late 1980s, research at the MEF and at other experimental forests has gravitated toward topics and projects that are supported extramurally, as real dollars (with consideration for inflation) for research have declined precipitously. As a result, scientists at most of the thriving experimental forests actively pursue extramural funding. Such pursuits are difficult because Federal scientists are not eligible to receive grants from some funding sources. Nevertheless, over the past decade, scientists at the MEF have received several grants, each in excess of US$1 million, from the Environmental Protection Agency (EPA), National Aeronautics and Space Administration (NASA), and the Department of Energy (DOE), as well as numerous smaller grants. This shift in funding sources has greatly expanded the audience served by Forest

Service Research and Development (R&D). Although the National Forest System still benefits from R&D study findings, the audience for our research findings has become more diverse, especially as the search for extramural funding expands research in new directions. Fortunately, this funding paradigm and the peer-review process ensure a research program that is directed toward compelling science. However, this new funding scenario taxes financial resources that are needed to measure and maintain long-term experiments. Currently, we invest our appropriated funds into the long-term measurements while we use extramural funding in short-term studies that answer specific questions, many times to better understand trends in long-term data and to develop science in emerging issues. However, based on current levels of appropriated funding, we are on the verge of needing to acquire extramural funding to maintain our long-term data collection.

Participation in Networks

One way for experimental forests to leverage their scientific potential is to share data with others via networks (Lugo et al. 2006). Well-designed networks allow examination of processes and responses across multiple ecosystems and varying scales. Scientific leaders at the MEF have long understood the value in contributing to and utilizing networks in our research. The MEF was one of three founding sites in the National Atmospheric Deposition Program/National Trends Network (NADP/NTN) (Miller 1980) and the MEF NADP site has been a test site for new monitoring instruments since its establishment. In the early 1990s, the MEF also became one of the first sites in the NADP subnetwork that quantifies mercury deposition, the Mercury Deposition Network (MDN) (Vermette et al. 1992, 1995a,b). During this same time, the MEF also became a pilot site for testing methods for the Long-Term Soil Productivity Network (LTSP) (Tiarks et al. 1993), the goal of which is to determine the effects of compaction and removal of organic matter on forest productivity across the United States and Canada (Powers et al. 2005). In early 2000, cooperative efforts between the USDA and Oregon State University led to the hydrological (HydroDB) and climate (ClimDB) databases (http://www.fs1.orstedulclimhy). The MEF also will participate in a water-chemistry database (tentatively named StreamChemDB) that is under development through collaboration among the Forest Service, National Center for Air and Stream Improvement, and Oregon State University.

The MEF has recently become part of the National Phenology Network (NPN) (Betancourt and Schwartz 2005), EcoTrends Project Network (Moran et al. 2008), United States–China Carbon Consortium (USCCC) (http://research.eeescience.utoledo.edu/lees/research/usccc/), and an internal Forest Service Experimental Forest and Range (EFR) Synthesis Network

(Lugo et al. 2006). The NPN was established to advance our understanding of how the phenology of plants, animals, and landscapes responds to environmental variation and climate change, and how these responses affect processes and phenomena in the biosphere. The EcoTrends Project was designed to promote and enable the use of long-term data to examine trends in climate, land cover, and habitat availability with important consequences for plant, animal, microbial, and human populations. The USCCC was established as a collaborative consortium between American and Chinese institutions that are interested in the use of eddy covariance to determine the role of managed ecosystems in global carbon and water cycles. The EFR Synthesis Network comprises 18 Forest Service experimental forests that form a national platform to synthesize studies of forest management impacts on water, changing climate, atmospheric chemistry, and invasive organisms. The MEF was considered as a possible relocatable site in the National Ecological Observation Network for the Great Lakes (GLEON), although it was not chosen in the first round of selections. It is likely that MEF will participate in GLEON at some level in the future.

Past Research

While this book reviews past research findings at the MEF, it is helpful to briefly summarize the research that informs our current philosophy and sets a firm foundation of a future research agenda (Table 3.1). The MEF conducted the first paired-watershed studies to determine the effects of experimental vegetation manipulations on stream water yield and chemistry from upland–peatland watersheds (Bay 1970a). Various forest management practices and effects on stream water yield and water chemistry have been studied (Bay 1968; Ohmann et al. 1978; Verry 1996; Grigal and Brooks 1997; Chapters 13 and 14). Treatments to experimental watersheds include upland clearcutting of aspen (Verry 1972), peatland strip, and clearcutting of black spruce (Verry 1981), prescribed fire in a harvested fen peatland (Knighton and Stiegler 1981), upland nitrogen fertilization (Perala 1983; Berguson and Perala 1988), conversion of upland aspen to a conifer forest, and use of cattle grazing as an alternative to herbicide application (Chapters 12 and 14). Field data and model results have been used to evaluate the effects of forest management practices on water yield and bankfull stream flow (Guertin and Brooks 1985; Verry 1987; Barten and Brooks 1988; Lu 1994; Chapter 15), which informed the development of forestry best management practices (BMPs) in Minnesota and the region (Verry 2004; Chapter 7). These data are also included in interregional and international data syntheses to assess forest harvest effects on water yield (Hornbeck et al. 1993; Sahin and Hall 1996; Stednick 1996; Brown et al. 2005; Guillemette et al. 2005).

While completing graduate research in 1959, Don Boelter became one of the first organic-soil scientists in the United States. His research and subsequent studies in the 1960s and 1970s quantified water storage and flow through peatland soils in relation to the physical properties of organic soils (Boelter 1964a, 1965, 1966, 1968, 1969; Nichols and Boelter 1984) and laid the foundation for later studies of organic soils at the MEF (Gafni and Brooks 1986a,b, 1990; Malterer et al. 1987; Malterer et al. 1992; Chapter 5). Boelter's initial research applied lab techniques that typically were used for mineral soils to study the water content and matric potential of organic soils (Boelter 1964b, 1965). This work was followed by field studies (Boelter 1965, 1968, 1975) and ultimately scaled up to explore undrained peatland hydrology (Boelter 1974a), water table manipulations (Boelter 1972a), and peatland drainage at the watershed scale (Boelter 1972b, 1974b). This research on the physical properties of peat later was used to develop and calibrate hydrological models, particularly the Peatland Hydrologic Impact Model (PHIM) (Guertin 1984; Guertin et al. 1987; Brooks and Kreft 1989; McAdams 1993; Chapter 15).

Various aspects of watershed energy budgets have been examined, including the link between evapotranspiration and the physical properties of peat (Bay 1966; Nichols and Brown 1980), effects of harvesting on net radiation fluxes (Brown 1972), albedo above black spruce canopies (Berglund and Mace 1972, 1976), and clearcut strips (Brown 1972). Later research was extended beyond point measurements of evapotranspiration to larger spatial scales, e.g., entire peatlands (Verma et al. 1995; Chapter 6).

The data collected on upland and peatland water levels at the MEF constitute one of the longest running daily records of groundwater in the Lake States; they reveal hydrologic patterns that can be discerned only from long-term data. Notably, these data have characterized the influence of interdecadal climatic variation on the regional groundwater table, and identified a surprising lack of synchrony between perched water table dynamics and the regional flow system (Bay 1967, 1970b; Nichols and Verry 2001; Chapters 2 and 13). Groundwater-well data from the MEF have been used to develop physical models of peatland hydrology (Sander 1976) and estimate deep seepage of groundwater from perched peatlands to the regional groundwater table (Nichols and Verry 2001; Verry 2003).

Over the decades, water, energy, and solute budgets have been calculated using the small-watershed approach by which major pools (e.g., soils and vegetation) and fluxes (precipitation inputs and watershed outfluxes) are measured (Chapters 7 and 8). Such watershed-scale mass balances have been published for hydrogen (Urban et al. 1987), dissolved organic carbon (Urban et al. 1989a; Kolka et al. 2001), major cations (Verry and Urban 1992; Urban et al. 1995), nitrogen (Urban and Eisenreich 1988; Verry and Urban 1992), phosphorus (Knighton and Stiegler 1981; Verry and Urban 1992), and sulfur (Urban et al. 1989b). To gain additional insight on the processes that control element pools and fluxes, biogeochemical transformations such as nitrification and denitrification (Urban and Eisenreich 1988), sulfate reduction, and sulfur reoxidation (Urban et al. 1989b) have been measured. In the field of mercury research, studies at the MEF have elucidated complex processes that affect the transport of mercury from atmospheric sources through the landscape (Kolka et al. 1999, 2001) and interactions among sulfur, organic carbon, and mercury (Skyllberg et al. 2000, 2006; Jeremiason et al. 2006; Mitchell et al. 2008a,b,c, 2009; Skyllberg 2008; Chapter 11).

Pioneering carbon-cycle studies at the MEF have formed a basis for our current focus on the effects of climate change on carbon storage in northern forest landscapes (Chapter 10). Quantifying sinks and sources of atmospheric carbon dioxide (CO_2) is an important area of carbon-cycle research at the MEF (Chapter 9), because storage of peatland carbon may become destabilized with climate change and potentially become a substantial source of CO_2 to the atmosphere, thus contributing to global warming. Previous studies have measured CO_2 fluxes using chamber and eddy covariance methods (Kim and Verma 1992), including the first use of eddy covariance during the early 1990s to measure methane and CO_2 emissions from peatlands (Kim and Verma 1992; Verma et al. 1992a,b, 1995; Clement et al. 1995). Data from the MEF and other northern Minnesota sites were the first to quantify methane emissions from natural peatlands (Harriss et al. 1985; Crill et al. 1988). The continually expanding data set from the MEF is included in regional and global scale assessments of methane emissions in peatlands (Shotyk 1988; Bartlett and Harriss 1993; Waldron et al. 1999; Wania et al. 2009). Several peatland landscape models have been developed that quantify the hydrological and biogeochemical processes controlling peatland carbon stores (Trettin et al. 1995, 2006; Zhang et al. 2002; Cui et al. 2005). Isotope studies of methane provide additional insight into the processes that control methane fluxes from peatlands (Quay et al. 1988; Stevens and Engelkeimer 1988; Kelly et al. 1992; Lansdown et al. 1992; Chanton et al. 1995; Hornibrook et al. 2000). A series of studies has measured environmental factors that control the spatial and temporal variability of methane emissions in peatlands (Dise 1992, 1993; Dise et al. 1993; Shurpali et al. 1993; Shurpali and Verma 1998; Dise and Verry 2001). Smemo and Yavitt (2007) documented anaerobic methane oxidation in freshwater wetlands. Studies at the MEF also have quantified the stabilization of organic carbon on soils (Fissore 2007; Fissore et al. 2008) and the hydrological transport of organic carbon from peatlands to downstream receiving waters (Gorham et al. 1985; Urban et al. 1989a; Kolka et al. 1999, 2001; Chapter 8).

Current Research

The current research agenda at the MEF builds on previous research and the monitoring legacy. Ongoing hydrological research is designed to determine how flowpaths through the landscape affect water and solute transport, and how vegetation change affects water yield. Studies have been initiated to determine how rainfall interception, evapotranspiration, and subsurface hydrology differ between the S2 control watershed and the S6 watershed, where the upland forest was converted from deciduous to conifer species in the 1980s (Chapter 13). To gain insight on the effect of various climate scenarios on the productivity of northern forests, we are exploring how climate variability and climate extremes affect annual tree growth in black spruce, aspen, and upland conifer stands.

Currently we have several major thrusts of carbon cycle research. In the last 3 years, we have reestablished the use of eddy covariance to measure CO_2 and methane at the Bog Lake fen, an open shrub, and grass fen. We hope to merge past data from studies in the 1990s with the current data to determine how peatland carbon pools and fluxes have changed at Bog Lake in the last 2 decades. We also measure soil respiration with an automated chamber system that analyzes samples from lawns, hummocks, and hollows in the immediate vicinity of the eddy covariance system, as well as a single chamber system that is deployed in the uplands of the S2 watershed.

In other carbon research, we also are exploring how dissolved organic matter (DOM) composition varies among hydrologic flow paths as well as the long-term trends of watershed total organic carbon (TOC) export. Stream TOC concentrations at the MEF have doubled since the 1980s (Chapter 8). Our long-term data sets are being used to assess the environmental or climatic factors that affect this increase. The studies of DOM composition are focused on characterizing the differences between peatland and upland sources that affect carbon, nitrogen, and mercury biogeochemical cycles.

A third major area of carbon research is aimed at scaling measurements to the landscape and regional scale (Bradford et al. 2009; Weishampel et al. 2009). Related research is assessing the influence on sample type and intensity, and distance between samples on scaling procedures for carbon pools and fluxes (Bradford et al. 2010). In 2009, a study was initiated to determine the decay rates of deadwood. Individual trees were cut in triplicate sites of aspen, red pine, and black spruce. Tree boles were propped up (simulating snags) or laid on the ground (simulating down woody debris) and cookies were cut to determine wood density at the time of felling. Future sampling over the next 20–30 years will allow us to determine decay rates.

We continue to build on the solid foundation of mercury-cycling research with four ongoing studies. In a large manipulative study on experimental and control areas, sulfate inputs were quadrupled across half of the S6 bog relative to ambient atmospheric deposition to study the effect of sulfate availability on methylmercury production (Jeremiason et al. 2006; Coleman-Wasik 2008; Chapter 11). Although the sulfate additions ended in 2008, we continue to monitor outflow and peatland pore waters for mercury to determine the rate of recovery. In a mesocosm study, enriched-abundance mercury isotope tracers were added along with various amounts of sulfate and different carbon sources (e.g., sugars and leaf-tissue leachates) to small chambers in a full factorial experiment to understand the controls on mercury methylation in peatlands. Other ongoing work integrates high-resolution topographic data from light detection and ranging (LiDAR) in a study of topographic controls on methylmercury production (Richardson et al. 2010, in press). In another new study, upland subsurface runoff will be collected to assess the effects of forest harvesting on methylmercury production through the use of small amounts of enriched-abundance isotope mercury tracers.

Past and current studies have contributed to our knowledge of biogeochemical hotspots in the landscape. Mitchell et al. (2008c) showed that the lagg zones along the bog edges where upland and peatland waters mix are biogeochemical hotspots for methylmercury production. Similarly, recently collected data show that lagg zones also are hotspots for the production of nitrous oxide, the possible result from denitrification.

Future Research at the MEF

Over the past decade, recommendations from blue ribbon panels on environmental science, hydrology, ecology, and climate change have identified a consistent set of research priorities. The National Research Council (2001) identified priorities in environmental research related to biogeochemical cycles, biological diversity and ecosystem functioning, climate variability, hydrologic forecasting, infectious disease and the environment, human uses of resources, land use and land cover change, and reinventing the use of resources (recycling). A subsequent panel assessed the state of knowledge about hydrologic consequences of changing forests and developed a list of 13 research needs for forest hydrology. The panel indicated the need to develop a landscape approach to forest hydrology, create a better understanding of the influence of forest disturbance, forest management, and climate change on water and chemical cycles, and to more fully understand cumulative watershed effects on water quality and quantity (National Research Council 2008).

The U.S. Geological Survey (2007) identified 10 year research priorities (2007–2017) that include understanding ecosystems and predicting ecosystem change, climate variability and change, and developing a nationwide water census. The National Science and Technology Council (2007) recommended research to support water availability and water quality. Priorities include identifying ways to use available water supplies more efficiently, developing and improving tools for water management, and also a national water census.

The Heinz Center, a nonprofit institution dedicated to improving the scientific and economic foundation for environmental policy, assessed the state of the nation's ecosystems in 2002 and 2008 (Heinz Center 2002, 2008). The assessment led to a report that identified data gaps that contribute to uncertainty in evaluating ecosystem health (Heinz Center 2006). The report listed 10 data gaps that related to key issues such as wetlands, remotely sensed land-cover data, human exposure to chemical contaminants (e.g., mercury in fish), nitrogen flows in rivers that produce hypoxia, ecosystem carbon storage, endangered and threatened species or communities, extent and impact of nonnative species, condition of plant and animal communities, condition of riparian areas and stream habitat, and use of groundwater levels to quantify groundwater depletion.

The Millennium Ecosystem Assessment (2005) recommended enhancing global monitoring networks to better assess trends in hydrology and land-use change, improve inventories of plant and animal species, provide greater understanding of drivers of ecosystem change at multiple scales, increase our understanding of nonlinear relationships in ecological processes that influence predicted thresholds of change, and collect additional data with which to evaluate ecosystem services, including the relationship between those services and human well-being.

A review by the U.S. Climate Change Science Program (USCCSP) identified 24 research needs (Lucier et al. 2006) in three categories: (1) feedbacks between ecological systems and climate change, (2) consequences of global change for ecological systems, and (3) sustaining and improving ecological systems in response to global climate change. The relevant research under (1) addresses how climate affects energy, water, and trace gas fluxes in terrestrial and aquatic ecosystems and how these changes will feedback on climate change. Changes would include perturbations to nitrogen and carbon cycles, effects on biological diversity and invasive species, and adverse effects on phenology, stomatal conductance, canopy growth, and albedo. The research questions under (2) include recommendations to better understand how changes in temperature and precipitation affect ecosystem carbon, water, and nutrient cycles and how those changes affect biological communities (e.g., aquatic food webs) in managed and unmanaged landscapes. Also included are questions related to ecosystem sensitivity and vulnerability to climate extremes, identifying processes and ecosystem types that are especially vulnerable to climate change or especially resilient, and how climate change affects ecosystem services. The research under (3) is directed at developing better management techniques at multiple scales that include considerations and trade-offs resulting from climate change. On the basis of these recommendations by the USCCSP, the National Research Council (2009) has recommended restructuring research on climate change in the United States, suggesting the development of a system that includes physical, biological, and social observations; a national assessment process to determine the risks and costs of climate change; and improved coordination among Federal agencies in providing tools for decision makers. In addition, recommendations include directed research aimed at better climate forecasting and understanding ecosystem vulnerabilities and their ability to adapt and mitigate climate change, and research focused on interactions among climate, human, and environmental systems.

When research priorities are reviewed, common themes center on how ecosystems respond to climate variability and change. Specifically, research is needed to understand how climate change affects carbon, nitrogen, and water cycles; biodiversity; spread of invasive species; and plant and animal phenology, and how changes in these parameters feedback to affect climate change. Other important areas of research include the effects of land-use change on ecosystems, interactions between land use and aquatic systems, pollutant cycling and human health, nitrogen flows, and groundwater dynamics, and how both climate and land-use change affect ecosystem services. To better assess these effects, the blue ribbon panels have recommended developing new networks to monitor water, land-use change, and plant and animal communities, making better use of current technology such as remote sensing, and developing new monitoring techniques.

The previous and current research direction at the MEF meshes well with research priorities identified by national and global scale assessments. Watershed hydrologic studies and greenhouse gas research at the MEF have contributed to our understanding land–surface interactions that affect global climate forcing. Northern peatlands have been identified as at risk due to climate change (Lenton et al. 2008). Research at the MEF will continue to assess the sensitivity and vulnerability of peatlands to climate change. Studies at the MEF and nearby systems also are measuring the trade-offs of managing forests for fiber production versus maximum ecosystem carbon sequestration. This research can be expanded across broader spatial scales and include additional species.

Understanding ecosystem responses to change at multiple scales encompasses a number of recommendations by the blue ribbon panels. Researchers at MEF are assessing carbon pools and fluxes at multiple scales and new studies will refine these assessments and extend to other chemical constituents such as DOM and mercury. We recently used LiDAR to map the vegetation and topography of the South Unit of the MEF. The accuracy and precision of the elevation data provided by LiDAR will increase our understanding of hydrological and chemical flow pathways and allow us to investigate the topographical controls and hotspots of biogeochemical cycling in peatland landscapes (Mitchell et al. 2008c). LiDAR canopy data will provide a new context for measuring vegetation and carbon pools on the MEF.

The gaged watersheds at the MEF have provided valuable insight into the effects of harvesting on water quality and quantity. We plan to extend our data and use MEF flow and chemistry data within the context of the larger Great Lakes region to determine the effects of land-use change on water resources. Although our two control watersheds (S2 and S5) have been used for comparisons of our manipulations at MEF, they have seldom been compared with other watersheds in the region. We plan to expand our small watershed research to determine the cumulative effects of forest management and natural disturbances on water yield and chemistry in larger landscapes. The scaling up of our watershed, carbon and mercury research will likely lead to developing collaborations with a new set of scientists proficient at using tools like remote sensing to help in the scaling process.

The mercury research program has contributed to our understanding of mercury cycles in northern forest landscapes, but much work remains. Although we have begun to tease apart the chemical and topographic controls on the mercury methylation process, we need to more fully understand how the sources of sulfate and DOM affect the methylation process and whether certain fractions of DOM are more important to methylation than others. There has been little research on the effects of climate change on mercury cycling. Climate change likely will affect hydrology, composition and productivity of plant communities, decomposition rates, and atmospheric inputs of mercury and other constituents. As a result, mercury currently stored in terrestrial systems may become more or less available for remobilization depending on changes in these processes. Investigating mercury speciation under elevated temperature, CO_2, and variable water tables could enhance our ability to predict the effect of climate change on mercury cycling.

The location of the MEF within the northern forest provides unique research opportunities related to the response of forests to climate change. For example, its location near the edge of the boreal forest places the MEF near the southern boundary of the range of many important plant species, notably black spruce. As a result, the MEF would be ideal for monitoring the long-term impacts of climate change on lowland conifer ecosystems. A comprehensive monitoring program targeted at capturing vegetation responses to climate change would strengthen the research portfolio at the MEF since many of our long-term measurements are currently focused on water or carbon cycling. A planned, large-chamber experiment will measure ecosystem responses to soil and air warming up to 9°C and doubled CO_2 concentrations in the black spruce bog in the S1 watershed. This experiment should reveal thresholds of change for peatland ecosystems under warming or elevated CO_2 or under the combined effects of both elevated temperature and CO_2.

As discussed earlier, the MEF is part of multiple networks, both national and international, which contribute to understanding ecosystems at larger scales when combined with other sites. Data collected at the MEF NADP and MDN sites and the entire network have advanced research in atmospheric and ecosystems science. One new network of sites is assessing the inputs of mercury in litterfall across the eastern United States. The MEF is now in its second year contributing to this effort. Although not currently part of the AmeriFlux Network, in the next few years we plan to add the eddy covariance data from the Bog Lake fen to the AmeriFlux Network of sites. As new networks develop, it is a priority for the MEF to contribute relevant information and expertise.

In our vision, the MEF will continue to contribute to priority basic and applied research through studies initiated by our scientists, university and other agency collaborators, graduate students, and postdoctoral scholars. The MEF has been and continues to be a test site for new monitoring technologies. A core activity is the extension of our long-term hydrological, meteorological, biogeochemical, and trace gas flux data series. In addition to continued monitoring that will extend our long-term records on forest harvest effects, we will conduct manipulative experiments aimed at issues that are important today and that likely will be important in the future (e.g., climate change and carbon). Based on our extensive long-term databases and rich history of research, we are well poised to address these important future issues.

References

Barten, P.K. and K.N. Brooks. 1988. Modeling streamflow from headwater areas in the northern Lake States. In *Modeling Agricultural, Forest, and Rangeland Hydrology*. ASAE Publication 07-88. American Society of Agricultural Engineers, pp. 347–356.

Bartlett, K.B. and R.C. Harriss. 1993. Review and assessment of methane emissions from wetlands. *Chemosphere* 26(1–4):261–320.

Bay, R.R. 1966. Evaluation of an evapotranspirometer for peat bogs. *Water Resources Research* 2(3):437–441.

Bay, R.R. 1967. Ground water and vegetation in two peat bogs in northern Minnesota. *Ecology* 48(2):308–310.

Bay, R.R. 1968. The hydrology of several peat deposits in northern Minnesota, USA. In *Proceedings of the Third International Peat Congress*. Quebec, Canada: National Research Council of Canada, pp. 212–218.

Bay, R.R. 1970a. Water resources research in Minnesota by the North Central Forest Experiment Station, USDA Forest Service. In *Proceedings of Conference on Ongoing Water Resources Research in Minnesota*, March 1970, Minneapolis, MN. St. Paul, MN: Water Resources Research Center, University of Minnesota, pp. 50–56.

Bay, R.R. 1970b. Water table relationships on experimental basins containing peat bogs. Symposium on the results of research on representative and experimental basins. In *Proceedings of the Wellington Symposium*, Paris, France, December 1970. UNESCO/International Association of Hydrological Sciences, Vol. 96, pp. 360–368.

Berglund, E.R. and A.C. Mace. 1972. Seasonal albedo variation of black spruce and *Sphagnum*-sedge bog cover types. *Journal of Applied Meteorology* 11(5):806–812.

Berglund, E.R. and A.C. Mace. 1976. Diurnal albedo variation of black spruce and *Sphagnum*–sedge bogs. *Canadian Journal of Forest Research* 6(3):247–252.

Berguson, W.E. and D.A. Perala. 1988. Aspen fertilization and thinning research results and future potential. In *Minnesota's Timber Supply: Perspectives and Analysis*, eds. A.R. Ek and H.M. Hoganson. Grand Rapids, MN: University of Minnesota, College of Natural Resources and Agricultural Experiment Station, pp. 176–183.

Betancourt, J.L. and M.D. Schwartz. 2005. Implementing a U.S. National Phenology Network. *EOS Transactions* 86(51):539–541.

Boelter, D.H. 1964a. Water storage characteristics of several peats in situ. *Soil Science Society of America Proceedings* 33:433–435.

Boelter, D.H. 1964b. Laboratory techniques for measuring water storage properties of organic soils. *Soil Science Society of America Proceedings* 28(6):823–824.

Boelter, D.H. 1965. Hydraulic conductivity of peats. *Soil Science* 100:227–231.

Boelter, D.H. 1966. Hydrologic characteristics of organic soils in Lakes States Watersheds. *Journal of Soil and Water Conservation* 21(2):50–53.

Boelter, D.H. 1968. Important physical properties of peat materials. In *Proceedings of the Third International Peat Congress*. Quebec, Canada: National Research Council of Canada, pp. 150–154.

Boelter, D.H. 1969. Physical properties of peats as related to degree of decomposition. *Soil Science Society of America Proceedings* 33:606–609.

Boelter, D.H. 1972a. Preliminary results of water level control on small plots in a peat bog. In *The Use of Peatland for Agriculture, Horticulture, and Forestry. Fourth International Peat Congress Proceedings*. Ontanienmi, Finland: International Peat Congress, Vol. 3, pp. 347–354.

Boelter, D.H. 1972b. Water table drawdown around an open ditch in organic soils. *Journal of Hydrology* 15(4):329–340.

Boelter, D.H. 1974a. The hydrologic characteristics of undrained organic soils in the Lake States. In *Histosols: Their Characteristics, Use, and Classification*. Madison, WI: Soil Science Society of America, pp. 33–46.

Boelter, D.H. 1974b. The ecological fundamentals of forest drainage peatland hydrology: Peatland water economy. In *The Coordinators' Papers and Discussions of the International Symposium on Forest Drainage*, ed. L. Heikurainen, Jyväskylä-Oulu, Finland, September 2–6, 1974. Helsinki, Finland, pp. 35–48.

Boelter, D.H. 1975. Methods for analyzing the hydrological characteristics of organic soils in marsh-ridden areas. In *International Symposium on the Hydrology of Marsh-ridden Areas. Proceedings of the Minsk Symposium*, Minsk, Byelorussia, June 1972. Paris, France: UNESCO/International Association of Hydrological Sciences, Vol. 105, pp. 161–169.

Bradford, J., P. Weishampel, M.L. Smith, R. Kolka, R.A. Birdsey, S.A. Ollinger, and M.G. Ryan. 2009. Detrital carbon pools in temperate forests: Magnitude and potential for landscape-scale assessment. *Canadian Journal of Forest Research* 39:802–813.

Bradford, J., P. Weishampel, M.L. Smith, R. Kolka, R.A. Birdsey, S.A. Ollinger, and M.G. Ryan. 2010. Carbon pools and fluxes in small temperate forest landscapes: Variability and implications for sampling design. *Forest Ecology and Management* 259:1245–1254.

Brooks, K.N. and D.R. Kreft. 1989. A hydrologic model for Minnesota peatlands. *Journal of the Minnesota Academy of Science* 55(1):113–119.

Brown, J.M. 1972. Effect of clearcutting a black spruce bog on net radiation. *Forest Science* 18(4):273–277.

Brown, A.E., L. Zhang, T.A. McMahon, A.W. Western, and R.A. Vertessy. 2005. A review of paired catchment studies for determining changes in water yield resulting from alterations in vegetation. *Journal of Hydrology* 310(1–4):28–61.

Chanton, J.P., J.E. Bauer, P.A. Glaser, D.I. Siegel, C.A. Kelley, S.C. Tyler, E.H. Romanowicz, and L. Lazrus. 1995. Radiocarbon evidence for the substrates supporting methane formation within northern Minnesota peatlands. *Geochimica et Cosmochimica Acta* 59(17):3663–3668.

Clement, R.J., S.B. Verma, and E.S. Verry. 1995. Relating chamber measurements to eddy correlation measurements of methane flux. *Journal of Geophysical Research, D, Atmospheres* 100(D10):21047–21056.

Coleman-Wasik, J.K. 2008. Chronic effects of atmospheric sulfate deposition on mercury methylation in a boreal wetland: Replication of a global experiment. MS thesis. St. Paul, MN: University of Minnesota.

Crill, P.M., K.B. Bartlett, R.C. Harriss, E. Gorham, E.S. Verry, D.I. Sebacher, L. Madzar, and W. Sanner. 1988. Methane flux from Minnesota peatlands. *Global Biogeochemical Cycles* 2(4):371–384.

Cui, J., C. Li, and C.C. Trettin. 2005. Analyzing the ecosystem carbon and hydrologic characteristics of forested wetland using a biogeochemical process model. *Global Change Biology* 11(2):278–289.

Dise, N.B. 1992. Winter fluxes of methane from Minnesota peatlands. *Biogeochemistry* 17(2):71–83.

Dise, N.B. 1993. Methane emission from Minnesota peatlands: Spatial and seasonal variability. *Global Biogeochemical Cycles* 7(1):123–142.

Dise, N.B. and E.S. Verry. 2001. Suppression of peatland methane emission by cumulative sulfate deposition in simulated acid rain. *Biogeochemistry* 53(2):143–160.

Dise, N.B., E. Gorham, and E.S. Verry. 1993. Environmental factors controlling methane emissions from peatlands in northern Minnesota. *Journal of Geophysical Research, D, Atmospheres* 98(D6):10583–10594.

Fissore, C. 2007. Biotic and abiotic controls on soil organic carbon quality along a paired pine and hardwood climosequence. PhD dissertation. Houghton, MI: Michigan Technological University.

Fissore, C., C.P. Giardina, R.K. Kolka, C.C. Trettin, G.M. King, M.F. Jurgensen, C.D. Barton, and S.D. McDowell. 2008. Temperature and vegetation effects on soil organic carbon quality along a forested mean annual temperature gradient in North America. *Global Change Biology* 14(1):193–205.

Gafni, A. and K.N. Brooks. 1986a. Hydrologic properties of natural versus mined peatlands. In *Advances in Peatlands Engineering*. Ottawa, Canada: National Research Council of Canada, pp. 184–190.

Gafni, A. and K.N. Brooks. 1986b. Tracing approach to determine groundwater velocity in peatlands. *Hydrological Science and Technology* 2(4):17–23.

Gafni, A. and K.N. Brooks. 1990. Hydraulic characteristics of four peatlands in Minnesota. *Canadian Journal of Soil Science* 70(2):239–253.

Gorham, E., S.J. Eisenreich, J. Ford, and M.V. Santelmann. 1985. The chemistry of bog waters. In *Chemical Processes in Lakes*, ed. W. Stumm. New York: John Wiley and Sons, pp. 339–363.

Grigal, D.F. and K.N. Brooks. 1997. Forest management impacts on undrained peatlands in North America. In *Northern Forested Wetlands: Ecology and Management*, eds. C.C. Trettin, M.F. Jurgensen, D.F. Grigal, M.R. Gale, and J.K. Jeglum. Boca Raton, FL: CRC/Lewis Publishers, pp. 379–396.

Guertin, D.P. 1984. Modeling streamflow response from Minnesota peatlands. PhD dissertation. St. Paul, MN: University of Minnesota.

Guertin, D.P., P.K. Barten, and K.N. Brooks. 1987. The peatland hydrologic impact model: Development and testing. *Nordic Hydrology* 18:79–100.

Guertin, D.P. and K.N. Brooks. 1985. Modeling streamflow response from Minnesota peatlands. In *Watershed Management in the Eighties*, eds. E.B. Jones and T.J. Ward. *Proceedings, Committee on Watershed Management, Irrigation and Drainage Division,*

American Society of Civil Engineers Symposium in Conjunction with the ASCE Convention, Denver, CO, April 30–May 1, 1985. New York: American Society of Civil Engineers, pp. 123–131.

Guillemette, F., A.P. Plamondon, M. Prevost, and D. Levesque. 2005. Rainfall generated stormflow response to clearcutting a boreal forest: Peak flow comparison with 50 world-wide basin studies. *Journal of Hydrology* 302(1–4):137–153.

Harriss, R.C., E. Gorham, D.I. Sebacher, K.B. Bartlett, and P.A. Flebbe. 1985. Methane flux from northern peatlands. *Nature* 315(6021):652–654.

Heinz Center. 2002. *The State of the Nations Ecosystems: Measuring the Lands, Waters and Living Resources of the United States*. New York: Cambridge University Press.

Heinz Center. 2006. *Filling the Gaps: Priority Data Needs and Key Management Challenges for National Reporting on Ecosystem Condition*. Washington, DC: Heinz Center.

Heinz Center. 2008. *The State of the Nations Ecosystems 2008: Measuring the Lands, Waters and Living Resources of the United States*. Washington, DC: Island Press.

Holscher, C.E. 1967. Forestry hydrology research in the United States. In *Forest Hydrology*, eds. W.E. Sopper and H.W. Lull. New York: Pergamon Press, pp. 99–103.

Hornbeck, J.W., M.B. Adams, E.S. Corbett, E.S. Verry, and J.A. Lynch. 1993. Long-term impacts of forest treatments on water yield: A summary for northeastern USA. *Journal of Hydrology* 150(2–4):323–344.

Hornibrook, E.R., F.J. Longstaffe, and W.S. Fyfe. 2000. Evolution of stable carbon isotope compositions for methane and carbon dioxide in freshwater wetlands and other anaerobic environments. *Geochimica et Cosmochimica Acta* 64(6):1013–1027.

Jeremiason, J.D., D.R. Engstrom, E.B. Swain, E.A. Nater, B.M. Johnson, J.E. Almendinger, B.A. Monson, and R.K. Kolka. 2006. Sulfate addition increases methylmercury production in an experimental wetland. *Environmental Science and Technology* 40(12):3800–3806.

Kelly, C.A., N.B. Dise, and C.S. Martens. 1992. Temporal variations in the stable carbon isotopic composition of methane emitted from Minnesota peatlands. *Global Biogeochemical Cycles* 6(3):263–269.

Kim, J. and S.B. Verma. 1992. Soil surface CO_2 flux in a Minnesota peatland. *Biogeochemistry* 18(1):37–51.

Knighton, M.D. and J.H. Stiegler. 1981. Phosphorus release following clearcutting of a black spruce fen and a black spruce bog. In *Proceedings of the Sixth International Peat Congress*, August 17–23, 1980, Duluth, MN. Eveleth, MN: W.A. Fisher Company, pp. 577–583.

Kolka, R.K., D.F. Grigal, E.A. Nater, and E.S. Verry. 2001. Hydrologic cycling of mercury and organic carbon in a forested upland-bog watershed. *Soil Science Society of America Journal* 65(3):897–905.

Kolka, R.K., D.F. Grigal, E.S. Verry, and E.A. Nater. 1999. Mercury and organic carbon relationships in streams draining forested upland/peatland watersheds. *Journal of Environmental Quality* 28(3):766–775.

Lansdown, J., P.D. Quay, and S.L. King. 1992. CH_4 production via CO_2 reduction in a temperate bog: A source of ^{13}C-depleted CH_4. *Geochimica et Cosmochimica Acta* 56(9):3493–3503.

Lenton, T.M., H. Held, E. Kriegler, J.W. Hall, W. Lucht, S. Rahmstorf, and H.J. Schellnhuber. 2008. Tipping elements in the Earth's climate system. *Proceedings of the National Academy of Sciences* 105:1786–1793.

Lu, S.-Y. 1994. Forest harvesting effects on streamflow and flood frequency in the northern Lake States. PhD dissertation. St. Paul, MN: University of Minnesota.

Lucier, A., M. Palmer, H. Mooney, K. Nadelhoffer, D. Ojima, and F. Chavez. 2006. Ecosystems and climate change: Research priorities for the U.S. Climate Change Science Program. Recommendations from the Scientific Community. Report on an Ecosystem Workshop, prepared for the Ecosystems Interagency Working Group. Special Series No. SS-92–06. Solomons, MD: University of Maryland Center for Environmental Science.

Lugo, A.E., F.J. Swanson, O.R. González, M.B. Adams, M.J. Palik, R.E. Thill, D.G. Brockway, C. Kern, R. Woodsmith, and R.C. Musselman. 2006. Long-term research at USDA Forest Service's experimental forests and ranges. *BioScience* 56(1):39–48.

Malterer, T.J., A.D. Cohen, and E.S. Verry. 1987. Comparison of fiber content determination by NRCC and proposed ASTM methods. In *Symposium '87 Wetlands/Peatlands*, eds. C.D.A. Rubec and R.P. Overend. Edmonton, Alberta, Canada, August 23–27, 1987. Ottawa, Ontario, Canada: International Peat Society, pp. 7–15.

Malterer, T.J., E.S. Verry, and J. Erjavec. 1992. Fiber content and degree of decomposition in peats: Review of national methods. *Soil Science Society of America Journal* 56(4):1200–1211.

McAdams, T.V.W. 1993. Modeling water table response to climatic change in a northern Minnesota peatland. MS thesis. St. Paul, MN: University of Minnesota.

Millennium Assessment Report. 2005. *Ecosystems and Human Well-being: Synthesis.* Washington, DC: Island Press.

Miller, J.M. 1980. National atmospheric deposition program: Analysis of data from the first year. In *Atmospheric Sulfur Deposition. Environmental Impact and Health Effects*, eds. D.S. Shriner, C.R. Richmond, and S.E. Lindberg. Ann Arbor, MI: Ann Arbor Science Publishers, pp. 469–476.

Mitchell, C.P.J., B.A. Branfireun, and R.K. Kolka. 2008a. Total mercury and methylmercury dynamics in upland–peatland watersheds during snowmelt. *Biogeochemistry* 90(3):225–241.

Mitchell, C.P.J., B.A. Branfireun, and R.K. Kolka. 2008b. Assessing sulfate and carbon controls on net methylmercury production in peatlands: An in situ mesocosm approach. *Applied Geochemistry* 23(3):503–518.

Mitchell, C.P.J., B.A. Branfireun, and R.K. Kolka. 2008c. Spatial characteristics of net methylmercury production hot spots in peatlands. *Environmental Science and Technology* 42(4):1010–1016.

Mitchell, C.P.J., B.A. Branfireun, and R.L. Kolka. 2009. Methylmercury dynamics at the upland-peatland interface: Topographic and hydrogeochemical controls. *Water Resources Research* 45:W02406, doi:10.1029/2008WR006832.

Moran, M.S., D.P.C. Peters, M.P. McClaran, M.H. Nichols, and M.B. Adams. 2008. Long-term data collection at USDA experimental sites for studies of ecohydrology. *Ecohydrology* 1(4):377–393.

National Research Council. 2001. *Grand Challenges in Environmental Sciences.* Washington, DC: National Academy Press.

National Research Council. 2008. *Hydrologic Effects of a Changing Forest Landscape.* Washington, DC: National Academy Press.

National Research Council. 2009. *Restructuring Federal Climate Change Research to Meet the Challenges of Climate Change.* Washington, DC: National Academy Press.

National Science and Technology Council. 2007. *A Strategy for Federal Science and Technology to Support Water Availability and Quality in the United States.* Washington, DC: Office of the President.

Nichols, D.S. and D.H. Boelter. 1984. Fiber size distribution, bulk density, and ash content of peats in Minnesota, Wisconsin, and Michigan. *Soil Science Society of America Journal* 48(6):1320–1328.

Nichols, D.S. and J.M. Brown. 1980. Evaporation from a *Sphagnum* moss surface. *Journal of Hydrology* 48(3–4):289–302.

Nichols, D.S. and E.S. Verry. 2001. Stream flow and ground water recharge from small forested watersheds in north central Minnesota. *Journal of Hydrology* 245(1–4):89–103.

Ohmann, L.F., H.O. Batzer, R.P. Buech, D.C. Lothner, D.A. Perala, A.L. Schipper, and E.S. Verry. 1978. Some harvest options and their consequences for the aspen, birch, and associated forest types of the Lake States. General Technical Report NC-48. St. Paul, MN: USDA Forest Service.

Perala, D.A. 1983. Growth response and economics of thinning and fertilizing aspen and paper birch. In *Silviculture of Established Stands in North Central Forests*, ed. J.C. Stier. Bethesda, MD: Society of American Foresters, pp. 100–110.

Powers, R.F., S.D. Andrew, F.G. Sanchez, R.A. Voldseth, D.S. Page-Dumroese, J.D. Elioff, and D.M. Stone. 2005. The North American long-term soil productivity experiment: Findings from the first decade of research. *Forest Ecology and Management* 220(1–3):31–50.

Quay, P.D., S.L. King, J.M. Lansdown, and D.O. Wilbur. 1988. Isotopic composition of methane released from wetlands: Implications for the increase in atmospheric methane. *Global Biogeochemical Cycles* 2(4):385–397.

Richardson, M.C., C.P.J. Mitchell, B.A. Branfireun, and R.K. Kolka. 2010. Analysis of airborne LiDAR surveys to quantify the characteristic morphologies of northern forested wetlands. *Journal of Geophysical Research, G, Biogeosciences,* 115:G03005, doi:10.1029/2009JG000972.

Sahin, V. and M.J. Hall. 1996. The effects of afforestation and deforestation on water yields. *Journal of Hydrology* 178(1–4):293–309.

Sander, J.E. 1976. An electric analog approach to bog hydrology. *Ground Water* 14(1):30–35.

Shotyk, W. 1988. Review of the inorganic geochemistry of peats and peatland waters. *Earth Science Review* 25:95–176.

Shurpali, N.J. and S.B. Verma. 1998. Micrometeorological measurements of methane flux in a Minnesota peatland during two growing seasons. *Biogeochemistry* 40(1):1–15.

Shurpali, N.J., S.B. Verma, R.J. Clement, and D.P. Billesbach. 1993. Seasonal distribution of methane flux in Minnesota Peatland measured by eddy correlation. *Journal of Geophysical Research, D, Atmospheres* 98(D11):20649–20655.

Skyllberg, U. 2008. Competition among thiols and inorganic sulfides and polysulfides for Hg and MeHg in wetland soils and sediments under suboxic conditions: Illumination of controversies and implications for MeHg net production. *Journal of Geophysical Research, G, Biogeosciences* 113:G00C03, doi:10.1029/2008JG000745.

Skyllberg, U., K. Xia, P.R. Bloom, E.A. Nater, and W.F. Bleam. 2000. Binding of mercury(II) to reduced sulfur in soil organic matter along upland-peat soil transects. *Journal of Environmental Quality* 29(3):855–865.

Skyllberg, U., P.R. Bloom, J. Qian, C.M. Lin, and W.F. Bleam. 2006. Complexation of mercury(II) in soil organic matter: EXAFS evidence for linear two-coordination with reduced sulfur groups. *Environmental Science and Technology* 40(13):4174–4180.

Smemo, K.A. and J.B. Yavitt. 2007. Evidence for anaerobic CH_4 oxidation in freshwater peatlands. *Geomicrobiology Journal* 24(7):583–597.

Stednick, J.D. 1996. Monitoring the effects of timber harvest on annual water yield. *Journal of Hydrology* 176(1–4):79–95.

Stevens, C.M. and A. Engelkeimer. 1988. Stable carbon isotopic composition of methane from some natural and anthropogenic sources. *Journal of Geophysical Research, D, Atmospheres* 93(D1):725–733.

Tiarks, A.E., R.F. Powers, D.H. Alban, G.A. Ruark, and D.S. Page-Dumroese. 1993. USFS long-term soil productivity national research project: A USFS cooperative research program. In *Proceedings of the 8th International Soil Management Workshop: Utilization of Soil Survey Information for Sustainable Land Use*, ed. J.M. Kimble, May 1993. Lincoln, NE: USDA Soil Conservation Service, National Soil Survey Center, pp. 236–241.

Trettin, C.C., M.F. Jurgensen, M.R. Gale, and J.W. McLaughlin. 1995. Soil carbon in northern forested wetlands: Impacts of silvicultural practices. In *Carbon Forms and Functions in Forest Soils*, eds. W.W. McFee and J.M. Kelly. Madison, WI: Soil Science Society of America, pp. 437–461.

Trettin, C.C., R. Laiho, and L.J. Minkkinen. 2006. Influence of climate change factors on carbon dynamics in northern forested peatlands. *Canadian Journal of Soil Science* 86:269–280.

Urban, N.R., S.E. Bayley, and S.J. Eisenreich. 1989a. Export of dissolved organic carbon and acidity from peatlands. *Water Resources Research* 25(7):1619–1628.

Urban, N.R. and S.J. Eisenreich. 1988. Nitrogen cycling in a forested Minnesota bog. *Canadian Journal of Botany* 66(3):435–449.

Urban, N.R., S.J. Eisenreich, and E. Gorham. 1987. Proton cycling in bogs: Geographic variation in northeastern North America. In *Effects of Atmospheric Pollutants on Forests, Wetlands, and Agricultural Ecosystems*, eds. T.C. Hutchinson and K.M. Meema. New York: Springer-Verlag, Vol. G16, pp. 577–598.

Urban, N.R., S.J. Eisenreich, and D.F. Grigal. 1989b. Sulfur cycling in a forested *Sphagnum* bog in northern Minnesota. *Biogeochemistry* 7(2):81–109.

Urban, N.R., E.S. Verry, and S.J. Eisenreich. 1995. Retention and mobility of cations in a small peatland: Trends and mechanisms. *Water, Air, and Soil Pollution* 79(1–4):201–224.

U.S. Geological Survey. 2007. Facing Tomorrow's Challenges–U.S. Geological Survey Science in the Decade 2007–2017. U.S. Geological Survey Circular 1309.

Verma, S.B., J. Kim, R.J. Clement, N.J. Shurpali, and D.P. Billesbach. 1995. Trace gas and energy fluxes: Micrometeorological perspectives. In *Soils and Global Change*, eds. R. Lal, J. Kimble, E. LeVine, and B.A. Stewart. New York: CRC Press, pp. 361–376.

Verma, S.B., F.G. Ullman, D. Billesbach, R.J. Clement, J. Kim, and E.S. Verry. 1992a. Eddy correlation measurements of methane flux in a northern peatland ecosystem. *Boundary-Layer Meteorology* 58(3):289–304.

Verma, S.B., F.G. Ullman, N.J. Shurpali, R.J. Clement, J. Kim, and D.P. Billesbach. 1992b. Micrometeorological measurements of methane and energy fluxes in a Minnesota peatland. *Suo* 43(4–5):285–288.

Vermette, S.J., S.E. Lindberg, and N. Bloom. 1995a. Field tests for a regional mercury deposition network—Sampling design and preliminary test results. *Atmospheric Environment* 29(11):1247–1251.

Vermette, S.J., M.R. Peden, S. Hamdy, T.C. Willoughby, L. Schroder, S.E. Lindberg, J.G. Owens, and A.D. Weiss. 1992. A pilot network for the collection and analysis of metals in wet deposition. In *The Deposition and Fate of Trace Metals in Our Environment*, eds. E.S. Verry and S.J. Vermette. Newtown Square, PA: USDA Forest Service, pp. 73–84.

Vermette, S.J., M.E. Peden, T.C. Willoughby, S.E. Lindberg, and A.D. Weiss. 1995b. Methodology for the sampling of metals in precipitation: Results of the national atmospheric deposition program (NADP) pilot network. *Atmospheric Environment* 29(11):1221–1229.

Verry, E.S. 1972. Effect of an aspen clearcutting on water yield and quality in northern Minnesota. In *Proceedings of a Symposium on Watersheds in Transition*, eds. S.C. Csallany, T.G. McLaughlin, and D. Striffler, Fort Collins, CO, June 19–22, 1972. Urbana, IL: American Water Resources Association, pp. 276–284.

Verry, E.S. 1981. Water table and streamflow changes after stripcutting and clearcutting an undrained black spruce bog. In *Proceedings of the Sixth International Peat Congress*, Duluth, MN, August 17–23, 1980. Eveleth, MN: W.A. Fisher Company, pp. 493–498.

Verry, E.S. 1987. The effect of aspen harvest and growth on water yield in Minnesota. In *Forest Hydrology and Watershed Management – Hydrologie Forestière et Aménagement des Bassins Hydrologiques*. Wallingford, U.K.: IAHS, Vol. 167, pp. 553–562.

Verry, E.S. 1996. Effects of forestry practices on physical and chemical resources. In *At the Water's Edge: The Science of Riparian Forestry Conference Proceedings*, Duluth, MN, June 19–20, 1995. St. Paul, MN: University of Minnesota Press, pp. 101–106.

Verry, E.S. 2003. Ground water and small research basins: An historical perspective. *Ground Water* 41(7):1005–1007.

Verry, E.S. 2004. Land fragmentation and impacts to streams and fish in the central and upper Midwest. In *Lessons for Watershed Research in the Future; A Century of Forest and Wildland Watershed Lessons*, eds. G.G. Ice and J.D. Stednick. Bethesda, MD: Society of American Foresters, pp. 129–154.

Verry, E.S. and N.R. Urban. 1992. Nutrient cycling at Marcell Bog, Minnesota. *Suo* 43(4–5):147–153.

Waldron, S., J. Lansdown, E.M. Scott, A.E. Fallick, and A.J. Hall. 1999. The global influence of the hydrogen isotope composition of water on that of bacteriogenic methane from shallow freshwater environments. *Geochimica et Cosmochimica Acta* 63(15):2237–2245.

Wania, R., I. Ross, and I.C. Prentice. 2009. Integrating peatlands and permafrost into a dynamic global vegetation model: 1. Evaluation and sensitivity of physical land surface processes. *Global Biogeochemical Cycles* 23:GB3014, doi:10.1029/2008GB003413.

Weishampel, P., R.K. Kolka, and J.Y. King. 2009. Carbon pools and productivity in a 1-km^2 heterogeneous forest and peatland mosaic in Minnesota, USA. *Forest Ecology and Management* 257(2):747–754.

Zhang, Y., C. Li, C.C. Trettin, H. Li, and G. Sun. 2002. An integrated model of soil, hydrology, and vegetation for carbon dynamics in wetland ecosystems. *Global Biogeochemical Cycles* 16(4):1–17.

4

Geology, Vegetation, and Hydrology of the S2 Bog at the MEF: 12,000 Years in Northern Minnesota

Elon S. Verry and Joannes Janssens

CONTENTS

Introduction

A clear understanding of geology and landscape setting is fundamental to the interpretation of water and solute movement among landscape forms. This understanding allows us to assess how land use affects water, soils, and vegetation as well as assess the fate of acids, nutrients, trace metals, and organic compounds deposited from the atmosphere. Pleistocene Glaciation and the Holocene accumulation of peat in ice-block depressions are two primary determinants of the landscape in the recessional moraine area of north-central Minnesota, which is the setting of the Marcell Experimental Forest (MEF). We examine these two processes using studies of lacustrine and peat sediment accumulation, ice-block paleobotany, mineral-soil development, and a review of Wisconsin Glaciation at the MEF.

An understanding of geology and soil development illuminates the physical link between water in "pothole" bogs and recharge of water in the lagg at their perimeter to groundwater aquifers. Regional paleobotany patterns reveal broad climate change as a third driver of landscape development, imposing its control on water levels, plant migration, interspecies competition, and the vegetation community—in short, a 12,000 year interpretation of Earth processes since a great ice sheet.

This chapter is an account of geologic, hydrologic, paleobotanical, and current botanical development in and around a single ice-block depression now occupied by a black spruce (*Picea mariana*)-*Sphagnum* bog on deep peat (Figure 4.1). Included with interpretations of the published literature are previously unpublished soil, water chemistry, piezometer and well records, topographic surveys, rigorous peat-core dating and pollen analysis, and a critical, regionally specific application of recent interpretations of Wisconsin Glaciation and climate change. The insights into environmental change over 26 millennia place our current evaluations of anthropogenic environmental change in perspective. In particular, the last 20,000 years, including Late-Wisconsin Glaciation and the Holocene (the last 12,000 years at this site), encompass fundamental geologic events that drive today's ecosystems and our interpretation of their function.

FIGURE 4.1
Oblique view of S2 bog on June 11, 1964 (photo by Bluford Muir, USDA Forest Service). The central peatland filling an ice block depression is forested with black spruce and the uplands with aspen and birch. The view is looking NNE. Numbers locate instrumentation: weir (1), weather station (2), bog well (3), upland wells (4–6), upland runoff plots (7, 8), soil pit A (9), soil pit B (10), piezometer nests (11–16), soil temperature stack (17), and neutron soil moisture tubes (18, 19). The arrow identifies the exit stream and the dashed line shows a picture-truncated watershed boundary.

General Setting and Contemporary Climate

The MEF is located in north central Itasca County, Minnesota (Lat. 47:31:52N, Long. 93:28:07W) at about 430 m above sea level (NAVD 1929). The Marcell Hills are in the Humid Temperate Domain (200), Warm Continental Division (210), Laurentian Mixed Forest Province (212), Northern Minnesota Drift and Lake Plains Section (212), and St. Louis Moraines Subsection (212Nb) of terrestrial ecoregions in North America (Keys and Carpenter, 1995).

In the North American aquatic classification (Maxwell et al., 1995), it is in the Nearctic Zone (North America), the Arctic-Atlantic Subzone (A), the Mississippi Region (A2), and at the very northern tip of the Upper Mississippi Subregion (A2a), several kilometers west of the tricontinental divide point for the Upper Mississippi, St. Lawrence, and Hudson Bay watersheds.

The climate is strongly continental (Chapter 2). The mean annual precipitation from 1961 to 2000 was 78 cm, and the annual temperature 3.3°C. Monthly precipitation ranges from 4 cm in February to 33 cm in August (Figure 4.2) and annual precipitation ranges from 412 to 946 mm. Monthly average temperatures are 16°C, 19°C, and 15°C in June, July, and August, and −12°C, −16°C, and −11°C in December, January, and February (Figure 4.3).

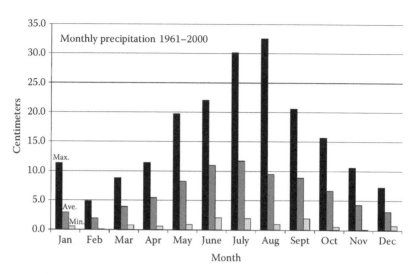

FIGURE 4.2
Monthly maximum, average, and minimum precipitation for the S2 meteorological station.

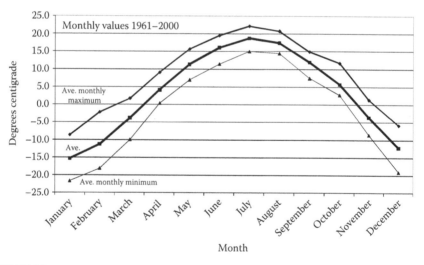

FIGURE 4.3
Monthly maximum, average, and minimum temperature for the S2 meteorological station.

Bedrock Setting

The MEF site is underlain by the Giants Range Batholith, a large complex of intrusions formed about 2.7 Ga before present (BP) (Figure 4.4). The bedrock lies some 40–50 m under late Wisconsin glacial drift (Oakes and Bidwell, 1968). The MEF and S2 bog lie in a recessional moraine complex (the Marcell

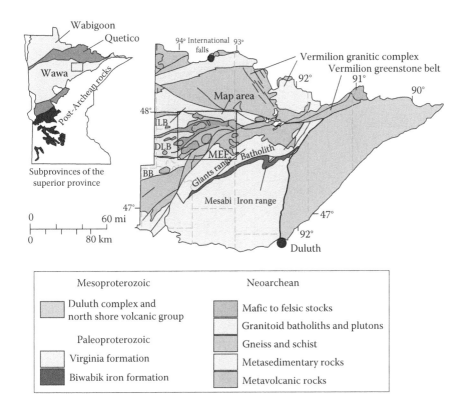

FIGURE 4.4
The MEF lies within the Bigfork USGS quadrangle (black rectangle) above the Giants Ridge, granitoid, batholith composed of tonalite to granodiorite, part of the Wawa subprovince of the Superior bedrock province (from Jirsa and Chandler, 2007). Black lines are major faults or NW trending dikes. ILB is Island Lake batholith, DLB is Dora Lake batholith, and BB is Bemidji batholith.

Hills) formed by Late-Wisconsin glaciation (26–12 ka BP), with elevation ranges of 10–30 m, mineral soil side slopes of 5%–40%, flat-topped hills, and peat- or lake-filled, ice-block depressions (Figure 4.5). Because this location is within 3 km of the Hudson Bay-Mississippi Continental Divide, we speculate that the bedrock surface may also be a topographic high.

Methods

Mapping

In 1966, Mark Hurd Aerial Surveys (Minneapolis, Minnesota) produced topographic maps with 4 ft contours for the MEF. We interpolated contour lines by hand at 1 m intervals. In 1982, we recorded a hollow and a hummock

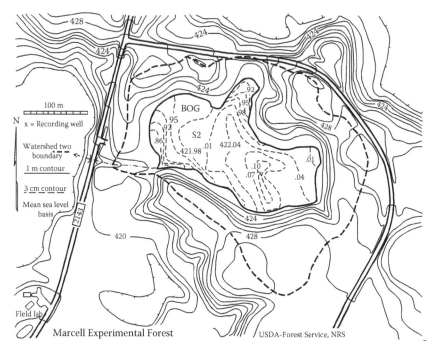

FIGURE 4.5
Surface contours of the uplands (1 m interval) and peatland (3 cm interval) in the S2 watershed.
A surface stream drains from the southwest tip of the bog.

surface elevation near each point on a 30.5 m grid over the S2 peatland (Verry, 1984). In 1988, we sounded the same grid for depth to the Koochiching till the surface lying below the peat deposit and mapped 1 m contours of the ice-block depression. We mapped only the hollow elevation of the peat surface and interpolated a series of 3 cm contour intervals. Detailed metric maps for watershed S2 were drawn and digitized in December 1994.

Peat Coring

On November 20, 1983, Herb Wright Jr. (emeritus professor, University of Minnesota) extracted a 782 cm-long peat core (MR2-8307) using a 10 cm-diameter piston corer at a site near the middle of bog S2 located 80° east of north and 33 m from the recording well. On April 20, 1994 at the same location, Janssens, Paul H. Glaser (senior research associate, University of Minnesota), Howard D. Mooers (professor, University of Minnesota), and Verry collected a 106 cm-long core (MR2-9401) using a Waadenaar corer. In addition, Janssens collected several short cores in S2 bog in July 1986. Data composites from the short cores were useful in documenting the post settlement peat horizon developing after about 1865.

Weather, Water, Soil, and Geologic Studies

Weather records were summarized for 1961–2000 at a weather station on the upland, 20 m north of bog S2. At the MEF technicians measured streamflow, bog and regional water table elevations continuously, and took soil moisture and water-chemistry samples periodically from 1966 to 2001 (Chapter 2). We derived the elevation, referenced to U.S. Geological Survey quad map elevations and benchmarks, of soil and peat horizons from logs for wells, piezometers, soil pits, and surface topography.

Soil descriptions are detailed in Paulson (1968) and Nyberg (1987). Detailed mineral- and organic-soil physical properties (including bulk density, carbonate content, hydraulic conductivity, color, texture, and degree of decomposition) were determined in two large, soil pits (about 3 m deep) in contiguous mineral and peat soils at two locations on the edge of the S2 bog (Brooks and Kreft, 1991; Tracy, 1997). Hydraulic conductivity was measured in the peat (Boelter, 1965; Gafni and Brooks, 1986, 1990) and in the A/E and Bt upland soil horizons of S2 (Tracy, 1997).

Sander (1971) carried out a local groundwater geology study from 1967 to 1969, which included a fully penetrating well for an aquifer pumping test (water transmissivity) and descriptions of geologic material, seismic refraction, resistivity, and residual Bouger gravity anomalies. Samples of surface and deep sand were collected in late September 2004 at 11 locations surrounding the S2 watershed to help differentiate between Rainy and Des Moines drift; both wet and dry soil color chips and hand textures were used.

Late Wisconsin Glaciation

Glacial Advances, Stable Margins, and Meltout over Stagnant Ice

Rainy/Itasca Lobe ice (from the north and northeast) stagnated north of the Itasca Moraine (lying along an east-west band near the southern shore of Leech Lake in north-central Minnesota; 100 km SW of MEF) after 18,000 calibrated years (cal y) BP. However, the Rainy/Itasca ice established stable margins northeast of Leech Lake about 3 km east of S2 bog (a north-south-oriented margin of the Rainy Lobe), and about 15 km to the north (an east-west-oriented margin of the Itasca Lobe). The ice sheet terminated at this stable margin for several thousand years. Meltwater and sediment released from the ice accumulated on the stagnant ice surface north and east of S2 bog, forming a large outwash fan.

Stagnant ice to the west and south of S2 bog melted over several thousand years, forming a topographic low. About 14,000 cal y BP, ice of the Des Moines Lobe advanced through this low from east to west along an axis extending through Grand Rapids, Minnesota and the large Lake Winnibigoshish

WSW of MEF. The northern margin of this advance, the St. Louis Sublobe, abutted against the Rainy/Itasca Lobe outwash fan about 15 km southwest of the S2 bog.

Meltwater and sediment then poured onto stagnant ice in the triangle-shaped, topographically low area bounded by the Itasca Lobe to the north, the Rainy Lobe to the east, and the St. Louis sublobe to the southwest. Meltwater escaped from the low area by means of a subglacial drainage channel passing through the Giant's Range near present-day Taconite, Minnesota. While the advance of the St. Louis Sublobe was short-lived, and its ice rapidly stagnated and melted, its advance allowed about 15 m of glaciofluvial and glaciolacustrine sediment to accumulate over stagnant ice at the S2 bog site.

After 13,000 cal y BP, the Rainy and Itasca Lobes retreated further to the north. Glaciofluvial sediment no longer accumulated at the S2 site but stagnant ice cored the topography (lay buried in the sediment) for thousands of years (to as late as about 11,000 cal y BP). As the buried ice melted gradually the overlying sediment slowly collapsed. Sediment slumped from high areas, accumulating in low areas. As the last ice melted, the sediment-filled areas remained as hills, while the relatively sediment-poor formerly high areas had collapsed to form depressions (kettles), one of which was the precursor to S2.

Concurrent with meltout of the last buried ice, large glacial lakes formed to the west and north of S2, most notably Lake Agassiz and its precursor Lake Koochiching. As the levels of these lakes fluctuated and finally drained, large areas of unvegetated lake bottom were exposed to wind erosion. Silt and very fine sand sediment eroded from the lake bottom and entrained by the prevailing westerly winds was deposited as a blanket of loess on the landscape to the south and east, including S2.

Sander (1971) describes sands beneath the till cap at S2 as brown. In addition, Rainy (reddish brown) and St. Louis Sublobe (yellowish brown) drifts are in the MEF area. The younger St. Louis Sublobe drift overlies the older Rainy drift in much of north-central Minnesota. Although the S2 watershed is overlain by a till cap of Koochiching drift origin, 11 samples of sand drift around the S2 watershed show a tongue of Rainy drift with no St. Louis drift. Sample sites with Rainy drift extend from at least 2.4 km north-northeast of the S2 watershed to 0.6 km south-southwest in a tongue about 0.6 km wide. The Rainy drift is overlain by St. Louis Sublobe drift in areas to the north, west, and south. The Rainy drift samples have a soil color description of 5YR 6/6 (dry) and 5YR 5/8 (wet) while the St. Louis Sublobe drift samples have a soil color description of 10YR 7/2 (dry and wet). The Rainy drift color is described as orange to bright reddish brown, while the St. Louis Sublobe drift is described as dull yellow orange. The Rainy sands are coarse, medium, and fine textured with few small-gravel pebbles while the Des Moines sands are coarse, medium to fine with a wide range of gravel and small cobble.

Table 4.1 lists the geologic materials documented at the MEF (Sander, 1971) and two possible interpretations of their drift origin (Meyer, 1986; Mooers,

TABLE 4.1

Geologic Materials Documented at the MEF (Shaded), Others in the Region, and Alternative Interpretations of Wisconsin Glaciation using Information from Mooers and Lehr (1997) or Meyer et al. (1986)

Geologic Material at the MEF or Near the S2 Bog	Elevation (m)	Mooers et al. Interpretation of Source Area	Mooers et al. Interpretation of Dates BP (ka)	Meyer et al. Interpretation of Source Area	Meyer et al. Interpretation of Dates BP (ka)
No Vermillion Phase drift at MEF		Rainy Lobe Vermillion Phase Labrador Center	12.3 13.7	Rainy Lobe	
Brown till, clay loam, with rock, and few slightly calcareous Cretaceous shale fragments	429	Red R.-Kooch. Lobe Riding Mtn. Prov. Labrador Center	11.9	Koochiching Lobe Riding Mtn. Prov.	
	424	Keewatin Center	12.2	Keewatin Center	
Brown Sand, fine to medium sands, with slight streaks of clay near the top, and some coarse sand with up to 13-mm pebbles near the bottom; noncalcareous	424	Rainy Lobe Bemis Phase Labrador Center North-central Ontario	13.2 14	Rainy Lobe Labrador Center	
	395		15		
Brown sand, medium to coarse sand with few pebbles up to 13 mm in top 1.0 m and bottom 1.2 m; noncalcareous	395	Rainy Lobe St. Croix Phase Labrador Center		Rainy Lobe Labrador Center	
	391	West of James Bay	17.5		
Gray sand, with pebbles and limestone fragments up to 6 mm in diameter	391	Itasca Lobe Outwash Hudson Provenance Labrador Center West Hudson Bay	18	Rainy Lobe Labrador Center	
	387				

(continued)

TABLE 4.1 (continued)

Geologic Materials Documented at the MEF (Shaded), Others in the Region, and Alternative Interpretations of Wisconsin Glaciation using Information from Mooers and Lehr (1997) or Meyer et al. (1986)

Geologic Material at the MEF or Near the S2 Bog	Elevation (m)	Mooers et al. Interpretation of Source Area	Mooers et al. Interpretation of Dates BP (ka)	Meyer et al. Interpretation of Source Area	Meyer et al. Interpretation of Dates BP (ka)
Gray till, compacted clay, sand, silt, pebbles and limestone fragments up to 13 mm in diameter	387	Itasca Lobe Hudson Provenance Labrador Center West Hudson Bay from the NNE	20	Winnipeg Lobe Browerville Phase Keewatin Center from the NNW	26
Greenstone and granite bedrock	380			(Pre-Late Wisc.)	30?
Other early Late-Wisconsin and pre Late-Wisconsin glacial drifts were not documented at the MEF	380				
Carbonate (<25%) and graywacke clasts called "Omars" (Mooers and Lehr, 1997)		Rainy Lobe Hewitt Phase Labrador Center East Hudson Bay	21 ka 23 ka	Rainy Lobe Hewitt Phase Labrador Center	
(Late Wisconsin)					
Carbonate (25%–40%), Cretaceous shale (Mooers and Lehr, 1997). Loamy-clay loam, pre-Cambrian pebbles (Meyer, 1986)		Winnipeg Lobe Browerville Phase Keewatin Center	23 ka	Winnipeg Lobe Browerville Phase Keewatin Center	>30
Silty clay, silty fine sand, coarse sand at base; gravel, cobbles and boulders, greenish gray, clay noncalcareous to moderately calcareous; rocky, hard, and compact. Alternates with Winnipeg Lobe (Meyer et al., 1986)				Old Rainy Lobe Various Phases Labrador Center	

Note: Elevations and dates are listed for the top and bottom of each major material described.

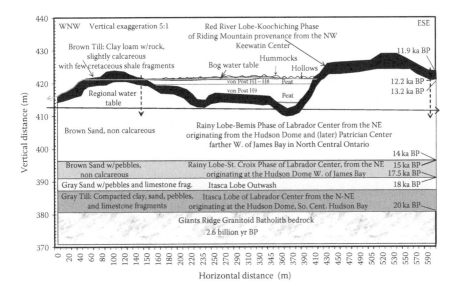

FIGURE 4.6

Longitudinal cross section of the S2 watershed (WNW to ESE) showing age and depth of major geologic and soil materials and the position of the bog and regional water tables during contemporary times. Note the slightly domed peat surface and parallel peatland water table. Vertical arrows are deep well locations.

1990; Mooers and Dobbs, 1993; Meyer, 1997; Mooers and Lehr, 1997; Mooers and Norton, 1997; Meyer et al., 1998). Figure 4.6 is a descriptive cross section drawn longitudinally through S2 bog and the entire S2 watershed using our interpretations based on various works of Mooers. It includes a surface deposit of loess and glacial flour at the bottom of the ice-block depression.

Beneath S2 Bog

The limestone fragments in the Itasca Lobe drift are the source of high calcium concentrations in groundwater (>20 mg L^{-1}) discharging from aquifers in the overlying Rainy Lobe sand into area streams and lakes. Rainy Lobe sands are brown and lack calcium carbonate except near the upper boundary where more recent slightly calcareous till overlies it.

Koochiching till caps the deep sandy drift of the Rainy Lobe, covering more than 75% of the MEF and ranging in thickness from 3 to 5 m; it was not covered with glacial meltwater or ice. An additional 20–30 cm of eolian loess was deposited over both the till cap and sandy drift. The loess is fine sandy loam in texture. It blew from the dry hills exposed after meltback of the ice (Nyberg, 1987). Beneath the bog peat the Koochiching till surface is mostly glacial flour formed directly beneath the grinding ice sheet, or washed into the tundra pond, beginning about 11,900 years ago.

Ice-Block Depression

The surface below the S2 bog is a muted contour of the original ice-block depression and hides much of the detail needed to explain the sequence of events resulting in the basin's current shape (Figure 4.7). The bottom contour is not the original ice-block depression, but the contour of the Koochiching till collapsed into an ice-block cavity in the underlying Rainy outwash. Nevertheless, it reveals the track of the original ice block (from the Rainy ice sheet) being dragged, pushed, and finally over-ridden by its mother sheet into the Rainy sands below.

It began when a wedge of ice broke from the leading edge of the Rainy Lobe glacier and rested near the outlet of S2, where the stream is today. It was perhaps 50 m wide at its base, 100 m wide at its top, and at least 20 m tall. We estimated its size from the width of the depression in the Rainy Lobe drift (Figure 4.6). It was a small block of ice compared to the enormous Laurentide ice sheet that rose 2,200 m into the air at maximum development some 18,000 years ago near its eastern Hudson Bay maximum. However, when the block broke away, perhaps 13,200 years ago, the ice was much thinner. There were tens of thousands of these ice-block depressions across the Lake States, New England, and the Canadian Provinces, giving rise to kettle (pothole) topography throughout.

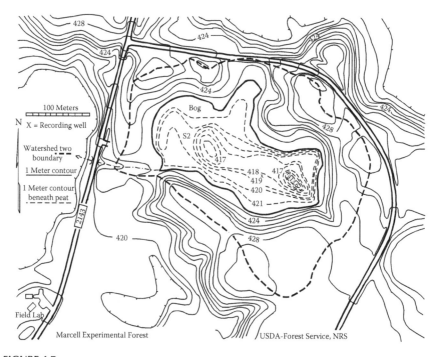

FIGURE 4.7
Contours at 1 m intervals of the uplands surface and the Koochiching drift that lies below the peatland in the S2 watershed.

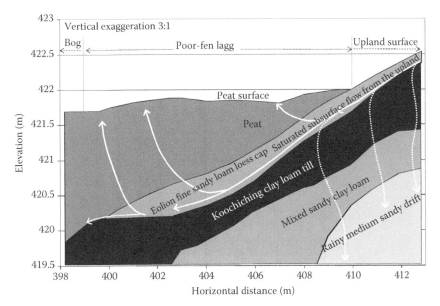

FIGURE 4.8

Soil boundaries and textures at the edge of the S2 bog near its eastern end. Arrows show saturated subsurface flow from the mineral soil upland flowing below the peat and exiting upward to the peat surface or downward as unsaturated flow into the Rainy Drift sands. Also, there is unsaturated flow through the Rainy sands directly beneath the uplands when the surface mineral soils saturate from snowmelt or large summer rains. The peat surface reflects hollow elevations only.

Revealing Sequence of Soil Horizons at Lagg Edge

A physical basis for chemical, hydrologic, and plant-biodiversity characteristics in the lagg is revealed when we interpret the story of horizons in a large soil pit near the east (or right) side of bog S2 (Figure 4.8). A deeper section showing the entire bog (see Figure 4.10) also uses our local climate interpretations to explain the accumulations of peat types within the S2 peatland.

Holocene Vegetation

Gross Composition of Core Segments

Figure 4.9 shows the gross composition of the long core from the S2 bog and indicates the slices examined for pollen and macrofossils. Relatively high bulk densities at the bottom reflect mineral entrainment in the algal gyttja on the now-filled lake bottom; low bulk densities at the top reflect the uncompacted *Sphagnum, Ericaceae* shrubs stems, and *Carex* roots in the acrotelm (active surface layer).

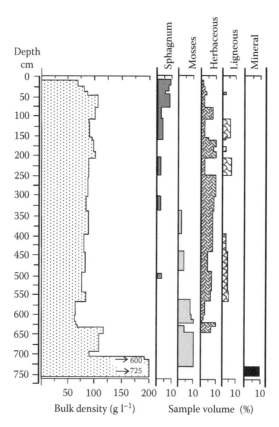

FIGURE 4.9
The gross composition of peat cores in the S2 bog: bulk density, *Sphagna*, other mosses, herbaceous, ligneous, and mineral composition values calculated as a percent of core volume within each visibly differentiated peat zone.

Dating of Core Segments and Peat Accumulation Rates

Eight 5 cm segments of the long core (MR28307) and one segment of the short core (MR29401) were radiocarbon dated and calibrated (Table 4.2). Calibrated year dates BP are plotted in Figure 4.10. The charcoal layer signifying post-fire establishment of the existing black spruce stand is dated at 1865 calendar year using tree rings.

Calendar years 1864 and 1610 are shown at a depth of 75 cm at the strong charcoal layer located there. The divergence between the long and short core also may reflect the loss of peat in fires (Figure 4.10). Prescribed fire in other peatlands at MEF have consumed 6–10 cm of peat; it is likely that hot fires during droughts can consume at least 15 cm of peat. The years 1863 and 1864 were two of the driest on record at St. Paul, MN; only 400 mm of annual precipitation fell in each year (Meyer, 1944). Fire likely burned the surface of the

TABLE 4.2

Calibrated Dates for Long-Core and Short-Core Samples at the S2 Bog

Sample	Lab No.	Depth of Top (cm)	Depth of Bottom (cm)	Radiocarbon Date (y BP)	Radiocarbon Error (y BP)	Calibrated Date, -2s (y BP)	Calibrated Date, Intercept (y BP)	Calibrated Date, -2s (y BP)
MR29401A	BEATA 72327	-95	-100	930	60	671	782	931
MR28307A	SI 6,481	-100	3	1,860	80	1,567	1,809	1,952
MR28307B	SI 6,482	-200	-205	3,475	70	3,564	3,699	3,900
MR28307C	SI 6,483	-300	-305	4,710	60	5,302	5,357	5,590
MR28307D	SI 6,506	-400	-405	5,265	55	5,917	6,016	6,186
MR28307E	SI 6,484	-500	-505	6,455	80	7,204	7,330	7,512
MR28307F	SI 6,485	-600	-605	8,265	85	8,986	9,245	9,441
MR28307G	SI 6,507	-645	-650	8,735	65	9,497	9,767	9,903
MR28307H	SI 6,486	-713	-718	9,675	90	10,474	10,931	11,001

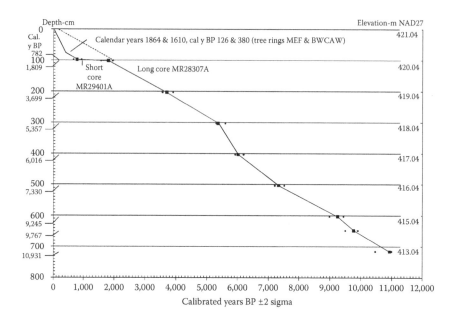

FIGURE 4.10
Calibrated y BP (and their ±2 sigma values) on the *x*-axis. Depth (cm) is on the left *y*-axis and the location of specific cores slices with cal y BP. Elevation (m NVGD 1929) is on the right *y*-axis.

S2 bog in the autumn of 1864. Seed within the surface peat or from surrounding unburned or standing burned black spruce on the peatland germinated in 1865. Tree cores on S2 show a 20 year range in dominant tree ages, suggesting a fully stocked stand by 1885.

The mass of peat accumulated (above the algal gyttja at a date of 9245 cal y BP) to the surface is 56.2 g m^{-2} y^{-1} (29.2 g C m^{-2} y^{-1}) is close to the overall rate for northern peatlands of 29 g C m^{-2} y^{-1} as calculated by Gorham (1991) (Figure 4.11). Accumulation rates from 18 cores in northern North America range from 16 to 80 g m^{-2} y^{-1} (8–41 g C cm^{-2} y^{-1}), and 12 sites within the Red Lake Peatland located 160 km northwest of the MEF are relatively high at 50–100 g m^{-2} y^{-1} (26–52 g C cm^{-2} y^{-1}). The Red Lake peatland sites developed over upwelling groundwater (Siegel and Glasser, 1987). Accumulation rates at the S2 bog are similar for all of the periods between dating ages except for the period 6000–7300 cal y BP, which accumulated carbon at the rate of 67 g C m^{-2} y^{-1}.

Major Changes in Upland Trees

Figure 4.12 shows pollen occurrence for upland trees and shrubs, upland herbs and ferns, wetland trees and shrubs, and aquatic plants. Four major

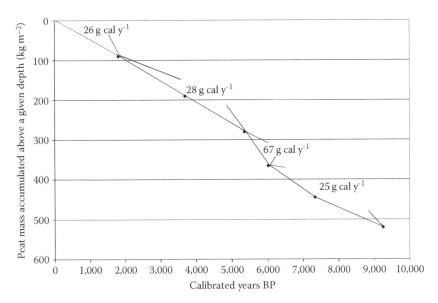

FIGURE 4.11
Accumulative mass of peat above a given depth in the S2 bog over the post-glacial period. Rates of carbon accumulation are shown for four periods.

vegetation associations are shown based on upland trees. The first association, 11,000–9,500 cal y BP, is dominated by *Picea*, which makes up 75%–30% of the segment volume. At the same time, *Pinus resinosa/Pinus banksiana* (red and jack pine) increases from 15% to 35% of the segment volume. A similar tree mixture at Steel Lake 106 km southwest of the MEF began its decline of *Picea* in favor of *P. resinosa/P. banksiana* at 11,200 and made the transition to *P. resinosa/P. banksiana* by 11,000 cal y BP as the climate became too warm and dry for *Picea* (Wright et al., 2004). At the MEF, the beginning of this decline is beyond the core depth, but the transition to the period dominated by *P. resinosa/P. banksiana* did not occur until 9500 cal y BP, some 1500 years later.

Between 9500 and 6500 cal y BP, *P. resinosa/P. banksiana* dominates the segment volumes (~40%) for 3000 years. Beginning at 6500 cal y BP, *Pinus strobus* (white pine) rises and dominates the pollen volume at 40% while *P. resinosa/P. baksiana* recedes to about 20% of the segment volume.

Major Changes in Wetland Vegetation

An examination of macrofossils with depth in the peat-core segments suggest seven vegetation phases in the S2 wetland (Figures 4.13 and 4.14). Changes in wetland vegetation occur more frequently and do not coincide with changes in upland vegetation (Figure 4.12).

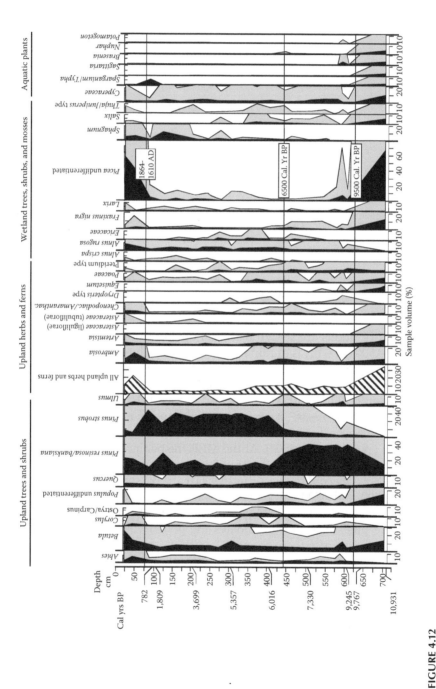

FIGURE 4.12

Pollen occurrence (percent of sample volume) in the long core from the S2 bog. Location of radiocarbon-dated (calibrated) core slices are shown on the *y*-axis.

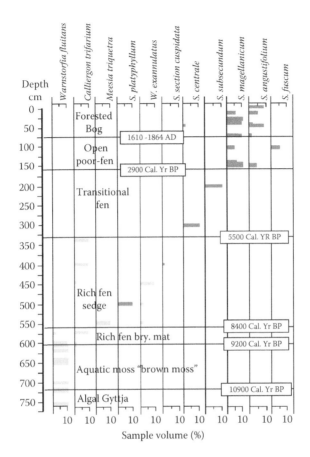

FIGURE 4.13
Sphagna and other mosses in the peat core of the S2 bog.

Discussion of Findings

11,900–11,800 Years BP

During the initial formation of a shallow pond, the Koochiching drift sank into the water-filled ice-block depression that eventually would become a small lake at the S2 bog site. The water level in the small lake began this period at an elevation of 422 m (near the top of the Koochiching till on the upland) and then dropped as the till sank to 420.2 m (see Figure 4.8 at the extent of the eolian deposit). Above the till, sediment of glacial flour and algal remains were slowly deposited. Algae are the only plant residues represented in both the bottom gyttja of the long core (MR2-8307) and in the

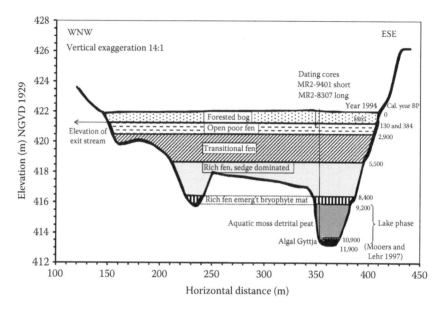

FIGURE 4.14

Major vegetation phases in the ice-block depression of S2 that filled with water and peat.

TABLE 4.3

Texture at the Organic-Mineral Interface at the Deepest Location in the S2 Depression

Elevation in (m)	Material Description	OM (%)	Sand (%)	Silt (%)	Clay (%)
416.7–416.5	Sapric peat and coprogenous gyttja	38	–	–	–
416.5–415.8	Silty clay loam	18	4	60	36
415.8–415.5	Silt loam	3	17	59	25
415.5–414.7	Loam	8	45	35	20
414.7–414.4	Loam	2	42	40	18
414.4–414.3	Clay loam	2	33	37	30

Source: Tracy, D.R., Hydrologic linkages between uplands and peatlands, PhD dissertation, University of Minnesota, St. Paul, MN, 1997.

Note: This bucket auger hole is deeper than the length of the long core shown in Figure 4.10.

auger hole sampled in the deepest part of the depression (Brooks and Kreft, 1991). Below the gyttja, the first 2.2 m into the Koochiching till, at the deepest part of the depression, averages 2% organic matter, 34% sand, 42% silt, and 23% clay (Table 4.3). Diameters in the sand fraction are fine to very fine (0.12–0.06 mm in diameter) and sand-particle edges under magnification are rounded indicating an eolian origin.

11,800–9,200 Years BP

The sedimentary profiles illustrate this period as a major zone of brown moss (*Warnstorfia fluitans*) near the base of the peat core, corresponding to Lacustrine Period II (10,900–9,200 cal y BP; Figures 4.13 and 4.14). The water level in the small lake dropped during this period as successively lower outlets to Lake Superior developed (Teller and Leverington, 2004). Perhaps the drop was slow at first as Figure 4.8 shows a possible wave-cut shelf in the Koochiching till at an elevation of 420.2 m. At 9200 y BP, the water level in the S2 depression had dropped a total of 6 m to the 416 m elevation. The southeast shore of glacial Lake Agassiz was 135 km north of S2 at 10,000 cal y BP, and 4,500 km north of S2 at 9,000 cal y BP (Viau et al., 2006).

Schwalb and Dean (2002) showed a drop in water levels at Williams and Shingobee Lakes 120 km SE of the MEF near the prairie boundary beginning at 9800 and extending to 7700 y BP (further lowering of lake levels occurred up to 5000 y BP). Donovan et al. (2002) showed that even in the droughts of the twentieth century, groundwater levels could drop by as much as 5 m in west-central Minnesota. Digerfeldt et al. (1992) reconstructed a drop in lake levels of 3–6 m at the Parkers Prairie sandplain in west-central Minnesota and Filby et al. (2002) reconstructed a drop in water level of 3.5 m in Williams Lake. The S2 bog is located only 3 km from the subcontinental divide (Hudson Bay, Mississippi Basins), so a drop in water level of 6 m from 11,000 to 9,200 cal y BP was possible (Figure 4.14).

The aquatic moss *W. fluitans* flourished during the end of this period (several centuries) when the water level was near 416 m. This species is the most acidophilous among its cogeners (Hedenäs, 2003; Hedenäs and Kooijman, 1996) and among most of the "brown mosses" typically found in more minerotrophic peatland mesohabitats. The water chemistry of the lake likely was more akin to today's upland subsurface or surface runoff than to the groundwater chemistry (Table 4.4) during this early period when the elevation of regional water tables ranged from 420 to 416 m.

The Lacustrine Period (11,900–9,200 y BP) included the aquatic mosses (Figure 4.13) but also floating-leaved and emergent aquatic plants: *Nuphar, Potamogeton, Brasenia, Sagittaria, Sparganium,* and *Typha* (Figure 4.12). Black ash (*Fraxinus nigra*) in shallow depressions and swales were at their Holocene climax during this initial tundra setting; their pollen accounted for 14% of the pollen sum. *Salix* (willow) also reached its Holocene maximum (6%) in this early lake period, as did *Larix* (eastern larch), but it accounted for less than 2% of the pollen sum. Two members of the Cupressaceae family, *Thuja* and *Juniperus,* represent less than 1% of the pollen but they too were at their Holocene maximum. *Thuja* likely is *Thuja occidentalis* (northern white cedar), which is well suited to wet and calcareous habitats. *Juniperus* probably is the shrubby, circumpolar *Juniperus communis* (common juniper) now common in northeastern and northwestern United States and Canada, and/or *Juniperus virginiana* (*eastern redcedar*) growing on cold dry ridge tops.

TABLE 4.4

Water Chemistry at Major Locations in the S2 Watershed on the MEF

Parameter	Units	Deep Groundwater, Number of Samples	Deep Groundwater, Mean	Deep Groundwater, Standard Deviation	Upland Subsurface Runoff, Number of Samples	Upland Subsurface Runoff, Mean
pH		76	7.43	0.29	496	6.28
Alkalinity	μeq/L	73	3144	314	484	62
Calcium	mg/L	74	41.60	2.65	486	3.44
Magnesium	mg/L	74	7.62	0.70	486	1.04
Sodium	mg/L	74	3.62	0.75	485	1.15
Potassium	mg/L	74	1.31	0.21	486	0.89
Ammonium-N	mg/L	69	0.08	0.07	418	0.05
Nitrate-N	mg/L	44	0.06	0.04	351	0.12
Sulfate	mg/L	73	3.13	1.03	485	2.56
Chloride	mg/L	73	1.03	0.47	467	0.49
Total phosphorus	mg/L	45	0.04	0.01	355	0.06
Phosphate-P	mg/L	54	0.017	0.006	395	0.042
Total organic carbon	mg/L	36	6.20	3.38	416	22.45
Total Kjeldahl nitrogen	mg/L	56	0.27	0.11	393	0.58
Carbon:nitrogen	Ratio	Ave.	23.0	—	—	38.7
Aluminum	mg/L	28	0.18	0.01	245	0.77
Iron	mg/L	61	3.12	4.77	412	0.60
Manganese	mg/L	25	1.38	0.01	234	0.03
Copper	mg/L	1	0.03	—	17	0.06
Zinc	mg/L	25	13.21[b]	2.87	234	0.29

[a] 1978–2001 National Atmospheric Deposition Program (access nadp.sws.uiuc.edu for complete weekly dataset).

[b] Zinc levels are elevated by galvanized plot borders or galvanized well pipe.

Upland Subsurface Runoff, Standard Deviation	Upland Surface Runoff, Number of Samples	Upland Surface Runoff, Mean	Upland Surface Runoff, Standard Deviation	Streamflow, Number of Samples	Streamflow, Mean	Streamflow, Standard Deviation	Mean Precipitation[a]
0.45	376	6.64	0.35	118	4.07	0.22	5.06
70	376	222	191	14	−73	54	—
1.00	366	5.56	2.24	116	2.32	0.87	0.20
0.29	366	1.27	0.54	116	0.83	0.31	0.03
0.52	365	0.38	0.29	115	0.53	0.21	0.07
0.74	366	4.92	3.61	116	0.92	0.64	0.04
0.10	349	0.30	0.58	109	0.14	0.21	0.33
0.21	364	1.14	2.47	96	0.13	0.28	1.05
2.43	367	2.04	2.03	114	1.22	2.41	1.03
4.07	361	0.93	1.43	111	0.41	0.26	0.08
0.14	285	0.41	0.85	85	0.08	0.10	—
0.12	308	0.343	0.264	89	0.035	0.074	—
13.35	296	23.07	32.98	99	49.83	22.47	—
0.26	290	1.86	235	95	1.15	0.81	—
—	—	12.4	—	—	43.3	—	—
0.29	137	0.3	0.19	50	0.64	0.31	—
0.42	296	0.21	0.25	98	1.14	0.71	—
0.28	127	0.03	0.02	46	0.05	0.03	—
0.05	7	0.09	0.13	11	0.05	0.03	—
1.08	127	4.89[b]	3.46	46	0.13	0.66	

Betula (birch) and undifferentiated *Populus* then, as now, were early invaders of disturbed landscapes (about 15% of the pollen sum). Oak (*Quercus*) and elm (*Ulmus*) were earlier invaders in smaller numbers (5% of the pollen sum). These may be associated with broad glacial river corridors assuming that the *Quercus* had a greater representation of white oaks (*Quercus alba* and *Quercus macrocarpa*) typical today on bottomlands and deep, well-drained, moist sites (Harlow and Harrar, 1958). The more abundant northern red oak (*Quercus rubra*), found today in association with sugar maple and white pine (and moister red pine sites), is typical of temperate forests on sandy loams and silty clay loams. The common upland trees were red and jack pine (*P. resinosa/P. banksiana*) beginning with pollen in 15% of the core volume and expanding rapidly to 42% of core volume by 9200 y BP.

Immediately after icemelt spruce (*Picea*), presumably both black (*P. mariana*) and white (*Picea glauca*), dominated the area with pollen in 75% of the pollen sum. This was reduced to 10% at the end of Lacustrine Period I (9200 BP) and remained below 2% until an intense fire in 1864 burned the peatland and, following harvest of white pine, *P. mariana* once again exceeded over 50%.

Balsam fir (*Abies balsamea*) was a minor component of the upland forest during the early lake period and remained under 5% of the pollen sum until immigrant harvest of the white pine in the late 1800s. The upland hazel shrubs (*Corylus americana* and *Corylus rostrata*) were early and persistent forest components, always occurring at 1%–2%. At the wet edges of the lake and shallow depressions within the upland, speckled alder (*Alnus rugosa*) was the only wetland shrub.

The upland was not a continuous forest as evidenced by the grass (*Poaceae*) and horsetail (*Equisetum*) pollen, which accounted for more than 10% of the core volume. Other upland herbs invading the glacial melt exposures included goosefoot (*Chenopodiacea*), aster (Asteraceae), ragweed (*Ambrosia*), amaranth (*Amaranthaceae*), and wormwood (*Artemisia*).

The entire Lacustrine Phase shown in the pollen record (Figure 4.14) suggests a widespread, large, and presumably gradual decline in the regional water table from 11,000 to 9,200 y BP. This is evidenced by decreases in tree pollen characteristic of wetter sites: *Picea, Thuja, Fraxinus,* and *Abies,* and increases in drier-site pines: *P. resinosa/P. banksiana* in particular but also small increases in the temperate and mesic *P. strobus*. Concomitant decreases in *Betula* and *Populus* may record early colonization by species adapted to a wide range of moisture regimes followed by declines owing to competition from the invading pines. Reductions in upland herb and fern pollen are strong and may reflect large forest areas nearing crown closure.

9200–8400 5Years BP

Water levels during this period remained at a pond-surface elevation of about 416 m. *Calliergon trifarium*, another aquatic brown moss, joined and succeeded by *W. fluitans* in what now were two small ponds, perhaps thinly joined, and then smothered by an emergent bryophyte mat at the water surface (see 416 and 417 contours in Figures 4.13 and 4.14). Pollen of submerged aquatics declined (*Brasenia* 4.9%–0%, *Nuphar* 4.9%–0.2%, *Potamogeton* 2.4%–0%), as did that of emergent aquatics (*Cyperaceae* 12.2%–0.6%, *Sagittaria* 2.4%–0%, *Sparganium/Typha angustifolium* 2.4%–0%).

The Rich-Fen Bryophyte Mat period is the shortest, only 800 years, *C. trifarium* peaked about 300 years into the period and appeared late in the Rich-Fen Phase. Sedges flourished in the next 500 years, poking through the bryophyte mat reaching their Holocene maximum. They persisted at 5%–10% of the pollen count for millennia until the latter part of the poor-fen period.

Wetlands with trees have sufficient drainage to a surface outlet to provide both air and water for growth in the near surface pore space. Wetlands without surface outlets collect water and retain it, giving rise to higher water levels and either sedge or shrub habitats (Verry, 2000). In the landlocked ponds at the bottom of S2, the floating plants *Brasenia* (water-shield) and *Potamogeton* (pondweed), increased. Most characteristically for the pond gyttja, *C. trifarium* smothered the surface and formed peat (Figure 4.13).

The bryophyte mat period reveals a landscape with dry uplands, aquatic moss, and sedge wetlands (mildly calcareous) with little drainage, and minor increases in treed wetlands with sufficient surface drainage to a small stream. The next major pollen reconstruction period (rich-fen sedge 8400–5500 cal y BP) reveals a drier period with longer summers when the local region reached its climatic optimum.

8400–5500 Years BP

A rich-fen sedge peat characterized this period. The peatland had a water pH of about 8 and associated calcium and magnesium values were near the high end of the groundwater chemistry dataset in Table 4.4 as inferred from plant species (Vitt et al., 1995). Early in the Rich-Fen period, the first identifiable *Sphagnum* species, the aquatic *Sphagnum platyphyllum* (an alkaline species), appeared (Figure 4.13). Sometimes known as *Sphagnum subsecundum* var. *platyphyllum* it grows in mineral-rich, open habitats in sedge mats marginal to eutrophic lakes and in the hollows of sedge fens (Crum, 1988). Also consistent with a calcium-rich environment is the first appearance of *Warnstorfia exannulatus* and additional mats of *C. trifarium* late in the period (Figure 4.13.). Toward and beyond the end of the rich-fen period, the pollen of *Ambrosia*, *Artemisia*, and *Chenopodiacea/Amaranthaceae* peaked between 7000 and 6000 y BP.

The MEF pollen record reveals an apparent warming and drying period during this transition from the Pleistocene to the Holocene. Over this early Aquatic Moss Phase there is a major reduction in the *Picea* pollen count representing cool/wet sites from 11,000 to 9,300 y BP (75%–5%, Figure 4.13). By contrast, *P. resinosa/P. banksiana* pollen count representative of warm/dry sites rises from 15% to 40% (Figure 4.14). Upland herbs and ferns also significantly decrease during this period (35%–15%) suggesting a forest succession explanation as well as an increasing amount of *P. resinosa/P. banksiana* invading the eolian and outwash sands on uplands.

The *P. resinosa/banksiana* pollen data at the MEF suggest that a warm/dry period achieved by 9200 cal y BP lasted until 6500 cal y BP. The next 500 years saw a transition to a cooler/wetter climate favoring *P. strobus* with pollen counts rising from 2% to 30%) over the warmer/drier climate that favors *P. resinosa/P. banksiana* when pollen counts fell from 40% to 12% (Figure 4.12).

There is an "acid anomaly" during this high pH period: *Sphagnum* section *Cuspidata sp.* occurs in 1.5% of the core pollen/spore count at a depth of 400 cm (Figure 4.13) and an elevation of 418.5 m. *Sphagnum* section *Cuspidata sp.* is rare in the Great Lakes Region, where it occurs as a submerged moss growing at the margins of acid lakes (Crum, 1988). The occurrence of this acidiphilous, aquatic *Sphagnum* with the alkaliphilous aquatic moss *C. trifarium* is curious. The 400 cm depth corresponds to 6000 cal y BP. The occurrence of *Sphagnum* section *Cuspidata sp.* is sandwiched in a bulk core description showing ligneous peat below it and aquatic mosses above it (Figure 4.13). Though relatively rare, thin acid bands of peat (pH 2.7–3.5) have been reported within alkaline peat profiles in Minnesota (Always, 1920), Michigan (Davis and Lucas, 1959), and Poland (Okruskzko, 1960). Acidity in organic soils is accounted for by the presence of organic compounds, exchangeable H, iron sulfide, and silicic acid. It is believed that these thin acid layers are formed under conditions similar to "cat clays" (cation clays) found in brackish coastal waters (Davis and Lucas, 1959). In these instances, the roots of reeds may contain as much as 3% sulfur in the form of bisulfides. When oxidized sulfuric acid is formed along with scattered crystals of calcium sulfate. The acid layer at the S2 peatland occurs at an elevation of 418.5 m, just about half meter above two deep pools in the bottom of S2 (Figure 4.14). It may be that the high lignin sedges accumulated on the shelf between the two pools (roughly the area between the 418 and 419 m contour lines in Figure 4.7). They also may be high in sulfur and during a period of high summer temperatures, relatively low rainfall, and low lake levels in the northern hemisphere oxidized to produce a thin acid layer in the otherwise alkaline peat.

How did calcium-rich groundwater rise into the S2 depression during the rich-fen, sedge-dominated period? First, the hydraulic conductivity of the Koochiching till, though extremely low (5×10^{-7} cm s^{-1}), is sufficient to allow regional groundwater to enter and exit the peat depression slowly over long periods. The hydraulic conductivity of the unweathered, silty clay layer (glacial flour) is even lower (1×10^{-8} cm s^{-1}), but over centuries can pass small amounts of water. Second, the hydraulic head on the regional water table at the S2 depression derives from the subcontinental divide 3 km north, where the contemporary water table elevation peaks near 418 m and fluctuates down to 417 m. The same regional water table range at the S2 depression is 414.8–413.8 m. In other words, the 4 m head drop over 3 km allows groundwater in the depression to rise to an elevation approaching 418 m. This elevation is just shy of the upper boundary of the Rich-Fen, Sedge-Dominated Peat Sediment Phase. Perhaps in the long Holocene after 9000 y BP water table elevations fluctuated through a range of 1.5 m as opposed to the contemporary 50 year record of 1 m.

Cyperaceae dominate the aquatic and wetland species in the peat sediment record. There also is one calciphile *Sphagnum* (*Sphagnum platyphylum*) and two calciphile aquatic mosses (*C. trifarium* and *W. fluitans*). *Salix* spp., *A. rugosa*, and *Alnus crispa* (its first occurrence) dominate the tall wetland shrubs. Short ericaceous shrubs also are strong in the peat record. Wetland trees strong in the record are *Fraxinus* and *Larix* with a notable near-absence of *Picea*. This is not a pond or lake environment, but a wet sedge meadow with ericaceous shrubs underlain by aquatic mosses and a calciphile *Sphagnum*. The sedge-shrub meadow is surrounded by alder and willow shrubs with sparse ash and tamarack trees, sitting in a landscape first dominated by dry site red and jack pine and then mixed with the temperate, moister, and dominant-crowned white pine.

Paludification is the process allowing the wet sedge meadow to accumulate peat. It occurs when there is a hydrostatic head (pressure field) above the meadow surface. Centuries of peat accumulation fill this energy field with pore space (about 8 cm per century), allowing groundwater under pressure to fill the pores at progressively higher elevations. The growth of sedge, *Ericaceae*, and moss continues upward. Hence, the peatland is a feature of the landscape that grows upward (and outward where the hills allow) on the death of its own vegetation and the pore structure that it creates in the organic soil.

5600–3000 Years BP

The overall picture of sediment during the Transitional-Fen Phase is nearly identical with that during the Rich-Fen Phase. The difference is the occurrence of circumneutral *Sphagnum* mosses, both *Sphagnum centrale* and *S. subsecundum*. *S. centrale* is typical of *Thuja* swamps (treed fens) where it occurs frequently today. It has been observed in a contemporary ice-block depression in outwash sands only 0.5 km distant and 7.5 m lower at an elevation of 414.5 m (Bay, 1967). The 414.5 m elevation at the S3 fen is coincident with the algal gyttja at the deepest part of S2 bog (Lacustrine, Calcareous Lake Phase) and also coincides with the contemporary upper limit of regional water table fluctuation beneath the S2 bog.

Groundwater in the S3 fen has a contemporary pH of 6.9–7.4, calcium of 17–20 mg L^{-1}, and magnesium of 3–4 mg L^{-1}. Groundwater beneath S2 bog has a contemporary pH of 6.8–8.0, calcium ranges from 37 to 47 mg L^{-1}, and magnesium ranges from 6 to 9 mg L^{-1} (Table 4.4). *S. subsecundum* grows in mineral-rich mats in open habitats, and in the hollows of poor-sedge fens (Crum, 1988). Thus, the Transitional-Fen Phase marks a subtle change in water chemistry from alkaline pH values near 8 to circumneutral values slipping below 7 (Vitt et al., 1995).

The hydraulic head arising at the subcontinental divide enables strong groundwater control on peatland water chemistry and water table elevation. Downward flowing groundwater flow lines begin at the divide and then flow deep through the Rainy sands into the Itasca Lobe ground moraine above the bedrock. There it dissolves the limestone fragments, pushed 20,000 years ago from south-central Hudson Bay. The flow lines then curve upward and fill the aquifer sands with water rich in calcium bicarbonate. This is the water flowing under S2 bog and discharging in large amounts to the surface at S3 fen. The groundwater control weakened when the peat surface in the S2 sedge meadow rose above the water table at the subcontinental divide (417.5 m) about 5500 cal y BP.

The paludification process continues upward and the surface water draining to the depression from the upland increases in importance relative to the slow lateral and upward seep of deep groundwater. The paludification process (about 6 cm per century during the Transitional-Fen Phase) accumulates an additional 1.6 m of peat, such that the peat surface rises above the subcontinental water table but retains (by chemical diffusion) a chemical groundwater memory during the process.

Annual peatland water table fluctuations (0.5 m) and severe drought fluctuations (1.2 m) (Verry, 1980) repeatedly mix the water in the undecomposed to moderately decomposed active peat layer (acrotelm) with water in the well-decomposed catotelm (Verry, 1984). During the Transitional-Fen Phase, groundwater still controlled the peat development through the power of a constant hydraulic head. This power gradually gave way to periodic processes including water table fluctuation, biologic mixing (vegetation growth, decay, and gas exchange), and deep frost penetration (0.1–0.5 m) that preferentially drives fine organic matter and elements to the bottom of freezing ice. Thus, frost formation fosters the transfer of material from the acrotelm to the catotelm (Verry, 1991).

At the end of the Transitional Fen Phase there is a clearer break with groundwater chronicled by new *Sphagna* such as *Sphagna magillanicum*, *Sphagna angustifolium*, and *Sphagna fuscum*, changes in water chemistry, and transition to a warmer continental climate.

2900–390 Years BP

There was a severe fire (a distinct charcoal layer) at the upper boundary in 1864, 130 years BP, but the layer may contain charcoal from an earlier fire documented north of Grand Marais, MN, in 1610 AD (384 y BP). An analysis of peat accumulation rates suggests how many years and how many centimeters of peat accumulation were consumed by the 1864 fire.

The amount of catotelm peat consumed by the 1864 fire are estimated by subtracting the years to reach the projected open poor-fen surface (413) from the middle of the last dated core slice (782). This yields a 369 y BP. A more reliable date estimated from St. Paul, MN, weather records and tree-ring cores from 1994 in the spruce at S2 bog is 1864 or 130 y BP. Thus, 369–130 suggests that the fire burned 239 years of catotelm peat. The lost years multiplied by the accumulation rate suggests the fire consumed 14 cm of catotelm peat. Because another severe fire occurred in 1610, the charcoal layer may represent two fires.

We derived our open poor-fen upper date of 390 y BP from peat accumulation rates extrapolated above the fire horizon depth of 75 cm (421.25 m elevation). The peat accumulation rate averaged over the last three dating slices is 0.55 mm y^{-1}. This rate applied to 22.5 cm of projected peat accumulation (decomposed and relatively dense catotelm peat) suggests that peat accumulated up to the pre-fire surface of the Open Poor-Fen Phase over a 413 year period beyond the center of the previous dating slice (782 y BP). This accumulation of catotelm peat extends from the middle of the last dating slice (421.07 m elevation, 97.5 cm depth) to just below the elevation of the contemporary stream outlet (421.29 m elevation, 75 cm depth). A less dense acrotelm peat of about 20 cm occurs above the catotelm boundary (Verry, 1984). Accordingly, we project the actual peat hollow surface at the end of the Open Poor-Fen Phase at virtually the stream-outlet elevation (421.5 m). This is deduced from the continued occurrence of *S. magellanicum* throughout the Open Poor-Fen Phase, the frequent occurrence of *S. angustifolium* above the 75 cm depth (Figure 4.12), and the peat humification values of 8 and 9 (on the von Post scale of 1:10; see Malterer et al., 1992) immediately below the charcoal horizon.

Characteristic of this Open Poor-Fen Phase is the occurrence (dominance) of acid *Sphagnum* species, particularly *S. magellanicum*, which first appeared about 2900 cal y BP and heralded the initiation of the Open Poor-Fen period. It was accompanied by *Sphagna recurvum* agg., identifiable as *S. angustifolium* at the top of the peat core (Figure 4.13) along with some transitional *S. fuscum*.

130/390–0 Years BP

Picea pollen rose sharply, from 2% to 60% of the upland pollen sum, as a forest of black spruce developed on the bog. Pollen of *Betula* also increased from 7% to 24%, presumably owing to the logging of pine and subsequent slash fires. White-pine pollen declined from 35% to 5%. The pollen of weedy invaders of open spaces, *Ambrosia sp.*, *Artemisia sp.*, and *Chenopodiaceae/Amaranthaceae sp.* increased by 5%–15% for the same reason. Fire history documents initiation of the aspen forest (*Populus tremuloides*) in 1917 and 1918 in the MEF.

The intense fire in 1864 probably burned peat accumulated in the narrow surface outlet on the southwest side of the S2 bog, and effectively lowered the water table at least 5 cm. This was sufficient to establish and grow a poor- to medium-site stand of black spruce (Verry, 1982, 2000). Bog development began immediately after the fire, fostered by unobstructed streamflow drainage, colonization by acidiophile *Sphagnum* species, decomposition of acid-producing *Sphagnum* moss (Gorham et al., 1985), minimal availability of bicarbonates from the deep peat (and little from the upland), and continuation of a wetter climate.

Since 1865, the S2 peatland has developed an *S. angustifolium/S. magellanicum* dome rising 18 cm from its edge to a high near the existing recording well (Figure 4.5). A small dome may have existed before the 1864 fire as well. This dome, built by *Sphagnum* accumulation above the elevation of the outlet stream, developed a localized groundwater flow net above the much deeper regional groundwater flow net (Ingram, 1978). Even before the acid, ombrotrophic peat in the Forested-Bog Phase rose an additional meter above the stream elevation, paludification had reversed the flow of water deep in the depression from vertically upward to vertically downward. Today the water table of the bog (422.2 m) is 4–5 m higher than that at the subcontinental divide (417.5 m).

Contemporary Water Chemistry in S2 Peatland

Extensive data on water chemistry for regional groundwater, upland, bog, streamflow, and precipitation show strong site differences (Table 4.4). Precipitation at the MEF is moderately impacted by industrial emissions. Sulfate concentration is slightly elevated (1.03 mg L^{-1}) and nitrate (1.05 mg L^{-1}) and ammonium (0.33 mg L^{-1}) more so. However, alkaline dust from the prairies yields high calcium (0.20 mg L^{-1}) and magnesium (0.03 mg/L) concentrations. Thus, pH is within the normal range for precipitation (5.06) as is chloride (0.08 mg L^{-1}) for midcontinental sites.

Although S2 streamflow pH reflects a mixing of the interior bog dome and the poor-fen lagg, it is acid (4.07) while calcium (2.3 mg L^{-1}) and magnesium (0.8 mg L^{-1}) are slightly elevated above bog waters in other cool and humid areas with no influence from windborne dust. Organic matter dissolved and suspended in the bog water yields high total phosphorus (TP) values (0.08 mg L^{-1}) and high total organic carbon (49.8 mg L^{-1}). Total Keljdahl nitrogen (TKN) also is high (1.15 mg L^{-1}), but the carbon to nitrogen ratio (C/N) of 43 is still very high compared with many natural systems.

The two components of upland water are surface and subsurface flow (Chapter 7). Upland surface water flows through the forest floor, whereas the subsurface component is water collected from the saturated A horizon of the mineral soil. Verry and Timmons (1982) described the chemical analyses and physical collection methods. Concentrations of calcium, magnesium, and alkalinity are higher in the surface flow than in the subsurface flow (Table 4.4). Uptake of these elements from the mineral-soil horizons by trees (*P. tremuloides*) and their subsequent release from decaying leaves is the source of elevated calcium and magnesium in the upland surface flow. Similarly, potassium, TKN, ammonium, TP, and phosphate-phosphorus (phosphate-P) are high in leaves and higher in upland surface flow than subsurface flow (Table 4.4). By contrast, sodium is higher in the subsurface flow, reflecting the quick release and leaching from leaves to the saturated A horizon. Total organic carbon and pH concentrations are similar between subsurface and surface flows. However, the C/N for subsurface flow is more than three times that in the surface flow. Aluminum is high in the subsurface flow and the peatland streamflow, reflecting acid leaching in both locations.

Regional groundwater (drawn from 3.2 cm diameter galvanized pipe, screened for 1 m, 15 m beneath the surface) is high in pH, alkalinity, calcium, magnesium, sodium, potassium, iron, and manganese relative to all other sites (Table 4.4) and provides the alkaline environment for the early lake and rich-fen environments deep in the S2 depression. These same concentrations are typical of the S3 fen 2 km north of the S2 bog. Phosphorous and nitrogen values are low as is TOC (6 mg L^{-1}), but the C/N of 23 is midway among all sites.

Contemporary Vegetation in the S2 Peatland

Tables 4.5 and 4.6 show the species present in single relevés (10×10 m^2) from both the bog dome and the marginal fen lagg. The bog relevé contains 19 species while the relevé from the lagg contains 35 species. Nomenclature follows Ownbey and Morley (1991) for vascular plants, Anderson (1990) for *Sphagnum*, Anderson et al. (1990) for other mosses, and Stotler and Crandall-Stotler (1977) for liverworts.

The overstory vegetation of the bog was almost entirely black spruce with some birches (*Betula papyrifera*) and tamarack (*Larix laricina*). *Ledum groenlandicum* was the dominant broad-leaved, evergreen woody shrub with five other ericaceous heaths present. *Carex trisperma* dominated the graminiods and *Smilacina trifolia* is the only common forb. Among the bryophytes, *S. magellanicum* and *S. angustifolium*—characteristic of low hummocks and cushions—had high cover values with another 14 species present (Table 4.6). Also common was *Pleurozium schreberi*, a feather moss usually found on taller, drier hummocks. The shrubby overstory of the lagg was dominated by *Alnus incana* with five other deciduous, broadleaf woody species. *C. trisperma* and *Carex disperma* were most common of the six graminiods, whereas *Calla palustris* was the chief forb among the 13 species present. *S. centrale* and *S. russowii* were the most common bryophytes in the relevé. When bryophyte species were enumerated during quantitative collections from the lagg, *Callicladium haldanianum* was the most frequent of the species collected followed by *Tetraphis pellucida*, *S. magellanicum*, *S. angustifolium*, and *S. centrale*. The bog plant list comprised 17 taxa on 2.8 ha while the lagg list comprised 55 taxa on just 0.4 ha (Table 4.6). The relative area-weighted species ratio (lagg:bog) is 23:1 and plant diversity in the lagg is three times greater than that in the bog.

TABLE 4.5

Vegetation of the S2 Bog and Lagg

Life Form	Species	Cover Scales[a] Bog	Cover Scales[a] Lagg
Needleleaf evergreen			
	Abies balsamea (<2 m)		+
	Picea mariana (10–35 m)	3	
	Picea mariana (<2 m)	+	+
Broadleaf deciduous			
	Acer rubrum (<2 m)	+	
	Alnus incana (<2 m)	1	
	Alnus incana (2–10 m)		1
	Amelanchier intermedia		+
	Betula papyrifera (2–10 m)	+	1
	Cornus stolonifera	+	
	Ribes glandulosum	1	
Broadleaf evergreen			
	Chamaedaphne calyculata	1	
	Gaultheria hispidula	1	+
	Kalmia polifolia	1	
	Ledum groenlandicum	3	+
	Vaccinium angustifolium	+	+
	Vaccinium oxycoccus	+	
Forbs			
	Aster puniceus	+	
	Calla palustris	2	
	Clintonia borealis		+
	Cornus canadensis		+
	Drosera rotundifolia	+	
	Dryopteris carthusiana		1
	Equisetum sylvaticum	+	
	Lycopus uniflorus	+	
	Lysimachia thyrsiflora		+
	Osmunda cinnamomea		+
	Rubus pubescens	1	
	Sarracenia purpurea	+	
	Smilacina trifolia	2	+
	Trientalis borealis		+
	Viola incognita		1
Graminoids			
	Calamagrostis canadensis		+
	Carex brunnescens	+	
	Carex disperma	1	
	Carex intumescens	+	
	Carex oligosperma	1	
	Carex pauciflora	1	
	Carex projecta		+
	Carex trisperma	2	1
	Eriophorum spissum	+	

TABLE 4.5 (continued)

Vegetation of the S2 Bog and Lagg

Life Form	Species	Cover Scales[a] Bog	Cover Scales[a] Lagg
Bryophytes			
	Calliergon giganteum		1
	Pleurozium schreberi	2	
	Polytrichum commune		1
	Sphagna angustifolium	3	
	Sphagnum centrale	2	
	Sphagnum magellanicum	3	
	Sphagnum russowii	2	

Note: The cover scores are those of the Braun-Blanquet scale.
[a] + = 5%, several individuals; 1 = 5%, many individuals; 2 = 5%–25%; 3 = 25%–50%.

TABLE 4.6

Additional Bryophyte Species Found in the Bog and Poor-Fen Lagg Habitats in the S2 Peatland, in Order of Frequency

Bog	Lagg	Lagg
Sphagnum magellanicum	*C. haldanianum*	*Campylium hispidulum*
Sphagna angustifolium	*Tetraphis pellucida*	*Climacium dendroides*
Pleurozium schreberi	*Sphagnum magellanicum*	*Dicranum polysetum*
Ptilidium pulcherrimum	*Sphagna angustifolium*	*Oncophorus wahlenbergii*
C. haldanianum	*Sphagnum centrale*	*Plagiomnium cuspidatum*
Dicranum polysetum	*Pleurozium schreberi*	*Plagiomnium ellipticum*
Dicranum undulatum	*Ptilium crista-castrensis*	*Pseudobryum cinclidioides*
Ptilium crista-castrensis	*Dicranum flagellare*	*Amblystegium serpens*
Aulacomnium palustre	*Brachythecium oedipodium*	*A. serpens* var. *juratzkanum*
Brotherella recurvans	*Thuidium recognitum*	*Anastrophyllum hellerianum*
Dicranum ontariense	*Plagiomnium ciliare*	*Brachythecium rivulare*
Lophocolea heterophylla	*Brachythecium salebrosum*	*Brachythecium velutinum*
Sphagna fuscum	*Calliergon cordifolium*	*Brotherella recurvans*
Pohlia sphagnicola	*Plagiomnium medium*	*Calliergon giganteum*
Cephalozia pleniceps subsp. *sphagnorum*	*Polytrichum commune*	*Dicranum undulatum*
Polytrichum strictum	*Sphagnum wulfianum*	*Hylocomium splendens*
Dicranum fuscescens	*Aulacomnium palustre*	*Hypnum pallescens*
	Jamesoniella autumnalis	*Orthotrichum speciosum*
	Lophocolea heterophylla	*Plagiothecium cavifolium*
	Plagiothecium denticulatum	*Polytrichum longisetum*
	Pohlia nutans	*Pylaisiella selwynii*
	Ptilidium pulcherrimum	*Rhytidiadelphus triquetrus*
	Brachythecium erythrorhizon	*Sanionia unciata*
	Bryohaplocladium microphyllum	*Sphagnum russowii*
	Plagiothecium laetum	*Sphagnum subnitens*
	Platygyrium repens	*Sphagnum warnstorfii*
	Sphagnum teres	*Warnstorfia fluitans*
	Brachythecium reflexum	

The greater plant diversity in the lagg is driven by the chemistry of water upwelling from the underlying A horizon of eolian loess beneath the peatland (Figure 4.8). At the time the relevés were established, pH was 4.03 in the bog and 5.28 in the marginal poor-fen lagg (compare to upland runoff pH of 6.5, Table 4.4). Specific conductivity ($K_{25°Ccorr}$) reduced by the conductivity attributable to H^+ ions indicated that both waters were dilute, the bog 17 and the lagg 23 μS cm^{-1}. The bog water is tea colored with an absorbance value ($Abs_{350nm\,1cm}$) of 0.77; the lagg water less so with a value of 0.31, reflecting the admixture of bog water and less colored water from the surrounding upland. Both the floristic and the water-chemistry data characterize the central peatland as a bog and the lagg as a poor fen according to the widely accepted classification of DuRietz (1949); see also Vitt et al. (1995).

Summary and Discussion

Accumulation of Peat Sediment

Dates in a long peat core (^{14}C dates) stretched over 610 cm and ranged from 783 y BP to 10,900 cal y BP. Simple division yielded an annual peat accumulation rate of 0.06 m century^{-1}. Overall, this was 56 g m^{-2} y^{-1} equivalent to 29 g C m^{-2} y^{-1} and similar to the North America average (Gorham, 1991).

Dates and pollen identification in the long core from the S2 bog identified seven peat sediment phases. The first was Lake Phase I (11,900–10,900 cal y BP), characterized by algal gyttja in the first 900 years. Lake Phase II (10,900–9,200 cal y BP) accumulated the remains of aquatic mosses (Amblystegiaceae, principally *W. fluitans*). Also prevalent were emergent lake aquatics (e.g., *Nuphar, Potamogeton, Brasenia, Sagittaria, Sparganium,* and *Typha*). During Lake Phase II, the water level in the ice-block dropped from an elevation of 420–416 m.

The Rich-Fen/Emergent-Bryophyte Mat Phase spanned only 800 years (9200–8400 cal y BP). A bryophyte mat smothered two small ponds near the bottom of the ice-block depression. The aquatic mosses *W. fluitans* and *C. trifarium* formed the mat. The water had high concentrations of calcium and magnesium and pH values around 8 (in both the Lake and Rich-Fen/Bryophyte Mat Phases). It occurred when the ice front was 390–470 km north fronted by the large glacial Lake Agassiz (Via et al., 2006).

The Rich-Fen/Sedge Phase encompassed 2900 years (8400–5500 cal y BP). *S. platyphyllum* (an alkaline *Sphagna* species), *C. trifarium*, the appearance of *W. exannulatus*, and abundant Cyperaceae characterize this long period of peat accumulation. Between 6300 and 5700 y BP there was a single 5–10 cm slice representing about a 75–150 year period containing *Sphagnum* section *Cuspidata* sp., an acid anomaly in a long period of alkaline water conditions caused by the oxidation of reed roots containing high concentrations of sulfide when water levels dropped to the level of the central flat in the bottom of S2 bog. Peat accumulation during this phase (2.4 m) responded to a piezometric water elevation set by a water table high at the subcontinental divide 3 km north of the S2 peatland. Limestone fragments in the mixed-texture Itasca ground moraine derived from the west side of Hudson Bay. Regional groundwater flowing in the deep sands above this layer dissolved the limestone and brought water into the S2 depression at pH values near 8.

The Transitional-Fen Phase encompassed 2600 years (5500–2900 cal y BP). Pollen of peatland origin during the Transitional-Fen Phase was virtually identical with that of the Rich-Fen/Sedge Phase, marked only by additions of the circumneutral *Sphagna*—*S. cenrale* and *S. subsecundum*. At the beginning of the Transitional-Fen Phase, the peatland was still under the influence of groundwater, but as further paludification occurred (accumulations of another 1.6 m), the groundwater influence waned and water pH values shifted from above 8 to just below 7. At the end of the phase, peat had risen to an elevation of 420.39 m and the depression began to drain to a stream. In the beginning of the phase, the pressure of constant, upwelling groundwater was a dominant force for peat accumulation. During this period, when the peat rose above the subcontinental divide water table at 418 m, the flow of water within the peat mass reversed from upward to downward. Flow reversals and the concurrent existence of local and regional groundwater flow systems are common in peatlands and paramount in understanding hydrology and biogeochemistry in peatlands (Devito et al., 1997). At the end of the Phase, the constant influence of regional, circumneutral, groundwater gave way to periodic processes controlling the chemistry of peat formation. These included water table fluctuation, biologic mixing (vegetation growth, decay, and gas exchange between soils and the atmosphere), and deep frost penetration. Frost penetration preferentially moves colloidal organic matter to the bottom of the frost layers, reaching as much a 45 cm below the surface hollows. At the end of the Transition-Fen Phase, there was a clear break with high-calcium, regional groundwater.

The Open Poor-Fen Phase (2900-130/390 y BP) is marked by the arrival of acid *Sphagna* species—*S. magellanicum, S. angustifolium,* and *S. fuscum*. This period accumulated peat (about 1.1 m), but intense fire during 1864 (based on St. Paul, MN weather records) burned 14 cm of peat. Heinselman (1973, 1996) also documented nearly a half million acres of fire-origin pine dating from 1863 and 1864 in the Boundary Waters Canoe Area Wilderness. He documented an earlier fire in 1610 north of Grand Marais, MN. Surface drainage to the stream was well established with no groundwater upwelling.

The Forested-Bog Phase (1864–2001) accumulated 70 cm of peat, including the construction of an 18 cm-high dome near the recording well toward the east end of bog S2. Built by *Sphagnum* mosses and *Ericaceae* roots, the dome rises above the stream elevation. Bog water acid derives from H-ion generation during organic-matter decomposition in the absence of significant bicarbonate ions (Gorham et al., 1985). The bog dome also contains its own shallow groundwater flow system on top of the 7.7 m of fen and algal gyttja once influenced by the flow strength and alkaline chemistry of the regional groundwater system. The additional 70 cm of hydraulic head induces minute amounts of vertically downward seepage through the bog bottom. However, much larger amounts of water leave the bog above the von Post H9 peat layer in the lagg as unsaturated flow through the thin Koochiching till and the Rainy River sands to the regional water table that is 7 m below.

Climate Change

The occurrence of upland pollen in the long peat core allowed us to interpret major changes in the regional climate at the MEF over the last 12,000 years. First is a transition-climate period (12,000–9,300 y BP). Its definition is hampered by the inability to slice algal gyttja, but change from the bottom of the core to the top of the Aquatic Moss Sediment Phase suggests an average annual temperature rising from −1°C to +5°C, and 1,000 mm of annual precipitation at 12,000 y BP dropping to 700 mm at 9,300 y BP.

A dry and warm phase persisted for 2700 years (9300–7400 y BP). During this period, *P. resinosa/P. banksiana* from the uplands dominated the long core with pollen counts that were 40% of the total pollen sum, the highest of any species in the Holocene. The local pollen record, other pollen work (Kutzbach et al., 1993), and CCM modeling suggest annual temperatures at 4°C–5°C and annual precipitation at 700 mm.

A transition climate from 7400 to 6100 y BP cooled from 4°C to 2°C and became wetter with 700–800 mm of annual precipitation. There was a major shift in upland tree pollen. The *P. resinosa/P. banksiana* component decreased from 40% to 15% and the *P. strobus* component increased from 8% to 35%. Then, the relative distribution of pines persisted through a stable period of nearly 3000 years from 6100 to 3200 y BP. A slight cooling period for 800 years from 3200 to 2400 y BP followed, with a subsequent shift to a short (200 year) warmer and drier period from 2000 to 1800 y BP. Changes in *P. resinosa/P. banksiana* and *P. strobus* pollen of 10%–20% followed suit.

The Little Ice Age from 600 to 150 y BP or about 1370–1840 saw the maximum amount of *P. strobus* pollen (40%). Average annual temperature cooled to 1°C and precipitation increased to 800 mm. For a brief period of 75–150 years, the acidiphilous aquatic *Sphagnum* section *Cuspidata* sp. occurred in the S2 bog pond-waters. In, 6000 y BP, significant remnants of the Keewatin and Labrador Ice Centers remained (Hamblin, 1989) but large expanses of ice in western and eastern Canada had melted, releasing enormous stress on the crust of North America and initiating rebound of the Earth's surface.

Peatland Ecology Perspective

Moore and Bellamy (1974) reviewed the evolution of mire ecology; treatises by Weber (1908) and Potonie (1908) conceptualize how peatlands develop from groundwater-fed basins that accumulate peat until the elevation exceeds the height of a surface outlet. Then the peat mass in the center is fed only by rainwater. Weber (1909) developed the hypothesis of terrestrialization based on a natural history of developing peat masses (ontogeny) near Hanover, Germany. The terms used to describe the three stages of mire development were: niedermoore or flachmoore (rich-fen), ubergangsmoore or zwischen-moore (poor fen), and hochmoore (moss or bog) for both Weber and Potonie. These terms simply mean low, medium, and high moors, referring to their elevation as they develop low to high. Mellin (1917) used a similar division for Swedish mires as riekarr, karr, and moss.

Kivinen (1935) summarized all the early chemistry of lake and moorwässer that laid a phytosociological foundation for the work of Witting (1947, 1949) to explain the occurrence of mire vegetation. It was accepted by DuRietz (1949) that water in the peat mass was distinctly different among the three mire types. Early work in Sweden by Sjörs (1948) followed, but he used the terms rich fen, poor fen, and moss. Later Sjörs (1950) made a direct correlation between water chemistry and vegetation in Swedish mires, but considered the change in mires and their vegetation as they built from low to high elevations as gradual rather than three distinct types.

Kulczynski (1949) also used three divisions for Polish mires—rehophilous or neidermoore, transition or ubergangsmoore, and ombrophilous or hochmoore. This was the first use of the term "transition." Earlier work was autecological in nature, while Kulczynski's work, which linked the terminology and classification of mires using the ontogenic, phytosociologic, and chemical approaches, was synecological in nature. Thunmark (1942) classified mire types in Sweden by pH and calcium concentrations, which Vitt et al. (1995) in Canada and Bridgham et al. (1996) in the United States further defined. In Europe, subsequent correlations between water chemistry and mire vegetation were made for England (Pearsall and Lind, 1941; Gorham, 1956a,b), Poland (Tolpa et al., 1967) and Ireland (Bellamy and Bellamy, 1967).

The work reported here for the S2 bog supports the three class divisions (i.e., rich fen, poor fen, and bog). However, it extends previous investigations to show how ice-block, watershed, and regional hydrology influence the amount and chemistry of water and eventually yield a peat mass based on peat phase development. It also considers not only the waning influence of groundwater under constant piezometric pressure but also the importance of a variety of periodic processes that wean dependence of the developing peat mass from alkaline groundwater to rainwater at the top of the bog dome. Although the water-flow and water-chemistry drivers of peat development gradually change as peat develops (Sjörs, 1950), the phytosociological character of rich fen, poor fen, and bog yield distinct phases of peat development interpreted in the study of paleobotany in peat cores.

In Fennoscandia mire terminology, lagg refers to the plant-rich peatland edge (Cajander, 1913; Sjörs, 1948; Kulczynski, 1949). Lagg always is described in terms of plant diversity. However, the detail of geologic and soil lithology at the edge of the S2 bog functionally defines a lagg that results from the movement of upland water through mineral soil horizons beneath the peat and upwelling to the lagg surface. It also shows and quantifies water movement at the lagg in three directions (Chapter 7). Water upwells to the lagg when upslope mineral soil horizons are saturated. Water seeps downward beneath the lagg to the regional water table daily. And, water forms surface flow (water tracks) originating from both the bog dome and the upland and sends it around the lagg to form first order streams leaving the peatland (Figures 4.7 and 4.8).

Lessons Learned

1. *Pleistocene, Wisconsin glaciation.* Matching glacial drift lithology to dates and regions of origin is a continuing process improved with new chemistry, cores, and insight. It includes details of contemporary soil surveys (e.g., shale fragments in the Koochiching till) and accounts for both regional (western Great Lakes) and distant geology.

2. *Topography.* Detailed maps of surface topography, soil horizons, ice-block bathymetry, and water table elevations that are tied to the same elevation datum greatly improve our ability to interpret Pleistocene and Holocene materials, dating, and climate.

3. *Core dating.* The precision of peat or lake sediment cores is derived from the thickness and cleanness of the slice, in our case 5–10 cm representing about 75–150 years; however, accuracy is a function of carbon dating with a variety of adjustments (± 250 years), tree rings (± 5 years), and climate or historic records (± 1 year).

4. Interpreting sediment cores. Records of moss and aquatic pollen preserved in the peat sediment core provide an insight to plant environmental conditions in the ice-block depression, suggesting identification of major peat sediment phases. Tree, shrub, and herbaceous pollen from the uplands provide an insight into major climate periods of transition or stability.

5. *Landscape hydrology.* Water table position, whether within the ice block depression or in the region, helps differentiate between environmental conditions within the depression giving rise to peat phases and climate periods within the region.

6. Hemisphere or regional? Local pollen records representing the regional climate temper our application of climate based on CCM models or interpreted between century and millennial periods in regional cores not subjected to inferences related to watershed hydrology.

7. *Linkages.* Linkages among the upland, lagg, bog, stream, and the regional water table define how a given watershed ecosystem functions. Topography, soil and glacial lithology, climate periods, peat phases, water chemistry, vegetation autecology and synecology, and hydrology refine our deductions of watershed and ecosystem function.

References

Always, F.J. 1920. Agricultural value and reclamation of Minnesota peat soils. Bulletin 188. St. Paul, MN: University of Minnesota Agricultural Experiment Station.

Anderson, L.E. 1990. A checklist of *Sphagnum* in North America north of Mexico. *The Bryologist* 93:500–501.

Anderson, L.E, H.A. Crum, and W.R. Buck. 1990. List of mosses of North America north of Mexico. *The Bryologist* 93:448–499.

Bay, R.R. 1967. Factors influencing soil moisture relationships in undrained forested bogs. In: *International Symposium on Forest Hydrology*, pp. 335–343. Pennsylvania State University, University Park, PA, Aug. 29–Sept. 10, 1965. Oxford: Pergamon Press.

Bellamy, D.J. and S.R. Bellamy. 1967. An ecological approach to the classification of the lowland mires of Ireland. *Proceedings of the Royal Irish Academy* 65B(6):237–251.

Boelter, D.H. 1965. Hydraulic conductivity of peats. *Soil Science* 100(4):227–231.

Bridgham, D.D., J. Pastor, J.A. Janssens, C. Chapin, and T.J. Malterer. 1996. Multiple limiting gradients in peatlands: A call for a new paradigm. *Wetlands* 16:45–65.

Brooks, K.N. and D.A. Kreft. 1991. Hydrologic Linkages between Uplands and Peatlands. Final Report. St. Paul, MN: Department of Forest Resources, University of Minnesota.

Cajander, A.K. 1913. Studien uber die Moore Finnlands. *Acta Forestalia Fennica* 2(3):1–208.

Crum, H. 1988. *A Focus on Peatlands and Peat Mosses.* Ann Arbor, MI: University of Michigan Press.

Davis, J.F. and R.E. Lucas. 1959. Organic soils, their formation, distribution, utilization and management. Special Bulletin 425. East Lansing, MI: Department of Soil Sciences and Agricultural Experiment Station, Michigan State University.

Devito, K.J., J.M. Waddington, and B.A. Branfireun. 1997. Flow reversals in peatlands influenced by local groundwater systems. *Hydrological Processes* 11:103–110.

Digerfeldt, G., J.E. Almendinger, and S. Björck. 1992. Reconstruction of past lake levels and their relation to groundwater hydrology in the Parkers Prairie sandplain, west-central Minnesota. *Palaeogeography, Palaeoclimatology, Palaeoecology* 94(104):99–118.

Donovan, J.J., A.J. Smith, V.A. Panek, D. Engstrom, and E. Ito. 2002. Climate-driven hydrologic transients in lake sediment records: calibration of groundwater conditions using 20th Century drought. *Quaternary Science Reviews* 21:605–624.

DuRietz, G.E. 1949. Huvudenheter och huvudgränser i Svensk myrvegetation. *Svensk Botanisk Tidskrift* 43:274–309.

Filby, S.K., S.M. Locke, M.A. Person, T.A. Winter, D.O. Rosenberry, J.L. Nieber, W.J. Gutowski, and E. Ito. 2002. Mid-Holocene hydrologic model of the Shingobee Watershed, Minnesota. *Quaternary Research* 58(3):246–254.

Gafni, A. and K.N. Brooks. 1986. Tracing approach to determine groundwater velocity in peatlands. *Hydrological Science and Technology* 2(4):17–25.

Gafni, A. and K.N. Brooks. 1990. Hydraulic characteristics of four peatlands in Minnesota. *Canadian Journal of Soil Science* 70:239–253.

Gorham, E. 1956a. The ionic composition of some bog and fen waters in the English lake district. *Journal of Ecology* 44:142–152.

Gorham, E. 1956b. On the chemical composition of some waters form the Moor House Nature Reserve. *Journal of Ecology* 44:377–384.

Gorham, E. 1991. Northern peatlands: Role in the carbon cycle and probable responses to climatic warming. *Ecological Applications* 1:182–195.

Gorham, E., S.J. Eisenreich, J. Ford, and M.V. Santelmann. 1985. The chemistry of bog waters. In: *Chemical Processes in Lakes*, ed. W. Stumm, pp. 339–363. New York: Wiley.

Hamblin, W.K. 1989. *The Earth's Dynamic Systems*, 5th edn. New York: Macmillan Publishing Co.

Harlow, W.M. and E.S. Harrar. 1958. *Textbook of Dendrology*, 4th edn. New York: McGraw-Hill Book Co.

Hedenäs, L. 2003. The European species of the *Calliergon–Scorpidium–Drepanocladus* complex, including some related or similar species. *Meylania* 28:1–117.

Hedenäs, L. and A. Kooijman. 1996. Phylogeny and habitat adaptations within a monophyletic group of wetland moss genera (Amblystegiaceae). *Plant Systematics and Evolution* 199:33–52.

Heinselman, M.L. 1973. Fire in the virgin forests of the Boundary Waters Canoe Area, Minnesota. *Quaternary Research* 3:329–82.

Heinselman, M.L. 1996. *The Boundary Waters Wilderness Ecosystem*. Minneapolis, MN: University of Minnesota Press.

Ingram, H.A.P. 1978. Soil layers in mires: function and terminology. *Journal of Soil Science* 29:224–227.

Jirsa, M.A. and V.W. Chandler. 2007 Bedrock geology of the Bigfork 30′ × 60′ quadrangle, northern Minnesota. Misc. Map Series Map M-176. Minneapolis, MN: University of Minnesota, Minnesota Geological Survey.

Keys, J.E. Jr. and C.A. Carpenter. 1995. *Ecological Units of the Eastern United States: First Approximation*. Atlanta, GA: USDA Forest Service.

Kivenen, E. 1935. Über electolytegehalt und reaction der moorwässer. *Maatouskoelatioksen Maatutkimusosasto Agrogeol. Julkaisuja*. 38. Helsingfors.

Kulczynski, S. 1949. Peat bogs of Polesie. *Memoirs Academy of Cracovie* 315:1–356.

Kutzbach, J.E., P.J. Bartlein, I.C. Prentice, W.F. Ruddiman, F.A. Street-Perrott, T. Webb III, and H.R. Wright Jr. 1993. Epilogue. In: *Global Climates Since the Last Glacial Maximum*, eds., H.W. Wright Jr., J.E. Kutzbach, T. Webb III, W.F. Ruddiman, F.A. Street-Perrott, and P.J. Bartlein, Chap. 20, pp. 536–542. Minneapolis, MN: University of Minnesota Press.

Malterer, T.J., E.S. Verry, and J. Erjavec. 1992. Fiber content and degree of decomposition in peats. *Soil Science Society of America Journal* 56(6):1200–1211.

Maxwell, J.R., C.J. Edwards, M.R. Jensen, S.J. Paustion, H. Parrott, and D.M. Hill. 1995. A hierarchical framework of aquatic ecological units in North America (Nearctic Zone). Gen. Tech. Report NC-176. St. Paul, MN: USDA Forest Service.

Mellin, E. 1917. Studier Över de noorlandska Handbibl, 7. Uppsala, Sweden.

Meyer, A.F. 1944. *The Elements of Hydrology*, 2nd edn, fifth printing, pp. 522. London: John Wiley & Sons.

Meyer, G.N. 1986. Subsurface till stratigraphy of the Todd County area, central Minnesota. Report of Investigations 34. St. Paul, MN: Minnesota Geological Survey.

Meyer, G.N. 1997. Pre-Late Wisconsinan till stratigraphy of north-central Minnesota. Report of Investigations 48. St. Paul, MN: Minnesota Geological Survey.

Meyer, G.N., C.J. Patterson, H.C. Hobbs, M.D. Johnson, and J.F.P. Cotter. 1998. Terrestrial record of Laurentide Ice Sheet reorganization during Heinrich events: Comments and Reply. *Geology* 26:667–668.

Moore, P.D. and D.J. Bellamy. 1974. *Peatlands*. New York: Springer-Verlag.

Mooers, H.D. 1990. Discriminating texturally-similar glacial tills in central Minnesota with graphical and multivariate techniques. *Quaternary Research* 34(2):133–147.

Mooers, H.D. and C.A. Dobbs. 1993. Holocene landscape evolution and the development of models for human interaction with the environment: An example from the Mississippi Headwaters region. *Geoarchaeology* 8(6):475–492.

Mooers, H.D. and J.D. Lehr. 1997. Terrestrial record of Laurentide Ice Sheet reorganization during Heinrich events. *Geology* 25(11):987–990.

Mooers, H.D. and A.R. Norton. 1997. Glacial landscape evolution of the Itasca/St. Croix moraine interlobate area including the Shingobee River headwaters area. In: *Hydrological and Biogeochemical Research in the Shingobee River Headwaters Area, North-central Minnesota*, ed., T.C. Winter, pp. 3–10. Water Resources Investigation Report 96-4215. Denver, CO: U.S. Geological Survey.

Nyberg, P.R. 1987. *Soil Survey of Itasca County, Minnesota*. 1987 0-493-527. QL3USDA-Soil Conservation Service. St. Paul, MN.

Oakes, E.L. and L.W. Bidwell. 1968. Water resources of the Mississippi headwaters watershed, north central Minnesota. USDI—Geological Survey. Hydrologic Atlas. HA-278.

Okruszko, H. 1960. Muck soils of the valley peat bogs and the chemical and physical properties. *Roczniki Nauk Rolniczych* 74-F1:5–89. Translation from U.S. Department of Commerce, Clearinghouse for Federal Scientific and Technical Information. 11 67-56119. Springfield, VA.

Ownbey, G.B. and T. Morley. 1991. *Vascular Plants of Minnesota. A Checklist and Atlas*. Minneapolis, MN: University of Minnesota Press.

Paulson, R.O. 1968. A soil survey: Marcell Experimental Forest. Itasca Co. Minnesota. Grand Rapids, MN: USDA Soil Conservation Service.

Pearsall, W.H. and E. Lind. 1941. A note on a Connemara bog type. *Journal of Ecology* 29:62–68.

Potonie, R. 1908. Aufbau und vegetation der norddeutschlands. *Engler. Bot. Jahrb.* 90, Leipzig, Germany.

Sander, J.E. 1971. Bog-watershed relationships utilizing electric analog modeling. PhD dissertation. East Lansing, MI: Michigan State University.

Schwalb, A. and W.E. Dean. 2002. Reconstruction of hydrological changes and response to effective moisture variations from North-Central USA lake sediments. *Quaternary Science Reviews* 21(202):1541–1554.

Siegel, D. and P. Glaser. 1987. Groundwater flow in a bog-fen complex, Lost River Peatland, northern Minnesota. *Journal of Ecology* 75:743–754.

Sjors, H. 1948. Myrvegetation i Bergslagen (Mire vegetation in Bergslagen). *Acta Phytogeographica Suecica* 21:1–299.

Sjors, H. 1950. On the relation between vegetation and electrolytes in north Swedish mires. *Oikos* 2:241–258.

Stotler, R. and B. Crandall-Stotler. 1997. A checklist of the liverworts and hornworts of North America. *The Bryologist* 80:405–428.

Teller, J.T. and D.W. Leverington. 2004. Glacial Lake Agassiz: A 5000 year history of change and its relationship to the $\delta^{18}O$ record of Greenland. *Bulletin of the Geological Society of America* 116:729–742.

Thunmark, S. 1942. Über rezente eisenocker und ihre mikroorganismengeme in-Schaften. *Bulletin of the Geological Institute of Uppsala*. Uppsala, 29:1–285.

Tolpa, S., M. Jasnowski, and A. Palczynski. 1967. System der genetischen klassifizierung der torfe mitteluropas. NO 76. Warsaw: Zesz. Probl. Post. Navk. Roln.

Tracy, D.R. 1997. Hydrologic linkages between uplands and peatlands. PhD dissertation. St. Paul, MN: University of Minnesota.

Verry, E.S. 1980. Streamflow and water table changes following stripcutting and clearcutting in an undrained black spruce bog. In: *Proceedings of the 6th International Peat Congress*, pp. 493–498, August 17–23, 1980. Duluth, MN. Helsinki: International Peat Society.

Verry, E.S. 1984. Microtopography and water table fluctuation in a *Sphagnum* mire. In: *7th International Peat Congress,* June 18–23, 1984, Dublin, Ireland. Helsinki: International Peat Society, Vol. 2, pp. 11–31.

Verry, E.S. 1991. Concrete frost in peatland and mineral soils: Northern Minnesota. In: *Proceedings of the International Peat Symposium Peat and Peatlands: The Resource and Its Utilization*, eds., D.N. Grubich and T.J. Malterer, August 19–22, 1991, Duluth, MN. Helsinki, Finland: International Peat Society, pp. 121–141.

Verry, E.S. 2000. Water flow in soils and streams: sustaining hydrologic function. In: *Riparian Management in Forests*, eds., E.S. Verry, J.W. Hornbeck, and C.A. Dolloff, pp. 99–125. Boca Raton, FL: Lewis Publishers.

Verry, E.S. and T.R. Timmons. 1982. Waterborne nutrient flow through an upland-peatland watershed in Minnesota. *Ecology* 63(5):1456–1467.

Viau, A.E., K. Gajewski, M.C. Sawada, and P. Fines. 2006. Millennial-scale temperature variations in North America during the Holocene. *Journal of Geophysical Research, D, Atmospheres* 111(D09102):1–12.

Vitt, D.H., S.E. Bayley, and T.L. Jin. 1995. Seasonal variations in water chemistry over a bog-rich fen gradient in Continental Western Canada. *Canadian Journal of Fisheries and Aquatic Sciences* 52:587–606.

Weber, C.A. 1908. Aufbau und vegetation der moore norddeutschlands. *Englers. Bot. Jahrb.* 90. Leipzig, Germany.

Weber, C.A. 1909. Das moor hanoverishce geschict blätter.

Witting, M. 1947. Katjonbestämningar I myrvatten. *Botaniska Notiser.* 287–311.

Witting, M. 1949. Kalsiumhalten I nagra nordsvenska myrvatten. *Svensk Botanisk Tidskrift* 43:715–739.

Wright, H.E., I. Stefanova, J. Tian, T.A. Brown, and F.S. Hu. 2004. A chronological framework for the Holocene vegetational history of central Minnesota: The Steel Lake pollen record. *Quaternary Science Reviews* 23(2004):611–626.

5

Physical Properties of Organic Soils

Elon S. Verry, Don H. Boelter, Juhani Päivänen,
Dale S. Nichols, Tom Malterer, and Avi Gafni

CONTENTS

Introduction

Compared with research on mineral soils, the study of the physical properties of organic soils in the United States is relatively new. Always (1920) and Anderson et al. (1951) considered the value and reclamation of peats in Minnesota and nationally. Davis and Lucas (1959) summarized organic soil formation, utilization, and management in Michigan; however, most of the literature on the detailed physical properties of peat was published from 1956 to 2003. This is true in Europe as well (Parent and Ilnicki 2003), except for a study by von Post (1922) who developed a field method for determining the degree of humification (decomposition) that is used widely outside of the United States (Box 5.1). A comprehensive series of studies on peat physical properties were conducted by Don Boelter (1959–1975), first at the Marcell Experimental Forest (MEF) and later throughout the northern Lakes States to investigate how to express bulk density (D_b, weight or volume basis), water retention characteristics, hydraulic conductivity (K), fiber content, specific yield (drainable porosity), and the degree of decomposition (pyrophosphate test). Juhani Päivänen, a graduate student from the University of Helskinki, spent nearly a year at the MEF to learn the techniques developed there. Upon returning to Finland, he sampled extensively in central Finland and developed a physical-property data set similar to that of Boelter. Together, the data of Boelter and Päivänen represent one of the largest examinations of the physical properties of peat. After 1975, physical-property studies continued at MEF including a detailed examination of fiber contents in Lake State organic soils, a comparison of international methods for physical properties,

BOX 5.1   VON POST FIELD-TEST PROCEDURE FOR H VALUE

Take samples about 5 m from prospective well sites. A Russian peat corer (Macaulay) or bucket auger may be used. Each horizon with a different H value should be evaluated. Take samples to at least 1.3 m or until a mineral-soil contact is found. Part of each sample can be retained for verification but the H value should be determined immediately using saturated soil or with water added if the soil is dry. Place enough soil to fill the hand when the fingers are gently curved against the palm. Gently bounce this egg-shaped soil until it just fits your hand. Add or remove soil to fill the gently curved pocket in your hand. Squeeze the sample as hard as you can. In the other hand catch the amorphous material and water squeezed between the fingers. Note the color and turbidity of the free water (water that is separate from any amorphous material). Thinning the water by opening the second hand facilitates the examination of color and turbidity.

Use the fingers of the second hand to scrape amorphous material from between the closed, squeezed fingers of the first hand and consolidate the amorphous material in the second hand. Consolidate the material in both hands by gently bouncing the material. Open both hands and compare the relative volume of the fiber material and the amorphous material. The relative volume of the amorphous material (in percent) is used to assign the von Post H value for whole and half values in the mid-range (Table Box 5.1). In Europe and Canada, use of half classes is common. Half classes in the mid-range are helpful in differentiating peat layers that are encountered frequently in partially drained areas. Table Box 5.1 shows half classes in the range of H4–H7.5.

TABLE BOX 5.1

von Post Field Evaluation Adapted for Hydraulic Conductivity in the Mid-Range (H4–H7.5)

von Post H Value	Volume Passing through Fingers (%)	Additional Description of Free Water Expressed to the Second Hand
1	0	Expressed water is clear to almost clear and yellow-brown in color. Slowly open the second hand and observe color as the water depth thins
2	0	
3	0	Water is muddy brown and retained fiber is not mushy
4	0	Very turbid, muddy water and retained fiber is somewhat mushy
4.5	1	Amorphous material primarily stays on outside of squeezed fingers
5	2–10	Use the volume of amorphous material passed. As with H4 and H4.5, water at the edges of the amorphous material is very turbid and muddy
5.5	11–25	
6	26–35	
6.5	36–45	
7	46–55	Water around the amorphous material is thick, soupy, and very dark
7.5	56–65	Water around the amorphous material is thick, soupy, and very dark
8	66–75	There is essentially no free water; it is all amorphous material
9	76–95	There is no free water associated with the amorphous material
10	95–100	

and a comparison of the piezometer and salt-dilution methods for hydraulic conductivity. In this chapter we present the peat physical properties, compare methods, and describe how to use physical peat data in lateral-extent equations to evaluate the effect of drainage in peatlands.

Early work at the MEF underpinned the U.S. Department of Agriculture's (USDA) interpretation and classification of organic soils in the United States by showing the range of fiber content in a variety of peats. In 1962, Boelter (1964a,b) sampled nine peatlands at MEF (Itasca County, Minnesota) and three peatlands in Koochiching County, Minnesota, to begin measurements of water content, bulk density, and water retention at 0.0 kPa (saturation) and at 0.5, 10, 20, 100, 200, and 1500 kPa. The 12 sites yielded 119 samples. Nichols and Boelter (1984) reported on samples from northern Minnesota, northern and central Wisconsin, and Upper Michigan, adding 57 samples from 26 sites. In total, 176 peat samples from 38 peatlands were collected. The botanical composition of the plant residues in the peats are listed in Table 5.2. Moss peats were predominately of *Sphagnum* origin, whereas herbaceous peats were predominately sedge (*Carex*) which also is known as reed-sedge or sedge peats.

Fiber content in organic soils is the fundamental characteristic that determines bulk density, water retention, hydraulic conductivity, and drainable porosity. The inclusion of mineral ash in excess of plant cellular ash components (e.g., windblown dust) also contributes to bulk density. The bulk density of a peat can be corrected for its ash content to quantify the bulk density of the organic portion (Nichols and Boelter 1984). When fiber content is divided by size category (>2.0, 1.0–2.0, 0.5–1.0, 0.25–0.5, 0.1–0.25, and <0.1 mm), it parallels the primary particles (sand, silt, clay) in mineral soils. Organic material less than 0.1 mm is subcellular and amorphous (not fiber); it affords strong cation exchange and water retention (Kwak et al. 1986), but strongly limits hydraulic conductivity because the percentage of amorphous material increases.

The degree of decomposition can be estimated by the amount of material (<0.1 mm) or the solubility of peat in sodium pyrophosphate solution (Farnham and Finney 1965). Lynn et al. (1974) used rubbed fiber content (rubbing small portions of a peat sample between thumb and fingers 10 times using moderate pressure) and the pyrophosphate index as the basis for classifying U.S. soils at the suborder level (Soil Survey Staff 1975). Lynn et al. (1974) sieved the rubbed peat through standard soil sieves using a gentle stream of water. This is similar to the method of Boelter (1964a,b) except that Lynn et al. used a bottom sieve of 0.15 mm. In this method, which is used to determine fiber content in the United States and Canada, the smallest fiber class is 0.25–0.15 mm and all material <0.15 mm is placed in the amorphous class. Sieves of 0.15 or 0.25 mm that are sealed across a centrifuge tube were used in the former Soviet Union (Tolonen and Saarenmaa 1979). All these national laboratory methods were compared using peat samples, most of which were from the MEF and central Minnesota (Malterer et al. 1992). Additionally, these authors characterized the von Post field method on the same samples (von Post 1922; von Post and Granlund 1926). The degree of humification or H value in the von Post method is based on the amount of peat and the color of water expressed from an egg-sized sample of peat squeezed in the hand. The von Post method defines 10 classes for degree of humification (H1–H10). This field-based method is used extensively outside the United States because the method is quick and, with practice, is consistent, more precise, and more accurate than sieving methods.

Bulk density of the organic portion of peats also reflects decomposition and is strongly related to the hydraulic conductivity of peats (Boelter 1965). Boelter used the borehole and piezometer methods to determine hydraulic conductivity at MEF. Both methods measure the rate of fall in the water level in the hole or in the piezometer as water flows into the soil around the hole or pipe opening. The peizometer method can be adjusted to measure a specific peat horizon of a given fiber content, bulk density, or degree of decomposition. In 1970, techniques developed at MEF were introduced to central

Finland by Päivänen (1973), who duplicated measurements of hydraulic conductivity at 28 sites with *Sphagnum* peat, 23 with sedge peat, and 29 with woody peat. These samples were collected from pristine and drained peatlands at the University of Helsinki's Hyytiälä Forestry Field Station. Earlier, he found relationships between bulk density and von Post H values for 316 samples collected from the same peatlands (Päivänen 1969).

The values for hydraulic conductivity may be biased because the water column in a borehole or piezometer creates a sloping water table that increases the hydraulic gradient relative to undisturbed peats. Gafni (1986) measured the hydraulic conductivity of peats by injecting small volumes of concentrated salt solutions into piezometers with a site at the MEF and three in the Cromwell area of St. Louis County, northeastern Minnesota. Salt movement was measured by detection in downslope wells or by dilution of salt in the injection well over time. When these methods were investigated at the MEF, point dilution was far more reliable than the trace method. Gafni also measured the von Post H value and related it to hydraulic conductivity (Gafni 1986; Gafni and Brooks 1990). A comparison of the piezometer and salt-dilution method revealed that the salt dilution is suitable for porous peats and the piezometer method is suitable for moderately to well-decomposed peats. However, the use of piezometers for porous soils greatly underestimates hydraulic conductivity. Earlier work in the United States and Finland includes that of Feustel and Byers (1930) and Huikari (1959), respectively. Examination of data on hydraulic conductivity in Finland and at the MEF revealed that neither methodology piezometer or point dilution is suitable for the entire range of peat decomposition (von Post H values 1–10). In the large-pored (H1–H6) peats K_{sat} is best determined by point dilution; in small-pored (H7–H10) peats, K_{sat} is best determined by the piezometer.

Parent and Illnicki (2003) summarized the chemical and physical properties of organic soils and peat materials. They included physical-property data for moorish soils (granular peats over well-decomposed horizons developed with prolonged drainage and cultivation), and long-term data on hydraulic conductivity and bulk density collected in Poland (e.g., Okruszko 1993).

In the United States, the physical properties of hydraulic conductivity, drainable porosity, and horizon thickness are used to determine the lateral effect of drainage. Lateral effect is the distance from a ditch or tile line where the soil is drained sufficiently to affect Federal or State protection of wetland status. Many wet sites were drained prior to the wetland regulation and protection specified in the 1974 Clean Water Act. However, the degree of drainage within a wetland was variable: some areas qualified as drained (within the lateral effect) while some areas beyond the lateral effect retained their protected status. The von Post H value correlates well with hydraulic conductivity, fiber content, and drainable porosity and is recommended as an alternative to laboratory or field determinations of hydraulic conductivity using piezometer methods.

In this chapter we discuss the important physical properties of peat as they have been determined and tested over 50 years, and show examples of how to use drainage equations to evaluate whether a wetland site is drained. The data set on peat physical properties developed at MEF and throughout the Lake States is the largest and most comprehensive in the United States.

Expression of Water Content and Bulk Density

Water content in mineral soils (%) is routinely expressed as the mass of water lost on drying to the oven-dry mass of the soil. The mass of water lost on drying is relatively small compared to the dry mass of the mineral soil (20%–35%). The mass of water in an organic soil is large compared to the oven-dry mass of the soil (organic fibers and some mineral ash), from 300% up to 3000%. Although the intrinsic meaning of water content is the same for mineral and organic soils when expressed this way, comparison of water content among organic soils is difficult when the range of values is undefined (i.e., <100% in mineral soils but essentially open-ended in organic soils). However, water content based on bulk saturated mass (mass of water at soil saturation divided by the total mass of water and wet soil) may be used to predict bulk density if allowances for gas volume are made (Laine and Päivänen 1982).

Boelter (1964a,b) used the bulk volume of a saturated sample for water content (volume basis) in organic soils in the United States. Water content (Wv) is the amount of water lost from the soil upon drying at 105°C (determined by mass loss) and expressed as the volume of water per unit volume of bulk soil. The volume of bulk soil is the volume of the soil sample removed in the field. It usually is sampled with a cylinder (sharpened at one end) of known volume. The cylinder is pushed gently to just below the peat surface, the extruded peat on top shaved off, and the cylinder is dug around on the outside and then detached at the lower end with a knife. Below the water table, a large-diameter caisson (about 60 cm) is evacuated of water as the peat is removed, and then a smaller volume cylinder (10 cm diameter by 10 cm long) is used to sample the peat. The volume of the peat is the volume of the cylinder when the peat is saturated (<0.33 kPa of water tension, which leaves the sample at full volume). Bulk density also is determined by dividing the total sample mass (oven dry) by the volume of the cylinder.

Laine and Päivänen (1982) sampled peats with a standard cylinder as part of a detailed method for calculating bulk density and total porosity that accounted for the volume of gases in the water. Rather than expressing the volume of the cylinder as m^3, they expressed the volume of the cylinder as the mass of water (Mg). The oven-dry mass of the peat and water divided by the mass of the water is another calculation for bulk density. This method reduces the variation in correlations between bulk density and degree of humification (von Post H value) but only when corrections for gas volume are made. It is suitable for samples below the water table (assuming that ash from windblown dust is low).

Boelter and Blake (1964) found a highly significant correlation between bulk density determined on a saturated-volume basis and on a dry-volume basis. Saturated-volume bulk densities ranged from 0.02 to 0.26 Mg m⁻³, though the standard error of estimate for a single value in the regression was 0.021 Mg m⁻³. Päivänen (1969) also found a high correlation between fresh, field-volume bulk density and laboratory-volume density as determined by the consistent packing of dried and ground samples. In subsequent work, Boelter and Päivänen consistently used the saturated or near saturated field volume of peat as the basis for bulk density (oven-dry mass per saturated volume; Mg m⁻³).

Correlation of Ash Content and Bulk Density

Both Nichols and Boelter (1984) and Päivänen (1973) correlated ash content as a function of bulk density. Both separated peat samples by botanical origin: *Sphagnum*, sedge (*Carex* or herbaceous), or woody. Lake States peats are from undrained peatlands while peats from central Finland are from peatlands drained for forestry. The results are plotted for woody peats (Figure 5.1) and for *Sphagnum* and sedge peats (Figure 5.2). The ash content of Lake States peat is about 5% higher in than that in woody peats of central Finland.

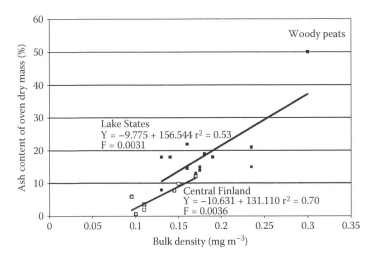

FIGURE 5.1
Relation of ash content to bulk density in the Lakes States and central Finland for woody peats as redrawn and regressed using data from Nichols and Boelter (1984) and Päivänen (1969). The F-value is the significance level of the regression.

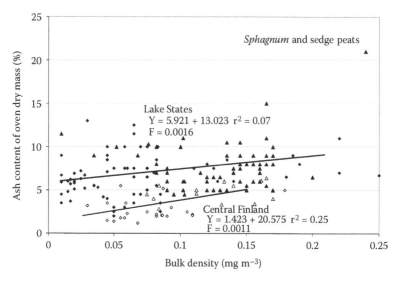

FIGURE 5.2

Relation of ash content to bulk density in the Lakes States and central Finland for *Sphagnum* and sedge peats as redrawn and regressed using data from Nichols and Boelter (1984) and Päivänen (1969). The open symbols are from central Finland (diamonds = *Sphagnum*; triangles = sedge); and the solid symbols are from the Lake States (diamonds = *Sphagnum*, triangles = sedge). The F-value is the significance level of the regression.

Higher ash contents in Lake States peat are traced to the continental position of the Lake States which are situated just east of large semiarid areas in the northern Great Plains. Mineral dust from these areas is routinely deposited in the Lake States. It is likely that peats in Manitoba and Ontario, Canada also have higher ash contents than peats in central Finland, which is just east of the slightly saline Gulf of Bothnia and south of the Arctic Circle. As such, precipitation in this region does not contain a large amount of mineral dust.

Correlation of Bulk Density and von Post Degree of Humification

Päivänen (1969) correlated bulk density with the von Post degree of humification for *Sphagnum*, *Carex* (sedge), and woody peats in central Finland (Figure 5.3). Similarly, a highly significant correlation between bulk density and degree of humification was found by Puustjärvi (1970), Karesniemi (1972), Päivänen (1973), Raitio and Huttunen (1976), Silc and Stanek (1977), Tolonen and Saarenmaa (1979), and Korpijaakko and Häikiö Leino (1981).

Most sedge and woody peats are found in fens fed by groundwater that is high in calcium and magnesium; they impart higher ash contents to ligneous

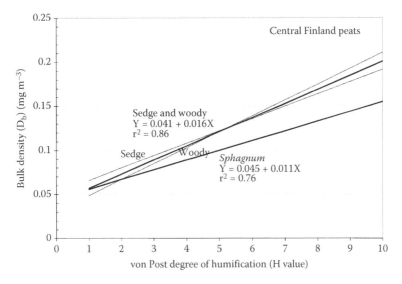

FIGURE 5.3
Relation between von Post degree of humification H value and bulk density, volume basis, for peats in central Finland as redrawn from Päivänen (1969). Sedge and woody peat regressions are not significantly different and are combined. (From Päivänen, J., *Acta Forest. Fenn.*, 129, 1, 1973.)

tree and sedge cells than to the nonligneous cells of *Sphagnum* peats occurring in poor fens or bogs (or in fens on hummocks, partially isolated from the groundwater below the hummocks). However, not all sedges are mesotrophic or eutrophic and high in lignins, some sedges that grow on poor fen sites are depauperate in minerals.

Available Water-Storage Capacity in Organic Soils

The handling of samples of organic soil can strongly affect available water-storage capacity (AWSC) and water retention characteristics. Boelter (1964a,b) measured AWSC at 3, 10, 33, and 1500 kPa of tension. Two sample-handling methods were compared: air-dried and ground peat samples and undisturbed field samples that were kept moist. AWSC at 3 kPa was about 13% less (by volume) in air-dried and ground samples about 5% less at 10 kPa, about 4% less at 33 kPa, and about the same at 1500 kPa. The irreversible shrinkage that occurs when organic soils are dried reduces the total porosity of the peat sample, no doubt altering the pore-size distribution such that more of the remaining pores are small and able to retain water only at higher tensions. The total loss of pore space with drying reduces total AWSC 5% at

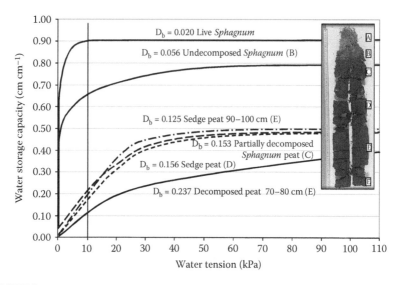

FIGURE 5.4
Available water storage capacity of undisturbed peats (S2 bog) at low water tensions for six peat samples with variable bulk density (D_b). Water storage capacity at 10 kPa is taken as field capacity in organic soils.

10 kPa taken as field capacity in peats. The use of moist field samples is recommended in determining water-storage capacity.

Boelter's first determinations of AWSC were with undisturbed, wet, peat core samples from the S2 bog. Six peat samples were taken from a single core. AWSC was measured on samples placed on large pressure plates or, for low tensions, on a single sample placed in a pressure cell that was the same diameter as the field-sampling core. In both instances, peat was seated on an asbestos slurry separated from the peat by cheesecloth. Changes in water content before and after application of the tension provided a measure of water-storage capacity (or a measure of specific yield). Water content was expressed as cm of water cm^{-1} of peat core thickness. In contemporary usage, drainable porosity is used synonymously with specific yield. From the outset at the MEF, basic science was undertaken to support applied science as related to soil, water, and forest management. This early work shows how fundamental water-storage characteristics in organic soils support field studies to calculate water balance, evaluate drainage and, years later, evaluate drainage lateral extent in wetlands. Figure 5.4 shows three major ranges in water storage or specific yield (live and undecomposed *Sphagnum*, three peats with D_b values of 0.125–0.156 Mg m^{-3}, and a well-decomposed peat with a D_b value of 0.237 Mg m^{-3}).

In mineral soils, field capacity (water content after 3 days of drainage from saturation) is represented by water content at 10 kPa for porous sands and 33 kPa for loams and clays. In organic soils, available water storage is taken at 10 kPa. The large-pored *Sphagnum* samples in Figure 5.4 (D_bs of 0.02 and 0.056 Mg m^{-3}) can store 0.75 or 0.90 cm of water cm^{-1}, that is, they have a specific yield or drainable porosity of 75%–90% of their volume. Likewise, more decomposed peats with a D_b of 0.125–0.156 Mg m^{-3} can store only 0.18–0.22 cm cm^{-1} and have a drainable porosity of 18%–22%. The most decomposed peat ($D_b = 0.237$ Mg m^{-3}) in Figure 5.4 can store only 0.12 cm cm^{-1} and has a drainable porosity of 12%. Water retention and drainable porosity of peat changes dramatically with degree of decomposition and bulk density, and must be accounted for by delineating horizons in natural settings. For instance, annual water budgets in peatlands account for changes in water storage from, say, the first day of the year when water tables are high to the last day of the year when, following a drought, water tables are low. The amount of actual water difference that left storage is dependent on the size of pores in the zone of water table fluctuation and their corresponding drainable porosity. On a daily basis, the water table response to precipitation is the inverse of drainable porosity (specific yield). A 1 cm precipitation addition to a raised-dome peatland (no immediate upland or groundwater input) with a water table in a horizon with a drainable porosity of 0.22 would have a water table response of 1/0.22 or 4.55 cm. A water table dropping 1 cm without precipitation would yield only 0.22 cm of water depth. Total porosity of the peat horizons in Figure 5.4 vary little (range of 84%–97%), but the size distribution varies greatly. The upper peats have many large pores and the deeper peats have many small pores. The distribution of pore space size is a principal correlate with degree of decomposition, bulk density, drainable porosity, and hydraulic conductivity.

Water Retention in Organic Soils

Samples from other peat cores in S2 bog were assessed for the amount of water retained under tensions of 0.5, 3.5, 10, 20, and 1500 kPa. Water retention in Figure 5.5 is on the same scale as that in Figure 5.4 (Boelter 1970). Although different peat samples (notice differences in bulk density) from the same peatland were used, water-retention curves (Figure 5.5) are nearly opposite of available water storage (Figure 5.4). These are large differences among peats with large differences in bulk density, but sedge peats with moderate decomposition and well-decomposed peats show similar water retention with different bulk densities.

Boelter (1970) illustrated the significant impact of bulk density on water retentions at 0.5, 10, and 1500 kPa (Figure 5.6). He also showed the impact of fiber content on water retention (Figure 5.7). It should be noted that

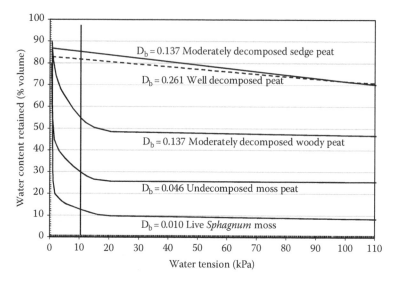

FIGURE 5.5
Water retention at various water tensions for a variety of peats from the S2 bog, MEF. Water tension at 10 kPa is considered field capacity. (Redrawn from Boelter, D.H., Important physical properties of peat materials, in *3rd International Peat Congress Proceedings*, Helsinki, Finland, 1970, pp. 150–154.)

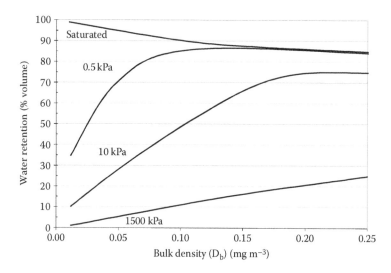

FIGURE 5.6
Water retention in peat as a function of bulk density. (Redrawn from Boelter, D.H., Important physical properties of peat materials, in *3rd International Peat Congress Proceedings*, Helsinki, Finland, 1970, pp. 150–154.).

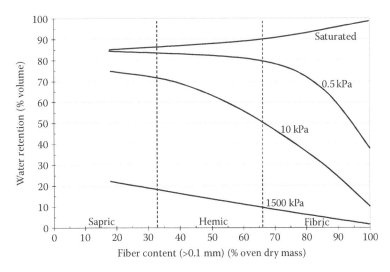

FIGURE 5.7

Water retention in peat as a function of fiber content. (Redrawn from Boelter, D.H., Important physical properties of peat materials, in *3rd International Peat Congress Proceedings*, Helsinki, Finland, 1970, pp. 150–154.) The vertical dashed lines illustrate USDA divisions for sapric (well decomposed), hemic (moderately decomposed), and fibric (least decomposed) categories used for soil-series classification (U.S. Soil Survey Staff 1975).

water-retention measurements at very low tensions require careful attention to sample volume and manometer column heights.

Figures 5.5 and 5.6 illustrate the dependence of water retention on bulk density and fiber content, particularly at 10 kPa of tension. Equations describing the relationship in Figures 5.6 and 5.7 are given in Table 5.1

TABLE 5.1

Regression Equations and R^2 Values for the Relationship between Water Retention (Y) at Saturation, 0.5, 10, and 1500 kPa and either Fiber Content (X) or Bulk Density (X)

Level of Saturation	Independent Variable (X)	Regression Equations by kPa of Tension	R^2
Saturated	Fiber content	$Y = 84.23 - 0.0279X + 0.00185X^2$	0.68
	Bulk density	$Y = 99.00 - 123.45X + 252.92X^2$	0.66
0.5 kPa	Fiber content	$Y = 52.45 + 1.5619X - 0.01728X^2$	0.69
	Bulk density	$Y = 39.67 + 638.29X - 2010.89X^2$	0.70
10 kPa	Fiber content	$Y = 67.91 + 0.4136X - 0.01064X^2$	0.80
	Bulk density	$Y = 2.06 + 719.35X - 1809.68X^2$	0.88
1500 kPa	Fiber content	$Y = 29.34 - 0.3420X + 0.00072X^2$	0.73
	Bulk density	$Y = 1.57 + 15.28 - 107.77X^2$	0.82

Source: Boelter, D.H., *Soil Sci. Soc. Am. Proc.*, 33(4), 606, 1969.

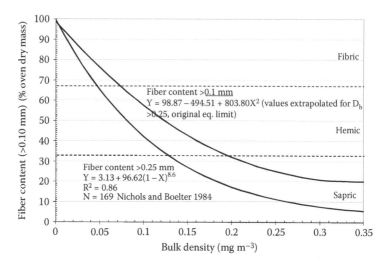

FIGURE 5.8
The relation between fiber content (>0.1 mm; Boelter 1969) and (>0.25 mm; Nichols and Boelter 1984) bulk density.

with R^2 values from 0.66 to 0.88. Figure 5.8 shows the correlation of fiber content (>0.1 mm) and bulk density. The correlations for other fiber content thresholds (>0.25, >0.50, >1.0, and >2.0 mm) are equally strong (Boelter 1969).

Fiber Content and Bulk Density in the Lake States

Nichols and Boelter (1984) measured fiber content and bulk density for samples taken from undrained peatlands in northern Minnesota, Wisconsin, and Michigan (Table 5.2). Moss peats are found throughout the entire range of fiber decomposition with most in the Fibric class. Sedge peats also found

TABLE 5.2

Numbers of Samples Used in Nichols and Boelter (1984) by Botanical Origin and USDA Fiber Class for 176 Peat Samples from the Lake States

Botanical Origin	Fibrists	Hemists	Saprist	Total
Moss peat (mostly *Sphagnum*)	48	21	9	69
Sedge peat (sedges, reeds, grasses)	9	53	4	66
Woody (at least one-third is wood remains)	0	1	13	14
Undetermined origin	0	7	20	27
Total	57	82	46	176

throughout the entire range with most in the Hemic class. Woody peats are rarely Hemists and most are in a Sapric class like peats of an undetermined origin. Note that bulk density tends to increase in the order of botanical origin shown in Table 5.3.

The most detailed relationship between fiber content (by size class) and field-volume bulk density (corrected for the mass of ash) is shown in Figure 5.9, which includes both unrubbed (field condition) and rubbed samples (USDA fiber-content protocol; Soil Survey Staff 1975). The bottom bars represent amorphous peat with bound water. The upper white bars

TABLE 5.3

Ash Content and Bulk Density of Lake States Peats

Type of Peat	Number of Samples	Mean Ash Content (%)	Range of Ash Contents (%)	Mean Bulk Density (Mg m⁻³)	Range of Bulk Densities (Mg m⁻³)
Moss	69	6.62 ± 0.61	2.2–13.4	0.067 ± 0.012	0.009–0.219
Herbaceous	66	8.54 ± 1.34	3.9–40.5	0.126 ± 0.010	0.010–0.201
Woody	14	18.47 ± 5.48	8.5–52.0	0.178 ± 0.029	0.126–0.314
Unidentified	27	12.67 ± 3.58	3.1–35.6	0.200 ± 0.030	0.071–0.374

Source: Nichols, D.S. and Boelter, D.H., *Soil Sci. Soc. Am. J.*, 48, 1320, 1984.

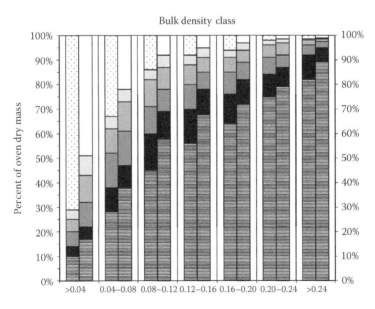

FIGURE 5.9

Distribution of fiber size for pristine (undrained) peats in the Lake States by volume-based bulk-density classes when corrected for ash content. (Redrawn from Nichols, D.S. and Boelter, D.H., *Soil Sci. Soc. Am. J.*, 48, 1320, 1984.)

(with small dots) represent water that runs from a saturated peat sample when picked up. The relatively narrow band of fiber size in the middle (<2.0 to >0.1 mm) represents peat horizons with values for bulk density, drainable porosity, and hydraulic conductivity that are critical for assessing the lateral effect of drainage. Peats with sufficient amorphous peat (<1.0 mm) to reach a bulk density of 0.18 or more is effectively an aquiclude. However, water level may vary upward or downward by several centimeters over a year.

In 1968, debate arose concerning the minimum size class for fiber to use for soil series classification. Original work by Farnham and Finney (1965) and Boelter (1964a,b, 1969) used tedious wet sieving to determine fiber content retained on a 0.1 mm sieve. However, the USDA Soil Conservation Service (Soil Survey Staff 1967) recommended a bottom-sieve size of 0.15 mm primarily to decrease time needed to carefully wet sieve highly decomposed samples. Because the national fiber categories for Histosol classification in the United States and Canada are large (Fibric, Hemic, Sapric), the size of the bottom sieve makes no significant difference to soil series description. Bottom sieve sizes of 0.1 and 0.25 mm and the three classification categories are compared in Figure 5.8. Similar curves can be drawn for any minimum size class.

Fiber content is a proxy measure of degree of decomposition as determined by a laboratory method. However, the widely accepted von Post H value also is directly dependent on the minimum fiber size. This is shown both in the strength of water bound to amorphous peat and water released from the peat matrix when peat is squeezed in the hand following field protocol (Box 5.1). For water-retention characteristics (Figure 5.7) and for field determination of degree of humification, 0.1 mm is an appropriate threshold for minimum fiber size. Plant fibers decompose to amorphous material that has a water-holding capacity similar to that of clays. This water cannot be removed by drainage or by squeezing because only prolonged or high-temperature drying will remove this water.

The von Post hand-squeeze method seems subjective; but the maximum pressure exerted by normal squeezing is about 138 kPa. In more decomposed peat (pores of less than 0.10 mm), holds water with a tension of 2800 kPa (Kwak et al. 1986). This 20-fold difference in tension marks the break where the first 10% of total peat volume, as amorphous peat, passes between the fingers (von Post H4: no peat passes between the fingers; H5: about one-tenth passes between the fingers).

Figure 5.10 illustrates how the amorphous material in peat (<0.1 mm) is the primary cause of increases in bulk density. The increase in bulk density is nearly linear as amorphous material (<0.1 mm) content increases. Except for undecomposed moss peat ($D_b < 0.04 \, \text{Mg m}^{-3}$), other fiber sizes show a minimal negative trend with increasing bulk density. Some fibers (0.1–2.0 mm) remain at high bulk densities (and have H values of 8–10). However, the large

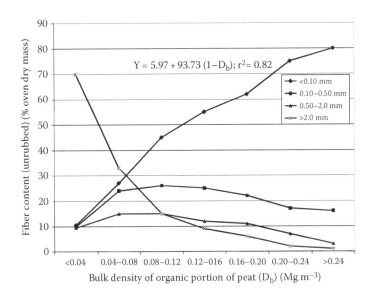

FIGURE 5.10
Percentage of fiber for four size categories versus bulk-density category. (From Nichols, D.S. and Boelter, D.H., *Soil Sci. Soc. Am. J.*, 48, 1320, 1984.)

amount of amorphous peat at these bulk densities and at high H values carries these remaining fibers (60%–80% of the bulk volume) through the fingers encased in the amorphous mass.

Fiber Content and Degree of Decomposition: An International Review

The debate about minimum fiber size (and its measurement) and use of field versus laboratory methods for degree of decomposition was prolonged by the adoption of a variety of methods by different nations since there are no internationally accepted methods. Malterer et al. (1992) reviewed national methods for degree of decomposition and fiber content and evaluated their precision and capacity to distinguish between classes of peat (von Post H value). Most samples were collected from S2 bog at the MEF (see H value description in Verry 1984) and from a site near Circle Pines in Anoka County, Minnesota. All 10 von Post classes were sampled (moss, sedge, and unidentified, well-decomposed peat).

The von Post degree of humification (Box 5.1), USSR humification index, and USDA pyrophosphate extract color are used to describe the degree of decomposition. The USSR method consists of centrifuging a small sample of peat through a screen sealed across a centrifuge tube. Screens of 0.25 mm (60 mesh) and 0.15 mm (100 mesh) were evaluated by Lishtvan and Kroll (1975). The volume of total sample and the volume passing the sieve are entered into one of five nomographs based on peatland trophic status and botanical origin of peat to yield a humification score (R%). The USDA pyrophosphate test consists of adding an aqueous sodium pyrophosphate solution and measuring the color intensity of the extract using the Munsel 10YR color chart (Soil Survey Staff 1975).

Fiber content is determined by the USSR centrifugation method (60 and 100 mesh) using the direct volume measurement of sample retained on the screen to total sample volume (in percent). In the ASTM fiber mass method, a moist unrubbed sample is soaked in a dispersing agent (sodium hexametaphosphate) and gently washed through a 0.15 mm sieve. This is similar to the method developed by Boelter who used a 0.1 mm sieve (Levesque and Dinel 1977). The original USDA method is determined by gently washing a peat sample through a 0.15 mm sieve; however, the percent fiber is calculated as the volume of sample retained divided by the volume of original sample. The USDA methods determines fiber volume by packing the original and retained fiber into a 35 cm^3 syringe with one-half of its length cut away and gently compressing with the plunger. Unrubbed and rubbed determinations are made in this way (Soil Survey Staff 1975). In a modified version, if more than 10% sapric material is retained on the screen, it is mixed with water, beaten with a whisk, and then washed through the sieve (Soil Conservation Service 1984). Coefficients of variation for the mean of 10 von Post H-value samples provide a measure of method precision (Table 5.4) and accuracy (Table 5.5). MUSSR 100-DFV was the most accurate fiber content method, distinguishing nine classes followed by MUSSR 60-DFV with eight classes. The ASTM method distinguished seven classes and the USDA methods distinguished four to six classes.

The sorting of fiber content methods into various accuracy categories likely relates to the method of determining fiber volume. The MUSSR method determines fiber volume directly in the centrifuge tube with volume calibration etched on a narrow portion of the tube. It "packs" material into this narrow tube portion with a constant centrifugal pressure sufficient to force amorphous material past the screen. The ASTM method determines percent fiber volume using the ratio of oven-dry mass of fibers passed divided by the oven-dry mass of the total peat sample.

In summary, the USSR centrifuge method with a nomograph (based on peatland site trophic condition and peat botanical origin) and the von Post field method are the most precise and accurate. Malterer et al. (1992) presented 42 regression equations and graphs comparing national methods. Staneck and Silc (1977) found that the von Post method was the most accurate for Ontario peats.

TABLE 5.4

Coefficients of Variation (%) for National Methods of Degree of Decomposition and Fiber Content

	Degree of Decomposition (%)						Fiber Content (%)					
	USSR 60N	USSR 100N	USDA 1974 pyrophosphate	MUSSR 60 DFV	MUSSR 100 DFV	ASTM 100 FW	USDA 1974 100 URFV	USDA 1974 100 RFV	MUSDA 1977 100 URFV	MUSDA 1977 100 RFV	
Field von Post H classes	8.0	5.2	5.0	23.6	7.2	5.2	11.3	12.7	21.7	15.8	22.6

Source: Malterer, T.J. et al., *Soil Sci. Soc. Am. J.*, 56, 1200, 1992.

Notes: 60- and 100-mesh are 0.15 and 0.25 mm sieve openings; N, nomograph; DFV, direct fiber volume; FM, fiber mass; U and R, unrubbed and rubbed peat samples. Values are means of 10 samples.

TABLE 5.5

Mean Values for Each Degree of Decomposition and Fiber Content Method for all 10 von Post H Classes

Degree of Decomposition Methods										
von Post class	1	2	3	4	5	6	7	8	9	10
von Post means	1.0	**2.4**	**2.7**	4.4	5.3	**6.5**	**6.7**	8.1	9.1	9.9
USSR 60 N	18.7	**26.0**	**27.0**	39.3	**44.7**	**46.1**	48.6	52.8	**58.8**	60.1
MUSSR 100 N	15.8	**21.7**	**23.3**	37.3	**40.2**	**40.5**	<u>42.5</u>	<u>43.1</u>	51.8	52.4
USDA pyrophosphate	6.6	**6.0**	**5.7**	2.6	**1.9**	**1.8**	**1.5**	**1.2**	*0.6*	*0.5*
Fiber content methods										
von Post class	1	2	3	4	5	6	7	8	9	10
MUSSR 60 DFV	85.6	**77.1**	**75.9**	58.4	49.9	**36.5**	**35.7**	23.3	19.5	11.6
MUSSR 100 DFV	89.1	**82.0**	**80.0**	61.7	56.8	49.2	45.8	36.9	29.8	2.34
ASTM 100 FW	71.4	**61.6**	**60.1**	37.5	**31.2**	**29.1**	27.2	**11.9**	**11.2**	7.7
USDA 100 URFV	95.6	**80.6**	**76.8**	**64.8**	**64.4**	**61.8**	**59.0**	**58.6**	44.6	30.4
USDA 100 RFV	84.0	**54.0**	**53.6**	<u>23.6</u>	<u>20.2</u>	<u>14.6</u>	<u>12.8</u>	4.8	2.1	0.1
MUSDA 100 URFV	94.8	**76.8**	**75.8**	58.0	43.8	**35.1**	**34.8**	<u>21.2</u>	<u>17.8</u>	<u>17.4</u>
MUSDA 100 RFV	86.4	**54.2**	**53.2**	25.2	18.4	**11.6**	**11.0**	<u>5.1</u>	<u>1.5</u>	0.0

Source: Malterer, T.J. et al., *Soil Sci. Soc. Am. J.*, 56, 1200, 1992.

Notes: The accuracy of each method is reflected in the number of von Post classes distinguished. Adjacent bold values (bold, underlined, or bold, underlined, and italicized) are not significantly different at the 0.05 level.

Hydraulic Conductivity

Early Work at the MEF: A Tabular Association with Drainable Porosity

Drainable porosity also known as water-yield coefficient, specific yield, AVWSC, and saturated hydraulic conductivity (K_{sat}) for MEF peats in 1965 were listed by the sampling depth and general peat type (Table 5.6). Hydraulic conductivity was measured by the piezometer method. Gafni (1986) measured effective porosity in the same MEF peatland as Boelter, and in a mined peatland near Cromwell, Minnesota (Table 5.6).

Hydraulic Conductivity and Fiber Content Correlations

Boelter (1969) provided a regression between hydraulic conductivity and fiber content (>0.1 mm; Figure 5.11). A similar relationship between hydraulic conductivity and bulk density was reported for Lake State peats (Boelter 1969) and peats in central Finland (Päivänen 1973; Figure 5.12). These relationships are reasonably close when D_b is at least $0.09\,Mg\,m^{-3}$. However, Boelter's relationship is more than twice Päivänen's for D_b less than $0.09\,Mg\,m^{-3}$. Both

TABLE 5.6

Drainable Porosity, Hydraulic Conductivity, and Bulk Density of Several Minnesota Peats

Peat Type	Degree of Decomposition	Sampling Depth (cm)	Hydraulic Conductivity (10^{-5} cm s^{-1})	Boelter's Drainable Porosity (cm cm^{-1})	Gafni's Drainable Porosity (cm cm^{-1})	Bulk Density (Mg m^{-3})
Sphagnum peat						
	Live, undecomposed moss[a]	10–0	[b]	0.85		0.100
	Live, undecomposed moss	0–10			0.63	
	Undecomposed moss	15–25	3810	0.60	0.58	0.040
	Undecomposed moss	45–55	104	0.53		0.520
	Moderately decomposed with wood	35–45	14	0.23		0.153
Woody peat						
	Moderately decomposed	35–45	496	0.32		0.137
	Moderately well decomposed	60–70	56	0.19		0.172
Sedge peat						
	Slightly decomposed	25–30	1280	0.57		0.069
	Moderately decomposed	79–80	0.70	0.12		0.156
Decomposed peat						
	Well decomposed	50–60	0.45	0.08		0.261
	Well decomposed, mined peat	130–140			0.21	

Source: Boelter, D.H., *Soil Sci.*, 100(4), 227, 1965; Boelter, D.H., Important physical properties of peat materials. In *Third International Peat Congress Proceedings*, Helsinki, Finland, 1970, pp. 150–154; Boelter, D.H., *Soil Sci. Soc. Am. Proc.*, 33, 1974.

[a] Sampled above the hollow elevation on a hummock.
[b] Drained too rapidly to measure.

equations use peats of different origin, but Päivänen's equation excluded samples in the upper 25 cm of the soil profile because this horizon had become well decomposed following drainage for forestry.

The piezometer methods for hydraulic conductivity are similar for Boelter and Päivänen. In each method, a narrow tube (3.2 cm) is driven into the peat near the middle of a peat horizon. The peat is augered from the tube and the hole augered another 10 cm below the tube. After the tube and hole are flushed with water pumped from the tube, the rate of water table rise

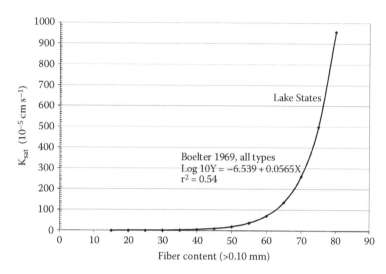

FIGURE 5.11
Relation between saturated hydraulic conductivity (K_{sat}) and fiber content (>0.1 mm) for all types of Lake States peat. (From Boelter, D.H., *Soil Sci. Soc. Am. Proc.*, 33(4), 606, 1969.)

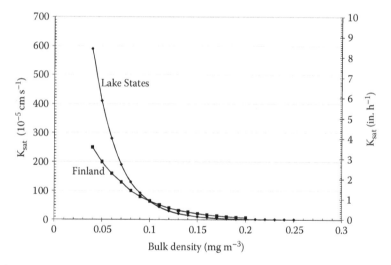

FIGURE 5.12
Relation between saturated hydraulic conductivity (K_{sat}) and bulk density (D_b) for all peat types in the Lake States (diamonds) and central Finland (squares). (From Boelter, D.H., *Soil Sci. Soc. Am. Proc.*, 33(4), 606, 1969; Päivänen, J., *Acta Forestalia Fennica*, 129, 1, 1973.)

is observed over time and hydraulic conductivity is calculated using the Kirkham equation (Frevert and Kirkham 1948; Kirkham 1951). In Päivänen's method, the water table was observed in a well hole near the piezometer. If the well water table dropped when the piezometer was pumped, the piezometer tube was not well sealed and was abandoned. While there is consistency

TABLE 5.7

Hydraulic Gradient and Groundwater Velocity in Four Northern
Minnesota Peatlands

Depth in Peat (cm)	Northern MN Peatland	Hydraulic Gradient (%)	Groundwater Velocity (10^{-5} cm s^{-1})
0–10			
	S2 bog	0.0526	1361
	Transitional fen	0.0425	1319
10–20			
	S2 bog	0.0526	4.1
	Transitional fen	0.0454	2.2
	Raised bog	0.0024	3.4
20–30			
	S2 bog	0.0524	2.8
	Transitional fen	0.0454	0.7
30–40			
	S2 bog	0.0543	0.4
40–50			
	S2 bog	0.0537	0.3
	Raised bog	2.2850	0.9
	Mined site	2.0950	0.8

Source: Gafni, A. and Brooks, K.N., *Can. J. Soil Sci.*, 70, 239, 1990.

between the K_{sat}:D_b relationships in the United States and Finland, actual K_{sat} values likely were underestimated when the peizometer method was used for undecomposed peats (H1–H6).

Hydraulic Gradient and Groundwater Velocity

Gafni and Brooks (1990) examined the correlation of hydraulic gradient measured with closely spaced well transects arranged on different azimuths around a central well and groundwater velocity measured by the point-dilution method. Groundwater velocities are listed in Table 5.7. The range of hydraulic gradients in peatlands is similar to the range of water-surface gradients in many Lake States streams and rivers (0.001%–3.0%). Generally, peatland hydraulic gradients range from 0.001% to 0.05% but steepen greatly where large, raised-bog domes develop with steep dome sides or where drainage ditches greatly steepen gradients. Groundwater velocities (0.49–0.016 m h^{-1}) decrease with depth (5–45 cm), reflecting peats with greater decomposition at depth.

Hydraulic Conductivity and Degree of Humification

Detailed work in Poland on heavily farmed agricultural land with organic soils is documented in Okruszko (1960, 1993) and Parent and Ilnicki (2003). The hydraulic conductivity of peats also has been investigated in Russia

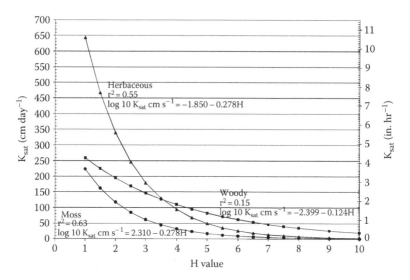

FIGURE 5.13

Relation between saturated hydraulic conductivity (K_{sat}) and von Post degree of humifica-tion (H value) using the piezometer method for central Finland peats. (From Päivänen, J., *Acta Forestalia Fennica*, 129, 1, 1973.)

(Ivanov 1953), Germany (Baden and Eggelsmann 1963), Great Britain (Rycroft et al. 1975a,b), and Canada (Stanek and Silc 1977). Päivänen (1973) developed regressions between K_{sat} and the von Post H value. Separate equations were developed for *Sphagnum*, *Carex*, and woody peats (Figure 5.13). The units are both cm h⁻¹ and inches h⁻¹ because inches h⁻¹ is used in web-based evaluations of reverse drainage equations. The relationship for woody peats are the least precise ($r^2 = 0.15$). *Carex* peats are more conducive to water flow than *Sphagnum* peats, though the relationships for both peat types tend to converge at H7 and are higher when the amount of amorphous material increases significantly.

The piezometer method for determining K_{sat} in saturated soils is used widely and assumes a hydraulic gradient (HG) of 1; however, peatlands have low water table slopes. In 1984 and 1985, Gafni (1986) used both point-dilu-tion and the well-to-well tracer methods (Gafni and Brooks 1990). He also measured the von Post H value. The tracer method did not detect movement of a salt injection at the source well, perhaps because the detection well was not aligned with the actual flow path. With point dilution, salt is injected into a piezometer and salt dilution is measured as groundwater travels through and past the well. Dilution is measured with a plot of electrical conductance over time (a correction for the natural diffusion rate of salt in the bog water is included). Groundwater velocity (GV) and the HG measured between transect wells, were used to solve for K_{sat} in the equation $K_{sat} = GV/HG$. Gafni found a significant relationship ($r^2 = 0.81$) between K_{sat} and von Post H value using 28 samples of *Sphagnum*, *Carex*, and woody peats combined

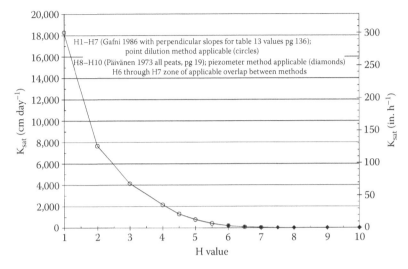

FIGURE 5.14

Relation between hydraulic conductivity (K_{sat}) and von Post H value using the point-dilution method (open circles) for estimating groundwater velocity and water table contour maps for estimating hydraulic gradient (H1–H7) (Gafni 1986), and the piezometer method (filled diamonds) for H8–H10 (Päivänen 1973). Values are from the respective equations and no combined equation is shown.

(for H1–H7; Gafni and Brooks 1990). This equation gave values for undecomposed H1 peat that were 150 times higher than those reported by Päivänen. However, at the H7 von Post value, K_{sat} values from both methods converged. The higher K_{sat} values measured by Gafni were largely due to the low, field-measured, water-slope values since GV/HG = K_{sat}.

Water levels in the transect wells were used to calculate HG and verify that it did not vary appreciably over the growing season when water levels declined. A water table contour map (Gafni 1986) revealed that well transects were not always perpendicular to the flow lines between water table contours. Water table slopes that were perpendicular to water table contour lines were higher, resulting in K_{sat} values that were 71%–86% less than Gafni's original equation relating K_{sat} to von Post H value. As before, the Gafni data still were 150 times higher for H1 values than those of Päivänen, but trended lower for H3–H7 where the two equations converged. The combination of the Gafni (1986) equation (H1–H7) corrected for steeper HG slopes and the Päivänen (1973) piezometer data for H8–H10 is shown in Figure 5.14.

Although hydraulic conductivity varies by more than 10 orders of magnitude, the values in Figure 5.15 vary only by 4 orders of magnitude. There are large differences in pore sizes from H1 to H10 (several μm to mm). In H1–H5 peats, pores are large, less than 10% of the peat volume in amorphous material, and the K_{sat} values are similar to sandy soils with 5% silt (25.4–30 cm h^{-1}; Table 5.11). As H values increase, K_{sat} values decline to about 0.25 cm h^{-1}.

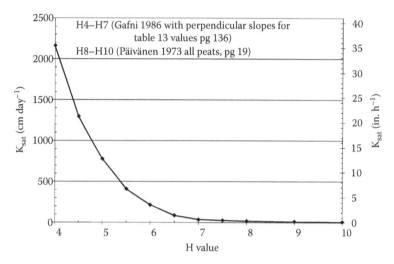

FIGURE 5.15

Relation between hydraulic conductivity (K_{sat}) and von Post H value using the point-dilution method for estimating groundwater velocity and water table contour maps for estimating hydraulic gradient (H4–H7) (Gafni 1986), and the piezometer method for H8–H10 (Päivänen 1973). Values are from the respective equations and no combined equation is shown.

Summary of Physical Properties of Organic Soil

Many laboratory values of K_{sat} for organic soil have been measured for soil series labeled fibric, hemic, and sapric, in the USDA soil classification system; these data are available in the U.S. General Soils Map (STATSGO2) inventory of soil data (NRCS 2006, http://soils.usda.gov/survey/geography/statsgo/description.html). However, laboratory values obtained with cylinders can be affected by trapped gasses or leaks that result in low or high values, respectively. Because hydraulic heads are greater in the laboratory than experienced in the field, K_{sat} values derived in the laboratory will be higher.

The data on peat physical properties collected at the MEF, across the Lake States, in Canada, and in central Finland show a consistent correlation among individual properties. Drainable porosity decreases from 0.60 to an asymptote of about 0.08. Bulk density increases from 0.04 to 0.24 Mg m^{-3} with complete decomposition and compaction (Boelter 1969). Bulk density values greater than 0.24 Mg m^{-3} indicate an admixture of mineral material with the organic material. Differences in bulk density among regions impart some variation to correlation data. Nonetheless, the higher ash contents yield higher bulk densities, primarily affect the bulk-density value, and have negligible effects on correlations of von Post H values with peat physical properties. Tables 5.8 through 5.10

TABLE 5.8

Peat Bulk Density Values and Drainable Porosity by von Post H Class

von Post Field Test by / Degree of Humification von Post H Class	Päivänen in Finland		Stanek and Silc in Ontario		Nichols and Boelter in Lake States	Boelter (1964) in Minnesota	Gafni (1986) in Minnesota
	Sphagnum with Ash D_b (mg mg⁻¹)	Sedge and Woody with Ash D_b (mg mg⁻¹)	Sedge with Ash D_b (mg mg⁻¹)	Woody with Ash D_b (mg mg⁻¹)	All Types without Ash D_b (mg mg⁻¹)	Drainable Porosity (cm cm⁻¹)	Peat Volume Passing Fingers (% Volume)
1	0.06	0.06	0.04	0.07	0.04	0.60	0
2	0.07	0.07	0.06	0.09	0.06	0.34	0
3	0.08	0.09	0.08	0.11	0.08	0.29	0
4	0.09	0.11	0.10	0.13	0.10	0.23	0
4.5	0.09	0.11	0.10	0.13	0.11	0.20	1–2
5	0.10	0.12	0.11	0.14	0.12	0.18	3–10
5.5	0.10	0.13	0.12	0.15	0.13	0.16	11–25
6	0.11	0.14	0.13	0.16	0.14	0.13	26–35
6.5	0.12	0.15	0.14	0.17	0.15	0.12	36–45
7	0.12	0.15	0.15	0.18	0.16	0.12	46–55
7.5	0.13	0.16	0.16	0.19	0.17	0.11	56–65
8	0.13	0.17	0.17	0.20	0.18	0.10	66–75
9	0.14	0.19	0.19	0.22	0.20	0.09	76–95
10	0.16	0.20	0.20	0.23	0.24	0.08	96–100

TABLE 5.9

Fiber Content, Water Retained at Saturation and at 10 kPa, and Drainable Porosity by von Post H Value Class

von Post Degree of Humification	Field Saturated Volume (%) >0.10 mm (Nichols and Boelter, Lake States)	Field Saturated Volume (%) >0.25 mm (Nichols and Boelter, Lake States)	Field Saturated Volume (%) >0.50 mm (Nichols and Boelter, Lake States)	Water Retained at Saturation (cm cm^{-1}) (Päivänen, Finland)	Water Retained at 10 kPa (cm cm^{-1}) (Päivänen, Finland)	Water Retained at 10 kPa (cm cm^{-1}) (Boelter, Lake States)	Drainable Porosity (Boelter 1964, Minnesota)	Peat Volume Passing Fingers (%) (Gafni 1986, Minnesota)
1	74	68	48	0.94	0.66	0.69	0.60	0
2	68	55	39	0.93	0.58	0.66	0.34	0
3	63	48	29	0.91	0.54	0.63	0.29	0
4	59	44	27	0.90	0.49	0.60	0.23	0
4.5	57	42	25	0.90	0.46	0.57	0.20	1–2
5	54	39	23	0.90	0.44	0.53	0.18	3–10
5.5	51	35	20	0.89	0.42	0.50	0.16	11–25
6	49	34	19	0.89	0.41	0.49	0.13	26–35
6.5	48	32	18	0.88	0.39	0.48	0.12	36–45
7	47	31	17	0.87	0.38	0.47	0.12	46–55
7.5	43	28	16	0.86	0.37	0.46	0.11	56–65
8	41	26	15	0.86	0.35	0.43	0.10	66–75
9	39	23	12	0.85	0.34	0.41	0.09	76–95
10	36	21	8	0.82	0.34	0.39	0.08	96–100

TABLE 5.10

Hydraulic Conductivity, Bulk Density, Fiber Content (>0.1 mm), and Drainable Porosity by von Post H Value Class

von Post H	K_{sat} (cm d^{-1})	Drainable Porosity (Decimal)	Peat Passing Fingers (% of Volume)
1	18,317	0.60	0
2	7,690	0.34	0
3	4,170	0.29	0
4	2,160	0.23	0
4.5	1,296	0.20	1–2
5	788	0.18	3–10
5.5	409	0.16	11–25
6	215	0.13	26–35
6.5	86	0.12	36–45
7	35	0.12	46–55
7.5	26	0.11	56–65
8	17	0.10	66–75
9	11	0.09	76–95
10	8	0.08	96–100

Poor fen near Bog Lake at Marcell

H1, H2, H3

Marcell s2 bog grading to poor fen with depth

H1, H2, H3, H4, H5, H6, H7, H8

Notes: Profiles at Marcell showing depth below surface in cm and H value for peat of all origins. The K_{sat} values are from Gafni (1986) and Päivänen (1973). The drainable porosity values are from Boelter (1964) and Gafni (1986)

list the most common values of bulk density, drainable porosity, fiber content, and hydraulic conductivity that are related to von Post H values and to the amount of material that passes between the fingers in the von Post H test.

Peats in Finland generally had 5% less ash content than those in the Lake States, where peats are denser because of dust from the Great Plains. In Ontario, woody peats are denser than sedge peats, probably reflecting the calcium and magnesium composition of wood cells. All peats are extremely light, only 0.002%–0.003% of mineral soil bulk densities. Ash content does not appreciably affect hydraulic conductivity or fiber content, but simply adds to the mass of volumetric peat samples without plugging pores.

Peats with H1–H5 values have large pores that drain quickly in 2 or 3 days. Boelter (1964) measured water table drawdown around a ditch in the S7 bog at the MEF. The S7 bog has peat similar to the S2 bog (Table 5.10). He found that the water table dropped rapidly to the interface between H5 and H6 peats before dropping farther with prolonged drainage. Note that K_{sat} values from H4 through H5.5 decreased from 89 to 17 cm h^{-1}. This pronounced drop in K_{sat} from free-flowing peats with no or low amounts of amorphous material (11%–25%) is similar to the drop in K_{sat} in mineral soils as they grade from sands to higher contents of silt or clay (e.g., 90% sand, 10% silt; Table 5.11). Applications of organic soil K_{sat} values to reverse-drainage equations for the determination of lateral extent nearly always include K_{sat} values for underlying mineral soils. Saturated hydraulic conductivity diminishes to about 0.25 cm h^{-1} in both mineral and organic soils. In organic soil, this occurs when the percentage of amorphous material exceeds two-thirds of the peat volume (Table 5.10). In mineral soils this occurs not in clay loams, not pure clay that typically has a K_{sat} of about 0.7 cm h^{-1}. The higher K_{sat} may be explained by the blocky structure of pure clays and the massive structure of clay loams.

Many peatlands in the Lake States are underlain by sandy outwash plains with a mixture of sand and silt. Various mixtures of sand and silt have K_{sat} values of 1.5–30.5 cm h^{-1}, a range similar to organic soil K_{sat} values in the range of H5–H7 (1.5–35.5 cm h^{-1}). Accurate determinations of weighted K_{sat} and drainable porosity values in lateral-extent equations depend on an accurate survey of K_{sat} and drainable porosity by horizon in both peat and mineral soil across the wetland of interest.

Evaluation of Wetland Drainage

On-site recording-well records provide the best estimates of lateral effect but may take several years to obtain. Models are an alternative used to evaluate the lateral extent of drainage. The Natural Resource Conservation Service (NRCS) provides hydrology tools for evaluating the lateral extent of wetland drainage. Hydraulic conductivity and drainable porosity are the primary parameters.

TABLE 5.11

Mineral Soil K_{sat} Values

Sand (%)	Silt (%)	Clay (%)	K_{sat} (cm d⁻¹)	Sand (%)	Silt (%)	Clay (%)	K_{sat} (cm day⁻¹)
100	0	0	1428	55	25	20	16
95	5	0	767	50	25	25	10
90	10	0	377	45	30	25	8
80	20	0	130	40	30	30	6
70	30	0	100	40	25	35	5
60	40	0	90	40	20	40	6
50	50	0	95	35	35	30	7
40	60	0	139	30	35	35	8
30	70	0	127	25	40	35	11
20	80	0	101	20	40	40	13
10	90	0	76	15	45	40	13
0	100	0	38	10	45	45	15
85	10	5	162	5	45	50	17
80	10	10	77	0	45	55	19
75	15	10	56	0	30	70	20
70	15	15	34	0	20	80	17
65	20	15	30	0	10	90	18
60	20	20	19	0	0	100	18

Source: Schaap, M.G., Rosetta version 1.2. Reverside, CA: USDA-ARS, U.S. Salinity Laboratory, 2000.

The von Post field-test procedure is suitable for assessing a wetland for lateral-effect drainage. In wetlands with horizons with different H values, both hydraulic conductivity and drainable porosity can be weighted by horizon depth for use in one of four drainage equations: Ellipse, Hooghoodt, van Schilgaarde, and Kirkham. In Minnesota, tests of efficacy at partially drained peatlands near Minneapolis–St. Paul has shown that the van Schilfgaarde equation best represents drainage recorded in long-term continuously recording wells. As with the other drainage equations, the van Schilfgaarde equation is used in reverse to determine the effectiveness of existing drains to lower the water table by 30 cm over 14 or more consecutive days. The area within a wetland where this occurs qualifies as a regulated wetland (water table remains at this level or higher) or a drained wetland (water table drops farther).

The NRCS uses a modified version of the van Schilfgaarde equation such that drainable porosity is replaced with an adjusted drainable porosity. The adjustment accounts for a small amount of water storage (s) on the surface (usually set to 0.025 cm). If surface roughness is ignored (s = 0), the equation is identical to the original van Schilfgaarde equation (Chapter 19 in USDA NRCS Staff 1997). A diagram adapted to show appropriate equation terms for both ditch and tile drainage is shown in Figure 5.16. Application to a

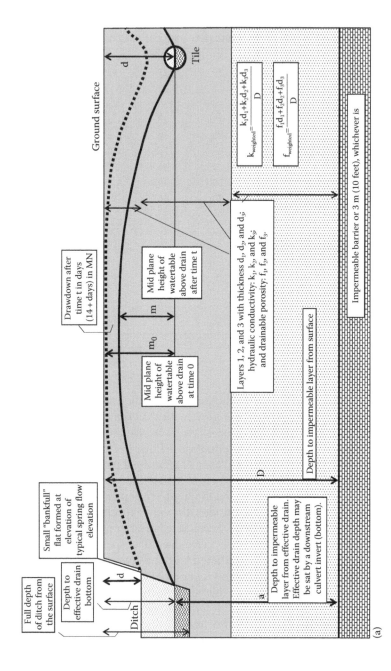

FIGURE 5.16
Important physical and hydrologic features of drainage systems for both ditched and tiled wetlands. In tiled systems, the value of d is taken at the center of the tile. In ditched systems, the value of d should be taken as the highest normal water level in the ditch that is maintained for 7–8 days during the growing season. This usually corresponds to the small flat developed within the ditch walls corresponding to approximately the 1.5 year recurrence interval flow (bankfull flat).

$$S = \sqrt{\frac{9Ktd_e}{f'[\ln m_o(2d_e + m) - \ln m(2d_e + m_o)]}}$$

The equation is first used with "a" in place of "d_e" to determine an estimated spacing S'. Equation variables are:

S = drain spacing, ft

K = hydraulic conductivity, ft day^{-1} (program takes ([in hr^{-1}] and converts to correct units)

d_e = equivalent depth from drainage feature to impermeable layer, ft

m = height of watertable above the center of the drain at midplane after time t, ft

m_0 = initial height of watertable above the center of the drain at t = 0, ft

t = time for watertable to drop from m_0 to m, days

a = depth from free water surface in drainage feauture to impermeable layer, ft

f' = drainable porosity adjusted for surface roughness, dimensionless (ft ft^{-1}), = f + (s(m$_o$ − m)$^{-1}$)

f = drainable porosity of the water conducting soil, dimensionless

s = water trapped on the surface by soil roughness, ft; s = 0.0083 ft (0.1 in) would be typical

S' = estimated drain spacing, ft

(b)

FIGURE 5.16 (continued)

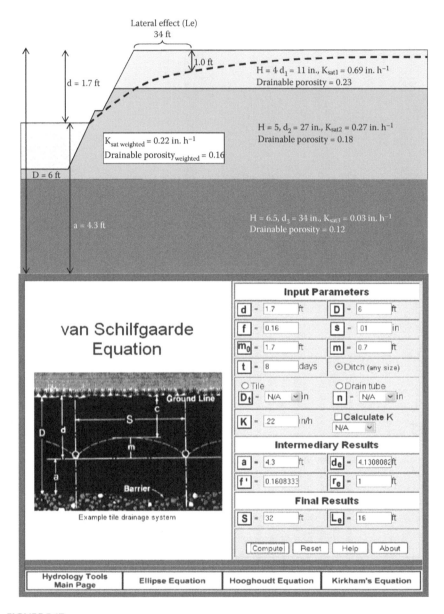

FIGURE 5.17

A ditched peatland in Minnesota showing the extent of lateral effect from the ditch edge (top). Lateral effect is the distance from the ditch edge (or tile center) where the water table is 30 cm below the ground surface. In Minnesota, that portion of a wetland site where the water table is within 30 cm of the surface for 8 or 9 days qualifies as a regulated wetland. If the water table is below 30 cm the land is not regulated and is considered drained. The Web site (USDA NRCS Staff 1997, http://www.wli.nrcs.usda.gov/technical/web_tool/Schilfgaarde_java.html, accessed June 1, 2010) displays only a tiled wetland (bottom), but Figures 5.16 and 5.17 (top) show how parameters are used for a ditched wetland too.

drained peatland in Minnesota is shown in Figure 5.17 and the range of lateral extent in organic soils with uniform peat soils and corresponding K_{sat} and drainable-porosity values shown in Table 5.12. von Post H1–H4 peats are fibric with no amorphous material; they drain at fast rates. The water table surface away from the drainage ditch tends to pass quickly below H1–H4 fibric peats; the primary impact of ditching is controlled by peat of H5 or greater and underlying layers of mineral soil.

Boelter (1972) measured water table drawdown away at S7 bog at the MEF and at a peatland drained for *Sphagnum* peat harvest near Floodwood, Minnesota. He found significant drawdown out to 5 m in late August at the S7 bog; H-value horizons were similar to those at the S2 bog. There, H1–H4 peats were in the horizon 0–30 cm below the hollows. A rapid progression of peat horizons followed: H5–H9 over a horizon of 30–115 cm. At the Floodwood peatland, there is no record of the H values; however, this peatland has relatively thick horizons of fibric peat near the surface. There, significant drawdown occurred to 30 m in both mid-June and late August. The values in Table 5.12 are estimates that assume that the peat is 3.3 m thick and has only one H value. A peat with one H value never occurs, but is useful for demonstrating the possible range of values.

Application of the van Schilfgaarde lateral-extent equation to drained or partially drained wetlands requires that K_{sat} and drainable-porosity values be weighted by horizon thickness (Figure 5.16). First determine the

TABLE 5.12

Range of Lateral Extent in Peatlands with the Average K_{sat} and Drainable Porosities of *Carex* Peat Shown for Each von Post H Value Class

H Value	K_{sat} (in. h^{-1})	K_{sat} (cm h^{-1})	Drainable Porosity	Lateral Extent (m)
1	300	762	0.60	105
2	126	320	0.34	82
3	68	173	0.29	65
4	35	89	0.23	52
4.5	21	53	0.20	44
5	12.7	32	0.18	36
5.5	6.7	17	0.16	27
6	3.5	9	0.13	22
6.5	1.4	4	0.12	14
7	0.56	1.4	0.12	9
7.5	0.43	1.1	0.11	8
8	0.28	0.7	0.10	7
9	0.18	0.5	0.09	6
10	0.12	0.3	0.08	6

Notes: Units are given in both English and metric units to correspond to units of in. h^{-1} in Figure 5.17.

TABLE 5.13

Lateral Extent Parameters for Well 8 at Village Meadows CRWP, Anoka County in Blaine, Minnesota, Ditch 53-62

Depth (cm)	H Value or Texture	K_{sat} (cm h^{-1})	Drainable Porosity (cm cm^{-1})
0–45	4	8.9	0.23
45–75	6	8.9	0.13
75–300	Sand = 70%, silt = 25%, clay = 5%	2.6	0.38
300		4.2	0.32

thickness of organic and mineral soil horizons on-site using a 60 m × 60 m sampling grid. Determine either von Post H value or soil texture at each location and for each horizon to estimate saturated hydraulic conductivity and drainable porosity for organic soils from Table 5.10. The procedure for horizon weighting for both organic and mineral soils should be modified if overlying horizon K_{sat} values exceed 15.2 cm h^{-1}. Horizons of rapid flow should not be included in the weighted K_{sat} or drainable-porosity estimates because their speed overwhelms the weighting procedure. Instead, assign a K_{sat} value of 5.0 or 8.9 cm h^{-1} for mineral or organic soils respectively (next value in the sequence of Table 5.10 and 5.11) to the thickness of these overlying horizons.

Detailed studies of drainable porosity and water table response to precipitation have been conducted in Anoka County, Minnesota (Emmons and Olivier Resources 2005). This is a partially drained peatland with organic soil overlying a sandy loam outwash. The U.S. Geological Survey (USGS) groundwater model, MODFLOW, was used to assess the lateral effect of ditches about 130 m apart. Soil horizons with estimated K_{sat} and drainable-porosity values are shown in Table 5.13 with peat values for H4 and H6 from Table 5.12. Weighted K_{sat} was estimated at 4.2 cm h^{-1} compared to the MODFLOW estimate of 4.3 cm h^{-1} at well 8. When these values were entered into the van Schilfgaarde model for lateral extent, a value of 9.8 m was obtained which was identical to the value obtained with MODFLOW.

Two points are emphasized for this comparison. First, the K_{sat} and drainable-porosity values for the MODFLOW model were calibrated manually until a good fit between the model and measured-well records was obtained. Second, horizons with the greatest thickness and greatest K_{sat} values dominated the results. Third, for overlying organic-soil horizons with K_{sat} values that exceeded 15.2 cm h^{-1}, the values immediately below were used because weighted values were biased due to the logarithmic function between H value and K_{sat} and because these horizons drained quickly.

Impact of Wetland Drainage

Wetland drainage significantly increases the growth of natural vegetation when the depth to the water table between ditches (m_0–m in Figure 5.16 or c in Figure 5.17) is increased from 5 to 25 cm (Figure 5.18). When the amount of drainage exceeds one-third of the watershed area, streamflow peaks can increase two to four times (Figure 5.19). Although the actual

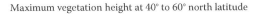

Maximum vegetation height at 40° to 60° north latitude

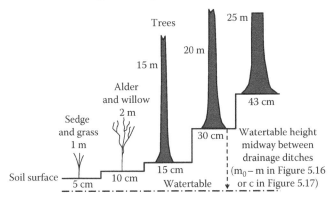

FIGURE 5.18
Response of natural vegetation growth (height) to wetland drainage. Depth to water table is measured midway between ditches.

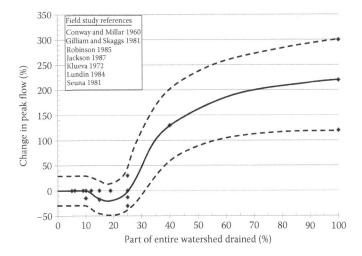

FIGURE 5.19
Change in streamflow peak flows as the percent of the entire watershed in drainage increases.

recurrence interval of the peak flows shown in Figure 5.19 is unknown, a similar response curve for upland watershed changes (Figure 13.13) indicates the recurrence interval affected varies from the 1 year to the 30 year storm and can significantly impact channel width and depth and accelerate channel erosion.

Conclusions

Research on the physical properties of organic soil should include both field and laboratory work to further refine correlations of degree of decomposition, bulk density, fiber content, water retention, drainable porosity, and hydraulic conductivity. Although fiber content is a suitable measure of fibric, hemic, and sapric peat categories in soil series classification, USDA laboratory methods are considerably less precise and less accurate than ASTM or USSR centrifuge methods or the von Post method (degree of humification). In the United States, the von Post field method should be given equal status with laboratory methods as it is elsewhere in the world. Few studies of the physical properties of organic soil have achieved the level of precision obtained by Boelter, Päivänen, Stanek and Silc, Okruszko, and Ilniki; such precision is required especially for various herbaceous peats and woody peats.

Field measurements of hydraulic conductivity are the most difficult to obtain. Piezometer methods greatly underestimate K_{sat} for high H1–H5 peats where there is no or little amorphous peat material because they assume a hydraulic gradient of 1. Point-dilution methods are erratic for H7–H10 peats where amorphous material exceeds half of the peat volume. Piezometer methods more accurately represent vertical water movement through peats that are 2–5 m thick with high HG in the vertical direction and are best for H7–H10 peats. Point-dilution methods are best for determining K_{sat} in H1–H6 peats because they more accurately represent actual water table HG in the horizontal direction. They may over estimate K_{sat} values by 30% or more if hydraulic gradients determined in well transects are not perpendicular to actual water table contour lines. Five or six radial transects are adequate to define actual water table contour lines and measure hydraulic gradients that are truly parallel to groundwater flow using the point-dilution method. The experiments and relationships discussed in this chapter and the modifications to the application of data on peat physical properties to lateral-extent equations allow a relatively quick determination of lateral extent when needed.

References

Always, F.J. 1920. Agricultural value and reclamation of Minnesota peat soils. University of Minnesota Agricultural Experiment Station Bulletin 188.

Anderson, M.S., S.F. Blake, and A.L. Mehring. 1951. Peat and muck in agriculture. Circular No. 888. Washington, DC: USDA.

Baden, W. and R. Eggelsmann. 1963. Zur Durchlässigkeit der moorböden. Z. f. *Kulturtechnik* und *Flurberein* 4:226–254.

Boelter, D.H. 1964a. Water storage characteristics of several peats in situ. *Soil Science Society of America Proceedings* 28(3):433–435.

Boelter, D.H. 1964b. Laboratory techniques for measuring water storage properties of organic soils. *Soil Science Society of America Proceedings* 28(6):823–824.

Boelter, D.H. 1965. Hydraulic conductivity of peats. *Soil Science* 100(4):227–231.

Boelter, D.H. 1966. Hydrologic characteristics of organic soils in Lake States watersheds. *Journal of Soil and Water Conservation* 21(2):51–52.

Boelter, D.H. 1969. Physical properties of peats as related to degree of decomposition. *Soil Science Society of America Proceedings* 33(4):606–609.

Boelter, D.H. 1970. Important physical properties of peat materials. In *Third International Peat Congress Proceedings*, Helsinki, Finland, pp. 150–154.

Boelter, D.H. 1972. Water table drawdown around an open ditch in organic soils. *Journal of Hydrology* 15(15):329–340.

Boelter, D.H. 1974. The hydrologic characteristics of undrained organic soils in the Lake States. Histosols: Their characteristics, use and classification. *Soil Science Society of America Proceedings* 33–46.

Boelter, D.H. and G.R. Blake. 1964. Importance of volumetric expression of water contents of organic soils. *Soil Science Society of America Proceedings* 28:176–178.

Clymo, R.S. 1983. Peat. In *Ecosystems of the World 4A: Mires, Swamp, Bog, Fen and Moor: General Studies*, ed. A.J.P. Gore, pp. 159–224. New York: Elsevier.

Conway, V.M. and Millar, A. 1960. The hydrology of some small peat covered catchments in the north Pennines. *Journal of the Institute for Water Engineering and Science* 14:415–424.

Davis, J.F. and R.E. Lucas. 1959. Organic soils: Their formation distribution, utilization, and management. Agricultural Experiment Station Special Bulletin 425. East Lansing, MI: Michigan State University, Department of Soil Science.

Emmons and Oliveir Resources. 2005. Memorandum: Comparative analysis of methodologies used to predict wetland impacts resulting from a repair of Anoka County Ditch 53-62 in Blaine, Minnesota. Emmons and Olivier Resources.

Farnham, R.S. and H.R. Finney. 1965. Classification and properties of organic soils. *Advanced Agronomy* 17:115–162.

Feustel, I.C. and H.G. Byers. 1930. The physical and chemical characteristics of certain American peat profiles. USDA Technical Bulletin 214.

Frevert, R.K. and D. Kirkham. 1948. A field method for measuring the permeability of soil below water table. *Highway Research Board Proceedings* 28:433–442.

Gafni, A. 1986. Field tracing approach to determine flow velocity and hydraulic conductivity of saturated peat soils. PhD dissertation. St. Paul, MN: University of Minnesota.

Gafni, A. and K.N. Brooks. 1990. Hydraulic characteristics of four peatlands in Minnesota. *Canadian Journal of Soil Science* 70:239–253.

Gilliam, J.W. and R.W. Skaggs. 1981. Drainage and agricultural development: Effects on drainage waters. In *Proceedings of Pocosins: A Conference on Alternative Uses of the Coastal Plain Freshwater Wetlands of North Carolina*, pp. 109–124. Stroudsburg, PA: Hutchinson and Ross Publishing.

Huikari, O. 1959. Kenttämittäustuloksia turpeiden vedenläpäisevyydestä. Referat: Feldmessungsergebnisse über die Wasserdurchlässigkeit von Torfen. *Commun. Inst. Forestalia Fennica* 51(1):1–26.

Ivanov, K.E. 1953. *Wetland Hydrology* (in Russian). Leningrad (St. Petersburg), Russia: Gidrometeoizdat Press.

Jackson, R.J. 1987. Hydrology of an acid wetland before and after draining for afforestation, western New Zealand. In *Proceedings of Forest Hydrology and Watershed Management*, Vancouver, British Columbia, eds. R.H. Swanson, P.Y. Bernier, and P.D. Woodard, Vol. 167, pp. 465–473. Wallingford, U.K.: International Association of Hydrological Sciences.

Kaila, A. 1956. Determination of the degree of humification of peat samples. *Journal of Scientific Agriculture Society* (Finland) 28:18–35.

Karesniemi, D. 1972. Dependence of humification degree on certain properties of peat. In *Proceedings, fourth International Peat Congress*, Vol. 2, pp. 273–282. Helsinki, Finland.

Kirkham, D. 1951. Seepage into drain tubes. I. Drains in subsurface stratum. *American Geophysical Union Transactions* 32:422–442.

Klueva, K.A. 1975. The effect of land reclamation by drainage on the regime of rivers in Byelorussia. In *International Symposium on the Hydrology of Marsh-Ridden Areas*, Vol. 105, pp. 419–438. Minsk, Byelorussia: UNESCO/International Association of Hydrological Sciences.

Korpijaakko, M. and J. Häikiö Leino. 1981. Vesipitoisuuden ja maaatuneisuuden vaikutus turpeen kuivatilavuuspanoon. Summary: Effect of water content and degree of humification on dry density of peat. *Suo* 32(2):39–43.

Kwak, J.C.T., A.L. Ayub, and J.D. Sheppard. 1986. The role of colloid science in peat dewatering: principles and dewatering studies. In *Peat and Water: Aspects of Water Retention and Dewatering in Peats*, ed. C.H. Fuschsman, pp. 95–118. New York: Elsevier.

Laine, J. and J. Päivänen. 1982. Water content and bulk density of peat. In *International Symposium of the International Peat Association Communications IV and II*, pp. 422–430. Minsk, Helsinki, Finland.

Levesque, M. and H. Dinel. 1977. Fiber content, particle-size distribution and some related properties of four peat materials in eastern Canada. *Canadian Journal of Soil Science* 57:187–195.

Lishtvan, I.I. and N.T. Kroll. 1975. *Basic Properties of Peat and Methods for Their Determination*. Minsk, Byelorusse: Nauka I Tekhnika.

Lundin, L. 1984. Torvmarksdikning Hydrologiska konsekvenser for Docksmyren (Peatland Drainage—Effects on the hydrology of the Mire Docksmyre). Department of Physical Geography Hydrology Division, University of Uppsala, Uppsala, Sweden.

Lynn, W.D., W.E. McKinzie, and R.B. Grossman. 1974. Field laboratory tests for the characterization of Histosols. In *Histosols: Their Characterizations, Use and Classification*, eds. Aandahl et al., pp. 11–20. SSSA Special Publication 6. Madison, WI: Soil Science Society of America.

Malterer, T.J., E.S. Verry, and J. Erjavec. 1992. Fiber content and degree of decomposition in peats: Review of national methods. *Soil Science Society of America Journal* 56:1200–1211.

Nichols, D.S. and D.H. Boelter. 1984. Fiber size distribution, bulk density, and ash content of peats in Minnesota, Wisconsin, and Michigan. *Soil Science Society of America Journal* 48:1320–1328.

Okruszko, H. 1960. Muck soils of valley peat bogs and their chemical and physical properties. *Roczniki Nauk Rolniczych* 74-F-1:5–89.

Okruszko, H. 1993. Transformation of fen-peat soils under the impact of draining. *Zeszyty Problemowe Postepów Nauk Rolniczych* 406:3–73.

Päivänen, J. 1969. The bulk density of peat and its determination. *Silva Fennica* 3(1):1–19.

Päivänen, J. 1973. Hydraulic conductivity and water retention in peat soils. *Acta Forestalia Fennica* 129:1–70.

Päivänen, J. 1982. Physical properties of peat samples in relation to shrinkage upon drying. *Silva Fennica* 16(3):247–265.

Parent, L.-E. and P. Illnicki. 2003. *Organic Soils and Peat Materials for Sustainable Agriculture*. Boca Raton, FL: CRC Press.

Puustjärvi, V. 1956. On the cation exchange capacity of peats and on the factors of influence upon its formation. *Acta Agriculturae Scandinavica* 6:410–449.

Puustjärvi, V. 1970. Degree of decomposition. *Suo* 14:48–52.

Raitio, H. and A. Huttunen. 1976. Turpeen maatuneisuuden määaritysmentelmistä. Summary: Methods of determining the humificiation degree of peat. *Suo* 27(1):19–23.

Robinson, M. 1985. The hydrological effects of moorland gripping: A reappraisal of the Moor House research. *Journal of Environmental Management* 21:205–211.

Rycroft, D.W., D.J.A. Williams, and H.A.P. Ingram. 1975a. The transmission of water through peat 1. Review. *Journal of Ecology* 63:535–556.

Rycroft, D.W., D.J.A. Williams, and H.A.P Ingram. 1975b. The transmission of water through peat 1. Field experiments. *Journal of Ecology* 63:557–568.

Schaap, M.G. 2000. Rosetta version 1.2. Riverside, CA: USDA-ARS, U.S. Salinity Laboratory.

Seuna, P. 1981. Long-term influence of forestry drainage on the hydrology of an open bog in Finland, Vol. 43, pp. 3–14. Publications of the Water Research Institute. Helsinki, Finland.

Silc, T. and W. Stanek. 1977. Bulk density estimation of several peats in northern Ontario using the von Post humification scale. *Canadian Journal of Soil Science* 51:138–141.

Soil Conservation Service. 1984. Procedures for collecting soil samples and methods of analysis for soil survey. Report No. 1. Washington, DC: USDA-SCS Soil Survey.

Soil Survey Staff. 1975. Soil taxonomy: A basic system of soil classification for making and interpreting soil surveys. *Agriculture Handbook 436*. Washington, DC: USDA-SCS.

Staneck, W. and T. Silc. 1977. Comparisons of four methods for determination of degree of peat humification (decomposition) with emphasis on the von Post method. *Canadian Journal of Soil Science* 57:109–117.

Tolonen, K. and L. Saarenmaa. 1979. The relationship of bulk density to three different measures of the degree of peat humification. In *Proceedings, International Symposium, Classification of Peat and Peatlands*, pp. 227–238. Hyytiälä, Finland. Helsinki, Finland.

USDA NRCS Staff. August 1997. Hydrology tools for wetland determination. In *NRCS Engineering Field Handbook*, Part 650, Chap. 19. Washington, DC: United States Department of Agriculture.

Verry, E.S. 1984. Microtropography and water table fluctuation in a *Sphagnum* mire. In *Proceedings of the 7th International Peat Congress*, Vol. 2, pp. 21–31. Dublin, Ireland: The Irish National Peat Committee/The International Peat Society.

von Post, L. 1922. Sveriges geologiska undersöknings torvinvenstering och några av dess hittills vaanna resultant. *Svenska Mosskulturfören. Tidskr.* 36:1–27.

von Post, L. and E. Granlund. 1926. Södra sveriges tarvtillgänger. Sverge Geologica Undersser Serial C335. *Årsbok* 19(2):1–127.

6

Scaling Up Evapotranspiration Estimates from Process Studies to Watersheds

Kenneth N. Brooks, Shashi B. Verma, Joon Kim, and Elon S. Verry

CONTENTS

Introduction

Watershed scientists involved with the establishment of the Marcell Experimental Forest (MEF) had the foresight to initiate research on evapotranspiration (ET) of peatlands at the onset. Bay (1966) emphasized the need for such fundamental hydrologic studies because of the prominence of the 6.1 million ha of organic soils in headwater watersheds of the northern Lake States. Bay's emphasis on ET was warranted given the dearth of research on evaporative losses from peatlands. Therefore, developing methods for determining ET in peatlands and peatland–upland forested watersheds became a key part of the research program at the MEF.

Quantifying ET of different soil-plant conditions has long been a problem confronting the science of hydrology (Calder 1982, 2005; Klemeš 1986). Klemeš (1986) cited Hall (1971) concerning the overall lack of emphasis on ET in the science of hydrology, stating "... about 80% of the hydrologic activity in the basin, the evapotranspiration, which is driven largely by radiation energy, is treated in 1%–2% of a typical hydrology textbook and the remaining 98%–99% is devoted to 20% of the activity governed by gravity and friction." Often the "big unknown" of the water balance of watersheds, ET could not be measured directly at the watershed scale and, as a consequence, was estimated frequently as the residual of the water budget in which all other components are measured or estimated (Brooks et al. 2003). The alternative "energy budget" approach for estimating ET, while theoretically sound, presents measurement challenges in determining the aerodynamic resistance associated with the large surface area and roughness of forested systems coupled with the difficulty of determining the physiological control that plants exert on transpiration (Calder 1982, 2005).

Evapotranspiration of peatlands usually should not be limited by the availability of water, because organic soils in peatlands are frequently at or near saturation. Therefore, ET should be governed by available energy, humidity, wind, and vegetative characteristics. Further, *Sphagnum* mosses that are common in most peatlands do not have stomata and act as wicks that efficiently transport shallow groundwater to the plant-evaporating surface (Nichols and Brown 1980). As a result, peatlands should evaporate and transpire water at or near the potential evapotranspiration (PET) rate as depicted by the Penman–Monteith relationship (Monteith 1965). Because of the importance of ET in peatlands, research at the MEF included several approaches to quantify ET, including lysimeter, growth-chamber, energy-budget, eddy-covariance, and water-budget methods. The succession of ET studies at the MEF that have contributed to our knowledge of peatland hydrology are summarized in this chapter. These research results have further been transformed into mathematical relationships of ET for hydrologic modeling. Modeling ET and other processes allows MEF research to be extended to other locations to simulate responses at different watershed scales and to simulate hydrologic response to changes in climate and anthropogenic disturbances.

Controlled Plot Experiments

Use of Lysimeters in Peatlands

A bottomless lysimeter or "evapotranspirometer" was developed to investigate rates of ET from saturated organic soils in peatlands of open- and closed-canopy bogs (Bay 1966). The evapotranspirometer consisted of a galvanized sheet metal cylinder 3.1 m in diameter and 0.9 m in depth that was driven into the peat surface until the top edge was approximately level with the top of the *Sphagnum* hummocks. Although the bottom of the evapotranspirometer is open to the soil, the hydraulic conductivities of the deeper herbaceous peats (>50 cm) are sufficiently low (Boelter 1964; Boelter and Verry 1977; Gafni and Brooks 1990) that little water would move into or out of the bottom of the evapotranspirometer. Thus, once inserted into the peat, the water level in the evapotranspirometer approximated the water table in the surrounding peatland and would represent the difference between precipitation (P) input and ET output over time. Taking into account the specific yield of peat, the change in water storage (ΔS) could be quantified by measuring the change in the water table in the evapotranspirometer and ET then determined as

$$ET = P - \Delta S, \tag{6.1}$$

Two evapotranspirometers were inserted into a treeless *Sphagnum* bog and weekly estimates of ET from mid-July through early November were compared with pan-evaporation rates and PET estimates from both Thornthwaite (Thornthwaite and Mather 1957) and Hamon methods (Russell and Boggess 1964). Prior to the experiment, checks on seepage from the evapotranspirometers were made by covering their surfaces with plastic sheeting to prevent ET losses and measuring change in water table; no measurable seepage was detected over the timescale of these measurements (Bay 1966). Measurements of water level in two evapotranspirometers were taken several times a week from July 13 to September 7, 1964 and calculated ET averaged 3.5 mm d^{-1}. By contrast, measurements from September 14 to November 2, 1964 averaged 1.5 mm d^{-1}. Regression analysis yielded coefficients of determination (r^2) and standard error of the estimates of 0.79 and 5.6 mm, 0.78 and 5.5 mm, and 0.83 and 4.8 mm for pan evaporation, Thorthwaite, and Hamon PET, respectively. On the basis of these results and its dependence only on air temperature data, the Hamon PET method was selected to determine PET in the Peatland Hydrologic Impact Model (PHIM) described in Chapter 15. Attempts to test two additional evapotranspirometers that were installed in a forested bog were unsuccessful due to problems related to measuring and encompassing larger trees and their roots in the evapotranspirometers.

Energy-Budget and Growth-Chamber Studies

Energy-budget studies were initiated at the MEF to better understand how energy is allocated in peatlands and how vegetative conditions in peatlands affect PET (Berglund and Mace 1972; Brown 1972). Berglund and Mace found that the albedo of a black spruce (*Picea mariana* [Mill.] B.S.P.) bog during the snow-free season ranged from 6% to 8% compared to 12%–16% for *Sphagnum*-sedge bogs. These results suggest that net radiation should be greater for ET in a watershed with black spruce cover than *Sphagnum*-sedge cover. A subsequent watershed experiment examined the water yield and ET response to clearcutting black spruce in the S1 bog at the MEF. The high rates of ET in the summer from *Sphagnum*-sedge vegetation in the clearcut areas were explained by factors other than net radiation (Verry 1981). The overall increase in summer ET could be attributed to increased transpiration rates resulting from the rapid increase in *Sphagnum*-sedge biomass and greater exposure of *Sphagnum*-sedge vegetation to wind and solar radiation following the removal of black spruce.

Nichols and Brown (1980) conducted a growth-chamber study of evaporation from *Sphagnum* under different water levels and temperatures. Peat cores of predominantly *Sphagnum* (*Sphagnum recurvum* Beau.) with an overstory of sedge (*Carex trisperma* Dew.) and cotton grass (*Eriophorum tenellum* Nutt.) were extracted from the field and placed in tubs in a growth chamber. Manometers measured changes in water level in the peat cores; these were compared to changes in water level in tubs that contained water. The controlled growth-chamber experiment indicated that

1. Evaporation rates from the peat cores were about twice those of free water surfaces at air temperatures ranging from 8.9°C to 25.3°C when water levels were maintained at the surface of the peat. Evaporation rates of peat surfaces ranged from 0.29 mm h^{-1} at 8.9°C to 0.56 mm h^{-1} at 25.3°C.

2. When water levels in the peat cores were lowered to 5–15 cm below the *Sphagnum* surface, evaporation rates of the peat cores still exceeded water-surface evaporation. Water readily wicks to the *Sphagnum* surfaces at these depths; the greater evaporating surface of the moss in contrast to the water surface explains the difference in evaporation rates. The ratio of moss to water evaporation ranged from 1.30 when the water level was at the surface to 1.59 at a depth of 5 cm and 1.51 at a depth of 15 cm.

3. Evaporation rates from the peat cores with the grass-sedge overstory intact did not differ from *Sphagnum* surfaces in which the vascular-plant overstory was removed.

4. From 78% to 131% of the net radiation received at the moss surfaces was used in evaporation in the growth-chamber environment; this was within the range of midday natural conditions at the peat bog.

The relationship between water level and evaporation rates in this experiment indicated that when water levels in a peatland drop to a depth of 15 cm, evaporation is not reduced. Boelter (1964) suggested that water does not rise more than 20 cm by capillary action in undecomposed peat. This work and that by Romanov (1968) and Boelter (1972) revealed that mosses become desiccated, and ET is reduced by 40% when the water level drops below a depth of 30 cm.

Evapotranspiration of an Open Bog: Studies at Bog Lake Fen

Evaportranspiration was measured at the Bog Lake fen at the MEF during 6 month periods in 1991 and 1992 (Kim and Verma 1996). The Bowen ratio-energy balance method was used in this open bog by measuring the latent heat flux λE (where E is the water vapor flux and λ is the latent heat of vaporization) as described by Verma (1990). Mean air temperature and humidity were measured using aspirated ceramic-wick psychrometers that exchanged between the heights of 1.0 and 2.0 m every 5 min (Hartman and Gay 1981). In 1992, these "exchanging" psychrometers equipped with Vaisala capacitive relative-humidity sensors (model HMP35C) were used; this eliminated the use of wicks and water reservoirs. Net radiation (R_n) above and below the canopy was measured with net radiometers (Radiation Energy Balance Systems, model Q6) positioned at 0.2 and 1.5 m above the peat surface at two locations. The peat heat flux (G) was measured with four heat transducers (Radiation Energy Balance Systems, model HFT-1) installed 0.50 m below the peat surface under hollows and hummocks at four locations. The average peat temperature above these heat transducers was measured with platinum resistance thermometers. The heat flux of surface peat (referenced to an average hollow bottom) was then estimated using the combination method (e.g., Kimball et al. 1976). The heat storage (W) in the *Sphagnum* hummocks and the standing-water column above the peat surface was estimated using the surface energy-balance equation by calculating the difference between (R_n and G) measured with the net radiometers and soil heat transducers and the sum of latent and sensible heat fluxes ($\lambda E + H$) measured independently by the Eddy-covariance technique. The best estimates of accuracy for the measurements of R_n, G, and $\lambda E + H$ for typical midday conditions are 30 W m^{-2}, 5 W m^{-2}, and 15–30 W m^{-2}, respectively. Hence, any errors in R_n, G, and $\lambda E + H$ will be implicit in the estimated values for W.

Incoming and outgoing solar radiation (R_s) and photosynthetically active radiation (Q_p) were measured with pyranometers (Eppley Laboratories, model PSP) and quantum sensors (LI-COR model LI-190SA), respectively. The profile of horizontal mean windspeed was measured with sensitive cup anemometers (Cayuga Development). Leaf-area index (L) was measured with an LAI-2000 area meter (LI-COR).

PET of an open water surface was calculated using the equation given by Penman (1948):

$$PET = \frac{\left[s(R_n - G - W) + \gamma E_a\right]}{\lambda(s + \gamma)}, \tag{6.2}$$

where
    s is the slope of the saturated vapor pressure–temperature curve
    γ is the psychrometric constant
    E_a is an empirical parameter that expresses the effect of windspeed and vapor–pressure gradient on PET, $E_a = 2.7(1 + u/100) D_a$
    u is the wind run (in kilometers per day)
    D_a equals the daily averaged vapor–pressure deficit in kPa measured above the canopy

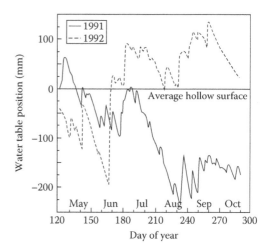

FIGURE 6.1
Seasonal distributions (Julian dates) of water table position referenced to the bottom of hollows. (Modified from Kim, J. and Verma, S.B., *Bound. Lay. Meteorol.*, 79, 243, 1996.)

Compared to the long-term averages, the mean air temperature in 1991 was about 2°C higher during the summer but about 1°C lower in autumn. In 1992, the summer was about 2°C cooler, and the autumn was near normal. Precipitation during the summer from June to August 1991 (245 mm) was 76% of the long-term mean value, and the water table always was at or below the hollow peat surface (typically, 80–230 mm below the surface). By contrast, summer rainfall in 1992 (457 mm) exceeded the long-term mean by 42%, resulting in an elevated water table and hollow bottoms that usually were waterlogged (Figure 6.1).

Daily ET exhibited a high degree of variability, with a range of 0.2–4.8 mm d^{-1} from mid-May to mid-October in both years (Figure 6.2). The growing-season average ET was 3.0 mm d^{-1}. These results are similar to the summertime ET values reported for a variety of wetlands within and outside of the

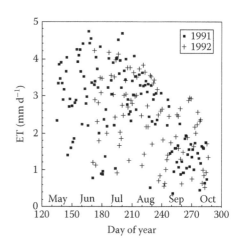

FIGURE 6.2
Seasonal distributions (Julian dates) of ET measured with the Bowen ratio-energy balance method. (Modified from Kim, J. and Verma, S.B., *Bound. Lay. Meteorol.*, 79, 243, 1996.)

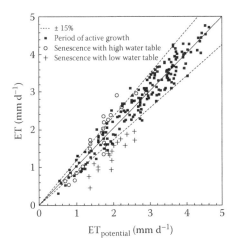

FIGURE 6.3

Relationship between the measured daily ET and PET. (Modified from Kim, J. and Verma, S.B., *Bound. Lay. Meteorol.*, 79:243, 1996.)

MEF (e.g., Chapman 1965; Sturges 1968; Boelter and Verry 1977; Lefleur and Roulet 1992).

The ET at the Bog Lake fen generally was within ±15% of the PET through the growing season (Figure 6.3). However, when the vascular plants became senescent and the water table was below the bottom of hollows, the measured ET was 5%–65% lower than the PET. When the water table was above the hollow bottom, ET was near the PET even during senescence. Water table dependence of ET/PET is depicted in Figure 6.4. Water table position becomes negative when it is below the hollow bottom. The ET/PET tended to increase with rising water table. Excluding the senescence period, the regression analysis indicated an r^2 of 0.42 (significant at $p = 0.05$). Advective enhancement of evaporation and surface geometry of *Sphagnum* could contribute to ET/PET greater than 1. Effective evaporating surface area of *Sphagnum* also could be a factor (e.g., Boelter and Verry 1977; Nichols and

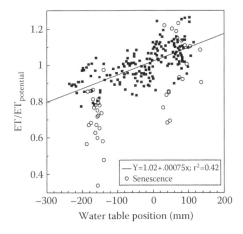

FIGURE 6.4

Measured daily ET as a fraction of PET in relation to water table position. (Modified from Kim, J. and Verma, S.B., *Bound. Lay. Meteorol.*, 79, 243, 1996.)

Brown 1980). However, it seems that the effective evaporative surface area of *Sphagnum* would be somewhat site-specific and would depend on the microtopographic distribution of hummocks and hollows and their position relative to the water table.

Kim and Verma (1996) used a dual-source modification of the Penman–Monteith equation (Massman 1992) to partition the measured ET into evaporation from the nonvascular *Sphagnum* surfaces and transpiration from vascular plants. The analysis indicated that about two-thirds of the ET was from evaporation when the *Sphagnum* surface was wet. Such an evaporative flux was expected because of vertical distribution of vascular plant leaves that had a small leaf-area index (0.4–0.7) and intercepted only about 30% of net radiation (R_n) during the day. Thus, the remainder of R_n was available for evaporation from *Sphagnum*. Evaporation decreased significantly as the *Sphagnum* surface dried out.

Evapotranspiration Estimates at Watershed Scale

Watershed Responses: Water-Budget Method

Water-budget studies on two watersheds with forested peatlands and upland aspen forests at the MEF compared estimated ET and PET as determined by the Thorthwaite PET method (Bay 1968). The S2 and S6 watersheds contained lake-filled bogs of 3.2 and 2.0 ha and upland aspen forests of 6.5 and 6.9 ha, respectively (Chapter 2). Six years of data at S2 were used, compared to 2 years of data at S6. ET was determined from

$$ET = P - Q - \Delta S - U, \qquad (6.3)$$

where
  P is the precipitation on the watershed
  Q the steamflow leaving the watershed
  ΔS the storage at the end of the period minus storage at the beginning of the period in the bog
  U the deep seepage from the watersheds, which was assumed to be negligible because of the low hydraulic conductivities of deeper peats

The water-budget study by Bay (1968) provided insight into watershed-scale ET responses. He found that

1. ET from May to November ranged from 87% to 121% of Thornthwaite PET; daily, ET ranged from 2.6 to 2.8 mm d^{-1} for S2 and 2.7 to 2.9 mm d^{-1} for S6.

2. For summer periods (June to September), ET was consistently lower than PET for both watersheds. ET averaged 2.7 and 3.0 mm d^{-1} in 1961 and 1963 on S2 when both summers were warm and dry. ET on S2 was 11% and 12% lower than PET in 1961 and 1963 when soil moisture may have limited ET. Alternatively, high temperatures could have inflated PET.

3. During the summer of 1966 (June to September), which was particularly wet, ET averaged 3.5 and 3.8 mm d^{-1}, which corresponded to 94% and 99% of PET for S2 and S6, respectively. Rainfall in September was three times the normal rate, resulting in ET exceeding PET during the month. Bay (1968) suggested that deep seepage could have occurred under these conditions.

Water budgets over 30 years for the S2 watershed and over 28 years for the S4 and S5 watersheds (Nichols and Verry 2001) provided estimates of "leakage" or groundwater recharge that improved estimates of annual watershed ET. Leakage amounted to 11.6% of precipitation (P) for S2 and 15.7% for S4 and S5; annual ET was 66.4% of annual P for S2 and 64.5% for S4 and S5. Watershed ET averaged 520 mm year^{-1} or 0.93 and 0.94 of PET (Thornthwaite method) for S2 and S4/S5, respectively. Assuming 180 days (May to November) when average daily temperatures were sufficient for significant ET, watershed ET ranged from 2.4 to 3.3 mm d^{-1} (average, 2.9 mm d^{-1}). These results are specific to watersheds with a till cap (clay loam texture at depth) and would vary in northern Minnesota, where deep sands or mixtures of till and sand occur.

Modeling Evapotranspiration at Watershed Scale

Hydrologic computer models provide the tool to extend research results to other locations and different scales and to examine watershed responses to different climatic and land-use conditions (Chapter 15). As ET is the dominant pathway by which water leaves peatlands and most upland forested watersheds in the northern Lake States (Bay 1969; Verry 1997), hydrologic models of these watersheds must adequately represent interception, surface evaporation, and transpiration processes. ET research at the MEF provided the basis for developing mathematical algorithms of ET for use in the PHIM (Guertin et al. 1987). The following section discusses in great detail the relationships used to calculate interception, evaporation, and transpiration that allow MEF research to be applied more broadly for the management of peatlands and upland forests across the northern Lake States. The other components of the PHIM are discussed in Chapter 15.

Interception losses for both peatlands and upland forests were calculated as a function of vegetative type, canopy coverage percentage, and PET. The forest canopy is represented as overstory (>3 m) and understory. The understory is divided into tall shrubs (1–3 m), low shrubs (<1 m), and herbaceous cover. Intercepted rainfall (I) is evaporated at the potential rate starting at the upper then to successive lower layers; this approach does not account for differences between PET within the canopy and above the canopy surface. Total I is the lesser of available evaporation rate and the storage capacity for each layer. Conceptually, interception storage capacity is treated as a bucket that must be filled before rainfall reaches the soil surface. Knowledge of the maximum storage capacity of the vegetative components is required as input to the model and generally estimated from the literature (e.g., Verry 1976; Zinke 1965).

Evaporation and transpiration relationships used in PHIM differed between peatlands and upland-forest components of watersheds. Peatland ET is linked directly to PET (using the Hamon or Thornthwaite method) and depth of the water table (Brooks et al. 1995; Guertin et al. 1987). A simple linear relationship between actual and potential ET (AET/PET) and water table depth is used in the model. The AET/PET typically is set at 1.0 when water table depth is between the soil surface at the level of the hollow of a *Sphagnum* surface and usually 15–20 cm depending on the type of vegetative cover (Figure 6.5). AET/PET declines linearly as water table depth increases, typically reaching a minimum value at depths of 40–80 cm as determined by the vegetative conditions. The ratio of AET/PET and water table depth is similar to that described by Verry (1997). Shortcomings of this method include the empirical nature of PET estimates that do not predict

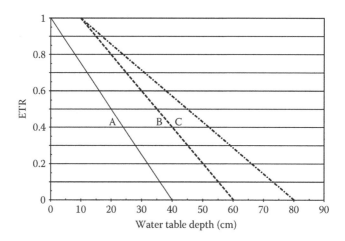

FIGURE 6.5
Generalized relationship of the evapotranspiration ratio (actual to potential evapotranspiration AET/PET) model for peatlands used in the Peatland Hydrologic Impact Model (PHIM): (A) for >25% *Sphagnum* moss, (B) mined areas, and (C) for sedges and <25% *Sphagnum* moss. (Modified from Guertin, D.P. et al., *Nord. Hydrol.*, 18, 79, 1987.)

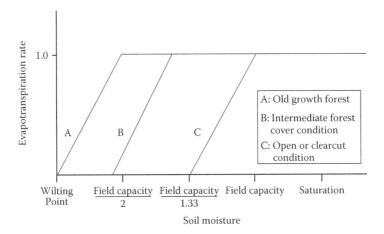

FIGURE 6.6

Relationship between evapotranspiration ratio (actual to potential evapotranspiration AET/ PET) and volumetric water content for mineral soil in upland forests: (A) mature forest, (B) intermediate forest cover condition, and (C) open or clearcut condition. (Adapted from Marker, M.S. and Mein, R.G., *Water Resour. Res.*, 23(10), 2001, 1987; Leaf, C.F. and Brink, G.E., Land use simulation model of the subalpine coniferous forest zone, Research Paper RM-50, USDA Forest Service, Rocky Mountain Station, Fort Collins, CO, 1975.). WP is wilting point, FC is field capacity, and SAT is saturated soil-moisture conditions.

the "true" PET. Methods of determining PET that account for advective energy input, and the aerodynamic and stomatal resistance of vegetative surfaces that are taken into account by the Penman–Monteith Equation (Monteith 1965; Thom 1975), would in some instances yield higher rates of PET than the methods of Penman, Thornthwaite, and Hamon. As a result, PET can be underestimated, explaining in part why AET may exceed PET as shown in Figure 6.4; such a possibility is not taken into account in the AET/PET relationship in the PHIM (Figure 6.5), which indicates an upper limit of 1.0.

In contrast to the AET/PET relationship used in the PHIM for peatlands, AET of upland forests (Figure 6.6) is a function of soil-moisture content (Guertin et al. 1987; Barten and Brooks 1988; Lu 1994; Brooks et al. 1995). The equation for estimating the AET is

$$AET_{1or2} = ROOTD_{1or2} * ETR_{1or2} * PET, \tag{6.4}$$

and

$$ETR = ETR_O + (ETR_F - ETR_O) * TRANSI, \tag{6.5}$$

where

    AET = actual evapotranspiration (cm h^{-1} or cm d^{-1})

    TRANSI = transpiration index; a species dependent index for deciduous trees; it starts at 0 during dormancy and increases from bud break until leaves mature in the spring, reaching 1.0 when leaves are fully mature (Federer 1976)

    ROOTD = cumulative fraction of rooting in soil zones to depth d (0–100 cm) estimated with Gale and Grigal's (1987) root-distribution equation where

$$ROOTD_1 = ROOTD1 = 1 - \beta^d, \tag{6.6}$$

and

$$ROOTD_2 = 1 - ROOTD_1, \tag{6.7}$$

The values of the mean distribution coefficient (β) are 0.95 for shade intolerant classes (species), 0.94 for midtolerant classes, 0.92 for tolerant classes, and 0.93 for bare soils.

$$ETR = \frac{AET}{PET}, \tag{6.8}$$

ETR is a function of volumetric water content (θ; Figure 6.6)

    For $\theta \geq \theta_L$, ETR = 1.0

    $\theta_L > \theta > \theta_Z$, ETR = a + b * θ

    $\theta \leq \theta_Z$, ETR = 0.0

where

    θ_L = limiting volumetric water content for old-growth forest $\theta_L \approx \theta_{FC}/2$, intermediate forest-cover condition $\theta_L \approx \theta_{FC}/2$, and open or clearcut condition $\theta_L \approx \theta_{FC}$ (Leaf and Brink 1975)

    θ_Z = volumetric water content at which AET drops to zero. For most cases, $\theta_Z \approx \theta_{WP}$, and open or clearcut condition $\theta_Z \approx \theta_{FC}/1.33$

    θ_{WP} = volumetric water content at wilting point

    θ_{FC} = volumetric water content at field capacity

    a and b are the interception and slope of the ETR function, respectively, determined from θ_Z and θ_L

PET = potential evapotranspiration (cm h^{-1} or cm d^{-1}), calculated with Thornthwaite or Hamon equations that are available in PHIM; other PET methods can be used with model adjustments.

The subscripts O and F represent open area and forested area, and subscripts 1 and 2 represent shallow subsurface soil layer (SSFL) and the lower root zone (LRZ), respectively.

The AET/PET relationships in PHIM have been tested by comparing simulated with observed flow volumes from peatlands and upland–peatland watersheds on the MEF and elsewhere (Guertin et al. 1987; Barten and Brooks 1988; McAdams et al. 1993; Lu 1994). Comparisons of both simulated and observed water table and flow-volume responses for peatlands provide valid tests for ET-simulation methods.

Conclusion

ET represents the major loss of water from peatlands and peatland–upland watersheds in northern Minnesota. Annual water budgets of MEF peatland–upland watersheds based on more than 28 years of observations show that average annual ET is 65%–66% of annual precipitation and 94%–95% of Thornthwaite PET. Studies using evapotranspirometers and the energy-balance method indicate that average daily ET of peatlands is 3–3.5 mm d^{-1} during the growing season, but can exceed 4.5 mm d^{-1}. Low PET in the spring and fall coupled with plant dormancy result in lower ET rates in those seasons. Peatland ET occurs near the potential rates when vascular plants are actively growing and when the water table is within the rooting zone, usually within a 15 cm depth from the bottom of the peatland hollows. The ratio of AET/PET in a peatland diminishes with increasing depth when the water table drops below 15 cm.

Because of these ET relationships, peatlands rarely produce significant amounts of streamflow during the growing season and usually contribute the majority of flow during the spring. Potential evapotranspiration as determined by Hamon and Thornthwaite methods provide "reasonable estimates" for modeling the water budgets of peatlands and peatland–upland watersheds. Ratios of AET to PET as a function of water table depth provide a method for tracking ET to model the hydrology of peatlands. The AET/PET for upland-forest components of watersheds is a function of soil-moisture content, seasonal transpiration relationships, and root distribution of trees with soil depth. The AET/PET relationships determined at the MEF have been applied in the PHIM, facilitating examination of the hydrologic response to climate change and the management of forest and peatland watersheds at different watershed scales.

References

Barten, P.K. and K.N. Brooks. 1988. Modeling streamflow from headwater areas in the northern Lake States. In *Modeling Agricultural, Forest, and Rangeland Hydrology*, pp. 347–356. Publ. 07-88. Chicago, IL: American Society of Agricultural Engineers.

Bay, R.R. 1966. Evaluation of an evapotranspirometer for peat bogs. *Water Resources Research* 2(3):437–441.

Bay, R.R. 1968. Evapotranspiration from two peatland watersheds. In *Geochemistry, Precipitation, Evapotranspiration, Soil-Moisture, Hydrometry*, ed., General Assembly of Bern, pp. 300–307. UNESCO/International Association of Hydrological Sciences.

Bay R.R. 1969. Runoff from small peatland watersheds. *Journal of Hydrology* 9:90–102.

Berglund, E.R. and A.C. Mace. 1972. Seasonal albedo variation of black spruce and *Sphagnum*-sedge bog cover types. *Journal of Applied Meteorology* 11(5):806–812.

Boelter, D.H. 1964. Water storage characteristics of several peats in situ. *Soil Science Society of America Proceedings* 28:433–435.

Boelter, D.H. 1972. Preliminary results of water level control on small plots in a peat bog. In *Proceedings of the 4th International Peat Congress*, Vol. 3, pp. 24–354, Otaniemi, Finland.

Boelter, D.H. and E.S. Verry. 1977. Peatland and water in the northern Lake States. Gen. Tech. Rep. NC-31. St. Paul, MN: USDA Forest Service North Central Forest Experiment Station.

Brooks, K.N., P.F. Ffolliott, H.M. Gregersen, and L.F. DeBano. 2003. *Hydrology and the Management of Watersheds*, 3rd edn. Ames, IA: Iowa State Press.

Brooks, K.N., S.-Y. Lu, and T.V.W. McAdams. 1995. User manual for the Peatland Hydrologic Impact Model (PHIM). Version 4.0. St. Paul, MN: Department of Forest Resources, University of Minnesota.

Brown, J.M. 1972. Effect of clearcutting a black spruce bog on net radiation. *Forest Science* 18(4):273–277.

Calder, I.R. 1982. Forest evaporation. In *Canadian Hydrology Symposium: 82, Hydrological Processes of Forested Areas*, pp. 173–193. Fredericton, N.B. Canada: National Research Council.

Calder, I.R. 2005. *Blue Revolution*, 2nd edn. London, U.K.: EARTHSCAN.

Chapman, S.B. 1965. The ecology of Coom Rigg Moss, Northumberland III. Some water relations of the bog system. *Journal of Ecology* 53:371–384.

Federer, C.A. 1976. Differing diffusive resistance and leaf development may cause differing transpiration among hardwoods in spring. *Forest Science* 22(3):359–364.

Gafni, A. and K.N. Brooks. 1990. Hydraulic characteristics of four peatlands in Minnesota. *Canadian Journal of Soil Science* 70(2):239–253.

Gale, M.R. and D.F. Grigal. 1987. Vertical root distributions of northern tree species in relation to successional status. *Canadian Journal of Forest Research* 17:829–834.

Guertin, D.P., P.K. Barten, and K.N. Brooks. 1987. The peatland hydrologic impact model: Development and testing. *Nordic Hydrology* 18:79–100.

Hall, W.A. 1971. Biological hydrological systems. In *Proceedings of 3rd International Seminar for Hydrology Professors*. West Lafayette, IN: Agriculture Experiment Station and National Science Foundation, Purdue University.

Hartman, R.K. and L.W. Gay. 1981. Improvements in the design and the calibration of temperature measurement system. In *Proceedings of 15th Conference on Agriculture and Forest Meteorology*. Boston, MA.

Kim, J. and S.B. Verma. 1996. Surface exchange of water vapour between an open *Sphagnum* fen and the atmosphere. *Boundary-Layer Meteorology* 79:243–264.

Kimball, B.A., R.D. Jackson, F.S. Nakayama, S.B. Idso, and R.J. Reginato. 1976. Soil heat flux determination: Temperature gradient method with computed thermal conductivities. *Soil Science Society of America Journal* 40:25–28.

Klemeš, V. 1986. Dilettantism in hydrology: Transition or destiny? *Water Resources Research* 22(9):177S–188S.

Leaf, C.F. and G.E. Brink. 1975. Land use simulation model of the subalpine coniferous forest zone. Research Paper RM-50. Fort Collins, CO: USDA Forest Service, Rocky Mountain Station.

Lefleur, P.M. and N.T. Roulet. 1992. A comparison of evaporation rates from two fens of the Hudson Bay Lowland. *Aquatic Botany* 44:59–69.

Lu, S.-Y. 1994. Forest harvesting effects on streamflow and flood frequency in the northern Lake States. PhD dissertation. St. Paul, MN: University of Minnesota.

Marker, M.S. and R.G. Mein. 1987. Modeling evapotranspiration from homogeneous soils. *Water Resources Research* 23(10):2001–2007.

Massman, W.J. 1992. A surface energy balance method for partitioning evapotranspiration data into plant and soil components for a surface with partial canopy cover. *Water Resources Research* 28:1723–1732.

McAdams, T.V.W., K.N. Brooks, and E.S. Verry. 1993. Modeling water table response to climatic change in a northern Minnesota peatland. In: *Management of Irrigation and Drainage Systems, Integrated Perspectives*, pp. 358–365. *Proceedings of the National Conference on Irrigation and Drainage*, Park City, Utah, July 21–23, 1993. New York: American Society of Civil Engineers.

Monteith, J.L. 1965. Evaporation and the environment. *Symposium of Society of Experimental Biology* 19:205–234.

Nichols, D.S. and J.M. Brown. 1980. Evaporation from a *Sphagnum* moss surface. *Journal of Hydrology* 48:289–302.

Nichols, D.S. and E.S. Verry. 2001. Stream flow and ground water recharge from small forested watersheds in north central Minnesota. *Journal of Hydrology* 245:89–103.

Penman, H.L. 1948. Natural evapotranspiration from open water, bare soil and grass. In *Proceedings of Royal Society of London Series A*, Vol. 193, pp. 120–145.

Romanov, V.V. 1968. Hydrophysics of bogs. Jerusalem: Israel Program for Scientific Translations (Translated from Russian).

Russell, R.L. and W.R. Boggess. 1964. Evaluation of the Hamon method of determining potential evapotranspiration. Forestry Note 108. University of Illinois Agricultural Experiment Station, Department of Forestry.

Sturges, D.L. 1968. Evapotranspiration at a Wyoming mountain bog. *Journal of Soil Water Conservation* 23(1):23–25.

Thom, A.S. 1975. Momentum, mass and heat exchange of plant communities. In *Vegetation and the Atmosphere*, ed. J.L. Monteith, Vol. 1, pp. 57–109. London, U.K.: Academic Press.

Thorthwaite, C.W. and J.R. Mather. 1957. Instructions and tables for computing potential evapotranspiration and the water balance. *Climatology* 10(3):311.

Verma, S.B. 1990. Micrometeorological methods for measuring surface fluxes of mass and energy. *Remote Sensing Reviews* 5(1):99–115.

Verry, E.S. 1976. Estimating water yield differences between hardwood and pine forests-an application of net precipitation data. Research Paper NC-128. St. Paul, MN: USDA Forest Service, North Central Forest Experiment Station.

Verry, E.S. 1981. Water table and streamflow changes after stripcutting and clearcutting an undrained black spruce bog. In *Proceedings of the Sixth International Peat Congress*, pp. 493–498, August 17–23, 1980, Duluth, MN. Eveleth, MN: W.A. Fisher Company.

Verry, E.S. 1997. Hydrological processes of natural, northern forested wetlands. In *Northern Forest Wetlands: Ecology and Management*, eds. C.C. Trettin, M.F. Jurgensen, D.F. Grigal, M.R. Gale, and J.K. Jeglum, pp. 163–188. New York: CRC Lewis Publishers.

Zinke, P.J. 1965. Forest interception studies in the United States. In *International Symposium on Forest Hydrology*, eds. W.E. Sopper and H.W. Lull, pp. 137–161. New York: Pergamon Press.

7

Watershed Hydrology

Elon S. Verry, Kenneth N. Brooks, Dale S. Nichols,
Dawn R. Ferris, and Stephen D. Sebestyen

CONTENTS

Introduction

Watershed hydrology is determined by the local climate, land use, and pathways of water flow. At the Marcell Experimental Forest (MEF), streamflow is dominated by spring runoff events driven by snowmelt and spring rains common to the strongly continental climate of northern Minnesota. Snowmelt and rainfall in early spring saturate both mineral and organic soils and feed water to streams and the regional groundwater aquifer. However, large rains in midsummer or late fall can cause significant stormflow as well as groundwater recharge persisting into January. Defining and measuring water flow in each pathway reveal its importance to evapotranspiration, streamflow, groundwater recharge, and to the movement of nutrients, minerals, heavy metals, and organics (Chapter 8).

Significant changes in climate at the MEF in the last 12,000 years and the results of these millennial steps are discussed in Chapter 4. During the Wisconsin Glaciation period, annual precipitation was half (400 mm) that at glacial retreat (800 mm). Annual precipitation has since ranged from 700 to

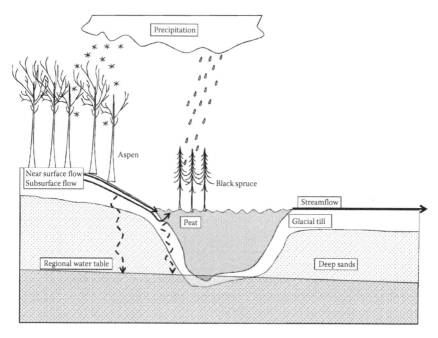

FIGURE 7.1

Net precipitation at perched watersheds in the MEF is rain or snow entering either the mineral or organic soil after interception through tree, shrub, and herbaceous layers. (Redrawn from Verry, E.S., Estimating water yield differences between hardwood and pine forests, Research Paper NC-128, USDA Forest Service, St. Paul, MN, 1976.)

800 mm and currently averages 782 mm. Mean temperature in July has varied 5°C since glacial retreat and now averages 18.9°C.

Once snow and rain reach the forests of MEF, interception, sublimation, throughfall, and stemflow are the first pathways to partition meteorological inputs between evaporation, storage, and runoff as mediated by forest type. Net precipitation (Figure 7.1; Verry 1976) is the rain or snowmelt that passes through the forest canopy and infiltrates through the litter to the forest floor and ultimately reaches the soil surface. Net precipitation is a key variable driving the amount of streamflow or groundwater recharge leaving the watershed.

Net Precipitation

In watersheds with deep sandy soils, net precipitation percolates directly to the regional groundwater system unless transpired or stored as soil water in the root zone. Fen peatlands in these watersheds receive net precipitation as a direct input to the regional water table, which usually flows from the fen as streamflow, but the direction of groundwater flow can reverse leading to groundwater recharge. In upland soils derived from glacial tills, net precipitation becomes (1) nearsurface flow occurring in the forest litter (generally because of underlying soil frost), (2) subsurface flow occurring in the saturated A and E horizons (often derived from eolian fine sands that overlie the tills), (3) deep seepage via unsaturated or saturated flow through the upland mineral soil into the regional water table, or (4) flow into bog peatlands at the wet edge (lagg). Flow from the peatland feeds streamflow in first-order channels.

Aspen (*Populus tremuloides*), red pine (*Pinus resinosa*), and black spruce (*Picea mariana*) are the major forest types at the MEF. The upland forests include an understory component dominated by the shrub hazel (*Corylus* sp.) and the herb bracken fern (*Pteridium* sp.). Bog peatlands have acid water (~pH 3.8) with depauperate flora characterized by low-*Ericaceae* shrubs and *Sphagnum* mosses. Tree species are primarily black spruce and tamarack (*Larix laricina*). Fen peatlands have near neutral pH water (~6.0–6.5) with a rich herbaceous flora, tall alder (*Alnus incana*) shrubs, and trees that include northern white cedar (*Thuja occidentalis*) along with black spruce and tamarack. Precipitation intercepted on the various vegetation layers is either evaporated or, on cold, sunny, and winter days, sublimated back to the atmosphere. Interception, stemflow, and throughfall were investigated at MEF with studies of hazel, aspen, and black spruce interception (Verry 1976) as well as literature data for red pine (Rogerson and Byrnes 1968).

In Lower Saxony, Germany, Delfs (1967) proposed a practical assessment of interception or net precipitation data: "Preliminary results suggest that it is chiefly interception which is responsible for the effect of beech and spruce stands on the water regime (streamflow) – the beech area discharged over 200 mm more than the spruce area." The hypothesis of Delfs was supported by findings of Swank and Miner (1968) who reported the effect on streamflow of converting a hardwood watershed to white pine (*Pinus strobus*) at the Coweeta Experimental Forest in North Carolina. Swank et al. (1972) concluded: "Increased interception loss occurs when hardwood-covered watersheds are converted to white pine, causing a significant reduction in streamflow (178 mm)."

At the MEF, the solution of 23 stemflow and throughfall equations for individual rainstorms and for annual snowfall, measured in the snowpack beneath each stand, yielded a nearly linear relationship between annual gross precipitation (rain plus snow) and annual net precipitation for each forest type and various basal areas (Figure 7.2).

The upland portion of S6 at MEF was converted from aspen to a mixture of red pine and white spruce (*Picea glauca*) in 1981. This provided a test of how well differences in net precipitation mirror differences in streamflow under hardwood and conifer forests. The S6 watershed contains 2.0 ha of black spruce bog and 6.9 ha of upland mineral soil. The mature aspen forest was harvested in the early winter and spring of 1980. Aspen regeneration was consumed by cattle grazing during the next three summers after which the upland was planted to a mixture of red pine and white spruce. When

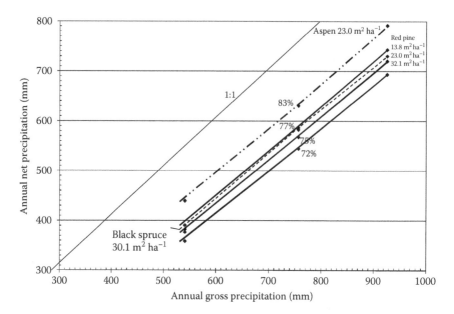

FIGURE 7.2

The relationship between net and gross annual precipitation for aspen (basal area of 23 m² ha⁻¹), red pine (three basal areas: 13.8, 23, and 32.1 m² ha⁻¹), and black spruce (basal area of 30.1 m² ha⁻¹) forests. Average precipitation at MEF is 783 mm. Annual net precipitation is 650 mm for aspen forests and 587 mm for red pine forests. (From Verry, E.S., Estimating water yield differences between hardwood and pine forests, Research Paper NC-128, USDA Forest Service, St. Paul, MN, 1976.)

conifers reached age 17, S6 produced 65 mm less streamflow than the watershed with mature aspen on the uplands (Chapter 13). Adjusted for a change in streamflow from just the upland area, there was a decrease in upland streamflow of 84 mm. The difference in net precipitation between mature aspen and red pine was 63 mm (Figure 7.2). The additional 18 mm reduction is attributed to transpiration from conifers, which retain their needles as opposed to aspen, which has leaves for about 150 days each year. Detailed net precipitation equations for aspen, spruce, and red pine are given in Verry (1976). These equations have been used to quantify the amounts of nutrients entering the forest floor as throughfall and stemflow (Timmons et al. 1977; Verry and Timmons 1977, 1982) and to partition the cycling of mercury and organic carbon through aspen and spruce forests (Kolka et al. 2001).

Water Flow in Mineral Soil

Frost affects the runoff of water from upland mineral soils to peatlands, and frost depth is inversely proportional to the snow depth in early winter. There is little frost when snow accumulates over 30 cm before air temperatures fall below –18°C. Net precipitation entering deep sands at the MEF satisfies soil-moisture storage and then percolates to the regional groundwater table via macropores and unstable wetting fronts. Water also may flow through the thawed upper sand horizons when soil frost at depth persists during snowmelt. The same is true of sandy loam and clay loam soils that comprise the glacial tills at the MEF. Vegetation cover type also affects frost occurrence due to canopy effects on snow depth. Annual variation in soil frost at the MEF is shown in Figure 7.3. In 1969, 1970, 1971, and 1996, little frost occurred when maximum snow depths exceeded 64 cm. The 1989 through 1995 sequence illustrates the inverse relationship between snow depth and frost depth.

Link between Uplands and Peatlands

The mineral soils of the glacial till upland and the organic soils of the peatland are linked at the lagg by water-flow pathways developed over millennia. Data collected on these pathways help us interpret how (1) upland water flows to the peatland, (2) upland and bog water mix in the lagg, (3) peatlands influence the generation of streamflow, and (4) the various components of the watershed contribute to deep, regional groundwater recharge via seepage.

Characteristics of the Upland Mineral-Soil Horizons in the Lagg

Water infiltrates the upland soil and flows downslope through the saturated A and E horizons. Figure 4.8 shows water flowpaths from the uplands to the peatland when the surface mineral soil horizons are saturated. In 1984, the soil horizons on the S2 upland were identified as the Warba Sandy Loam soil series in the Itasca County Soil Survey (Nyberg 1987). The A, E1, E2, E/B, and B/E horizons developed in an eolian loess cap of fine sandy loam that covers the glacial Koochiching clay loam till (Table 7.1). These horizons are sandy or fine sandy loam, medium to strongly acid in reaction, and friable with moderate to medium platy structure in the E horizons (Nyberg 1987; Tracy 1997). The eolian loess cap fell on the S2 uplands and some fine sand swirled into the open water of a tundra pool that covered the bottom of the ice block depression about 9300 cal years BP as suggested by pollen correlates of climate variables (Table 4.3). In subsequent millennia, peat filled the ice-block depression and eventually covered (paludified) the eolian deposits from an elevation of 420.2–422.0 m. The persistence of the A and E horizons under the peat allows a direct conduit for subsurface flow from the upland into the peatland lagg. When soil pits are excavated at the outside edge of the

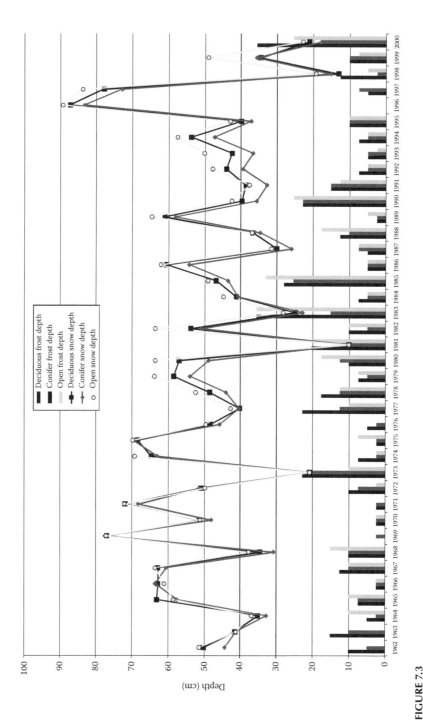

FIGURE 7.3
Annual variation in the depth of soil frost (bars) and maximum snow depth (lines) under deciduous (aspen), open (grasses), and conifer (black spruce on organic soils with some red pine on sands) forests.

TABLE 7.1

Description of the Warba Fine Sandy Loam Soil Series in Pit A at Eastern End of the S2 Bog

Horizon	Depth (cm)	Dry Color	Moist Color	Texture	Comment
0	1–0				Leaves and twigs
A	0–2	10YR 4/1	10YR 3/1	Sandy loam	Medium acid
E1	2–10	10YR 6/2, 7/2	10YR 4/2	Sandy loam	Strongly acid
E2	10–15	10YR 7/2	10YR 5/3	Sandy loam	Strongly acid
E/B	15–23	10YR 6/2		Fine sandy loam	Strongly acid
B/E	23–29			Fine sandy loam	Strongly acid
Bt1	29–54	10YR 4/3	10YR 4/4	Clay loam	Many distinct clay films (dry color)
Bt2	54–79	10YR 3/3, 3/2	10YR 4/3	Clay loam	10YR 3/2 clay films (dry color)
Bt3	79–98	Many faint 10YR 4/4 clay films	10YR 4/3	Loam	Common 10YR 4/4 and 7.5YR 3/2 root channel fillings; 10YR 8/2 filamentous carbonate masses
C	>98 (to sand)	No films	10YR 5/3	Loam	10YR 8/2 filamentous carbonate masses

Note: Information is for the mineral soil above and east of bog edge.

lagg, water flows freely from the interface of the peat and the underlying fine sandy loam (loess cap).

The Bt1 horizon has a moderate fine to moderate, angular, blocky structure that is firm, hard, and strongly acidic. Clay films coat the ped faces and sand sticks to the top of the clay boundary. The Bt2 horizon has a moderate coarse, prismatic structure with clay films, organic stains on the surface of peds, and a medium acidity. The Bt3 has a coarse subangular structure with few roots, organic stains, and a neutral pH. The C-horizon is massive, sandy clay loam, has few roots, and contains shale fragments in 1%–2% of its volume. These soils effervesce slightly with 10% HCl and are mildly alkaline (Nyberg 1987; Brooks and Kreft 1991; Tracy 1997).

Water moves through the upland soils into the deep sands beneath as evidenced by the filamentous carbonate masses following cracks in the Bt3 and C horizons. However, the amount of water loss is not sufficient to wash all the amorphous carbonate crystals from the soil. By contrast, all the mineral soil beneath the peatland lagg (Figure 4.8, horizontal meter scale 399.5–409.5 m) is free of carbonate masses (Brooks and Kreft 1991) and is acid in reaction. The eolian loess cap is up to 1.5 m beneath the peat surface (see 420 m contour in Figure 4.8).

Characteristics of Organic Soil Horizons at the Lagg

Tracy (1997) instrumented five sites around the perimeter of the S2 bog with piezometer nests. At each site, three nests were established on a transect from the upland mineral soil through the lagg and into the bog. Three piezometers were open to selected horizons within each type of profile except on of some mineral sites where only two piezometers were installed. Sand-point piezometers were also installed in the lagg areas to determine hydraulic pressure within the upper meter of the underlying sand. Piezometers screens were placed at elevations of 422.5, 421.5, and 419.6 m. In all cases, when bog water levels were normal to high (and no large amount of water came as sub-surface flow from the upland), flow lines constructed between lines of equi-potential water levels in the piezometers showed water at the outside edge of the lagg flowing downward and outward from the peatland. This water flows through the eolian loess horizon beneath the peatland lagg, through the Koochiching till that is stretched thin at the shoulders of the ice block depression, and finally through the mixed sandy clay loam layer (Figure 4.8) into the deep sands of the Rainy Lobe drift.

Water flows freely in the upper, less decomposed, organic soil horizons in S2. Values of field-measured hydraulic conductivity in the S2 bog are related to von Post degree of decomposition and range from 2.2×10^{-1} (von Post H1) to 1×10^{-6} (von Post H9) cm s^{-1} (Päivänen 1973; Gafni 1986; Gafni and Brooks 1990; Chapter 5).

Water arrives at the lagg from both the upland and the central dome of the bog (Figures 4.8, 7.4, and 7.5). The water-flow lines illustrated in Figure 7.4 occur when the upland mineral soils are below field capacity (unsaturated) and do not contribute water to the lagg. Figure 7.5 is a plan view of flow lines at the water table surface from the bog dome to the lagg. Flow through the lagg coalesces into a stream that flows from the southwest corner of the peatland.

Streamflow Generation at the S2 Bog

We separated the water leaving the bog as streamflow into a portion originating in the bog and a portion originating in the upland using a hydrograph-separation technique (Verry and Timmons 1982; Nichols and Verry 2001; Figure 7.6). Near surface and subsurface flow from the upland are combined as upland flow in Figure 7.6.

Figure 7.6 depicts the desynchronization of flow in a mixed basin (upland/peatland), an example of the variable source areas for streamflow. This variable source area concept was developed for streams by Hewlett (1961) to estimate the area of land directly contributing to streamflow as a saturation to the surface expands outward and upstream as precipitation increases. In upland/peatland mixed basins, the peatland water table is closer to the surface, making the peatland the primary source area for streamflow during late spring and summer. In a 4 cm rainstorm in July 1967, the S2 bog water

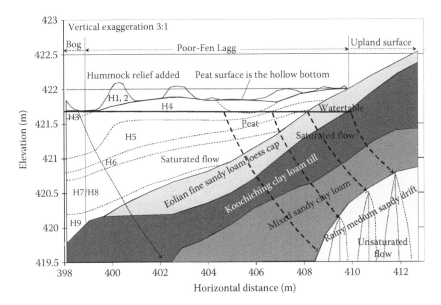

FIGURE 7.4

Organic (von Post H values) and mineral-soil layers beneath the lagg on the east end of S2. Water-flow lines represent a no-flow period from the upland; the water table in the bog slopes slightly downward from left to right.

table and the stream leaving the bog responded immediately to rainfall, but the peak flow from the upland was delayed 1–2 h depending on aspect (Verry and Kolka 2003). During snowmelt in early spring, the peatland and upland contributed equally to streamflow. However, streamflow from the peatland precedes and exceeds flow from the upland in this till-capped peatland basin (Figure 7.7).

Deep Seepage in the S2 Watershed

Some studies of water balance and nutrient yield on small watersheds have documented the tightness of the watershed and show or assume that virtually all liquid water leaves via streamflow (e.g., Likens and Bormann 1977). Despite this common assumption, some subsurface water flows to the regional water table. Penman (1963) stated that deep seepage "is frequently ignored altogether in catchment studies in the quiet hope that it is in fact, zero."

Nichols and Verry (2001) documented significant amounts of deep seepage from 10 to 50 ha watersheds on the MEF. The amount of deep seepage varies with the portion of glacial clay loam till and deep outwash sand in each watershed, the temporal distribution of precipitation, and air temperature.

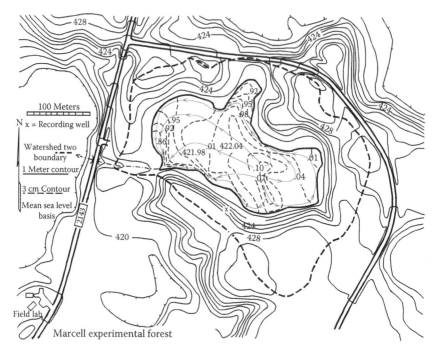

FIGURE 7.5

Much of the surface water leaving the S2 bog is from the dome. It flows from the dome to the poor-fen lagg (fine-lined arrows on the bog surface) where the water leaves the bog as a stream.

On average, deep seepage constitutes about 40% of the total water yield (streamflow plus seepage to groundwater). Adjacent watersheds (S4 and S5) on the North Unit of the MEF contain 15% deep outwash sands (Rainy Lobe drift) and 85% Koochiching till cap. Deep seepage ranges from 5 to 20 cm year^{-1} and averages 13 cm year^{-1}. The S2 watershed is completely covered with the Koochiching till cap; deep seepage ranges from 5 to 15 cm year^{-1} and averages 9 cm year^{-1}. Figure 7.8 shows the relationship using only annual precipitation (November–October) for 30 years on the S2 watershed. Nine centimeter of deep seepage across the entire surface area of the S2 watershed amounts to 8845 m^3 of water.

We estimated seepage to groundwater under watershed S2 using saturated hydraulic conductivity of the major material layers in the basin. We divided the surface area into upland, lagg, and bog and used average hydraulic-conductivity values for the soil material (K_{sat} values) to estimate seepage from each landscape unit. The critical impeding layers in the watershed include the Koochiching till covering the uplands, where the impeding layer is a series of clay-rich Bt horizons, the unweathered, and neutral-gray glacial flour (0.9 m thick; 10% very fine sand, 60% silt, and 30% clay) beneath both the bog and the lagg zone at the edge of the bog. Beneath the lagg above an elevation of 419.8 m, the stretched (and thus thinner) Koochiching till is mixed with

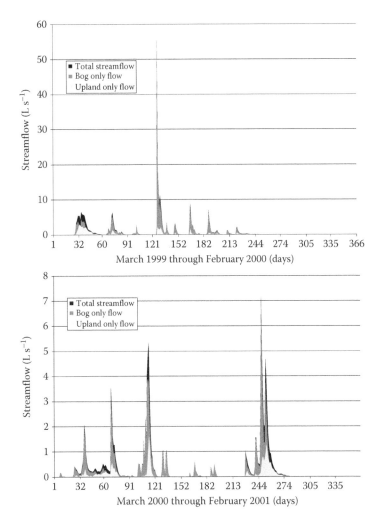

FIGURE 7.6
Annual hydrographs with upland and bog contributions separated from the total watershed streamflow. The year 1999 (top) illustrates a normal spring snowmelt peak and an extraordinary storm event on July 4, 1999. The year 2000 (bottom) illustrates an extraordinary wet period in the fall; note that the bog contribution precedes and exceeds the upland contribution.

sand on its bottom side (Figure 4.8). Table 7.2 shows K_{sat} values measured at the MEF by Tracy (1997) for mineral soils along with pertinent literature values for organic-soil K_{sat} values (see Tables 5.10 through 5.12). We used 16 years (1985–2000) of hydrograph separation data for S2 combined with rates of hydraulic conductivity and the area of the upland, lagg, and bog to calculate the amount of seepage for each watershed component (Figure 7.7).

In addition to the rate of seepage (K_{sat}), the time that free water is available to percolate must be considered. In the bog and lagg this is virtually

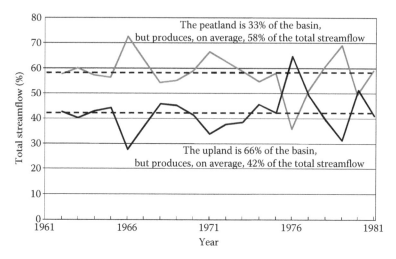

FIGURE 7.7
Relative contribution of the peatland and upland to streamflow over a 20 year period. Data for individual years were accumulated from daily data. The relative role of upland and peatland reverses in extreme drought years (e.g., 1976 and less so in 1980).

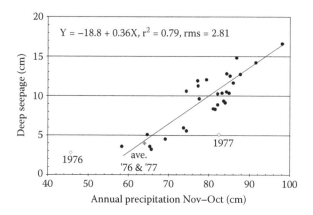

FIGURE 7.8
Deep seepage from S2 as a function of annual precipitation. The consecutive years of 1976 (dry with 45 cm) and 1977 (wet with 83 cm) were averaged due to pronounced hysteresis. (Modified from Nichols, D.S. and Verry, E.S., *J. Hydrol.*, 245, 89, 2001.)

every day (365 days year^{-1}) except for years with severe droughts, such as 1976. In the upland, the number of days when the A and E horizons (and thus the Bt horizons) are saturated is highly variable. We estimated that deep seepage through the upland soil occurred on any day from April to June when the upland hydrograph exceeded 0.028 L s^{-1}, a threshold of saturation in the Bt horizon (Figure 7.6). In the months of high evapotranspiration that required substantial filling of the soil moisture deficit in the A, E, and B

TABLE 7.2

Hydraulic Conductivity (K_{sat}) Values for MEF Samples and from the Literature

Description	Bulk Density (g cm^{-3})	K_{sat} (cm s^{-1})	K_{sat} (cm year^{-1})	Source
Mineral Bt, field	1.59	1.7×10^{-6}	54	Tracy, 1997, S2
Bt, field	1.59	1×10^{-6}	32	Tracy, 1997, S2
Bt, field	1.59	7×10^{-7}	22	Tracy, 1997, S2
Bt, lab	1.59	1.6×10^{-6}	50	Tracy, 1997, S2
Bt, lab	1.59	7×10^{-7}	22	Tracy, 1997, S2
Mixtures of sand, silt, clay	—	5×10^{-6}	160	Todd, 1959
Glacial till	—	5×10^{-7}	16	Todd, 1959
Stratified clays	—	2×10^{-7}	6	Todd, 1959
Unweathered clay, high	—	1×10^{-7}	3	Todd, 1959
Unweathered clay, mid	—	3×10^{-8}	1	Todd, 1959
Unweathered clay, low	—	1×10^{-9}	0.04	Todd, 1959

horizons (July through October), we estimated that deep seepage occurred on any day when the upland hydrograph exceeded 0.567 L s^{-1}. These rates are confirmed by the measurement of water in tanks that collect subsurface runoff from plots (Chapter 2). The number of days available for deep seepage through the upland soils ranged from 53 to 134.

We used the 30 year average water budget for S2 (Nichols and Verry 2001) and average or rounded K_{sat} values (Table 7.2) to estimate the relative amount of deep seepage from the upland, bog, and lagg (Table 7.3). A sum of water-shed-component seepage rates to groundwater totaled 9916 m^3 year^{-1}. This is 12% more than that based on groundwater table recession curves. An agreement by two methods within 12% on S2 contrasts with errors associated with more general methods for estimating groundwater seepage that exceed 50% (Winter 1981). Table 7.3 also shows the relative importance of each watershed component to groundwater seepage.

The upland portion of the watershed accounts for slightly less seepage than its area (59 vs. 67%). The lagg accounts for 12 times the seepage its area would suggest (38 vs. 4%), and the bog accounts for only a tenth of what its area would suggest (3 vs. 29%). Plausible groundwater flow lines based on this analysis are shown in Figure 7.9.

The upper active layer of the peatland referred to as the acrotelm has a von Post degree of decomposition ranging from H1 to H8 and includes a low dome of *Sphagnum* moss peat (Figure 7.5). This layer has a separate, thin, subsurface water flow system that discharges water radially, downslope from the bog water table to the lagg and the stream (Figure 7.5), but also discharges water via shallow arcs to the peatland lagg (solid flow lines in the upper peat in Figure 7.9).

The deeper, hydrologically inactive zone is the catotelm that consists of highly decomposed von Post H9 peat. Such H9 *Sphagnum* peat transmits

TABLE 7.3

Deep Seepage for Each Component of the S2 Watershed

Watershed Component	Area (ha)	Area (%)	K_{sat} (cm s^{-1})	K_{sat} (cm Year^{-1})	Saturated (Days)	Effective K_{sat} (cm year^{-1})	Annual Seepage (m^3)	Annual Seepage (%)
Upland	6.48	67	1.14×10^{-6}	36	91	9	5832	59
Lagg	0.4	4	3×10^{-6}	95	365	95	3800	38
Bog	2.84	29	3×10^{-8}	1	365	1	284	3
Total watershed	9.72	100	—		—		9916[a]	100

[a] About 8845m^3 based on water table recession and an aquifer specific yield of 0.2 (Nichols and Very 2001).

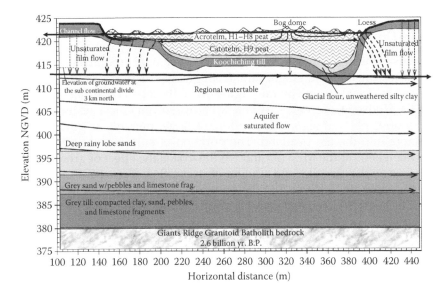

FIGURE 7.9
Plausible water-flow lines (solid lines show saturated flow; dashed lines show unsaturated flow) illustrate the relationship between the bog water table and the regional groundwater table when there is no upland subsurface flow. When upland subsurface flow occurs, unsaturated flow from uplands through the unsaturated zone is similar to the unsaturated flow that originates from the lagg. The dome peaks along the distance scale at 320 m.

21 cm water year^{-1}. However, a layer of unweathered, smooth, silty clay beneath the peat transmits less than 1 cm water year^{-1}. Although it is effectively an aquiclude, this layer does connect the saturated deep peat of the bog with the saturated regional aquifer.

Concrete Frost and Water Movement of Colloidal Organic Matter

Another mechanism of water and organic-carbon flow in organic soils is related to concrete frost in both bog and fen organic soils, although the process is more important in bog soils. When concrete frost forms in organic soils, colloidal organic matter is excluded preferentially from the ice crystal lattice and moves progressively deeper to the bottom of the concrete frost. The upper part of concrete frost is clear and the bottom layer is dark brown. The process of freezing moves colloidal organic matter and associated elements on nearly an annual basis from the acrotelm (where organic matter regularly contributes to dissolved organic carbon and nutrients leached in streamflow) to the catotelm (where they can be stored for millennia).

Streamflow from Upland–Peatland Watersheds

The high-pH water in fens is derived from the regional groundwater aquifer. It dissolves calcium and magnesium from gravel-size pieces of limestone in the deepest glacial ground moraine (gray till for the Itasca Lobe that originates from the Hudson Bay Dome Ice Center). The dissolved limestone imparts the near neutral pH and signals the inflow of large amounts of groundwater discharge into the fen peatland. The duration of surface streamflow from two watersheds at the MEF (S3 and S5) illustrates the effect of groundwater discharge on the duration and amount of streamflow (Figure 7.10). S3 and S5 have the same surface area, but the groundwatershed for S3 is 10 times the size of its surface watershed (Sander 1976). Because water from the regional groundwater aquifer discharges into the S3 fen, streamflow is nearly constant 70% of the time and an order of magnitude higher than streamflow from S5. Streamflow at S3 is perennial. Streamflow from the S5 watershed has a steep flow-duration curve with no flow about 25% of the time during

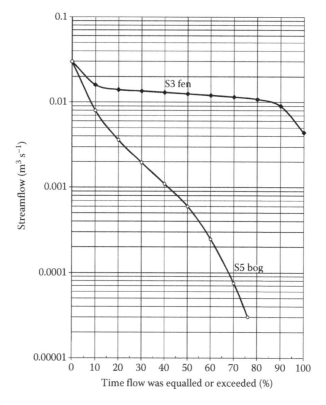

FIGURE 7.10

Streamflow-duration curves for the S3 watershed (groundwater-fed fen, solid circles) and watershed S5 (bog peatland, open circles). Both watersheds have the same surface area.

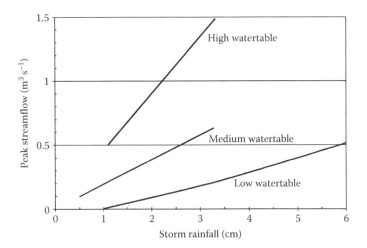

FIGURE 7.11
Water table position in the peatland at the time of a rainstorm affects the magnitude of the streamflow peak flow for the same size of rainstorm. Streamflow from peatlands with high water table is most responsive rainfall. (From Bay, R.R., *J. Hydrol.*, 9, 90, 1969.)

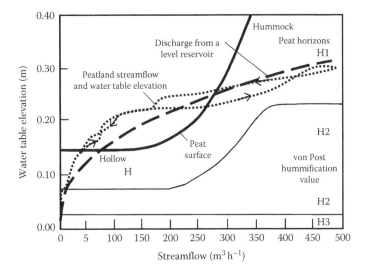

FIGURE 7.12
Peatland streamflow and measured water table elevation (returning dashed line) are imposed on a cross section of the hollow/hummock microtopography showing the porous von Post H1–H3 peat horizons. The direction of the arrow shows the progression of the hydrograph with hysteresis loop at the high flows. The measured stormflow response in the peatland is nearly identical with the rating curve for a level reservoir (heavy dashed line).

summer months and winter. Note that the maximum streamflow rate for both watersheds is the same (flow-duration curves converge in upper left in Figure 7.10), because annual streamflow peaks are driven largely by spring snowmelt occurring in the surface watershed.

Stormflow from rain in a bog or fen watershed is dependent on the position of the water table in the peatland at the time of the rain. Bay (1969) reported this phenomenon (Figure 7.11) and is supported by the hydrographs in Figure 7.6.

The largest peak streamflow ($486\,m^3\,h^{-1}$) from rain (17 cm) was caused by a large storm cell over a 35 h period on July 2 and 3, 1979. Water table response in the hummock and hollow topography of the S2 bog illustrates a hysteresis effect on the rising and falling leg of the hydrograph and its near match to the rating curve for a level reservoir (Figure 7.12). This similarity of reservoir and peatland stormflow responses is not surprising given that both peatlands and reservoirs have flat surfaces. The loop in the peatland response curve also occurs in reservoirs and is caused by the movement of a wedge of water through the reservoir (Verry et al. 1988). This proof of concept has been used to model rainstorm response in peatlands (Chapter 15).

Summary

- Intermittent streamflow from bog watersheds occurs when rainfall or snowmelt is rapidly routed at times of high water tables in bogs and laggs. Most of the annual stream water yield occurs in response to spring snowmelt. Although peakflows may be similar between fen and bog watersheds, the streamflow from the fen varies over about one order of magnitude in contrast to bog streamflow that varies up to four orders of magnitude.

- Mineral soils on uplands and organic soils in peatlands have distinct effects on streamflow from peatland watersheds. Studies at the MEF have quantified the relative magnitude of water movement along various flowpaths for watersheds with large central bogs and partitioned hydrological outflows to show that evapotranspiration (52 cm year^{-1}) on average accounts for the largest output of water followed by streamflow (17 cm year^{-1}) and deep seepage (9 cm year^{-1}).

- The magnitude and timing of streamflow from peatland watersheds at the MEF is a function of the hydrogeological setting. The surface elevation of perched bogs is at least several meters above the regional groundwater table. Saturated organic soils maintain a permanent hydraulic gradient that results in slow, but constant deep seepage of water to the regional aquifer from bogs and laggs. Streamflow from fens with outlet streams is augmented by the inflow of groundwater from the regional aquifer in addition to precipitation inputs on upland soils. In contrast to bog watersheds that lose significant amounts in the annual water budget to deep seepage, groundwater inflow to fens may cause annual stream-water yields that are orders of magnitude larger than water yields from bog watersheds.

- Although laggs around bogs in perched watersheds occupy a small fractional area of the landscape, laggs are particularly important to the hydrology of a watershed. Laggs receive rainfall and snowmelt melt inputs from direct inputs as well as bog and upland soils. Flow through the lagg routes these waters to outlet streams when water tables are high. In addition, most of the deep seepage for the entire watershed occurs through soils that lie beneath laggs.

References

Bay, R.R. 1969. Runoff from small peatland watersheds. *Journal of Hydrology* 9(1):90–102.

Brooks, K.N. and D.R. Kreft. 1991. Hydrologic linkages between uplands and peatlands. Final Report to Water Resources Research Center, University of Minnesota, College of Natural Resources, St. Paul, MN.

Delfs, J. 1967. Interception and stemflow in stands of Norway spruce and beech in West Germany. In *International Symposium of Forest Hydrology*. Penn. State Univ. New York: Pergamon Press, pp. 179–185.

Gafni, A. 1986. Field tracing approach to determine flow velocity and hydraulic conductivity of saturated peat soils. PhD dissertation. St. Paul, MN: University of Minnesota.

Gafni, A. and K.N. Brooks. 1990. Hydraulic characteristics of four peatlands in Minnesota. *Canadian Journal of Soil Science* 70:239–253.

Hewlett, J.D. 1961. Soil moisture as a source of base flow from steep mountain watersheds. Research Paper 132. Asheville, NC: USDA Forest Service. Southeastern Forest Experiment Station.

Kolka, R.K., D.F. Grigal, E.A. Nater, and E.S. Verry. 2001. Hydrologic cycling of mercury and organic carbon in a forested upland-bog watershed. *Soil Science Society of America Journal* 65(3):897–905.

Likens, G.E. and H.P. Bormann. 1977. *Biogeochemistry of a Forested Ecosystem*. New York: Springer Verlag.

Nichols, D.S. and E.S. Verry. 2001. Stream flow and ground water recharge from small forested watersheds in north central Minnesota. *Journal of Hydrology* 245(1–4):89–103.

Nyberg, P.R. 1987. *Soil Survey of Itasca County, Minnesota*. St. Paul, MN: USDA Soil Conservation Service.

Päivänen, J. 1973. Hydraulic conductivity and water retention in peat soils. *Acta Forestalia Fennica* 129:1–70.

Pennam, H.L. 1963. Vegetation and hydrology. Technical Communication No. 53. Harpenden, UK: Commonwealth Bureau of Soils.

Rogerson, T.L. and W.R. Byrnes. 1968. Net rainfall under hardwoods and red pine in central Pennsylvania. *Water Resources Research* 4(1):55–57.

Sander, J.E. 1976. An electric analog approach to bog hydrology. *Ground Water* 14(1):30–35.

Swank, W.T. and N.H. Miner. 1968. Conversion of hardwood-covered watersheds to white pine reduces water yield. *Water Resources Research* 4:947–954.

Swank, W.T., N.B. Goebel, and J.D. Helvey. 1972. Interception loss in loblolly pine stands of the South Carolina Piedmont. *Journal of Soil and Water Conservation* 27(4):160–1264.

Timmons, D.R., E.S. Verry, R.E. Burwell, and R.F. Holt. 1977. Nutrient transport in surface runoff and interflow from an aspen-birch forest. *Journal of Environmental Quality* 6(2):188–192.

Todd, D.K. 1959. *Groundwater Hydrology*. New York: Wiley.

Tracy, D.R. 1997. Hydrologic linkages between uplands and peatlands. PhD dissertation. St. Paul, MN: University of Minnesota.

Winter, T.C. 1981. Uncertainties in estimating the hydrologic balance of lakes. *Water Resources Bulletin* 17(1):82–115.

Verry, E.S. 1976. Estimating water yield differences between hardwood and pine forests. Research Paper NC-128. St. Paul, MN: USDA Forest Service.

Verry, E.S., K.N. Brooks, and P.K. Barten. 1988. Streamflow response from an ombrotrophic mire. In *Proceedings, International Symposium on the Hydrology of Wetlands in Temperate and Cold Regions*. Helsinki, Finland: International Peat Society/The Academy of Finland, pp. 52–59.

Verry, E.S. and R.K. Kolka. 2003. Importance of wetlands to streamflow generation. In *Proceedings, First Interagency Conference on Research in the Watersheds*, ed. K.G. Renard, S.A. McElroy, W.J. Gburek, H.R. Canfield, and R.L. Scott. Benson, AZ, October 27–30, 2003. Tucson, AZ: USDA Agricultural Research Service, pp. 126–132.

Verry, E.S. and D.R. Timmons. 1977. Precipitation nutrients in the open and under two forests in Minnesota. *Canadian Journal of Forest Research* 7(1):112–119.

Verry, E.S. and D.R. Timmons. 1982. Waterborne nutrient flow through an upland-peatland watershed in Minnesota. *Ecology* 63(5):1456–1467.

8

Element Cycling in Upland/
Peatland Watersheds

**Noel Urban, Elon S. Verry, Steven Eisenreich,
David F. Grigal, and Stephen D. Sebestyen**

CONTENTS

Introduction

Studies at the Marcell Experimental Forest (MEF) have measured the pools, cycling, and transport of a variety of elements in both the upland and peatland components of the landscape. Peatlands are important zones of element retention and biogeochemical reactions that greatly influence the chemistry of surface water. In this chapter, we summarize findings on nitrogen (N), sulfur (S), carbon (C), major cations, and other biogeochemically important elements in uplands and peatlands. We have organized this chapter to describe processes that affect the transport and storage of elements in watersheds. First, we address a primary route for entry into ecosystems, atmospheric deposition. We then describe compositional changes as water passes through canopies in upland forests before water and solutes infiltrate soils. We also track changes in chemistry as precipitation infiltrates organic soils in bogs, upland runoff passes into the laggs that surround bogs, and groundwater passes through fens. We then examine solute concentrations and yields as waters exit watersheds at the downstream end. Finally, we summarize the cycling of elements within the watersheds.

Atmospheric Deposition

Monitoring of atmospheric deposition at the MEF began in the early 1970s with measurements of bulk deposition in open sites and measurement of throughfall and stemflow in forested sites (Verry and Timmons, 1977). The MEF was one of the original sites of the National Atmospheric Deposition Program (NADP). The chemistry of wet-only precipitation has been monitored continuously there since 1978. This long record of atmospheric inputs has been invaluable in formulating budgets and informing process studies. Verry (1983) was the first to summarize the solute composition of wet-only precipitation at the MEF. The dominant cations were ammonium (NH_4^+), calcium (Ca^{2+}), and hydrogen ion (H^+), and the dominant anions were nitrate (NO_3^-) and sulfate (SO_4^{2-}; Figure 8.1). The mean pH from 1978 through 1980 was 4.9, and variability in H^+ concentration most closely tracked variations in NO_3^- concentrations. The nearly equal concentrations of NH_4^+ and NO_3^- reflect the remote, rural location distant from both industrial and agricultural sources. Little of the SO_4^{2-} is associated with sea salt at this site, but

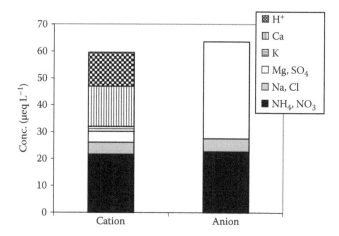

FIGURE 8.1

Average (volume-weighted mean) ionic composition of wet-only precipitation as measured at the MEF for 1978–1980, Verry, 1983). Concentrations are expressed as charge equivalents. The Na^+ concentration reported by Verry was reduced because it was more than 50% higher than the 3-year mean of values now reported by the NADP for 1978–1980.

Verry reported that SO_4^{2-} concentrations were not correlated with H^+ concentration and hence probably not derived only from sulfur dioxide (SO_x) emissions.

There have been significant changes in precipitation chemistry during the NADP monitoring (1978 to present). Volume-weighted mean annual concentrations of SO_4^{2-} have declined, closely paralleling a decline in chloride (Cl^-) concentrations (Figure 8.2). The pH has increased by about 0.3 units, but this decrease in H^+ concentration is five times smaller than the decrease in SO_4^{2-} concentration on an equivalent basis. The decrease in SO_4^{2-} concentration is most strongly correlated with decreases in sodium (Na^+) and Cl^- concentrations, though significant correlations are also observed among SO_4^{2-}, Ca^{2+}, and magnesium (Mg^{2+}) concentrations (Table 8.1). These correlations support the contention that soil dust is a significant source of the SO_4^{2-} deposited at

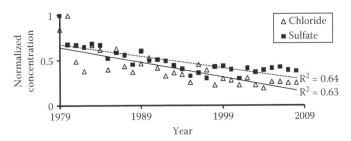

FIGURE 8.2
Historical changes in precipitation at the MEF. Volume-weighted mean concentrations of Cl^- and SO_4^{2-} were normalized to the maximum value in the period of record to put both on the same scale. Solid lines show the least-squares regression; the upper line is for SO_4^{2-}, and the lower is for Cl^-. Both regressions are statistically significant ($p < 10^{-6}$).

TABLE 8.1

Correlation Coefficients for Volume-Weighted Mean Nutrient Concentrations in Wet-Only Precipitation Collected at the MEF NADP Station (1978–2008)

	Ca^{2+}	Mg^{2+}	K^+	NH_4^+	NO_3^-	Cl^-	SO_4^{2-}	pH
Ca^{2+}	1							
Mg^{2+}	0.89[a]	1						
K^+	0.68[a]	0.64[a]	1					
NH_4^+	0.38[a]	0.17	0.19	1				
NO_3^-	0.60[a]	0.45[a]	0.48[a]	0.43[a]	1			
Cl^-	0.52[a]	0.51[a]	0.38[a]	−0.06	0.71[a]	1		
SO_4^{2-}	0.66[a]	0.68[a]	0.42[a]	0.17	0.75[a]	0.80[a]	1	
pH	0.10	−0.02	0.01	0.43[a]	−0.32	−0.58[a]	−0.55[a]	1

[a] Significant at $p < .05$. Na^+ was excluded due to significant skew and lack of normality in the distribution.

this site (Urban et al., 1989c); agricultural lands are several hundred kilometers to the south and southwest. During this period, a trend toward decreasing deposition of NO_3–N (Figure 2.3b) at the MEF and elsewhere throughout the upper Midwest is consistent with a decrease in N oxide emissions (McDonald et al., 2010).

Comparison of bulk precipitation (wet plus dry deposition) collected in the open and below canopies of aspen on uplands and black spruce in peatland showed that some elements increased in concentration and others decreased while passing through the canopy. Inorganic N is taken up by the canopy while phosphorus (P), organic N, Ca^{2+}, Mg^{2+}, K^+, and Na^+ are enriched in throughfall (Verry and Timmons, 1977). Some of this enrichment results from washoff of dry deposition from the canopy, but, for K^+, the increased fluxes in throughfall represent leaching from plant foliage. For N, there may be conversion of inorganic N to organic N by epiphytic bacteria, lichens, or algae; in the black spruce stand, there was no enrichment in total N after passing through the canopy. Collection of dry deposition by foliage and subsequent washoff by precipitation has been reported for other substances, including SO_4^{2-} (Verry, 1986), mercury (Hg) (Kolka et al., 1999b), and trace metals (e.g., Lindberg et al., 1986). Because of the greater leaf area index of conifers (dominant in the peatlands), the enrichment of these substances in throughfall into the peatlands is greater than in the uplands that are characterized by hardwood (aspen) stands (Kolka et al., 1999b). That deposition of most cations increased more upon passage through the aspen canopy compared to the black spruce canopy suggests that much of the increase results from leaching from deciduous vegetation.

Atmospheric deposition of organic carbon has been poorly characterized (e.g., Willey et al., 2000). Studies at the MEF have provided some of the few measurements in the region. Results showed that wet deposition of organic carbon is about seven times less than the yield of organic carbon from the MEF watersheds. Kolka et al. (1999b) reported an organic carbon flux in bulk precipitation of 1.2 g C m^{-2} year^{-1} for 1995. This flux is within the range (0.2–1.7 g m^{-2} year^{-1}) reported from other locations (Dillon and Molot, 1997; Willey et al., 2000). Considerable organic carbon is leached from tree canopies; the flux (throughfall and stemflow) below aspen canopies was 7.6 g C m^{-2} year^{-1} vs. 8.5 g C m^{-2} year^{-1} below a spruce canopy. Whether this organic matter is exported, respired, or stored within the ecosystems is not known.

The measurements of wet-only deposition and bulk deposition at the MEF provide some of the few data available to estimate atmospheric deposition of organic N (cf. Neff et al., 2002). Verry and Timmons (1977) reported bulk deposition in the open of 0.23 g organic N m^{-2} year^{-1} for a typical year; Urban (1983) reported wet-only deposition of organic N of 0.05 g m^{-2} year^{-1} in 1982. The discrepancy suggests that dry deposition of organic N is much larger than wet deposition, perhaps on the order of 0.2 g N m^{-2} year^{-1}.

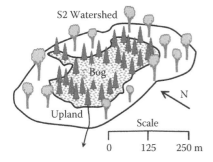

FIGURE 8.3
Schematic of the S2 watershed. The uplands comprise two-thirds of the watershed. (Adapted from Kolka, R.K. et al., *J. Environ. Qual.*, 28, 766, 1999a.)

Throughfall studies generally focus on the overstory, but shrubs, herbaceous species, and bryophytes likely modify precipitation chemistry. To determine the significance of denitrification in the S2 wetland (Figure 8.3), Urban et al. (1989b) measured NO_3^- at successive depths below the moss surface. They observed that NO_3^- was essentially removed from the percolating rain water, leaving little to be denitrified in anaerobic regions of the peat.

Processing of Elements in the Uplands of the MEF Watersheds

The bogs within the MEF are perched rather than raised, so laggs receive runoff from the surrounding mineral-soil uplands (Chapter 7). The uplands support mixed hardwood forests in various stages of succession (Chapters 2 and 12). Timmons et al. (1977) reported on water and element fluxes in runoff from the upland of the S2 watershed from 1971 to 1973. Runoff from the upland in this watershed occurs as surface runoff (primarily within the organic O horizon) and as interflow in mineral soil horizons above the less permeable B2t horizon due to clay. The composition of the runoff entering the wetlands is determined by the flow path.

Water flowpaths in the uplands vary seasonally. Surface runoff is confined largely to snowmelt. From 1971 to 1973, snowmelt accounted for 46%–68% of total annual water loss from the S2 upland, but it supplied 93%–100% of the total annual runoff. Some surface runoff occurs in response to large rain events but does not usually contribute a significant amount to the annual surface runoff.

Elements can be placed in four categories based on the manner in which they cycle within and pass through the uplands (Figure 8.4). Category 1 elements (e.g., Na^+) include substances that pass largely unaltered through the upland; bars showing flux ratios in Figure 8.4 are small, because the

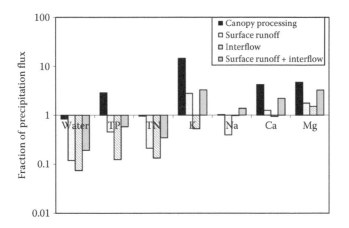

FIGURE 8.4
Processing of elements within the S2 upland. The y axis shows the ratio of each flux (through-fall + stemflow, surface runoff, interflow, and surface runoff + interflow) to the flux in bulk deposition; values greater than one indicate that this flux is larger than the input to the system from atmospheric deposition (i.e., a net release of the element from the catchment occurred); values less than one indicate the element was retained in the system. The canopy processing bar includes both throughfall and stemflow. Data for interflow and surface runoff represent a 3 year average (1971–1973; Timmons, et al., 1977). Fluxes in bulk precipitation, throughfall, and stemflow are from Verry and Timmons (1977).

ratio of fluxes remains close to 1. Category 2 elements are leached from the canopy in throughfall and stemflow (large flux ratio shown by the black bar in Figure 8.4), but then are strongly retained in the soil (flux ratios less than one for remaining three bars). Total P falls into this category; the upland is a net sink for this substance. Category 3 substances (e.g., total N) are not leached from the aspen canopy (small black bar) and are also strongly retained in the soil (all other flux ratios below one), such that the upland is a stronger sink than for Category 2 elements. Category 4 elements (e.g., K^+, Ca^{2+}, and Mg^{2+}) have flux ratios greater than 1; they are highly enriched in throughfall and stemflow relative to bulk precipitation. Unlike N and P, these elements are also enriched in upland runoff relative to inputs from bulk precipitation. The uplands act as a net source for these substances. Part of the export of these substances from the uplands is a result of chemical weathering of the mineral soils. Plants augment weathering and affect the seasonality of the element export. The cations are extracted by plants from the mineral soils, and a fraction of the uptake is transported into the foliage; the cations are then leached from the canopy and flushed from the uplands upon decomposition of the leaf litter. Potassium and Ca^{2+} are flushed primarily in surface runoff, but Mg^{2+} also has a large flux in interflow. The peatlands receive significant inputs of Category 4 elements in runoff from the uplands.

Processing of Elements in the MEF Peatlands

Element Export from the MEF Peatlands

Concentrations in Outflow

Hydrology has a major influence on seasonal patterns in element cycling. The major hydrologic event of the year at these sites is snowmelt, which generates the peak flows for the year (Julian day 125 in Figure 8.5b). During snowmelt, water retention time in the wetlands is at an annual minimum, and element uptake is correspondingly low. Although most precipitation occurs in summer months (Chapter 2), high evapotranspiration rates during these months result in little or no runoff to streams. In autumn, streamflow increases again as evapotranspiration decreases in response to plant senescence, reduced solar radiation, and falling temperatures.

Element concentrations in the MEF streams reflect this hydrologic cycle. Biologically active substances (NH_4^+, NO_3^-, and SO_4^{2-}) exhibit peak

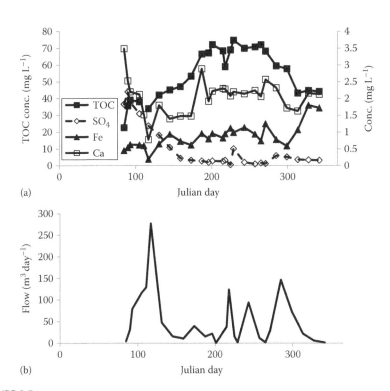

(a)

(b)

FIGURE 8.5
Concentrations for several solutes including TOC, Fe, Ca^{2+}, and SO_4^{2-}–S for 1981 (a); and seasonal cycle of water flow in 1981 (b).

concentrations during early snowmelt and minimum concentrations in summer (Figure 8.5a). Spring snowmelt generally is the only time of year when NO_3^- is measurable in the outflow from S2. Concentrations are highest in summer for substances generated by peat decomposition or bound to the dissolved organic matter (DOM) from peat decomposition (e.g., organic C, organic N, organic S, aluminum [Al], and iron [Fe]). Other substances exhibit intermediate patterns. As shown in Figure 8.5a, Ca^{2+} shows high concentrations during early snowmelt, but concentrations also increase during summer low-flow periods.

Streamflow draining the MEF watersheds reflects processes both in the uplands and in the peatlands. For instance, the increase in Ca^{2+} concentrations in autumn in the S2 stream likely reflects leaching from leaf litter in the uplands (Figure 8.5). Comparison of volume-weighted mean annual concentrations in streamflow and precipitation (Figure 8.6) shows the net effect of all processes in the watershed. Some ions (NO_3^-, NH_4^+, and SO_4^{2-}) are retained by the watershed while others (Ca^{2+}, Mg^{2+}, K^+, and H^+) are "produced" within the watershed. The uptake of Ca^{2+}, Mg^{2+}, and K^+ from the mineral soils into the upland plants and subsequent litterfall, leaching, and flow into the wetland was discussed previously as was the net retention of N by the upland (Timmons et al., 1977; Verry and Timmons, 1982). Figure 8.6 also indicates

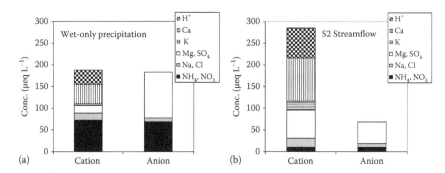

FIGURE 8.6

Comparison of volume-weighted mean ion concentrations in atmospheric deposition and streamflow (S2) for 1981. Concentrations are expressed in charge equivalents per liter. Because much of the precipitation is evapotranspired, solute concentrations are much higher in streamflow than in precipitation even for substances with identical fluxes in precipitation and streamflow. To facilitate seeing the changes in fluxes, precipitation concentrations have been multiplied by 3.98, the quotient of annual precipitation (77.65 cm), and streamflow (19.5 cm) in 1981. Cationic charges are increased as they pass through the watershed due to the release of Ca^{2+}, Mg^{2+}, and K^+ from the mineral soils and also due to the generation of free H^+ in the wetland. Sodium and Cl^- pass through the watershed nearly unaltered (export/input = 132% for Na^+, and 99% for Cl^-). Measured anionic charge equivalents are greatly reduced due to retention of NO_3^- in both the upland and peatland and retention of SO_4^{2-} in the peatland. The apparent charge imbalance indicates the quantity of dissociated organic anions (humic and fulvic acid anions); organic anions protonated at the pH of the bog water (~50% of total organic anions, Urban et al., 1989a) as well as those bound to Fe and Al (~12% of total organic anions) do not appear in the charge balance shown.

the generation of free acidity (H^+) by the peatland, and the large imbalance in ionic charge in the stream indicates the presence of high concentrations of organic anions. High concentrations of DOM generated in the peatlands and its associated acidity are important characteristics of streamflow from the MEF watersheds.

Research at the MEF has elucidated many features of the export from peatlands of DOM, which is generally measured as dissolved organic carbon (DOC). DOM is an important water-quality parameter that is intimately linked with biogeochemical cycles of several elements and with important ecosystem processes. DOM in surface waters affects light penetration (e.g., Fee et al., 1996; Jackson and Hecky, 1980), metal speciation and mobility (e.g., Lazerte, 1991), nutrient availability (e.g., Jones et al., 1988; Koenings and Hooper, 1976), alkalinity and pH (e.g., Oliver et al., 1983; Urban et al., 1989a), and toxicity and bioavailability of organic contaminants (e.g., Capel and Eisenreich, 1990). Colored organic matter is one of the primary controls on the penetration of potentially harmful UV radiation into lakes (e.g., Jerome and Bukata, 1998; Schindler et al., 1996; Smith et al., 1998). Carbon and energy are moved from terrestrial systems through surface waters in the form of DOM. Respiration of DOM in surface waters returns the carbon to the atmosphere (Cole et al., 1994; Kling et al., 1991). Many lakes are net sources of carbon dioxide to the atmosphere because of the organic carbon flow from their watersheds (Cole et al., 2000; Dillon and Molot, 1997; Schindler et al., 1997a). Concentrations of DOM derived from watersheds are high enough to cause 20%–40% of the lakes in northern Europe, eastern Canada, and the northeastern and upper Midwest of the United States to be brown colored, with color greater than 50 platinum–cobalt units (Gorham et al., 1986; Lillie and Mason, 1983; Overton et al., 1986; Rogalla, 1986; Wright, 1983). Clearly, DOM export from the MEF watersheds is important not only within the watersheds but also for receiving waters.

Awareness of water color, its association with organic compounds, and its production in wetlands predates the MEF and can be traced to at least the early nineteenth century (e.g., De Luc, 1810, cited in Garham, 1953). Early investigators had no method for measuring concentrations of DOM, and so there are few early paradigms for the factors regulating DOM concentrations in wetlands (see review in Gorham et al., 1985). De Luc (cited in Gorham, 1953) noted in 1810 that a "brown peat tint" first appeared in wetland waters at the stage when *Sphagnum* invaded. Methods for quantifying color were developed in the late nineteenth century (Hazen, 1892, 1896; Richards and Ellis, 1896), and Thompson et al. (1927) noted that the intensity of the yellow color in wetland waters increased with the stage of plant development.

Because of the disproportionate contribution of DOM by wetlands relative to their surface area, numerous authors have suggested that DOM concentrations in rivers and lakes can be predicted on the basis of the percentage of the watershed occupied by wetlands. Engstrom (1987) demonstrated that the color of Labrador lakes could be explained largely on the basis of the area of wetland in the watershed. Similarly, DOM from wetlands was the primary contributor to color and natural acidity in Nova Scotian lakes (Gorham et al., 1986) and streams (Gorham et al., 1998). Eckhardt and Moore (1990) and Koprivnjak and Moore (1992) found that DOM concentrations in Canadian streams can be predicted based on the fraction of the watershed occupied by wetlands. Gergel et al. (1999) found that DOM concentrations in streams were predicted better by the percentage of watershed in wetlands than DOM concentrations in lakes; lakes appeared to be influenced more strongly by nearshore wetlands than by wetlands higher in the watershed. It is not known whether the predictive relationships in the individual studies can be extrapolated to larger geographic areas (cf. Aitkenhead et al., 1999) or whether the predictive capabilities can be refined by distinguishing among different types of wetlands that have different DOM concentrations (cf. Frost et al., 2006).

Work at the MEF helped document the relationship between acidity of peatland drainage and DOM. A link between peatland acidity and organic acids was hypothesized in the early 1900s (e.g., Ramaut, 1954; Skene, 1915; Thompson et al., 1927). In 1980, Hemond (1980) showed that the acidity of one bog was due to dissolved humic and fulvic acids (see also McKnight et al., 1985). The generality of this conclusion was demonstrated by Gorham et al. (1985) and Urban et al. (1987a) in a survey of North American peatlands, among which the S2 bog featured prominently. The buffering capacity and acidity of DOM have been characterized in several studies (e.g., Oliver et al., 1983; Tipping et al., 1988); titrations of waters from the S2 watershed were consistent with the model proposed by Oliver (Urban, 1987; Urban et al., 1989). The pH of individual peatlands results from the titration of the humic and fulvic acids generated in the peatlands by the bases entering peatlands from atmospheric deposition, groundwater discharge, upland runoff, or anion uptake in the peatland (Urban, 1987; Urban and Bayley, 1986; Urban et al., 1987a). Gorham et al. demonstrated the titration of organic acids that occurs across the spectrum from bogs to calcareous fens and the bimodal distribution of pH in peatlands that results from the low buffer capacity of organic acids in the pH range of 4.5–5.5 (e.g., Gorham et al., 1984, 1985; Mullen et al., 2000). In many locations, there is a large effect of wetland drainage on the alkalinity of receiving waters (e.g., Bishop et al., 2000; Driscoll and Bisogni, 1984).

Research at the MEF documented the magnitude of DOC export from peatlands (Urban et al., 1989), systematic differences in DOM concentrations and export among different types of wetlands (Urban et al., 1989), variations in DOM composition, and geographic patterns in DOM concentrations (and associated acidity) in peatland waters (Gorham et al., 1985; Urban et al., 1987b). From 1981 to 1985, export of DOC ranged from 9 to 28 g C m^{-2} year^{-1} for S2 and 9 to 43 g C m^{-2} year^{-1} for S6 (Urban et al., 1989a). The export of DOC represents 5%–10% of net primary production in the peatlands, a significant component of the carbon budget. A similar rate of DOC export (19 g C m^{-2} year^{-1})

was measured for a bog in watershed 239 of the Experimental Lakes Area (ELA) in western Ontario, Canada, and estimated for bogs across northeastern North America ranged from 5 to 20 g C m^{-2} year^{-1} (Urban et al., 1987a). These rates of DOC export from peatlands are large relative to export rates from upland (unsaturated soils) forests in the same area (<1–6 g C m^{-2} year^{-1}). The differences in the export rates among three wetlands, two at the MEF and one at ELA, reflected differences in water yields. The high carbon export rates from peatlands explain why the areal extent of wetlands is a good predictor of organic carbon loading to lakes and streams.

Concentrations of DOC in peatland waters range from 5 to more than 60 mg L^{-1} (Gorham et al., 1985). The DOC concentrations in outflow from both the S2 (flow-weighted means of 31–75 mg L^{-1} for 1981–2008) and S6 (flow-weighted means of 31–47 mg L^{-1}) peatlands exceed the mean concentration reported for 28 peatlands across northeastern North America (Gorham et al., 1985). Higher DOC concentrations in midcontinental peatlands reflect the higher ratios of evapotranspiration to precipitation that result in lower water flow rates and less dilution of DOC by rain (Urban et al., 1987a). Watershed studies in Ontario have shown that droughts reduce water and DOC yields from the watersheds (D'arcy and Carignan, 1997; Moore et al., 1998; Schiff et al., 1998; Schindler et al., 1996).

Research at the MEF also has documented other effects of the export of DOM from peatlands. The high concentrations of DOM in bog waters result in a large binding capacity of bog waters for minor and trace metals. Binding sites on DOM compete with binding sites on the solid phase (peat), and can result in significant loss of trace metals from peatlands that experience significant water flow out of the system. In the S2 peatland, only about 35% of lead (Pb) inputs were retained (Urban et al., 1990). Leaching of ^{210}Pb from peat below the water table rendered ^{210}Pb-dating inaccurate at this site as well as at numerous other peatlands throughout North America (Urban et al., 1990). Aluminum in outflow from the MEF watersheds also is primarily (>85%) organically bound (Helmer et al., 1990). There are strong correlations between export of Hg and DOC from the MEF watersheds (Chapter 11; Kolka et al., 1999a, 2001), and, across the landscape, there is a correlation between the percentage of wetlands in watersheds and stream Hg concentrations (Grigal, 2002). Retention of Hg in the S2 watershed is only about 80% efficient largely due to the export of Hg bound to particulate organic matter and DOM in the outflow (Grigal et al., 2000). Enrichment of the DOM with organic S may contribute to the loss of Hg from the watershed (cf. Drexel et al., 2002; Ravichandran, 2004).

Given the importance of DOM to biogeochemical processes within the MEF watersheds as well as in receiving waters, any changes in export of DOM over time may have significant biogeochemical repercussions. Historical changes in DOC concentrations in lakes, streams, and rivers have been reported in several areas of the world and variously attributed to climate change and changes in acid deposition and land use (Hruska et al., 2009; Lepisto et al., 2008; Monteith et al., 2007; Schindler et al., 1997b). Data

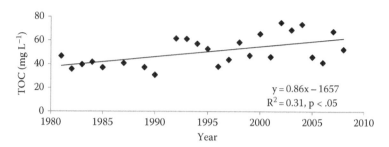

FIGURE 8.7
Historical increase in volume-weighted concentrations of TOC in outflow from the MEF S2.

from the S2 watershed also indicate that concentrations of DOC and total organic C (TOC) in the outlet stream have increased over the past 30 years (Figure 8.7). The statistically significant trendline (linear regression, $p < .05$) suggests that concentrations have increased by more than 50% (15–25 mg L^{-1}) since 1980. Although concentrations have increased significantly, fluxes have not; the trend toward increasing fluxes over time is not statistically significant (linear regression, $p = .14$), and fluxes in the most recent 10 years are not significantly different (t-test, $p > .05$) than fluxes in the first 10 years of the record.

Decreasing inputs of sulfuric acid in rain probably do not cause an increase in TOC concentrations in the S2 stream. It is thought that the trend in England, Scandinavia, and northeastern North America of increasing DOC concentrations in rivers, streams, and lakes over the past 20 years (e.g., Lepisto et al., 2008; Worrall and Burt, 2004; Worrall et al., 2003) is due to a decrease in atmospheric deposition of sulfuric acid (De Wit et al., 2007; Evans et al., 2006; Monteith et al., 2007). This hypothesis reinvokes the theory of Krug and Frink (1983) that mineral acid deposition inhibits the release of organic acids into surface waters by increasing ionic strength and decreasing pH (see Evans et al., 2008). However, at the MEF, there has been no significant change in concentrations of SO_4^{2-} or NO_3^-, and pH has declined in the stream outflow over the 30 year record despite a decrease in the atmospheric deposition of both SO_4^{2-} and NO_3^-. An increase in DOC solubility as a result of higher pH has not occurred here. Because there has been no change in SO_4^{2-} or NO_3^- concentration in the outflow, the change in TOC cannot be attributed to changing ionic strength driven by declines in concentrations of acid anions. These two ions are strongly retained by the watershed; it is possible that a decrease in retention of these anions has caused a decrease in alkalinity generation within the watershed. Such a change in alkalinity generation and export, if it has occurred, is hidden in the "anion deficit" of the outflow; changes in this unmeasured concentration of organic anions could cause changes in ionic strength, but such a change has not been observed. There has been a slight increase in electrical conductivity over the 30 year record, which suggests that ionic strength has not declined at this site, certainly not by the amount necessary to induce an increase in DOC concentrations (cf. Hruska et al., 2009). Concentrations and yields of other major ions also are buffered biogeochemically by the watershed such that changes in precipitation inputs have minor effects on the stream draining the watershed. For instance, the historical reduction in Ca^{2+} deposition in precipitation (14% decrease from 1980 to 2008) has not resulted in a decrease in concentrations or fluxes of Ca^{2+} in the S2 outflow. Atmospheric deposition of Na^+ and Cl^- has declined by 60%–70% over the past 30 years; this decline would cause at most a decrease in ionic strength of 3% if it was observed in the streamflow. However, there has been no trend in streamflow concentrations and yields since 1980. The interquartile ranges in annual streamflow, TOC export, and volume-weighted mean concentrations of TOC and several major ions have increased over 30 years. Hence, it seems that acid rain is not the causative factor at this site, but it is not clear whether the increased variability and increased concentrations of TOC are a response to climate change.

Variations among Wetland Types

Studies at the MEF have helped clarify the dependence of dissolved solute composition on water flowpaths. There is a large difference in the water chemistry of streams draining the two types of wetlands (perched bogs and groundwater fens) in the gaged watersheds at the MEF. Although both uplands and wetlands comprise the watersheds, streamflow is generated primarily from the wetland portions of the watersheds (Verry and Kolka, 2003). The relative contributions of peatlands and uplands to the solutes in the streamflow vary by solute (e.g., Kolka et al., 2001; Verry and Timmons, 1982).

A major distinction among the watersheds is that the groundwater fen (S3) has high concentrations of solutes derived from mineral dissolution (e.g., Ca^{2+}, Mg^{2+}, HCO_3^-, and Si) relative to the perched bogs (S1, S2, S4, S5, and S6). A longer contact time of the groundwater with soil minerals allows greater dissolution than occurs in the upland runoff into the perched bogs. The bogs have higher concentrations of organically bound solutes (e.g., DOC, dissolved organic N [DON], dissolved organic phosphorus, Al, Fe, and Hg) as a result of lower water flow rates (less dilution of substances released from the peat) and lower ash content of the peat in the bogs. Higher ash content in fen peat implies higher concentrations of base cations, iron oxides, and aluminum oxides that would limit the release of organic matter into fen waters. The higher concentrations of inorganic ions in fen waters lead to higher conductivity while the higher organic-matter content of bog waters leads to higher color and lower pH. This contrast is illustrated in Table 8.2 with data from Verry (1975) and Kolka et al. (1999a).

TABLE 8.2

Comparison of Fen and Bog Water Solute Concentrations[a]

Solute Category or Source	Solute Perched Bog Streamflow	Groundwater Fen Streamflow	Fen:Bog
Mineral dissolution			
Ca^{2+}	2.4 mg L^{-1}	16.6 mg L^{-1}	6.9
Mg^{2+}	0.97 mg L^{-1}	2.88 mg L^{-1}	3.0
Alkalinity	~0	54.2 mg L^{-1} as $CaCO_3$	>54
Si	2.7 mg L^{-1}	4.9 mg L^{-1}	1.8
Specific conductance	51 µS cm^{-1} (25°C)	125 µS cm^{-1} (25°C)	2.45
Organically bound			
DOC[b]	46.6 mg L^{-1}	4.2 mg L^{-1}	0.09
	0.69 mg L^{-1}	0.33 mg L^{-1}	0.48
TDP[c]	0.19 mg L^{-1}	0.09 mg L^{-1}	0.47
Al	0.79 mg L^{-1}	0.16 mg L^{-1}	0.20
Fe	1.35 mg L^{-1}	0.98 mg L^{-1}	0.72
Hg[b]	11.6 ng L^{-1}	1.4 ng L^{-1}	0.12
pH	3.6	6.5	
Color	303	100	
Atmospherically derived			
SO_4^{2-}	4.6 mg L^{-1}	6.0 mg L^{-1}	1.30
NO_3^-	0.20 mg L^{-1}	0.10 mg L^{-1}	0.50
Cl$^-$	0.7 mg L^{-1}	0.4 mg L^{-1}	0.57

[a] All data except DOC and Hg are from Verry (1975). Values represent means of concentrations measured from 1968–1972.
[b] Data from Kolka et al. (1999a) includes the bogs at S1, S2, S4, and S5, and the S3 fen. Mean of the volume-weighted means for 1994 and 1995.
[c] TDP, total dissolved phosphorus.

Element Budgets

Element budgets have been measured and reported multiple times for the MEF watersheds (e.g., Urban and Bayley, 1986; Urban and Eisenreich, 1988; Urban et al., 1989c, 1990, 1995; Verry, 1975; Verry and Timmons, 1982; Verry and Urban, 1992). Budgets for the S2 watershed for the four major cations S, P, N, and C are summarized in Table 8.3. Most of these data were collected from 1971 to 1973 and from 1981 to 1984. Here, the areal basis for these budgets is the wetland (bog plus lagg) portion of the watershed. Inputs include atmospheric deposition (bulk deposition or separate estimates of wet and dry) and upland runoff; streamflow is the only output considered. These budgets differ from those of Verry and Urban (1992), who used throughfall plus stemflow as the measure of atmospheric deposition. For elements not leached from foliage, throughfall may provide a

TABLE 8.3

Element Budgets for the S2 Watershed (g m^{-2} Year^{-1}). Accounting for Gas Emissions Leads to Lower Retention of C and N (in Parentheses)

	Atmospheric Deposition[a] (g m^{-2} Year^{-1})	Surface Runoff (g m^{-2} Year^{-1})	Interflow (g m^{-2} Year^{-1})	Outflow (g m^{-2} Year^{-1})	% Retention (g m^{-2} Year^{-1})
Carbon[b]	1.2	0.49	1.75	33.6	−879% (25%)
Nitrogen[c]	1.04	0.16	0.17	0.6	56% (38%)
Phosphorus	0.048	0.022	0.004	0.046	38%
Sulfur	0.49[d]		0.3	0.46	42%
Calcium	0.44[d]	0.88	0.44	0.91	48%
Magnesium	0.08[d]	0.25	0.14	0.38	19%
Sodium	0.20[d]	0.09	0.14	0.24	45%
Potassium	0.18[d]	0.6	0.08	0.61	28%

[a] Bulk deposition from Verry and Timmons (1977).
[b] Values from Kolka et al. (1999a) for 1993–1995.
[c] Values from table 9 in Urban and Eisenreich (1988).
[d] Values from Urban et al. (1995) for years 1971–1973 and 1981–1984.

better estimate of dry deposition than bulk deposition. For elements with gas exchange (C and N), this budget is incomplete and the second value in parentheses for percent retention accounts for gas fluxes (Chapter 10) into and out of the wetland.

The major conclusion to be drawn from comparison of the element budgets is that retention of many of the elements (N, P, S, Ca^{2+}, and Na$^+$) is similar (40%–50%), but retention of K$^+$ and Mg^{2+} is lower by 20%–30%. The macronutrients N, P, and S are all similarly conserved (~40%) while gas emissions of C lead to lower retention (25%). Differences among elements are not due to the differing importance of inputs from the upland as large interannual differences in upland inputs cause little change in the magnitude of element export (Urban et al., 1995). The major mechanism for retention of all the elements is burial of partially decomposed vegetation and, to a lesser extent, retention within an aggrading vegetation compartment; differences among elements likely reflect differential decomposition rates of different components of the organic matter. Greater incorporation of Ca into wood and slower decomposition of wood may explain the greater retention of Ca^{2+} relative to Mg^{2+}. Magnesium exhibits a net export during the summer months of some years (Urban et al., 1995), which is further evidence of low Mg^{2+} retention in vegetation. The greater retention of Na$^+$ relative to K$^+$ may reflect the rapid release and recycling of K$^+$ within the wetland (Buttleman and Grigal, 1985); the pool of K$^+$ in living vegetation represents more than 100 years of atmospheric inputs while the amount of Na$^+$ in vegetation represents only 2 years of atmospheric deposition.

Element Cycling within the MEF Peatlands

Nitrogen

One of the major strengths of the MEF is the length of the historical record. This strength is apparent in the 30 year record of atmospheric deposition of N. Data from this site document the decrease in atmospheric deposition of NO_3^- that has occurred over this 30 year period. Deposition of NO_3^- has decreased at a rate of about 1% per year or 30% over the period of record. The decline observed at the MEF (Figure 2.3b) is consistent with that observed throughout the upper Midwest (McDonald et al., 2010) and results from the decrease in emissions of N oxides over this period (U.S. Environmental Protection Agency, 2000). The 30% decrease in wet deposition of NO_3^- has resulted in about a 7% decrease in the total N deposition.

Earlier studies on N cycling in the MEF watersheds (David et al., 1988; Grigal, 1991; Urban and Eisenreich, 1988) allowed predictions of the likely effects of this decrease in N deposition. Given that the MEF peatlands are N-limited (Bridgham et al., 2001; Urban and Eisenreich, 1988), a decrease in N inputs likely leads to a decrease in primary production, litter quality, and overall rates of N cycling through these systems. Surveys of peatlands across gradients of increasing N deposition have shown these responses to changes in N inputs (Turunen et al., 2004). Elevated rates of atmospheric deposition of N tend to favor growth of vascular plants over that of bryophytes (Berendse et al., 2001; Heijmans et al., 2002a,b; Limpens et al., 2003; Monique et al., 2001). High rates of nitric acid deposition in British moors are thought to have contributed to the loss of bryophytes, acidification of the moors, and inhibition of microbial enzymes (Lee and Stewart, 1978; Lee and Tallis, 1973; Lee et al., 1987). There is some evidence that high N deposition rates may threaten the survival of some of the carnivorous plant species characteristic of peatlands (Gotelli and Ellison, 2002). Large inputs of the limiting nutrient may cause supplies of other nutrients to become inadequate and to switch the limiting nutrient from N to either P or K (Hoosbeek et al., 2002). Nitrogen deposition never was high enough to observe many of these effects at the MEF; maximum historical rates were about $1\,g\,N\,m^{-2}\,year^{-1}$. It is not known whether the composition of vegetation at the MEF is changing in response to decreased N inputs.

Studies at the MEF showed that NO_3^- inputs are unlikely to be denitrified and, therefore, are unlikely to affect rates of other forms of anaerobic respiration. In bogs receiving low rates of N deposition, NO_3^- is quickly taken up by the vegetation and does not reach anaerobic strata where it might be denitrified (Heijmans et al., 2002a; Nykanen et al., 2002; Urban and Eisenreich, 1988; Urban et al., 1989b). Consequently, there is no evidence of inhibition of methane emissions or of SO_4^{2-} reduction by elevated inputs of NO_3^- (Dise and Verry, 2001); in contrast, N inputs appear to stimulate methane emissions from some wetlands (Aerts and De Caluwe, 1999; Nykanen et al., 2002). The low pH in bogs inhibits nitrification (Bridgham et al., 2001), and hence the internal cycling of N is largely as NH_4^+ (Figure 8.8a). The internal cycling

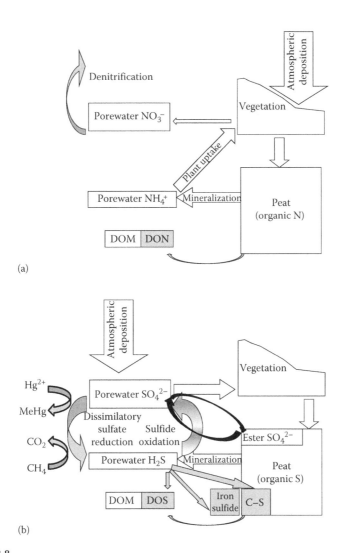

(a)

(b)

FIGURE 8.8

Comparison of N (a) and S (b) cycles in the S2 peatland at the MEF. Both elements are similar in having large fluxes into the vegetation, passage of this material to the upper layers of peat, and subsequent mineralization of the peat. Because of the inhibition of nitrification at the low pH of the bog, N is cycled back to the vegetation in the reduced form while S is first reoxidized to SO_4^{2-}. Because the vegetation is very efficient at capturing NO_3^- inputs to the bog, very little denitrification occurs. By contrast, SO_4^{2-} reduction is considerable. Sulfide generated via microbial SO_4^{2-} reduction or mineralization of organic matter is reoxidized, precipitated with Fe, or reacted with organic matter. Sulfate reduction inhibits the release of methane, but, in the presence of Hg, promotes Hg methylation.

of NH_4^+ in the S2 bog is about eight times larger than the N input from atmospheric deposition (Figure 8.8a). The absence of nitrification and low rates of denitrification in bogs suggests that N_2O emissions from these ecosystems should be low; this hypothesis is supported by measurements by Huttunen et al. (2002), Nykanen et al. (2002), and Regina et al. (1996).

Work in the MEF watersheds suggests that peatlands are not unique in rates of cycling of N but are unique in exporting considerable organic N. Although the high C:N ratios of the peat might be expected to limit N mobility in peatlands (e.g., Fenn et al., 1998; Lovett et al., 2002), the S2 bog is inefficient at retaining N. Recent studies have suggested that N retention in ecosystems can be predicted on the basis of the soil C:N ratio (e.g., Lovett et al., 2002). The high C:N ratios in bog peat (30–60, e.g., Malmer and Holm 1984; Urban and Eisenreich, 1988) lead to efficient retention of inorganic N; retention of inorganic N is more than 90% in the S2 watershed. However, the low rates of microbial activity in acidic peat may limit the extent to which NH_4^+ may be immobilized, and the loss of DOM engendered by anaerobic decomposition results in a total N retention efficiency of only 45% in the S2 bog. In bogs with adequate water flow, the export of organic N in the DOM offsets the high efficiency with which inorganic N is retained. Export of DON by peatlands is important at both the landscape and watershed scales. As has been widely reported for DOC (e.g., Engstrom, 1987; Frost et al., 2006; Koprivnjak and Moore, 1992), DON concentrations in rivers and streams in New England were strongly correlated with the areal extent of wetlands in the watershed (Pellerin et al., 2004).

Comparison of the MEF peatland watersheds with raised bogs on the extensive blanket peatlands in Minnesota suggests that the perched bogs in the MEF are less nutrient stressed than nearby raised bogs (Grigal, 1991). The fraction of plant N uptake supplied by atmospheric deposition was higher (12% vs. 15%) on raised than on perched bogs. The fraction of plant N uptake supplied by mineralization in the acrotelm was correspondingly lower in the raised bogs, but the fraction of N mineralized in the acrotelm of each type of peatland was similar (1.5% year^{-1}). The turnover time of N in the vegetation, about 25% shorter (3.8 vs. 4.8 years) in the raised bogs, is further evidence of the greater nutrient stress in these systems.

Sulfur

Studies of the biogeochemistry of S in peatlands at the MEF have provided important insights into its cycling and interactions with other elements. Sulfur is a macronutrient and the S cycle within peatlands is similar to cycling of N in many respects (Figure 8.8b). Research on the characterization of S cycling through the vegetation of the S2 watershed (Urban et al., 1989c) is among the most thorough for any peatland. Atmospheric deposition (0.4–0.5 g S m^{-2} year^{-1}) is less than the annual uptake of S by the vegetation (~1.3 g S m^{-2} year^{-1}) at this site. Thus, there is a large internal cycle of S within the S2 wetland. In part because of the large uptake and recycling by the vegetation, there is a net retention of S within peatlands; in the S2 peatland, about 40% of inputs is retained on an annual basis. Uptake of SO_4^{2-} by the vegetation is not as rapid as uptake of NO_3^- (Urban and Bayley, 1986; Urban et al., 1989b); accordingly, retention of inorganic S inputs (~60%) is not as efficient as retention of inorganic N (>90%). Export of atmospherically deposited SO_4^{2-} also has been observed in other peatlands (Novak et al., 2005b). Because a fraction of the organic matter is not decomposed but buried in peat, there is a large burial of organic S within peatlands. However, because of significant mineralization of organic S and uptake of this mineralized S by vegetation, there is significant vertical mobility of S within peat, and accumulation rate profiles of S in peat cannot be interpreted as historical records of atmospheric S deposition (Novak et al., 2003, 2005a; Urban et al., 1989c). Also, S inventories in peat are not proportional to S loadings to the peatland (Novak et al., 2003).

A second cycle of S through the vegetation and shallow peat (Figure 8.8b) occurs concurrently with the cycle described previously. As for N, any SO_4^{2-} in precipitation that is not taken up by the vegetation may be used as an electron acceptor for anaerobic respiration (dissimilatory SO_4^{2-} reduction). Unlike N, substantial quantities of inorganic sulfur (SO_4^{2-}) penetrate through the vegetation to the water table and are available to be reduced via dissimilatory SO_4^{2-} reduction in the S2 bog (Urban and Bayley, 1986). In many peatlands, high rates of microbial SO_4^{2-} reduction have been measured by the addition of SO_4^{2-} or radiolabeled $^{35}SO_4^{2-}$ (Bayley et al., 1987; Chapman and Davidson, 2001; Keller and Bridgham, 2007; Novak et al., 2003; Vile et al., 2003; Wieder and Lang, 1988; Wieder et al., 1987), including SO_4^{2-} additions to the S6 wetland at the MEF (Jeremiason et al., 2006). Sulfate reduction can account for 0%–25% of the anaerobic oxidation of organic matter in peatlands (Keller and Bridgham, 2007; Vile et al., 2003). The fate of the microbially produced H_2S varies among peatlands. In minerotrophic peatlands or those rich in iron, formation of acid-volatile sulfide (FeS) and pyrite (FeS_2) is considerable (Novak and Wieder, 1992; Novak et al., 2003; Wieder and Lang, 1986, 1988). In peatlands, low in iron, most of the H_2S reacts with the peat and DOM to form C-bonded S (Brown, 1986; Keller and Bridgham, 2007; Novak et al., 1994; Urban et al., 1989c). Rates of SO_4^{2-} reduction greater than the annual deposition of SO_4^{2-} to peat bogs have been documented (Novak et al., 2005b); rates of oxidation of sulfide are substantial in peatlands (Bayley et al., 1986; Bottrell et al., 2007; Heitmann and Blodau, 2006; Novak et al., 2005b), allowing an internal cycle (Figure 8.8) to play a significant role in C oxidation (Keller and Bridgham, 2007; Vile et al., 2003). The factors controlling the rates and pathways for the reoxidation of sulfide in peatlands are poorly documented. Internal cycling of S does not appear to be as significant in upland mineral soil systems (Eimers et al., 2004).

A third cycle of S within peatlands about which relatively little is known entails the formation and hydrolysis of ester sulfates (Figure 8.8b). Formation of ester sulfates upon addition of $^{35}SO_4^{2-}$ to peat has been documented

(Chapman and Davidson, 2001; Novak et al., 2003), and sulfate esters have contributed 6%–30% of total S in peat profiles (Novak et al., 2003; Prietzel et al., 2007). Aryl sulfatase activity in peat was lower in English peatlands receiving high rates of acid and pollutant deposition (Press et al., 1985). Research is needed to determine whether this cycle is quantitatively important in most peatlands.

Work at the MEF documented the significance of the export of organic S from peatlands. DOM in the S2 stream was enriched in S in summer months; C:S dropped from 300 in spring to 100 in summer (Urban et al., 1989c). Export of organic S accounted for one third of the S exported from the S2 watershed. The abiotic formation of dissolved organic S by reaction of H_2S with DOC in peatlands was documented by Heitmann and Blodau (2006). Spectroscopic studies have shown oxidation states of −2 to +6 for S in humic acids from peatlands (Morra et al., 1997; Prietzel et al., 2007). Organic S may be important for the binding and transport of trace metals including Hg in bog waters (Qian et al., 2002; Xia et al., 1999; Yoon et al., 2005).

Interactions of S and Hg cycling have been well documented for the MEF wetlands. Wetlands have long been known as significant sources of total and methyl Hg to lakes and streams (Dennis et al., 2005; Hurley et al., 1999; Shanley et al., 2005). It also is widely recognized that Hg methylation occurs in association with SO_4^{2-} reduction (e.g., Goulet et al., 2007; King et al., 2001). Postulating a linkage between S transformations in peatlands and Hg export, researchers at MEF demonstrated and quantified this linkage. They artificially applied SO_4^{2-} to a portion of the S6 wetland and compared the concentrations of total and methyl Hg between the amended and control portions of the wetland (Jeremiason et al., 2006). Results showed that SO_4^{2-} additions caused increased concentrations of methyl Hg in porewaters and an increase in export of methyl Hg following SO_4^{2-} applications (Coleman-Wasik, 2008; Chapter 11).

Observations at the MEF also pointed to interactions between C and S cycling in wetlands. The potential for inhibition of methane production in wetlands due to competition for C substrates between methanogenic and SO_4^{2-} reducing bacteria had been postulated by Wieder et al. (1990). This hypothesis was tested at the Bog Lake peatland, an open, nutrient-poor fen within the MEF (Dise and Verry, 2001). Application of NH_4SO_4 to field plots lowered methane emissions by 30%. These results have since been replicated at other locations (Gauci and Chapman, 2006; Gauci et al., 2002, 2004). The mechanism behind the suppression of methane fluxes remains under investigation. Although studies have documented a stimulation of SO_4^{2-} reduction in response to SO_4^{2-} additions (Gauci and Chapman, 2006), laboratory studies have not always shown inhibition of methanogenesis by SO_4^{2-} addition (Vile et al., 2003). Anaerobic oxidation of methane by SO_4^{2-}-reducing bacteria, a phenomenon well documented in marine sediments, does not appear to be important in peat bogs (Smemo and Yavitt, 2007).

References

Aerts, R. and H. De Caluwe. 1999. Nitrogen deposition effects on carbon dioxide and methane emissions from temperate peatland soils. *Oikos* 84:44–54.

Aitkenhead, J.A., D. Hope, and M.F. Billett. 1999. The relationship between dissolved organic carbon in stream water and soil organic carbon pools at different spatial scales. *Hydrological Processes* 13:1289–1302.

Bayley, S.E., R.S. Behr, and C.A. Kelly. 1986. Retention and release of S from a freshwater wetland. *Water, Air, and Soil Pollution* 31:101–114.

Bayley, S.E., D.H. Vitt, R.W. Newbury, K.G. Beaty, R. Behr, and C. Miller. 1987. Experimental acidification of a *Sphagnum*-dominated peatland: First year results. *Canadian Journal of Fisheries and Aquatic Science* 44:194–205.

Berendse, F., N. Van Breemen, H. Rydin, A. Buttler, M. Heijmans, M.R. Hoosbeek, J.A. Lee et al. 2001. Raised atmospheric CO_2 levels and increased N deposition cause shifts in plant species composition and production in *Sphagnum* bogs. *Global Change Biology* 7:591–598.

Bishop, K.H., H. Laudon, and S. Koehler. 2000. Separating the natural and anthropogenic components of spring flood pH decline: A method for areas that are not chronically acidified. *Water Resources Research* 36:1873–1884.

Bottrell, S.H., R.J.G. Mortimer, M. Spence, M.D. Krom, J.M. Clark, and P.J. Chapman. 2007. Insights into redox cycling of sulfur and iron in peatlands using high-resolution diffusive equilibrium thin film (DET) gel probe sampling. *Chemical Geology* 244:409–420.

Bridgham, S.D., K. Updegraff, and J. Pastor. 2001. A comparison of nutrient availability indices along an ombrotrophic-minerotrophic gradient in Minnesota wetlands. *Soil Science Society of America Journal* 65:259–269.

Brown, K.A. 1986. Formation of organic sulphur in anaerobic peat. *Soil Biology and Biochemistry* 18:131–140.

Buttleman, C.G. and D.F. Grigal. 1985. Use of the Rb/K ratio to evaluate potassium nutrition of peatlands. *Oikos* 44:253–256.

Capel, P.D. and S.J. Eisenreich. 1990. Relationship between chlorinated hydrocarbons and organic carbon in sediment and porewater. *Journal of Great Lakes Research* 16:245–257.

Chapman, S.J. and M.S. Davidson. 2001. ^{35}S-sulphate reduction and transformation in peat. *Soil Biology and Biochemistry* 33:593–602.

Cole, J.J., N.F. Caraco, G.W. Kling, and T.K. Kratz. 1994. Carbon dioxide supersaturation in the surface waters of lakes. *Science* 265:1568–1570.

Cole, J.J., M.L. Pace, S.R. Carpenter, and J.F. Kitchell. 2000. Persistence of net heterotrophy in lakes during nutrient addition and food web manipulations. *Limnology and Oceanography* 45:1718–1730.

Coleman-Wasik, J.K. 2008. Chronic effects of atmospheric sulfate deposition on mercury methylation in a boreal wetland: Replication of a global experiment. MS thesis. St. Paul, MN: University of Minnesota.

D'arcy, P. and R. Carignan. 1997. Influence of catchment topography on water chemistry in southeastern Quebec Shield lakes. *Canadian Journal of Fisheries and Aquatic Science* 54:2215–2227.

David, M.B., D.F. Grigal, L.F. Ohmann, and G.Z. Gertner. 1988. Sulfur, carbon, and nitrogen relationships in forest soils across the northern Great Lakes States as affected by atmospheric deposition and vegetation. *Canadian Journal Forest Research* 18:1386–1391.

De Wit, H.A., J. Mulder, A. Hindar, and L. Hole. 2007. Long-term increase in dissolved organic carbon in streamwaters in Norway is response to reduced acid deposition. *Environmental Science & Technology* 41:7706–7713.

Dennis, I.F., T.A. Clair, C.T. Driscoll, N. Kamman, A. Chalmers, J. Shanley, S.A. Norton, and S. Kahl. 2005. Distribution patterns of mercury in Lakes and Rivers of northeastern North America. *Ecotoxicology* 14:113–123.

Dillon, P.J. and L.A. Molot. 1997. Dissolved organic and inorganic carbon mass balances in central Ontario lakes. *Biogeochemistry* 36:29–42.

Dise, N.B. and E.S. Verry. 2001. Suppression of peatland methane emission by cumulative sulfate deposition in simulated acid rain. *Biogeochemistry* 53:143–160

Drexel, R.T., M. Haitzer, J.N. Ryan, G.R. Aiken, and K.L. Nagy. 2002. Mercury(II) sorption to two Florida Everglades peats: Evidence for strong and weak binding and competition by dissolved organic matter released from the peat. *Environmental Science & Technology* 36:4058–4064.

Driscoll, C.T. and J.J. Bisogni. 1984. Weak Acid/base systems in dilute acidified lakes and streams of the Adirondack region of New York State. In: *Modeling of Total Acid Precipitation Impacts*, ed. J.L. Schnoor. Boston, MA: Butterworth Publishers, pp. 53–72.

Eckhardt, B.W. and T.R. Moore. 1990. Controls on dissolved organic carbon concentrations in streams, Southern Quebec. *Canadian Journal of Fisheries and Aquatic Science* 47:1537–1544.

Eimers, M.C., P.J. Dillon, and S.L. Schiff. 2004. A S-isotope approach to determine the relative contribution of redox processes to net SO_4 export from upland, and wetland-dominated catchments. *Geochimica et Cosmochimica Acta* 68:3665–3674.

Engstrom, D.R. 1987. Influence of vegetation and hydrology on the humus budgets of Labrador lakes. *Canadian Journal of Fisheries and Aquatic Science* 44:1306–1314.

Evans, C.D., P.J. Chapman, J.M. Clark, D.T. Monteith, and M.S. Cresser. 2006. Alternative explanations for rising dissolved organic carbon export from organic soils. *Global Change Biology* 12:2044–2053.

Evans, C.D., D.T. Monteith, B. Reynolds, and J.M. Clark. 2008. Buffering of recovery from acidification by organic acids. *Science of the Total Environment* 404:316–325.

Fee, E.J., R.E. Hecky, S.E. Kasian, and D.R. Cruikshank. 1996. Effects of lake size, water clarity and climatic variability on mixing depths in Canadian Shield lakes. *Limnology and Oceanography* 41:912–920.

Fenn, M.E., M.A. Poth, J.D. Aber, J.S. Baron, B.T. Bormann, D.W. Johnson, A.D. Lemly, S.G. McNulty, D.F. Ryan, and R. Stottlemeyer. 1998. Nitrogen excess in North American ecosystems: Predisposing factors, ecosystem responses, and management strategies. *Ecological Applications* 8:706–733.

Frost, P.C., J.H. Larson, G.A. Lamberti, C.A. Johnston, K.C. Young, P.A. Maurice, and S.D. Bridgham. 2006. Landscape predictors of stream dissolved organic matter concentration and physicochemistry in a Lake Superior river watershed. *Aquatic Science* 68:40–51.

Gauci, V. and S.J. Chapman. 2006. Simultaneous inhibition of CH_4 efflux and stimulation of sulphate reduction in peat subject to simulated acid rain. *Soil Biology and Biochemistry* 38:3506–3510.

Gauci, V., N. Dise, and D. Fowler. 2002. Controls on suppression of methane flux from a peat bog subjected to simulated acid rain sulfate deposition. *Global Biogeochemical Cycles* 16: doi 10.1029/2000GB001370.

Gauci, V., D. Fowler, S.J. Chapman, and N.B. Dise. 2004. Sulfate deposition and temperature controls on methane emission and sulfur forms in peat. *Biogeochemistry* 71:141–162.

Gergel, S.E., M.G. Turner, and T.K. Kratz. 1999. Dissolved organic carbon as an indicator of the scale of watershed influence on lakes and rivers. *Ecological Applications* 9:1377–1390.

Gorham, E. 1953. Some early ideas concerning the nature, origin and development of peatlands. *Journal of Ecology* 41:257–273.

Gorham, E., S.E. Bayley, and D.W. Schindler. 1984. Ecological effects of acid deposition upon peatlands: A neglected field in acid-rain research. *Canadian Journal of Fisheries and Aquatic Science* 41:1256–1268.

Gorham, E., S.J. Eisenreich, J. Ford, and M.V. Santelmann. 1985. The chemistry of bog waters. In: *Chemical Processes in Lakes*, ed. W. Stumm. New York: John Wiley & Sons, p. 339.

Gorham, E., J.K. Underwood, F.B. Martin, and J.G. Ogden. 1986. Natural and anthropogenic causes of lake acidification in Nova Scotia. *Nature* 324:451–453.

Gorham, E., J.K. Underwood, J.A. Janssens, B. Freedman, W. Maass, D.H. Waller, and J.G. Ogden. 1998. The chemistry of streams in southwestern and central Nova Scotia, with particular reference to catchment vegetation and the influence of dissolved organic carbon primarily from wetlands. *Wetlands* 18: 115–132.

Gotelli, N.J. and A.M. Ellison. 2002. Nitrogen deposition and extinction risk in the northern pitcher plant, *Sarracenia purpurea*. *Ecology* 83:2758–2765.

Goulet, R.R., J. Holmes, B. Page, L. Poissant, S.D. Sicilliano, D.R.S. Lean, F. Wang, M. Amyot, and A. Tessier. 2007. Mercury transformations and fluxes in sediments of a riverine wetland. *Geochimica et Cosmochimica Acta* 71:3393–3406.

Grigal, D.F. 1991. Elemental dynamics in forested bogs in northern Minnesota. *Canadian Journal of Botany* 69:539–546.

Grigal, D.F. 2002. Inputs and outputs of mercury from terrestrial watersheds: A review. *Environmental Reviews* 10:1–39.

Grigal, D.F., R.K. Kolka, J.A. Fleck, and E.A. Nater. 2000. Mercury budget of an upland-peatland watershed. *Biogeochemistry* 50:95–109.

Hazen, A. 1892. A new color-standard for natural waters. *American Chemical Journal* 14:300–310.

Hazen, A. 1896. The measurement of colors of natural waters. *Journal of the American Chemical Society* 18:264–275.

Heijmans, M.M.P.D., H. Klees, W. De Visser, and F. Berendse. 2002a. Effects of increased nitrogen deposition on the distribution of ^{15}N-labeled nitrogen between *Sphagnum* and vascular plants. *Ecosystems* 5:500–508.

Heijmans, M.M.P.D., H. Klees, W. De Visser, and F. Berendse. 2002b. Response of a *Sphagnum* bog plant community to elevated CO_2 and N supply. *Plant Ecology* 162:123–134.

Heitmann, T. and C. Blodau. 2006. Oxidation and incorporation of hydrogen sulfide by dissolved organic matter. *Chemical Geology* 235:12–20.

Helmer, E.H., N.R. Urban, and S.J. Eisenreich. 1990. Aluminum geochemistry in peatland waters. *Biogeochemistry* 9:247–276.

Hemond, H.F. 1980. Biogeochemistry of Thoreau's Bog, Concord, Mass. *Ecological Monographs* 50:507–526.

Hoosbeek, M.R., N. Van Breemen, H. Vasander, A. Buttler, and F. Berendse. 2002. Potassium limits potential growth of bog vegetation under elevated atmospheric CO_2 and N deposition. *Global Change Biology* 8:1130–1138.

Hruska, J., P. Kram, W.H. Mcdowell, and F. Oulehle. 2009. Increased dissolved organic carbon (DOC) in Central European streams is driven by reductions in ionic strength rather than climate change or decreasing acidity. *Environmental Science & Technology* 43:4320–4326.

Hurley, J.P., J.M. Benoit, C.L. Babiarz, M.M. Shafer, A.W. Andren, J.R. Sullivan, R. Hammond, and D.A. Webb. 1999. Influences of watershed characteristics on mercury levels in Wisconsin rivers. *Environmental Science & Technology* 29:1867–1875.

Huttunen, J.T., H. Nykanen, P.J. Martikainen, J. Turunen, and O. Nenonen. 2002. Fluxes of nitrous oxide on natural peatlands in Vuotos, an area projected for a hydroelectric reservoir in northern Finland. *Suo* 53:87–96.

Jackson, T.A. and R.E. Hecky. 1980. Depression of primary productivity by humic matter in lake and reservoir waters of the boreal forest zone. *Canadian Journal of Fisheries and Aquatic Science* 37:2300–2317.

Jeremiason, J.D., D.R. Engstrom, E.B. Swain, E.A. Nater, B.M. Johnson, J.E. Almendinger, B.A. Monson, and R.K. Kolka. 2006. Sulfate addition increases methylmercury production in an experimental wetland. *Environmental Science & Technology* 40:3800–3806.

Jerome, J.H. and R.P. Bukata. 1998. Tracking the propagation of solar ultraviolet radiation: Dispersal of ultraviolet photons in inland waters. *Journal of Great Lakes Research* 24:666–680.

Jones, R.I., K. Salonen, and H. De Haan. 1988. Phosphorus transformations in the epilimnion of humic lakes: Abiotic interactions between dissolved humic materials and phosphate. *Freshwater Biology* 19:357–369.

Keller, J.K. and S.D. Bridgham. 2007. Pathways of anaerobic carbon cycling across an ombrotrophic-minerotrophic peatland gradient. *Limnology and Oceanography* 52:96–107

King, J.K., J.E. Kostka, M.E. Frischer, F.M. Saunders, and R.A. Jahnke. 2001. A quantitative relationship that demonstrates mercury methylation rates in marine sediments are based on the community composition and activity of sulfate-reducing bacteria. *Environmental Science & Technology* 35:2491–2496.

Kling, G.W., G.W. Kipphut, and M.C. Miller. 1991. Arctic lakes and streams as gas conduits to the atmosphere: Implications for tundra carbon budgets. *Science* 251:298–301.

Koenings, J.P. and F.F. Hooper. 1976. The influence of colloidal organic matter on iron and iron-phosphorus cycling in an acid bog lake. *Limnology and Oceanography* 21:684–696.

Kolka, R.K., D.F. Grigal, E.S. Verry, and E.A. Nater. 1999a. Mercury and organic carbon relationships in streams draining forested upland/peatland watersheds. *Journal of Environmental Quality* 28:766–775.

Kolka, R.K., E.A. Nater, D.F. Grigal, and E.S. Verry. 1999b. Atmospheric inputs of mercury and organic carbon into a forested upland/bog watershed. *Water, Air, and Soil Pollution* 113:273–294.

Kolka, R.K., D.F. Grigal, E.A. Nater, and E.S. Verry. 2001. Hydrologic cycling of mercury and organic carbon in a forested upland-bog watershed. *Soil Science Society of American Journal* 65:897–905.

Koprivnjak, J.F. and T.R. Moore. 1992. Sources, sinks and fluxes of dissolved organic carbon in subarctic fen catchments. *Arctic and Alpine Research* 24:204–210.

Krug, E.C. and C.R. Frink. 1983. Acid rain on acid soil: A new perspective. *Science* 221:520–525.

Lazerte, B.D. 1991. Metal transport and retention: The role of dissolved organic carbon. In: *3rd International Nordic Symposium on Humic Substances*. Finnish Humus News, pp. 71–82.

Lee, J.A. and J.H. Tallis. 1973. Regional and historical aspects of lead pollution in Britain. *Nature* 245:216–218.

Lee, J.A. and G.R. Stewart. 1978. Ecological aspects of nitrogen assimilation. *Advances in Botanical Research* 6:1–43.

Lee, J.A., M.C. Press, S. Woodin, and P. Ferguson. 1987. Responses to acidic deposition in ombrotrophic mires. In: *Effects of Air Pollutants on Forests, Agriculture and Wetlands*, ed. T.C. Hutchinson and K.M. Meema. New York: Springer-Verlag, pp. 549–560.

Lepisto, A., P. Kortelainen, and T. Mattsson. 2008. Increased organic C and N leaching in a northern boreal river basin in Finland. *Global Biogeochemical Cycles* 22:GB3029.

Lillie, R.A. and J.W. Mason. 1983. *Limnological Characteristics of Wisconsin Lakes*. Technical Bulletin #138. Madison, WI: Wisconsin Department of Natural Resources.

Limpens, J., F. Berendse, and H. Klees. 2003. N deposition affects N availability in interstitial water, growth of *Sphagnum* and invasion of vascular plants in bog vegetation. *New Phytologist* 157:339–347.

Lindberg, S.E., G.M. Lovett, D.D. Richter, and E.W. Johnson. 1986. Atmospheric deposition and canopy interactions of major ions in a forest. *Science* 231:141–145.

Lovett, G.M., K.C. Weathers, and M.A. Arthur. 2002. Control of nitrogen loss from forested watersheds by soil carbon:nitrogen ratio and tree species composition. *Ecosystems* 5:712–718.

Malmer, N. and E. Holm. 1984. Variation in the C:N quotient of peat in relation to decomposition rate and age determination with Pb-210. *Oikos* 43:171–182.

McDonald, C.P., N.R. Urban, and C. Casey. 2010. Modeling historical trends in Lake Superior total nitrogen concentrations. *Journal of Great Lake Research* 36(4):715–721.

McKnight, D., E.M. Thurman, R. Wershaw, and H. Hemond. 1985. Biogeochemistry of aquatic humic substances in Thoreau's Bog, Concord, Massachusetts. *Ecology* 66:1339–1352.

Monique, M.P., P.D. Heijmans, F. Berendse, W.J. Arp, A.K. Masselink, H. Klees, W. De Visser, and N. Van Breeman. 2001. Effects of elevated carbon dioxide and increased nitrogen deposition on bog vegetation in the Netherlands. *Journal of Ecology* 89:268–279.

Monteith, D.T., J.L. Stoddard, C.D. Evans, H.A. deWit, M. Forsius, T. Hogasen, A. Wilander et al. 2007. Dissolved organic carbon trends resulting from changes in atmospheric deposition chemistry. *Nature* 450:537–540.

Moore, T.R., N.T. Roulet, and J.M. Waddington. 1998. Uncertainty in predicting the effect of climatic change on the carbon cycling of Canadian peatlands. *Climatic Change* 40:229–245.

Morra, M.J., S.E. Fendorf, and P.D. Brown. 1997. Speciation of sulfur in humic and fulvic acids using X-ray absorption near-edge structure (XANES) spectroscopy. *Geochimica et Cosmochimica Acta* 61:683–688.

Mullen, S.F., J.A. Janssens, and E. Gorham. 2000. Acidity of and the concentrations of major and minor metals in the surface waters of bryophyte assemblages from 20 North American bogs and fens. *Canadian Journal of Botany* 78:718–727.

Neff, J., E. Holland, F. Dentener, W. Mcdowell, and K. Russell. 2002. The origin, composition and rates of organic nitrogen deposition: A missing piece of the nitrogen cycle? *Biogeochemistry* 57: 99–136.

Novak, M. and R.K. Wieder. 1992. Inorganic and organic sulfur profiles in nine *Sphagnum* peat bogs in the United States and Czechoslovakia. *Water, Air, and Soil Pollution* 65:353–369.

Novak, M., R.K. Wieder, and W.R. Schell. 1994. Sulfur during early diagenesis in *Sphagnum* peat: Insights from $\delta^{34}S$ ratio profiles in ^{210}Pb-dated peat cores. *Limnology and Oceanography* 39:1172–1185.

Novak, M., M. Adamova, and J. Milicic. 2003. Sulfur metabolism in polluted *Sphagnum* peat bogs: A combined 34S-35S-210Pb study. *Water, Air, and Soil Pollution: Focus* 3:181–200.

Novak, M., M. Adamova, R.K. Wieder, and S.H. Bottrell. 2005a. Sulfur mobility in peat. *Applied Geochemistry* 20:673–681.

Novak, M., M. Stepanova, I. Jackova, F. Buzek, E. Prechova, M.A. Vile, S.H. Bottrell, and R.J. Newton. 2005b. Isotope systematics of sulfate-oxygen and sulfate-sulfur in six European peatlands. *Biogeochemistry* 76:187–213.

Nykanen, H., H. Vasander, J.T. Huttunen, and P.J. Martikainen. 2002. Effect of experimental nitrogen load on methane and nitrous oxide fluxes on ombrotrophic boreal peatland. *Plant and Soil* 242:147–155.

Oliver, B.G., E.M. Thurman, and R.L. Malcolm. 1983. The contribution of humic substances to the acidity of colored natural waters. *Geochimica et Cosmochimica Acta* 47:2031–2036.

Overton, W., P. Kanciruk, L.A. Hook, J.M. Eilers, D.H. Landers, D.F. Brakke, D.J. Blick, R.A. Linthurst, M.D. Dehaan, and J.M. Omernik. 1986. Characteristics of lakes in the eastern United States. Vol. 2, *Lakes Sampled and Descriptive Statistics for Physical and Chemical Variables*. US Environmental Protection Agency/600/4–86/0076. Washington, DC.

Pellerin, B.A., W.M. Wollheim, C.S. Hopkinson, W.H. McDowell, M.R. Williams, C.J. Vörösmarty, and M.L. Daley. 2004. Role of wetlands and developed land use on dissolved organic nitrogen concentrations and DON/TDN in northeastern U.S. rivers and streams. *Limnology and Oceanography* 49:910–918.

Press, M.C., J. Henderson, and J.A. Lee. 1985. Arylsulphatase activity in peat in relation to atmospheric deposition. *Soil Biology & Biochemistry* 17:99–103.

Prietzel, J., H. Knicker, J. Thieme, and M. Salome. 2007. Sulfur K-edge XANES spectroscopy reveals differences in sulfur speciation of bulk soils, humic acid, fulvic acid, and particle size separates. *Soil Biology & Biochemistry* 39: 877–890

Qian, J., U. Skyllberg, W. Frech, W.F. Bleam, P.R. Bloom, and P.E. Petit. 2002. Bonding of methyl mercury to reduced sulfur groups in soil and stream organic matter as determined by X-ray absorption spectroscopy and binding affinity studies. *Geochimica et Cosmochimica Acta* 66:3873–3885.

Ramaut, J. 1954. Modifications de pH apportees par la tourbe et le *Sphagnum* secs aux solutions salines et a l'eau bidistillee. *Bulletin de la Classe Des Sciences*, Bruxelles, Series 5, 40.

Ravichandran, M. 2004. Interactions between mercury and dissolved organic matter—A review. *Chemosphere* 55:319–331.

Regina, K., H. Nykanen, J. Silvola, and P.J. Martikainen. 1996. Fluxes of nitrous oxide from boreal peatlands as affected by peatland type, water table level and nitrification capacity. *Biogeochemistry* 35:401–418.

Richards, E.H. and J.W. Ellis. 1896. The coloring matter of natural waters, its source, composition and quantitative measurement. *Journal of the American Chemical Society* 18:68–81.

Rogalla, J.A. 1986. Empirical acidification modeling for Lakes in the Upper Great Lakes Region. MS thesis. St. Paul, MN: University of Minnesota.

Schiff, S., R. Aravena, E. Mewhinney, R.J. Elgood, B. Warner, P.J. Dillon, and S.E. Trumbore. 1998. Precambrian Shield wetlands: Hydrologic control of the sources and export of dissolved organic matter. *Climatic Change* 40:167–188.

Schindler, D.W., P.J. Curtis, B.R. Parker, and M.P. Stainton. 1996. Consequences of climate warming and lake acidification for UV-B penetration in North American boreal lakes. *Nature* 379:705–708.

Schindler, D.E., S.R. Carpenter, J.J. Cole, J.F. Kitchell, and M.L. Pace. 1997a. Influence of food web structure on carbon exchange between lakes and the atmosphere. *Science* 277:248–251.

Schindler, D.W., P.J. Curtis, S.E. Bayley, B.R. Parker, K.G. Beaty, and M.P. Stainton. 1997b. Climate-induced changes in the dissolved organic carbon budgets of boreal lakes. *Biogeochemistry* 36:9–28.

Shanley, J.B., N.C. Kamman, T.A. Clair, and A. Chalmers. 2005. Physical controls on total and methylmercury concentrations in streams and lakes of the northeastern USA. *Ecotoxicology* 14:125–134.

Skene, M. 1915. The acidity of *Sphagnum* and its relation to chalk and mineral salts. *Annals of Botany* 29:65–87.

Smemo, K.A. and J.B. Yavitt. 2007. Evidence for anaerobic CH_4 oxidation in freshwater peatlands. *Geomicrobiology Journal* 24:583–597.

Smith, R.E.H., J.A. Furgal, and D.R.S. Lean. 1998. The short-term effects of solar ultraviolet radiation on phytoplankton photosynthesis and photosynthate allocation under contrasting mixing regimes in Lake Ontario. *Journal of Great Lakes Research* 24:427–441.

Thompson, T.G., J.R. Lorah, and G.B. Rigg. 1927. The acidity of the waters of some Puget Sound Bogs. *Journal of the American Chemistry Society* 49:2981–2988.

Timmons, D.R., E.S. Verry, R.E. Burwell, and R.F. Holt. 1977. Nutrient transport in surface runoff and interflow from an aspen-birch forest. *Journal of Environmental Quality* 6:188–192.

Tipping, E., C.A. Backes, and M.A. Hurley. 1988. The complexation of protons, aluminum and calcium by aquatic humic substances: A model incorporating binding-site heterogeneity and macroionic effects. *Water Research* 22:597–611.

Turunen, J., N.T. Roulet, and T.R. Moore. 2004. Nitrogen deposition and increased carbon accumulation in ombrotrophic peatlands in eastern Canada. *Global Biogeochemical Cycles* 18:GB3002.

U.S. Environmental Protection Agency. 2000. National air pollutant emission trends. Washington, DC: US Environmental Protection Agency.

Urban, N.R. 1983. The nitrogen cycle in a forested bog watershed in northern Minnesota. MS thesis. St. Paul, MN: University of Minnesota.

Urban, N.R. 1987. The nature and origins of acidity in bogs. PhD dissertation. St. Paul, MN: University of Minnesota.

Urban, N.R. and S.E. Bayley. 1986. The acid-base balance of peatlands: A short-term perspective. *Water, Air, and Soil Pollution* 30:791–800.

Urban, N.R. and S.J. Eisenreich. 1988. The nitrogen cycle of a *Sphagnum* bog. *Canadian Journal of Botany* 66:435–449.

Urban, N.R., S.J. Eisenreich, and E. Gorham. 1987a. Proton cycling in bogs: Geographic variation in northeastern North America. In: *The Effects of Air Pollutants on Forests, Wetlands and Agricultural Ecosystems*, ed. T.C. Hutchison and K.M. Meema. New York: Springer-Verlag, pp. 577–598.

Urban, N.R., S.J. Eisenreich, and E. Gorham. 1987b. Proton cycling in bogs: Geographic variation in northeastern North America. In: *The Effects of Air Pollutants on Forests, Wetlands and Agricultural Ecosystems*, ed. T.C. Hutchison and K.M. Meema. New York: Springer-Verlag, pp. 577–598.

Urban, N.R., S.E. Bayley, and S.J. Eisenreich. 1989a. Export of dissolved organic carbon and acidity from peatlands. *Water Resources Research* 25:1619–1628.

Urban, N.R., S.J. Eisenreich, and S.E. Bayley. 1989b. The relative importance of denitrification and nitrate assimilation in mid-continental bogs. *Limnology and Oceanography* 33:1611–1617.

Urban, N.R., S.J. Eisenreich, and D.F. Grigal. 1989c. Sulfur cycling in a forested *Sphagnum* bog in northern Minnesota. *Biogeochemistry* 7:81–109.

Urban, N.R., E.S. Verry, and S.J. Eisenreich. 1995. Retention and mobility of cations in a small peatland. *Water, Air, and Soil Pollution* 79:201–224.

Urban, N.R., S.J. Eisenreich, D.F. Grigal, and K.T. Schurr. 1990. Mobility and diagenesis of Pb and Pb-210 in peat. *Geochimica* et *Cosmochimica Acta* 54:3329–3346.

Verry, E.S. 1975. Streamflow chemistry and nutrient yields from upland-peatland watersheds in Minnesota. *Ecology* 56:1149–1157.

Verry, E.S. 1983. Precipitation chemistry at the Marcell Experimental Forest in north central Minnesota. *Water Resources Research* 19:454–462.

Verry, E.S. 1986. The fate of ions in till, granite, and organic soil landscapes in Minnesota. Grand Rapids, MN: USDA Forest Service.

Verry, E.S. and D.R. Timmons. 1977. Precipitation nutrients in the open and under two forests in Minnesota. *Canadian Journal of Forest Research* 7:112–119.

Verry, E.S. and D.R. Timmons. 1982. Water-borne nutrient flow through an upland-peatland watershed in Minnesota. *Ecology* 63:1456–1467.

Verry, E.S. and N.R. Urban. 1992. Nutrient cycling at Marcell Bog, Minnesota. *Suo* 43:147–153.

Verry, E.S. and R.K. Kolka. 2003. Importance of wetlands to streamflow generation. In: *First Interagency Conference on Research in the Watersheds*, ed. K.G. Renard, S.A. McElroy, W.J. Gburek, H.E. Caufield, and R.L. Scott. Benson, AZ: USDA, Agricultural Research Service, 126–132.

Vile, M.A., S.D. Bridgham, and R. Kelman Wieder. 2003. Response of anaerobic carbon mineralization rates to sulfate amendments in a boreal peatland. *Ecological Applications* 13:720–734.

Wieder, R.K. and G.E. Lang. 1986. Fe, Al, Mn, and S chemistry of *Sphagnum* peat in four peatlands with different metal and sulfur input. *Water, Air, and Soil Pollution* 29:309–320.

Wieder, R.L. and G.E. Lang. 1988. Cycling of inorganic and organic sulfur in peat from Big Run Bog, West Virginia. *Biogeochemistry* 5:221–242.

Wieder, R.K., G.E. Lang, and V.A. Granus. 1987. Sulphur transformations in *Sphagnum*-derived peat during incubation. *Soil Biology & Biochemistry* 19:101–106.

Wieder, R.K., J.B. Yavitt, and G.E. Lang. 1990. Methane production and sulfate reduction in two Appalachian peatlands. *Biogeochemistry* 10:81–104.

Willey, J.D., R.J. Kieber, M.S. Eyman, and G.B. Avery. 2000. Rainwater dissolved organic carbon: Concentrations and global flux. *Global Biogeochemical Cycles* 14:139–148.

Worrall, F. and T. Burt. 2004. Time series analysis of long-term river dissolved organic carbon records. *Hydrological Processes* 18:893–911.

Worrall, F., T. Burt, and R. Shedden. 2003. Long term records of riverine dissolved organic matter. *Biogeochemistry* 64:165–178.

Wright, R. 1983. Predicting acidification of North American lakes. Acid Rain Research Series, Norwegian Institute for Water Research.

Xia, K., U.L. Skyllberg, P.R. Bloom, E.A. Nater, W.F. Bleam, and P.A. Helmke. 1999. X-ray absorption spectroscopic evidence for the complexation of Hg(II) by reduced sulfur in soil humic substances. *Environmental Science & Technology* 33:257–261.

Yoon, S.-J., L.M. Diener, W.F. Bleam, P.R. Bloom, and E.A. Nater. 2005. X-ray absorption studies of CH_3Hg^+-binding sites in humic substances. *Geochimica et Cosmochimica Acta* 69:1111–1121.

9

Ecosystem Carbon Storage and Flux in Upland/Peatland Watersheds in Northern Minnesota

David F. Grigal, Peter C. Bates, and Randall K. Kolka

CONTENTS

Introduction

Carbon (C) storage (the amount of C in the system at a given time) and fluxes (inputs and outputs of C per unit time) are central issues in global change. Spatial patterns of C storage on the landscape, both that in soil and in biomass, are important from an inventory perspective and for understanding the biophysical processes that affect C fluxes. Regional (e.g., Grigal and Ohmann 1992; Homann et al. 1995; Johnson and Kern 2003; Kulmatiski et al. 2004; Simmons et al. 1996) and national estimates of C storage (Birdsey and Heath 1995; Birdsey and Lewis 2003; Smith et al. 2006) are uncertain because they are based on a specific subset of sites or on extrapolation of broad-based means. The patterns and processes affecting ecosystem C vary considerably among landscapes, limiting extrapolation across broad climatic–geomorphic regions.

Estimates of C storage for specific watersheds or other forested tracts (e.g., Arrouayes et al. 1998; Bell et al. 2000; Thompson and Kolka 2005) provide metrics to evaluate these broader estimates. The catena concept, firmly established in soil science, embodies the view that, in humid regions, hydrologic processes active on hillslopes lead to considerable differences in soil characteristics. Although the concept is firmly established, quantification of changes in soil C with landscape position, especially in forests, is lacking.

This is especially true in more recently glaciated regions such as the northern Great Lakes States, northern New York, and much of south-central Canada, where climate and a poorly developed drainage network have led to abundant peatlands, occupying 10%–30% of most basins. Quantification of the C stored in these immature landscapes, the pools in which it is stored, and the relationship of that storage to landscape features will aid in predicting the magnitude of C response to various scenarios of global change.

Carbon storage is ultimately the summation of the flux of C into and out of ecosystems. Attention has been focused on detailed measurements of net primary productivity, but less attention has been paid to simultaneous C losses. Numerous data relating to C flux in selected ecosystems have been collected over the past several decades, but those empirical data have not been linked. Such a linkage can both provide an estimate of net C flux and indicate where additional research is needed.

Objectives

The primary objective of this work is to better understand the mechanisms responsible for C sequestration at landscape scales. Specific objectives were to

1. Quantify the magnitude and spatial patterns of ecosystem C storage across landscapes at the USDA Forest Service's Marcell Experimental Forest (MEF) in northern Minnesota. Two approaches were used and the results compared; the *categorical* approach extrapolated published estimates of C storage for similar systems and components, and the *continuous* approach used functional relationships developed from an intensive set of plot measurements.

2. Estimate C flux for two common forest types on the MEF: aspen-birch on uplands and black spruce on peatlands. A conceptually simple model/spreadsheet considered C inputs via plant growth and losses via decomposition using functions based wholly on literature data. Results were compared to measures of C storage at various stand ages from the literature and from the MEF.

Carbon Storage

The *categorical* approach to quantify ecosystem C storage was based on categories of vegetation and soils of MEF from conventional mapping (at scales of 1:15,840 and 1:24,000, respectively). Those estimates were applied to all occurrences of each category on the MEF. The *continuous* approach was based on functional relationships predicting C storage, developed primarily using linear and nonlinear regression with plot measurements, mapping, and landscape attributes, developed from a digital elevation model (DEM) as independent variables. These relationships were then applied to each cell of a GIS database describing the MEF. Although reports of use of the continuous approach to estimate C storage (with different sets of variables than used here) have been optimistic (Mueller and Pierce 2003; Stutter et al. 2009), the categorical approach is much less costly in terms of both time and resources. If results from the two approaches are similar and if the goal is simply estimates of C storage per se, the categorical approach could be applied more widely.

A detailed description of the methods and results of the inventory of ecosystem C at the MEF is available (Grigal 2009). They provide a background for the overview presented here.

Methods

Field

The location and characteristics of the MEF are described elsewhere (Chapter 2). In 1992 and 1993, descriptions and samples of soils, vegetation, and topography were collected at 25 m increments along 20 transects crossing topographic and vegetation boundaries of the MEF. All sample points (596 points) were included in a reconnaissance dataset, and about one-third of those points (219 points) were elected randomly for intensive sampling. In addition, 30 points (including 10 intensively sampled points) were revisited to assess variability in data collection.

The following data were collected at all sample points:

- Topography: slope gradient, aspect, profile curvature (perpendicular to the contour), planar curvature (parallel to contour), and slope position.
- Vegetation: species and dbh (diameter at breast height) for all "in" trees > 2.5 cm dbh using a 10-BAF prism; standing dead trees were also measured and assigned to one of four condition classes (ranging from recently dead to standing bole without branches); percent cover and average height of tall shrubs (SHs); and percent cover of low SHs and forbs.
- Mineral soils (when present): thickness of both forest floor and of the A horizon.
- Organic soils (when present): depth of organic soils (peat) was measured using a McCauley peat auger, and degree of humification was estimated using the von Post method (Malterer et al. 1992; Chapter 5).

At the 219 intensive points, soil samples were collected for laboratory analyses. For mineral soils, samples of the forest floor, a composite sample of the upper 25 cm of mineral soil, and a single sample from the 25–100 cm layer were collected. For peat, samples were collected from the 0–50 cm layer, 50–100 cm layer and from each additional meter thereafter.

The 1992–1993 inventory of the MEF did not include complete sampling of coarse woody debris (CWD); standing dead trees were included but logs (fallen branches and boles) were not. In 2003, logs were inventoried following the methods outlined in Duvall (1997), based in part on techniques described by Van Wagner (1968). Log CWD was sampled at 25 m intervals along 28 transects that roughly followed the same bearings and lengths as the 1992/1993 transects. No attempt was made to relocate sample points from the earlier inventory. Each sample point (n = 546) served as the center of a 10 m line-transect, oriented along the bearing of the overall transect, and logs (≥2.5 cm in diameter at the plane of intersection between the transect and the log) were measured. Stand basal area (BA) was measured with a 10-BAF prism centered at each point. Overall, transect bearings, and hence those of CWD transects, varied across MEF. Because of concern that a single orientation could bias results, about 6% of the points were resampled with transects perpendicular to the original bearing. Paired t-tests indicated no significant differences between log mass from the two sets of data.

Laboratory

All samples of forest floor, mineral soil, and peat were kept cool in the field and sent to the laboratory within 48 h. Upon receipt, mineral and peat samples were frozen until further processing. The moist mass of the forest-floor samples was determined, a subsample was removed to determine water content, and the remainder was frozen. After thawing, subsamples of all materials were analyzed for loss on ignition (LOI) by ashing at 450°C for 12 h. Total C was determined on about 20% of the samples using a LECO CR-12 analyzer (David 1988).

Geospatial Data

A raster GIS database with 10 m cells was developed for MEF, including a DEM, vegetation cover type, and soils from an order II soil survey. The DEM was created from an existing topographic map with 1.2 m elevation contours derived from a ground survey. Cover type was interpreted from 1:15,840 color-infrared airphotos, dated May 1990. Twelve forested types (red pine, jack pine, balsam fir, white spruce, black spruce, tamarack, northern white-cedar, northern hardwoods, lowland hardwoods, black ash, aspen, and birch) and 10 miscellaneous types (upland grass, upland brush, lowland grass, lowland brush, beaver pond, marsh, muskeg, permanent water, stagnant spruce, and stagnant tamarack) were recognized. The forested types were assigned to one of three dbh size classes. Forested types in the smallest class were further characterized by the percent canopy cover and in the two larger classes by the estimated merchantable volume (in cords per acre). Soil-map unit delineations were digitized from a 1:24,000 Natural Resources Conservation Service (NRCS) order II soil survey for Itasca County, MN (Nyberg 1987). Sixteen soil-mapping units were recognized.

Numerical and Statistical

Categorical C Estimates

Vegetation Type Estimates of aboveground biomass were derived for each cover type. Estimates for the larger dbh classes of the forested types were based on extrapolation of the mapped volume to total biomass by vegetation type, using data from Grigal and Bates (1992). Biomass estimates for the smallest class of the aspen-birch forest type were primarily based on Perala (1972), with comparisons for reasonableness from Silkworth (1980) and Alban and Perala (1992). For the smallest diameter class of the conifer forest types, including pine, spruce-fir, lowland conifers, and black spruce, biomass was estimated with data from Berry (1987) and Methven (1983). Biomass for hardwoods other than aspen-birch was the geometric mean of those for aspen-birch and conifers. Biomass estimates for the non-forest cover types were based on a variety of studies in Minnesota (Bell et al. 1996; Connolly-McCarthy and Grigal 1985; Perala 1972; Swanson and Grigal 1991).

Belowground biomass estimates were simply a ratio of the aboveground estimates. For forested types, the root:shoot was 0.3 (Alban and Perala 1994; Santantonio et al. 1977; Whittaker and Marks 1975). For shrub types (upland brush, lowland brush, and muskeg), it was 1.3 (Alban and Perala 1994; Grigal et al. 1985; Johnston et al. 1996). For the herbaceous types (upland grass, lowland grass, and marsh), the ratio was 2.0 (Abrahamson 1979; Ashmun et al. 1985; Grigal et al. 1985; Gross et al. 1983; Nihlgard 1972; Paavilainen 1980; Remezov and Pogrebnak 1969). In all cases, living biomass was converted to C using a ratio of 1:0.475 (Raich et al. 1991).

Soil-Mapping Unit Soil C was estimated for each soil-map unit. Because tabulated data in the published soil survey report (Nyberg 1987) are relatively general, more detailed data from 73 pedons, representing 16 taxonomic units, were used. Data were primarily collected from the soil characterization database of the NRCS (Soil Survey Staff 1997) and from the University of Minnesota Department of Soil, Water, and Climate. Other sources included Alban and Perala (1990), Balogh (1983), Grigal et al. (1974), and Kolka (1993). Organic C was computed for both the upper 25 cm and the upper meter of these pedons and then extrapolated to the soil-mapping units based on the taxonomic composition of each unit (Nyberg 1987).

Continuous C Estimates

Laboratory   The LOI and C data for forest floor, mineral soil, and peat from MEF and similar data gathered from another site in Minnesota in a companion study (Bell et al. 1996, 2000) were combined, and simple linear regressions were developed for each sample type relating LOI (%) and C (%).

Mineral soil bulk density (D_b) was estimated from LOI using the approach of Federer et al. (1993), where soil D_b is a function of the proportion of organic material in the soil and the D_b of the organic (or C) and mineral fractions. The relationship was developed from the data collected in a transect of forested sites across the north-central United States (Ohmann and Grigal 1991). For materials high in organic matter, such as forest floor and peat, soil D_b is simply a function of the organic fraction and its D_b (Gosselink et al. 1984). Using the measured forest-floor thickness, LOI, and mass from the intensively sampled points, the D_b of the organic material was determined using nonlinear least-squares regression (SYSTAT Inc. 2007). The D_b of the organic material in peat was similarly determined using an extensive database of peat LOI and D_b collected in MEF and other peatlands in northern Minnesota (Buttleman 1982).

Vegetation   Overstory (tree) biomass was based on the tree dbh collected at each point. For aboveground biomass, we used estimation equations from Alemdag (1983, 1984). Those equations use both dbh and total height as independent variables, so a relationship between dbh and height for trees on the MEF was developed. For belowground biomass, an estimation equation was based on data from Santantonio et al. (1977), Whittaker and Marks (1975), and Alban and Perala (1994).

Although understory strata (below the canopy) are a relatively minor proportion of total stand biomass in closed forest (about 5%; Ohmann and Grigal 1985a; Swanson and Grigal 1991), they may constitute the majority of mass as canopies become more open. Biomass estimation equations for understory strata were developed using data from two comprehensive studies from northern Minnesota, one of uplands (211 stands by Ohmann and Grigal 1985b) and the other of peatlands (235 stands by Swanson, 1988). Because these studies included only data from one mesic hardwood (i.e., northern hardwood) stand, data for seven mesic hardwood stands from east-central Minnesota were added (Suhartoyo 1991). The data were aggregated into three understory groups (ground cover, forbs, and shrubs) and an overstory group (trees) and further grouped into 10 vegetation types (aspen-birch [AB], black spruce [BS], lowland conifers [LC], lowland hardwoods [LH], pine [PI], spruce-fir [SF], upland hardwoods [UH], open bog [OB], shrub [SH], and grass [GR]), based on the majority of overstory biomass (Ohmann and Grigal 1985b) or on the physiognomic group (Swanson 1988).

Vegetation types for points with no or minimal tree cover were based on field notes. For example, residual trees in recently cut areas may not be indicative of the original or regenerating type. Field notes helped assign the appropriate type in those cases and in the shrub, open bog, and grass types.

Understory biomass was estimated by linear regression as a function of the biomass of the conifer and the deciduous overstories, with vegetation type as a dummy variable. Cover of forbs and the cover and height of the tall SHs were used to adjust the estimates.

Aboveground biomass of standing dead trees (snags) was calculated using the same functions as those for live trees, and the resulting mass was adjusted for the decay class. Log volumes were calculated (Van Wagner 1968), and biomass and C content was estimated by using tabulations of density and C concentration as a function of decay class (Duvall 1997). Log CWD for each point in the original (1992/1993) inventory was estimated by extrapolation of the data collected in 2003. Points were placed into vegetation groups, and a regression equation was developed using the continuous variables from each point and with group as a dummy variable.

Soil   Forest floor had been quantitatively sampled and LOI determined at each intensive point. Mass per unit area was calculated and converted to C per unit area using the equation relating C to LOI. Extrapolation of the forest-floor C (FFC) mass from the intensive to the reconnaissance points was based on the collected data. The measured aspect was transformed to more closely reflect biological importance, with a maximum of 1 at 225° (SW) and a minimum of 0 at 45° (NE) (Beers et al. 1966). A randomly chosen sub-sample of 10% of the data from intensive points (15 of the 152 mineral soil points) was used to evaluate the uncertainty of the final estimates.

Determination of C mass of mineral soils requires C concentration, D_b, and horizon thickness. For the intensive points, LOI data were used to estimate C concentration and D_b, and C mass was calculated for the 0–25 cm and the 26–100 cm soil layers. As with forest floor, those data were extrapolated to the reconnaissance points using the same subset of data but including thickness of forest floor and A-horizon as additional independent variables. A subsample of 10% of the data (the same points used for evaluating forest-floor uncertainty) was used to evaluate the uncertainty of the final estimates.

For peat, both C concentration and D_b were expressed as functions of LOI, and C mass was computed as a direct function of LOI and thickness. Depth had been measured at both intensive and reconnaissance points where peat occurred, and C mass was calculated for all those points.

Landscape Attributes   The C data from the inventory and landscape attri-butes were used to estimate C over the entire landscape. The landscape attri-butes were calculated from the DEM for the 10×10 m cell associated with each inventory point using Arc/Info. Primary attributes included elevation, aspect, flow accumulation, plan/profile curvature, and slope steepness, and secondary attributes included both compound topographic index and stream power index. An aspect code was calculated from aspect (Beers et al. 1966).

The C data for each inventory point, both intensive and reconnaissance, and their map attributes (as categorical variables) were merged with the landscape attributes to create a single database (Table 9.1). Some categorical

TABLE 9.1

Data for Each Inventory Point, Landscape Attributes Calculated from 10 m DEM Interpolation of 1.2 m Contour Data for the MEF

Attribute	Description
POINT	Transect point
NORTHING	Location
EASTING	Location
ELEV	Elevation (m)
ASP	Aspect direction (0°–360°)
CTI	Compound topographic index
PLCURV	Curvature measured in the plan direction
PROCURV	Curvature measured in the profile direction
SCA	Specific catchment area (also known as flow accumulation)
SPI	Stream power index
SLOPE	Slope (degrees)
LEGEND_NUM	Numerical code for combination of overstory, size, and density
OVER_TYPE	Overstory type number
OVER_SIZE	Overstory size
OVER_DENS	Overstory density
SPECIES	Species group or cover type—alpha
MUSYM	Soil-mapping unit symbol
LVL	Level of field sampling; 1 = reconnaissance, 2 = intensive
FFC	Forest floor C
SURFC	Mineral C (0–25 cm)
SUMINC	Mineral C (0–100 cm)
SOILC	Sum FFC + SUMINC
PEATDPTH	Peat depth (cm)
PEATC	Peat C to mineral substrate
PEAT100	Peat C to 100 cm
FF MEAS	Forest floor mass was determined (1) or estimated (0)
MIN MEAS	Mineral soil mass was determined (1) or estimated (0)
COVTYP	Cover type based on measured basal area
C/D	Conifer or deciduous basal area majority?
AGTREC	Aboveground tree C
ROOTC	Belowground tree C
SHRBAGC	Estimated shrub aboveground C
FORBAGC	Estimated forb aboveground C
GRNDC	Estimated ground layer aboveground C
AGVEGC	Sum AGTREC + SHRBAGC + FORBAGC + GRNDC
LOGC	Estimated log CWD C
SNAGC	Standing dead tree C
CWDC	Sum LOGC + SNAGC
ASPCOD	Computed aspect code
LNCTI	Natural log (CTI + 2)

(*continued*)

TABLE 9.1 (continued)

Data for Each Inventory Point, Landscape Attributes Calculated from 10 m
DEM Interpolation of 1.2 m Contour Data for the MEF

Attribute	Description
LNSCA	Natural log SCA
LNSPI	Natural log (SPI + 2)
SOILCODE	Coalescence of soil-mapping units into four categories
FORTYP	Coalescence of overstory type into 10 categories
OVSZCOD	Coalescence of overstory size into four categories
OVDNCOD	Coalescence of overstory density into four categories

variables were further grouped. Soil-mapping units were combined into four
groups based on a combination of physiography and associated surface soil
texture. Cover types were combined into 10 groups based on predominance
of overstory biomass as described previously. An additional group, coded as
0 for nonforest types, was added to overstory size; overstory density, based
on estimated volume, was combined into four groups, including 0 for non-
forest types (Table 9.1).

Because of the continuous vegetation cover at the boundaries of MEF, some
inventory points had been established inadvertently outside MEF and had no
associated landscape attributes. Those points were eliminated from the analyses, leaving 541 points with a full set of data. Fifty points were chosen randomly to serve as a check, and 491 points were used to develop estimation
equations.

Analysis Extensive screening was conducted to determine the best overall
set of independent variables to use to estimate C. This screening included
analyses of variance (ANOVA) to determine the relative influence of the categorical variables (soil and cover type) on C, stepwise regression to determine
the influence of the continuous variables (landscape attributes), and scat-
terplots. Based on the results, data were separated into peatland and non-
peatland to estimate soil-related C data and into forested and nonforested to
estimate vegetation-related C.

Although peatland occurrence should presumably be established by the soil
mapping, that occurrence has uncertainty through simple errors or lack of
sufficient map resolution (minimum map unit size). Similarly, there is uncertainty in plot locations as determined by GPS. Even without error, these uncertainties can amount to 7–15 m (National Imagery and Mapping Agency, http://
www.geocomm.com/channel/gps/news/nimagps2/, accessed February 19,
2010). As a result, mineral-soil data were collected from some plots whose
coordinates indicated that they were located on peatland, and peat data were
collected on sites mapped as upland. Logistic regression, a mix of categorical
and continuous variables used to predict probabilities of a binary response
variable, was used to predict the probability that any point was likely to be

peatland (i.e., quantify the uncertainty). Elevation, a continuous variable, and cover type and soil-mapping unit, and categorical variables (Table 9.1) were used as independent variables.

A similar situation occurred with vegetation mapping, and a similar rationale led to the use of logistic regression to estimate the probability that a data point was forested (i.e., had a forest canopy). The criterion for presence or absence of a canopy was approximately $5 \, m^2 \, ha^{-1}$ BA of living trees, equivalent to about $7.5 \, Mg \, ha^{-1}$ of aboveground tree C. Regression was based on the categorical variables overstory dbh class, overstory density class, and soil group.

The final estimation equations for C in vegetation-related ecosystems components—aboveground tree, root, aboveground shrub, aboveground forb, aboveground ground cover, sum of all aboveground vegetation, down log CWD, standing dead CWD, and sum of all CWD—were based on a common set of independent variables; for some dependent variables, the data were divided into nonforested and forested categories. The estimation equations for soil-related C used a different but common set of independent variables following the separation of points into peatland (in which case, peat C to mineral substrate was calculated) or nonpeatland (with calculation of FFC, surface [0–25 cm] mineral soil C, total mineral soil C [0–100 cm], and sum of forest floor plus mineral soil C to 100 cm). Belowground C of SHs was based on a root:shoot = 1.3 and of forbs = 2.0 (sources as described earlier).

The final equations were applied to each cell of the 10 m GIS database (approximately 100,000 cells). Statistical tests were conducted to determine the significance of differences in C estimates among mapped vegetation types and among soil-mapping units.

Results

Categorical C Estimates

Vegetation Type

Total aboveground tree biomass of closed-canopy forested types ranged from about $500–1000 \, kg \, m^{-3}$ of merchantable volume. No estimates of understory biomass were made in closed-canopy types. Total aboveground biomass estimates for the forested size class 1 (<12.5 cm dbh) ranged from 2 to $80 \, Mg \, ha^{-1}$ and from 7 to $35 \, Mg \, ha^{-1}$ for the nonforest vegetation types. After conversion of biomass to C, the results indicated an average of about $51 \, Mg \, ha^{-1}$ of C stored in the vegetation of MEF (Table 9.2). This result has an undefined uncertainty.

Soil-Mapping Unit

As described, data from 73 pedons representing 16 taxonomic units were used to estimate C content of the soil-mapping units on the MEF. Where D_b for a horizon was missing, it was computed using estimates of the mineral (D_{bm}) and the C fractions (D_{bC}) from the remainder of the database

TABLE 9.2

Estimates of Biomass C for Cover Types Mapped on the MEF Based on Categorical Analysis

Cover Type	Area (ha)	C (Mg)	C (Mg ha^{-1})
Aspen	488	32,765	67.1
Balsam fir	3	193	56.0
Birch	26	917	35.3
Black ash	3	51	17.8
Black spruce	76	1,445	19.1
Jack pine	24	1,638	68.3
Lowland brush	17	346	20.2
Lowland grass	2	23	13.5
Lowland hardwoods	18	1,279	69.1
Marsh	18	179	10.0
Muskeg	59	609	10.4
Northern hardwoods	64	5,475	86.1
Northern white-cedar	7	185	27.9
Permanent water	51		
Red pine	36	1,047	29.5
Stagnant spruce	10	194	19.1
Stagnant tamarack	7	151	20.3
Tamarack	28	390	13.9
Upland grass	3	39	13.5
White spruce	26	142	5.5
Sum	966	47,069	51.5[a]

Note: Data include both aboveground and belowground C.
[a] Mass per area-weighted mean, not including area of permanent water.

(Federer et al. 1993). Nonlinear least squares yielded $D_{bm} = 1.73\,Mg\;m^{-3}$ and $D_{bC} = 0.100\,Mg\;m^{-3}$ ($n = 99$, $R^2 = 0.98$, $s_{y \cdot x} = 0.11\,Mg\;m^{-3}$). For organic soils (C > 15%), $D_{bC} = 0.07\,Mg\;m^{-3}$ ($n = 46$, standard error of the mean, SE = 0.004 Mg m^{-3}). Where C data were missing for some subsurface mineral horizons, a general relationship of C to depth was developed using data from Ohmann and Grigal (1991). Mass of organic C was computed for both the upper 25 cm and the upper meter of the pedons and thence for the soil-mapping units (Table 9.3). Uncertainty can be roughly quantified by the pooled SE of estimates for units represented by three or more pedons ($n = 12$). Based on the SEs (13% for the 0–25 cm layer and 11% for the 0–100 cm layer 68 degrees of freedom, d.f.), Fisher's least significant difference (LSD) among units is about 24%.

The estimated C for mapping units tended to be bimodal, as expected in a landscape with a mix of mineral soils and peat (Histosols). Mineral soils generally had less than 50 Mg ha^{-1} C in the surface 25 cm while Histosols had greater than 150 Mg ha^{-1} (Table 9.3). Similarly, mineral soils tended to have about 100 Mg ha^{-1} C in the upper meter while Histosols had greater than

TABLE 9.3

Organic Carbon Content of the Soil-Mapping Units of the MEF Based on Detailed Soil Characterization Data for Individual Taxonomic Units within Mapping Units

Soil-Mapping Unit	Area (ha)	Organic Carbon 0–25 cm (Mg[a])	Organic Carbon 0–100 cm (Mg[a])	Organic Carbon 0–25 cm (Mg ha^{-1})	Organic Carbon 0–100 cm (Mg ha^{-1})
Borosaprists, depressional	51	10,095	38,477	198.9	758.0
Cathro muck	7	1,252	5,295	173.3	733.1
Greenwood peat	48	8,321	29,158	173.5	607.9
Loxely peat	4	1,290	3,338	358.1	926.8
Menahga and Graycalm soils	135	4,193	8,954	31.1	66.4
Menahga loamy sand	71	2,269	5,467	31.9	76.8
Mooselake and Lupton mucky peats	35	5,809	25,505	164.7	722.9
Nashwauk fine sandy loam	159	7,924	17,991	50.1	113.6
Nashwauk–Menahga complex, 1%–10%	45	2,214	5,203	48.9	115.0
Nashwauk–Menahga complex, 10%–25%	96	4,397	9,831	46.1	103.1
Sago and Roscommon soils	8	1,248	3,009	149.9	361.3
Seelyeville–Bowstring association	42	7,996	30,484	188.7	719.2
Warba fine sandy loam, 1%–8%	153	4,576	12,346	30.0	81.0
Warba fine sandy loam, 10%–25%	54	1,644	4,418	30.4	81.7
Sum	964	63,228	199,475	69.7[b]	220.0[b]

[a] Total C mass in each unit.

[b] Mass per area-weighted mean, not including area of permanent water.

600 Mg ha^{-1} in that depth (Table 9.3). A minor exception to this generality was the Sago and Roscommon mapping unit, where the major taxonomic components were Histic and Mollic Aquepts, respectively. These wet mineral soils with surface organic accumulations grouped with the Histosols in the surface 25 cm, but with lower organic matter in deeper horizons fell between mineral and organic soils at 100 cm (Table 9.3).

When the data were extrapolated to the entire landscape of MEF, soil C, even in the 0–25 cm layer (Table 9.3), was higher than that of any vegetation strata, both above- and belowground (Table 9.2). When deeper soil layers were considered, differences were even greater. This demonstrates the importance of soils, especially, peat, in influencing landscape C storage.

Continuous C Estimates

Vegetation

A wide range of species and sizes of trees were used for the height estimation equation. Based on linearized least squares, and including a correction for bias (Beauchamp and Olson 1973), the best-fit relationship was

$$HT = 2.03 * dbh^{0.71}, \quad n = 173, \quad r^2 = 0.99, \tag{9.1}$$

where height (HT) is in m and dbh is in cm. The expression relating root mass to tree dbh, an average of the literature sources, was

$$ROOT = 0.031 * dbh^{2.39}, \tag{9.2}$$

where ROOT is root biomass in kg and dbh is in cm. Goodness-of-fit statistics are not relevant for Equation 9.2 because it is simply an average of other relationships.

The understory estimation equations, where biomass of the understory was a function of the conifer and the deciduous overstory biomass, with vegetation type as a dummy variable, were significant (herbs, $R^2 = 0.67$, $s_{y \cdot x} = 575$ kg ha^{-1}; SHs, $R^2 = 0.32$, $s_{y \cdot x} = 2200$ kg ha^{-1}; ground cover, $R^2 = 0.50$, $s_{y \cdot x} = 1550$ kg ha^{-1}; $n = 436$ for all cases).

There was no need to extrapolate vegetation C estimates from intensive to reconnaissance points, because tree data had been collected on all points. The data indicated a mean biomass C, both above- and belowground, of about 51 Mg ha^{-1} (Table 9.4). This is virtually identical to the areally weighted mean based on the vegetation cover-type map (Table 9.2). The agreement between the results of the two methods increases confidence in those results.

Soil

Laboratory/Statistical   Simple linear regressions were developed for forest floor, mineral soil, and peat where x = LOI (%) and y = C (%). Because the y-intercept was only marginally significant for forest floor (p = .043) and not significantly different than 0 for peat (p = .175), both regressions were rerun forcing the intercept through 0. The final results were

$$Min\ C\ (\%) = 0.50 * LOI\ (\%) - 0.15, \quad n = 229, \quad r^2 = 0.95, \quad s_{y \cdot x} = 0.13, \tag{9.3}$$

TABLE 9.4

Estimates of Biomass C for All Sampled Points on the MEF.
Biomass Includes Both Aboveground and Belowground Trees,
Shrubs, Forbs, and Ground Cover

Vegetation Type	Number of Observations	C, Mean (Mg ha^{-1})	C, Standard Error (Mg ha^{-1})
Aspen-birch	415	53.33	2.03
Black spruce	29	40.68	3.38
Grass	14	6.05	0.21
Lowland conifers	23	35.37	5.61
Lowland hardwoods	22	52.40	7.74
Open bog	16	6.14	0.16
Pine	20	90.21	5.90
Spruce-fir	18	48.67	5.97
Shrub	2	9.72	0.00
Upland hardwoods	36	54.24	5.02
All	595	50.61[a]	1.62

[a] Mean of all observations.

$$FF\ C\ (\%) = 0.484 * LOI\ (\%),\ n = 129,\ r^2 = 0.84,\ s_{y \cdot x} = 3.5, \qquad (9.4)$$

$$Peat\ C\ (\%) = 0.55 * LOI\ (\%),\ n = 82,\ r^2 = 0.91,\ s_{y \cdot x} = 3.2. \qquad (9.5)$$

These relationships are similar to others reported for the north-central United States. For the same LOI, however, relationships spanning the North Central States yield higher estimates for both forest floor and mineral soil (about 8% and 0.2% C, respectively, or about 25% of the observed mean) (David 1988). LOI (equivalent to organic matter) apparently is lower in C at MEF than at other locations in the region.

Data collected in a transect of forested sites across the north-central United States (Ohmann and Grigal 1991) yielded D_{bo} of 0.11 Mg m^{-3} and D_{bm} of 1.63 Mg m^{-3} (R^2 between observed and predicted $D_b = 0.70$, n =172). In coarse-textured soils of New England, D_{bo} of 0.10 Mg m^{-3} and D_{bm} ranged from 1.45 to 2.19 Mg m^{-3} (Federer et al. 1993), and, in Quebec, D_{bo} of 0.12 Mg m^{-3} and D_{bm} of 1.40 Mg m^{-3} (Tremblay et al. 2002). For forest floor, D_{bo} of 0.052 Mg m^{-3}, with R^2 between observed and predicted forest-floor mass of 0.46, n of 164. This compares favorably to estimates by Gosselink et al. (1984) (D_{bo} of 0.054 Mg m^{-3} after conversion of C to organic matter). In part, the relatively low-R^2 for D_{bo} of forest floor arose because it was a function of both forest-floor density and thickness and variations in both contributed to uncertainty. Within-plot variation in thickness was high, with a pooled coefficient of variation (CV) of about 42% based on three measurements per point on 164 points.

The estimate of D_{bo} for surface peat (<50 cm depth) was 0.069 Mg m^{-3} with an R^2 of 0.11 for a sample size of 157. For subsurface peat (>50 cm depth), the estimate of D_{bo} was 0.092 Mg m^{-3} with an R^2 of 0.23 for a sample size of 104. The D_{bo} for surface and subsurface peat were significantly different and reflect increasing density of subsurface peats due to compaction, as also noted by Grigal et al. (1989). High variability in the relationship between LOI and peat D_b is commonly observed. For example, Nichols and Boelter (1984) found an r^2 of 0.17 for the relationship between ash (equivalent to 1 − LOI) and D_b for 176 samples of peat from 38 peatlands from the north-central United States.

Forest Floor   Based on the quantitative sampling and the LOI and C data, the mean C mass of forest floor on the intensively sampled points was 5.3 Mg ha^{-1} (Table 9.5). In the preliminary analysis of the extrapolation of the FFC mass from the intensive to the reconnaissance points, planar curvature (prob. = 0.97), position on slope (prob. = 0.22), and deciduous tree BA (prob. = 0.16) were not significant predictor variables. The remaining variables were significant, and the final regression had an R^2 of 0.64 (Table 9.5). The correlation between the observed and predicted C for the subsample used to evaluate uncertainty was r^2 = 0.30 (Table 9.5). A paired t-test indicated no significant difference between the observed and predicted FFC of that subsample (prob. = 0.45, n = 15).

TABLE 9.5

C Mass of Soil Variables from Intensive Points and Basis of Extrapolation to Reconnaissance Points Sampled on the MEF

Descriptor	Units	Forest Floor	Mineral C, 0–25 cm	Mineral C, 0–100 cm	Peat[a]
All points					
n		152	161	161	113
Mean	Mg ha^{-1}	5.31	34.36	115.23	962.60
Standard deviation	Mg ha^{-1}	3.75	9.62	35.99	751.57
Basis of equations					
n		137	146	146	
Mean	Mg ha^{-1}	5.27	34.56	116.70	
Standard deviation	Mg ha^{-1}	3.64	9.58	36.16	
r^2		0.64	0.42	0.29	
$s_{y \cdot x}$	Mg ha^{-1}	2.23	7.41	31.12	
Check of equations					
n		15	15	15	
Mean	Mg ha^{-1}	5.72	32.43	100.94	
Standard deviation	Mg ha^{-1}	4.80	10.22	31.93	
r^2		0.30	0.60	0.45	
$s_{y \cdot x}$	Mg ha^{-1}	4.17	6.68	24.55	

[a] No extrapolation was made for peat; all points were sampled to mineral soil.

Based on the extrapolation, forest-floor mass increased with an increase in conifer tree BA and with a decrease in slope gradient. Less expected, warmer aspects (toward the southwest) had greater forest-floor mass; the greatest mass was found where profile curvature was convex, less where flat, and least where it was concave. Although some of these relationships are contrary to accepted paradigms, they may have rational explanations. For example, the decline of forest floor-mass as the profile curvature becomes more concave may be associated with increased incorporation of C into the surface mineral soil at those positions.

Mineral Soil   Measured mineral soil C to 100 cm from the intensive points was 115 Mg C ha^{-1} (Table 9.5). The extrapolation of the mineral soil C mass from the intensive to the reconnaissance points was based on the same variables used for extrapolation of FFC. In the preliminary analysis, profile curvature (prob. = 0.25 for surface soil and prob. = 0.51 for 0–100 cm), planar curvature (prob. = 0.16 for surface and prob. = 0.11 for 0–100 cm), deciduous tree BA (prob. = 0.47 for surface and prob. = 0.92 for 0–100 cm), aspect code (prob. = 0.69 for surface and prob. = 0.72 for 0–100 cm), and slope gradient (prob. = 0.11 for surface and prob. = 0.82 for 0–100 cm) were not significant predictor variables. The remaining variables were significant, and the final regression for surface soil had an R^2 of 0.42 (Table 9.5). For the subsample used to evaluate uncertainty, the correlation between the observed and predicted C was $r^2 = 0.60$ (Table 9.5); a paired t-test indicated no significant difference between the observed and predicted C (prob. = 0.51, n = 15). For the sum of mineral C to 100 cm, the final regression had an R^2 of 0.29 (Table 9.5), the correlation between the observed and predicted C for the subsample was $r^2 = 0.45$ (Table 9.5), and a paired t-test again indicated no significant difference between the subsample observed and predicted C (prob. = 0.15, n = 15).

Mineral C mass increased with A thickness and was lowest in upper slope positions, increasing downslope. Mineral-soil C mass also increased with forest-floor thickness and decreased with increase in conifer-tree BA. The increase with forest-floor thickness is interesting because there was no relationship between forest-floor thickness and A thickness ($r^2 = 0.01$). The decrease in mineral C mass with increase in conifer BA may be related to a general tendency for conifers to occur on coarser-textured soils than broadleaf deciduous trees.

Peat   As described, C content was a function of LOI and thickness. Carbon mass of surface peat (<50 cm depth) was 3.80 Mg ha^{-1} cm^{-1} and of subsurface peat (>50 cm depth) was 5.06 Mg ha^{-1} cm^{-1}. Because depth had been determined at all points where peat occurred, both intensive and reconnaissance, there was no need to extrapolate data (Table 9.5).

CWD

Points from the CWD inventory were aggregated into one of nine groups (CWD groups) based on the plurality of live-tree BA. These groups were nearly identical to the 10 vegetation types used to develop estimation equations for understory biomass, but the open CWD group (O) included the grass and open bog types; neither had live-tree BA. Log biomass was statistically significant among the groups [$F(8537) = 3.689$, prob. < 0.001]. Although differences were significant, mean separation tests indicated considerable overlap. In addition, relatively high-log mass in the open and shrub groups indicated that many of those points had been recently forested.

A subset of 25 points was removed randomly from the data to test the uncertainty of the final equation used to extrapolate measured log CWD to the points in the original inventory. The final equation for that extrapolation used CWD group as a dummy variable and a set of continuous variables including standing dead-tree BA and density, total and deciduous live-tree BA, live-tree density, and average live-tree diameter. The resulting equation ($R^2 = 0.21$, $n = 521$, $s_{y \cdot x} = 17.4 \, \mathrm{Mg \, ha^{-1}}$) had relatively high uncertainty. Points with high observed mass were especially underpredicted because of the weighting of the equation by a large number of points with minimal CWD. To accommodate some of that underprediction, a simple quadratic equation with 0 intercept was fitted with predicted CWD as the independent variable and observed CWD as the dependent variable. The relationship was not a major improvement ($R^2 = 0.24$, $n = 521$, $s_{y \cdot x} = 16.9 \, \mathrm{Mg \, ha^{-1}}$), but it reduced some of the error in the high estimates and explained 33% of the variation in the 25 observation test dataset.

The resulting estimates of log C mass for the original MEF inventory points, when summarized by vegetation type, ranged from 5 to 20 Mg ha^{-1} (Table 9.6). Those estimates are also similar to the measured C from the 2003 inventory summarized by CWD group (Figure 9.1).

Landscape Attributes

Peatland Probability   The parameters in the logistic model for predicting the probability that a point was peatland (elevation, presence of a wetland vegetation type, and presence of an organic soil-mapping unit) were significant (prob. < 0.03), and the model was significantly different than using a constant to predict peat occurrence (Chi-square $= 198$, 3 d.f., and prob. < 0.001). A point was considered to be peatland if the predicted probability was >0.5. Cohen's kappa (Rosenfield and Fitzpatrick-Lins 1986) was used to measure agreement between the observed and predicted occurrences. Values range from 0 (when agreement is no better than chance) to 1.0 (when agreement is perfect), with >0.75, indicating strong and <0.40 indicating poor agreement (SYSTAT Inc. 2007). Kappa was 0.64 for the data used to develop the model and 0.56 for the 50 observation test dataset.

Forested Probability   The parameters in the logistic model predicting probability of a point being forested (overstory dbh class, overstory density class,

TABLE 9.6

Estimated C Mass of CWD from Inventory Points on Marcell Experimental Forest

Vegetation Type	Number of Observations	Snag, Mean (Mg ha⁻¹)	Snag, Standard Error (Mg ha⁻¹)	Log, Mean (Mg ha⁻¹)	Log, Standard Error (Mg ha⁻¹)
Aspen-birch	415	10.3	0.6	12.2	0.5
Black spruce	29	11.7	2.4	9.3	2.7
Grass	14	0.7	0.4	6.4	0.1
Lowland conifers	23	6.9	1.9	5.2	1.4
Lowland hardwoods	22	12.2	3.6	19.0	3.5
Open bog	16	1.5	1.1	7.4	0.7
Pine	20	15.0	2.7	10.7	2.1
Spruce-fir	18	13.4	3.3	10.4	2.4
Shrub	2	0.0	0.0	6.4	0.0
Upland hardwoods	36	13.4	1.6	11.9	1.5
ALL	595	10.3[a]	0.5	11.6[a]	0.4

[a] Mean of all observations.

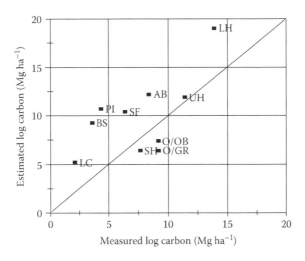

FIGURE 9.1

Mean measured C mass of logs by CWD group (x axis), based on a 2003 survey of 546 plots, compared to mean estimated C mass by vegetation type (y axis) based on application of estimation equations to data from 595 other plots. Vegetation groups/types are AB, aspen-birch; BS, black spruce; LC, lowland conifers; LH, lowland hardwoods; O/GR, open/grass; O/OB, open/open bog; PI, pine; SF, spruce-fir; SH, shrubs; and UH, upland hardwoods. The generally good relationship between the two variables is indicated by the 1:1 line.

and soil group) were significant (prob. < 0.06), and the model was significantly different than no prediction (Chi-square = 147, 9 d.f., prob. < 0.001). Kappa for the data used for the model was 0.51 and for the 50 observation test dataset was 0.35.

Estimation Equations   The C data from the inventory, used to extrapolate ecosystem C over the entire MEF, were highly variable (Table 9.7). For the 443 nonpeat points, the CV was nearly 80% for FFC and about 25% for mineral soil C (SUMINC). For the 98 peat points, the CV for peat C to mineral substrate (PEATC) was similar to that for FFC, about 80%. Vegetation-related C was similarly highly variable. For all sample points (n = 541, with n = 443 for mineral soil and n = 98 for peat), the CV was nearly 100% for CWD and 80% for vegetation C (Table 9.7).

A limited number of variables was used in the final estimation equations (equations in Grigal 2009, at http://www.nrs.fs.fed.us/ef/marcell/pubs/proceedings/accessed August 13, 2009). These included the categorical variables from the vegetation and soil mapping, and the landscape variables of slope, aspect (coded from NE to SW), elevation, specific catchment area (the number of cells contributing flow to a specific cell), and compound topographic index (wetness index, a function of both slope, and catchment area). In general, the relationships used to predict soil-related C did not have high-explanatory power (Table 9.7), explaining only about 15% of the variation in both FFC and mineral-soil C, and about 38% of that in peat C (n = 401 for mineral points and n = 90 for peat points). The relationships only explained about 7% of the variation in both FFC and mineral-soil C for the check data, but about 55% of that in peat C (n = 42 for mineral points, n = 8 for peat points) (Table 9.7). It is clear that soil C is not easily predictable over the MEF despite the large suite of variables we had measured (Table 9.1).

The continuous relationships had a higher explanatory power for vegetation-related C than for soil C (Table 9.7). Relationships explained about 61% of the variation in aboveground tree C and root C, 24% of the variation in aboveground shrub C, 43% in aboveground forb C, and 51% in ground layer C (n = 491). Using the check data, the relationships explained about 65% of the variation in aboveground-tree C and root C, 16% of the variation in aboveground-shrub C, 37% in aboveground-forb C, and 23% in ground-layer C. The sum of aboveground-vegetation C had 58% of the variation explained in the estimation dataset and 66% in the check data (Table 9.7). Vegetation-related C is better predicted than is soil C. As with soil C, CWD C generally was not well predicted by the estimation equations. Relationships explained about 28% of the variation in snag C, 13% of the variation in log C, and 24% of the variation in their sum (n = 491; Table 9.7). Using the check data, the relationships explained about 37% of the variation in snag C, less than 1% of the variation in log C, and 14% of the variation in their sum (Table 9.7).

TABLE 9.7

Descriptive Statistics and Details of Estimation Equations

C Pool	All Points			Basis of Equations					Check of Equations				
	n	Mean (Mg ha⁻¹)	Standard Deviation (Mg ha⁻¹)	n	Mean (Mg ha⁻¹)	Standard Deviation (Mg ha⁻¹)	R^2	$s_{y \cdot x}$	n	Mean (Mg ha⁻¹)	Standard Deviation (Mg ha⁻¹)	R^2	$s_{y \cdot x}$
Forest floor	443	4.00	3.19	401	4.08	3.27	0.146	3.10	42	3.24	2.24	0.069	2.18
Mineral, 0–25 cm	443	34.10	10.55	401	34.35	10.85	0.146	10.29	42	31.78	6.72	0.060	6.60
Mineral, 0–100 cm	443	112.52	28.96	401	112.79	29.48	0.174	27.47	42	109.89	23.65	0.067	50.44
Forest floor + 100 cm mineral	443	116.52	29.13	401	116.87	29.66	0.161	27.65	42	113.13	23.76	0.039	50.49
Peat	98	966.60	785.39	90	977.50	804.70	0.379	709.74	8	843.80	544.80	0.549	395.36
Aboveground tree	541	38.43	33.05	491	39.19	32.93	0.613	20.50	50	31.01	33.60	0.660	19.80
Belowground tree	541	9.57	7.56	491	9.74	7.47	0.606	4.70	50	7.96	8.27	0.653	4.92
Aboveground shrub	541	1.03	0.38	491	1.04	0.38	0.242	0.34	50	1.00	0.39	0.163	0.36
Aboveground forb	541	0.34	0.24	491	0.33	0.23	0.429	0.18	50	0.39	0.30	0.374	0.24
Aboveground ground layer	541	0.64	0.53	491	0.64	0.54	0.505	0.39	50	0.65	0.42	0.230	0.37
Sum vegetation aboveground	541	40.45	32.66	491	41.20	32.54	0.612	20.30	50	33.03	33.26	0.660	19.60
Log CWD	541	11.75	9.83	491	11.75	9.91	0.127	9.46	50	11.79	9.13	0.006	9.19
Standing dead tree	541	10.46	11.64	491	10.64	11.84	0.267	10.14	50	8.75	9.47	0.367	7.62
Log + standing dead	541	22.21	20.65	491	22.39	20.98	0.213	18.64	50	20.54	17.17	0.144	16.05

Application to MEF

The estimation equations were applied to the approximately 100,000 10 m cells in the MEF GIS database. ANOVA invariably indicated that differences in C among categories of soils or vegetation exceeded those arising by chance, due in part to the large number of cells in each class. The significance of the differences among estimates was evaluated by Fisher's LSD (at prob. = 0.05). Nearly, all the vegetation types differed from one another in live-vegetation C, with highest C in the pine type (Figure 9.2). This is consistent with the categorical estimates of vegetation C (Table 9.2). Most of the pine types on the MEF are plantations, and their uniform stocking apparently leads to high vegetation C. Ironically, the lowest vegetation C also was in conifer plantations in the upland spruce-fir type (Figure 9.2). Nearly, all these plantations were young when sampled with associated low tree C. However, the soil was the largest contributor to ecosystem C differences among vegetation types (Figure 9.2). Types on peat, including lowland conifers and hardwoods, open, SHs, and black spruce, all had high C (Figure 9.2).

When viewed at a landscape scale, the distribution of vegetation C at the MEF shows expected spatial variation (Figure 9.3). However, this variation

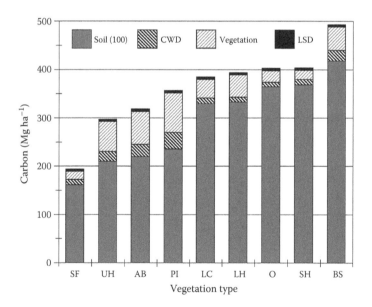

FIGURE 9.2
Estimates of ecosystem C mass for MEF, categorized by vegetation type. Types are AB, aspen-birch; BS, black spruce; LC, lowland conifers; LH, lowland hardwoods; O, open; PI, pine; SF, spruce-fir; SH, shrubs; and UH, upland hardwoods. "Soil (100)" includes sum of forest floor and mineral soil or organic soil (peat) to 100 cm depth, "Vegetation" includes above- and belowground living vegetation, and "CWD" includes both snags and logs. Fisher's least significant difference (prob. = 0.05), based on sum of components, indicated (LSD). With the exception of young SF plantations, the major difference is between peatland (LC, LH, O, SH, and BS) and upland ecosystems.

Carbon mass—North and South Units, Marcell Experimental Forest

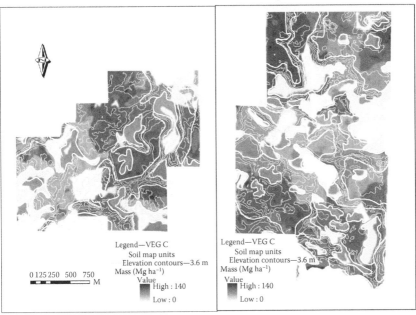

FIGURE 9.3
Landscape distribution of above- and belowground vegetation C at MEF. Distribution based on estimation equations using mapped and landscape variables that were applied to individual cells of a GIS database with 10 m resolution. Left panel, North Unit; right panel, South Unit. Note the continuous gradation from low-C open peatlands and recently cut areas to fully stocked forests. Scale from 0 to 140 Mg ha⁻¹.

becomes a simple nuance when the C content of the entire ecosystem, either to a depth of 100 cm or to the entire peat depth, is considered. The resulting map is nearly "black and white," with much higher C in peatland ecosystems than in upland systems (Figure 9.4); contrast is even greater when the entire depth of peat is included. As shown here and in other studies, peatlands are important C reservoirs in northern landscapes (Bell et al. 2000; Gorham 1991; Johnston et al. 1996).

Comparisons

With Other Studies

As described earlier, the categorical and continuous C estimates for vegetation (mean for total aboveground and belowground) are similar (51 Mg ha⁻¹; Tables 9.2 and 9.4, respectively). Because these means include nonforested types, they are lower than reports for forests. As a point of comparison, the continuous estimate for aspen-birch, the dominant cover type on MEF (53 Mg ha⁻¹; Table 9.4), can be compared to other estimates for aspen-birch, including 67 Mg ha⁻¹ (Smith et al. 2006), 69 Mg ha⁻¹ (Grigal and Ohmann

Carbon mass–North and South Units, Marcell Experimental Forest

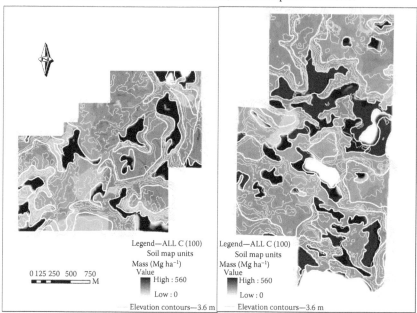

FIGURE 9.4

Landscape distribution of ecosystem C, including vegetation, CWD, and soil to 100 cm depth at MEF. Distribution based on estimation equations using mapped and landscape variables that were applied to individual cells of a GIS database with 10 m resolution. Left panel, North Unit; right panel, South Unit. Note the sharp contrast between high C peatlands and uplands. Scale from 0 to 560 Mg ha^{-1}.

1992), 80 Mg ha^{-1} (Weishampel et al. 2009), and an average of 51 Mg ha^{-1} for a chronosequence and 81 Mg ha^{-1} for a 63-year-old stand (Ruark and Bockheim 1988). The MEF estimate included recently harvested stands with low C. The estimate for pine on MEF, 90 Mg ha^{-1}, falls between other reports for pine in the Northern Lake States, 65 Mg ha^{-1} (Grigal and Ohmann 1992) and 107 Mg ha^{-1} (Smith et al. 2006).

The estimates of CWD, about 22 Mg C ha^{-1} evenly divided between snags and logs (Table 9.6), fall within the broad range of literature values. On the basis of the data from 778 plots, Chojnacky et al. (2004) estimated log CWD in the North Central States at 7.3 Mg ha^{-1}; C yield tables estimate a CWD mass of 19.6 Mg ha^{-1} for upland forest types in the Northern Lake States (Smith et al. 2006). Another study on MEF estimated 9.3 Mg C ha^{-1} CWD in aspen-birch and 23.1 Mg ha^{-1} in upland conifers (Weishampel et al. 2009). Other studies have similar estimates of about 9 Mg C ha^{-1} in aspen-birch stands (Alban and Perala 1992; Gower et al. 1997; Ruark and Bockheim 1988).

The FFC (slightly more than 5 Mg ha^{-1}) (Table 9.5) is lower than other reports from similar systems. For example, Smith and Heath (2002) indicated an average FFC mass of about 20 Mg ha^{-1} for northern upland forests, including aspen-birch, pine, and northern hardwoods. They also reported a C:mass of 0.38. At MEF, the C:mass was 0.33 (SE = 0.006 and n = 152); the MEF forest floor is less C-rich than average northern forests. Grigal and Ohmann (1992), reporting on five upland-forest cover-types across the Great Lakes States, found average FFC mass of about 16 Mg ha^{-1}. Although average LOI was higher in the MEF data (68%) than in their dataset (55%), the ratio of C:LOI was lower for the MEF data (0.48 versus 0.61). This led to slightly less C per unit mass of forest floor at MEF than in the broader dataset. The most important reason for the difference in C mass between the two datasets is that the average forest-floor thickness across the Great Lakes States was about 40 mm (Ohmann and Grigal 1991) compared to 17 mm for the MEF.

There are other reports of lower FFC mass for northern forests. The average FFC mass at the MEF is slightly lower than those for forested sites from a similar study in east-central Minnesota (8.0 Mg ha^{-1}, Bell et al. 1996) and from a study using a different sampling design at the MEF (8.2 Mg ha^{-1}, Weishampel et al. 2009) and is similar to that for a chronosequence of aspen forests in northern Wisconsin (4.8 Mg ha^{-1}, Ruark and Bockheim 1988).

A comparison of the estimates of mineral soil C at the MEF with data from the literature is constrained by differences in depth of reporting, though a common denominator for many reports is C to 100 cm. The mean C content of the soil orders that are areally most important on the MEF, Alfisols and Entisols (85 Mg ha^{-1}, based on the categorical approach), is similar to that based on large databases (56 Mg ha^{-1}, Guo et al. 2006; 70 Mg ha^{-1}, Kern 1994; 87 Mg ha^{-1}, Johnson and Kern 2003). The mean of all intensive sampling points on mineral soils (115 Mg ha^{-1}; Table 9.5) is near that estimated for all cells of the 10 m GIS database that fell on Alfisols or Entisols mapping units (110 Mg ha^{-1}). This is consistent with the mean from five upland forest cover types across the Great Lakes States (106 Mg ha^{-1}, Grigal and Ohmann 1992), for Entisols and Inceptisols in hardwood forests in Rhode Island (144 Mg ha^{-1}, Davis et al. 2004) or north-central New York (135 Mg ha^{-1}, Galbraith et al. 2003), and Alfisols, Entisols, and Inceptisols in Danish forest soils (105 Mg ha^{-1}, Vejre et al. 2003). These values are considerably higher than mineral soil C for forested sites from a similar study in east-central Minnesota (46 Mg ha^{-1}, Bell et al. 1996), but those soils were almost exclusively sandy Entisols. In general, our estimates fall well within the range of other estimates from the literature.

The C mass in peat at the MEF (about 960 Mg ha^{-1}, the mean of sampling points falling on peat; Table 9.5) is similar to that for Histosols from large databases (975 Mg ha^{-1}, Guo et al. 2006; 843 Mg ha^{-1}, Kern 1994; 832 Mg ha^{-1}, Johnson and Kern 2003). However, the estimate for the MEF is for mass from the surface to contact with the mineral substrate, which averaged about 200 cm, whereas the estimates from large databases were to a 100 cm depth. At the MEF, peat to 100 cm contained approximately 445 Mg C ha^{-1}, less than that from a companion study in east-central Minnesota (630 Mg ha^{-1} to 100 cm, Bell et al. 1996) and lower than the mean of the organic soil-mapping units on the MEF based on the categorical estimate (745 Mg ha^{-1} to 100 cm; Table 9.3). Average D$_b$ and C content of peat from an extensive review of the literature (Gorham 1991) yields an estimate of 580 Mg ha^{-1} to 100 cm. It appears that the peatlands at MEF contain relatively less decomposed peat than other sites in the literature, resulting in a combination of lower C concentration and D$_b$, both of which result in lower C mass per unit depth or area. The lower estimate for the MEF is similar to that from a comprehensive study of 10 northern Minnesota peatlands, evenly divided between bogs and fens, with organic material ranging from hemic to fibric (Grigal and Nord 1983). The average C mass in the surface 100 cm of those peatlands was 470 Mg ha^{-1}, with Fisher's LSD based on within-peatland variance of 68 Mg ha^{-1}. The continuous estimate for MEF peatlands is not different from this average, though the categorical estimate is higher. Uncertainty in the C content of peatlands in the MEF is important because of their dominance in the overall C inventory.

Between the Two Estimates

A relevant question is whether the machinations described here were worth it. In other words, did the data collection and statistical manipulations that produced the continuous estimates of C, as described here, provide better estimates than the much less intensive categorical mapping and application of existing data?

The categorical estimates of C were about 15% greater than those using continuous variables. The areally weighted estimates of vegetation C were nearly identical; the difference between the two approaches was due to the difference in soil C (to 100 cm), primarily related to the estimates of peat C. The categorical estimate of peat C (745 Mg ha^{-1} to 100 cm) was more than 1.5 times the continuous estimate (445 Mg C ha^{-1}). Within-stand variation reported or computed from studies of mature aspen-birch and black spruce stands can be used to calculate Fisher's LSD and evaluate the significance of the difference in the estimates from the two approaches. For vegetation C, the difference between estimates from the two approaches (0.9 Mg ha^{-1}) is much less than the LSD (51 Mg ha^{-1}). Even in the case of soil C, the difference between estimates is less than the LSD (42 Mg ha^{-1} versus 68 Mg ha^{-1}, respectively). Statistically, the two approaches to estimating ecosystem C do not differ.

It appears that landscape-level estimates of ecosystem C might be economically feasible using literature data and vegetation and soil mapping, which exists for most systems in the north-central United States. Any ecosystem sampling should be concentrated on peatlands because of both their large C mass and the variation in that mass as reported in the literature. This comparison via overall averages and LSDs does not speak to landscape patterns of C. No attempt was made in this study to develop landscape patterns of C mass based on the categorical approach, but a spatial integration of the vegetation and soil mapping almost certainly would yield patterns similar to those from the continuous approach.

Carbon Flux

Quantification of the C storage at the MEF is one step in understanding the C budget of its ecosystems, but C storage is the summation of C fluxes into and out of ecosystems. A simple spreadsheet/model (Carbon FLuX—CFLX) was developed to describe C fluxes for two common forest types on MEF: aspen-birch on uplands and black spruce on peatlands (70% of the inventory points were assigned to the aspen-birch type and 5% to black spruce). CFLX annually computes the size of major C pools by tracking the annual cohort of each pool over the simulation period. Some pools change as a function of time (stand development) and others as a function of inputs and outputs. CFLX is simply an accounting tool for even-aged stands; it does not attempt to mechanistically simulate growth processes, interactions among pools, effects of climate, or other factors important to forest development. Although it is a spreadsheet, CFLX is coded in FORTRAN. It was not calibrated for a specific stand or site; rather, it generically describes average stand behavior. The functional forms are embedded in the coding, but CFLX was designed, so that the constants can be easily altered to produce alternative scenarios or application to specific sites or conditions.

Major Pools and Central Relationships

The major C pools considered by CFLX differed between the two forest types (Table 9.8). Carbon flux was estimated by quantifying a number of central

TABLE 9.8

Major C Pools Used to Estimate C Flux in the Aspen-Birch and Black Spruce Forest Types at the MEF

C Pool	Aspen-Birch	Black Spruce
Overstory foliage	X	X
Overstory wood	X	X
Fine woody debris	X	X
Coarse woody debris	X	X
Understory herbs	X	
Tall shrubs	X	
Tree seedlings	X	X
Forest floor	X	
Mineral soil to 50 cm	X	
Low shrubs		X
Moss		X
Roots		X
Peat accretion		X
Residual peat		X

relationships using data from the literature. The data sources and the general approaches used to define key relationships are described in Appendix 9.1 at the end of this chapter. In nearly all cases, nonlinear regression was used to estimate constants (SYSTAT Inc. 2007).

Basal Area versus Time

The fundamental basis of CFLX is a logistic function describing BA change over time,

$$BA = a/\big(1+\exp(b*(T-c))\big), \tag{9.6}$$

where
  BA is stand basal area (BA) in $m^2\,ha^{-1}$
  T is stand age in years
  a, b, and c are constants

The constant a is the maximum BA at infinite time, b is the rate at which the function approaches maximum BA (a), and c can be considered a lag term related to time to establishment of a new stand. Constants were derived from a set of empirical yield tables for Minnesota. Results indicated that a tended to increase with site quality, b nominally decreased (becoming more negative, but not a strong trend), and c decreased (Tables 9.9 and 9.10). The increase in the asymptote with site quality is logical, and the decrease in the lag may be related to a quicker occupancy of a site following disturbance. More negative values of b are associated with more rapid rise toward near asymptotic values.

Biomass versus Basal Area

The relationship between biomass (or C) and BA was expressed as the power function

$$BIOM = a*BA^b, \tag{9.7}$$

where
  BIOM is stand aboveground biomass in $kg\,ha^{-1}$
  BA is stand basal area (BA) in $m^2\,ha^{-1}$
  a and b are constants

Constants were estimated using data from about 450 aspen-birch stands (Table 9.9) and more than 100 black spruce stands (Table 9.10).

Diameter versus Time

The change in average stand diameter over time also is described by the logistic relationship (Equation 9.6), with diameter as the dependent variable.

TABLE 9.9

Functional Relationships Used in the Aspen-Birch Version of CFLX

Dependent Variable	Independent Variable	Function	Constant, a	Constant, b or k	Constant, c or p	R^2	n	S_{yx}	Notes
Basal area	Time	Logistic	24.962	−0.055	14.206		5,019		Mean of six site classes
Biomass	Basal area	Power	2,185.073	1.223		0.765	447	33,198.042	
Diameter	Time	Logistic	22.232	−0.064	31.305		920		Mean of six site classes
ln Herb mass	Ln overstory mass	Linearized power	12.899	−0.580		0.515	348	0.788	Plus a dummy variable for type
ln Shrub mass	Ln overstory mass	Linearized power	9.688	−0.240		0.243	313	1.077	Plus a dummy variable for type
Herb mass	Overstory mass	Power	270,410	−0.580					Exponentiation of ln herb
Shrub mass	Overstory mass	Power	27,351	−0.240					Exponentiation of ln shrub
Litterfall	Basal area/latitude	Modified power	80,984	0.443	−1.246	0.458	33	523.000	With latitude
Litterfall	Basal area	Power	661.758	0.443					At 47.3° N
Tall shrub density	Age	Neg. exponential		−0.202	0.950		34		
Herb decay rate	Time	Mod. neg. exponential	1	−1.811	1	0.996	4	0.047	
Leaf decay rate	Time	Mod. neg. exponential	1	−0.492	0.722	0.906	20	0.060	
Fine wood decay rate	Time	Mod. neg. exponential	1	−0.208	1				Duvall and Grigal (1999)
Residual forest floor decay rate	Time	Mod. neg. exponential	1	−0.078	1				Alban and Perala (1982)
Standing snags	Time	Neg. exponential	1	−0.145			62		
Snag decay rate	Time	Mod. neg. exponential	1	−0.037		0.829	3	0.114	
Log decay	Time	Neg. exponential	1	−0.070		0.984	6	0.058	

TABLE 9.10

Functional Relationships Used in the Black Spruce Version of CFLX

Dependent Variable	Independent Variable	Function	Constant, a	Constant, b or k	Constant, c or p	r²	n	S_yx	Notes
Basal area	Time	Logistic	22.254	-0.056	21.243	0.980	651	20.929	
Biomass	Basal area	Power	3,795.365	1.015		0.644	117	30,083.731	
Diameter	Time	Logistic	18	-0.020	60				
Moss mass	Overstory mass	Power	13,473	-0.082		0.622	6	394.757	
Low shrub plus herb mass	Overstory mass	Power	203,041	-0.409		0.645	6	1,047.291	
Low shrub mass	Overstory mass	Power	283,755	-0.447		0.643	6	1,098.919	
Tree seedling mass	Overstory mass	Power	263,747	-0.667		0.812	6	108.097	
Litterfall	Basal area/latitude	Modified power	1.036E+12	1.826	-6.736	0.927	18	476.398	With latitude
Litterfall	Basal area	Power	5.417	1.826					At 47.3° N
Accumulating peat decay	Time	Mod. neg. exponential	1	-0.286	0.723	0.765	28	0.057	
Decay peat 0–25 cm	Time	Mod. neg. exponential	1	-0.0005000	1				
Decay peat 25–50 cm	Time	Mod. neg. exponential	1	-0.0001000	1				
Decay peat 50–100 cm	Time	Mod. neg. exponential	1	-0.0000500	1				
Decay peat 100–200 cm	Time	Mod. neg. exponential	1	-0.0000100	1				
Decay peat 200–300 cm	Time	Mod. neg. exponential	1	-0.0000010	1				
Standing snags	Time	Neg. exponential	1	-0.106					
Snag decay rate	Time	Mod. neg. exponential	1	-0.010	1	0.899	3	0.062	
Log decay	Time	Mod. neg. exponential	1	-0.071	0.774	0.953	6	0.073	

For aspen-birch, the constants tended to change less with site class than was the case with BA. Although average diameter increased with site quality, the asymptotes a and c changed little, and there was a small increase in b (less negative) (Table 9.9). For black spruce, constants resulted in a slowly increasing diameter to about 15 cm at 150 years (Table 9.10).

Stand Density versus Time

The change in stand density over time, as trees per hectare, is important to stand dynamics; however, no separate function was developed to describe it. Stand density is computed within CFLX from changes in BA and stand diameter with time.

Understory Mass

Mass of understory strata was based on the nonlinear function

$$UBIOM = a * BIOM^b,$$ (9.8)

where
   UBIOM is aboveground biomass for a vegetation stratum in kg ha^{-1}
   BIOM is aboveground tree biomass in kg ha^{-1}
   a and b are constants

The relatively low-explanatory power for aspen-birch (Table 9.9) is tolerable, because understory C pools are relatively small; less than 2% of aboveground biomass for stands in the database. Over half of the variation in black spruce understory strata was explained (Table 9.10).

Substrate

A major difference between the application of CFLX to aspen-birch compared to black spruce is the treatment of the substrate. The substrate for aspen-birch is the forest floor and mineral soil, and for black spruce it is peat. In both cases, substrate C over time is the sum of initial conditions and inputs (positive) and losses (negative). The forest floor (aspen-birch) or peat (black spruce) existing before the simulation begins is carried as subpools without inputs, only losses, whereas forest floor or peat produced during the simulation has both inputs and losses. Although CFLX includes a C pool for mineral soil, the size of that pool does not change.

Inputs

Inputs to forest floor or accumulating peat include litterfall from overstory and SHs, annual herb turnover, and coarse and fine woody debris. Inputs to peat also include moss growth and root turnover and mortality.

Overstory and Shrub Litterfall

Although overstory foliar mass can be estimated by biomass estimation equations, crown size and thus foliage mass are affected by stand density (Grigal and Kernik 1984). Because of the centrality of BA change over time in CFLX, estimates of overstory litterfall were based on a relationship with stand BA,

$$LIT = a * BA^b, \qquad (9.9)$$

where
LIT is litterfall in kg ha^{-1}
BA is stand BA in m^2 ha^{-1}
a and b are constants

The relationship for aspen-birch indicates that overstory litterfall increases with BA at a much slower rate (b = 0.44) than total overstory biomass (b = 1.22) (Table 9.9), reflecting the proportionally decreasing leaf mass with stand development. Conversely, black spruce overstory litterfall increases at a more rapid rate with BA (b = 1.83) than total overstory biomass (b = 1.02) (Table 9.10). Because black spruce stocking often is low on ombrotrophic peatlands, as BA increases, the trees may develop proportionally larger crowns with greater litterfall.

For both aspen-birch and black spruce, litterfall from SHs and tree seedlings was based on foliage:total biomass and foliage turnover (Equation 9.8; Table 9.11). Herb biomass (C) (from Equation 9.8) also was added to the substrate annually.

TABLE 9.11

Ratios Used in Black Spruce Version of CFLX to Convert among Biomass Components

Component	Ratio	Multiplier
Low shrub leaves	0.3	Low shrub biomass
Low shrub litter	0.5	Low shrub leaves
Herb litter	1	Herb biomass
Seedling needles	0.5	Seedling biomass
Seedling litter	0.15	Seedling needles
Low shrub turnover	0.4	Low shrub wood mass
Seedling turnover	0.2	Seedling wood mass
Moss production	1.750	Mg C ha^{-1} year^{-1}
Fine root turnover	1.75	Litterfall
Tree root mortality	0.3	Aboveground mortality
Seedling root mortality	0.3	Aboveground mortality
Low shrub root mortality	1.85	Aboveground mortality

Fine Woody Debris

The fine woody inputs to the substrate have three sources: overstory litter-fall and mortality of SHs and young trees. The proportion of woody twigs in aspen-birch litterfall is relatively uniform (20.2%, SE = 1.5%; Grigal and Homann 1994; Grigal and McColl 1975; Van Cleve and Noonan 1975) and similar to that in black spruce litterfall (17%, SE = 1.9%; Grigal et al. 1985). Therefore, annual inputs of fine woody debris via litterfall were considered to be those proportions.

For aspen-birch, estimates of additions to forest floor by the woody frac-tion of shrub mortality were based on shrub demography as described by a negative exponential function

$$STEMS = a * exp(-b * A), \tag{9.10}$$

where
STEMS is tall shrub density in stems per hectare
A is shrub age in years
a and b are constants (Balogh and Grigal 1987) (Table 9.9)

In the case of black spruce stands, data from a detailed study of low shrub and tree seedling mortality in black spruce (Grigal et al. 1985) were used to estimate additions to peat (Table 9.11).

Moss and Roots

Growth and accumulation of moss, predominantly *Sphagnum*, is a significant part of the C cycle in peatland black spruce forests, and growth was assumed to be 1.75 Mg C ha^{-1} year^{-1} (Table 9.11). Root turnover and mortality are also major inputs to accumulating peat (Table 9.11).

CWD

Tree mortality is the source of CWD. In CFLX, mortality is assumed to be chronic, not episodic, and is described by changes in stand BA and tree diameter with time (i.e., self-thinning). Mortality first creates standing dead trees (snags; Tables 9.9 and 9.10), and those snags subsequently transfer to the ground surface (fall) based on a simple negative exponential (Gore et al. 1985),

$$Y = exp\ (-k * T), \tag{9.11}$$

where Y is the proportion of dead snags remaining at time T in years (Tables 9.9 and 9.10).

Losses

After materials reach the substrate, decomposition (mass loss) begins. The classic functional expression of loss of biomass (or C) is a negative exponential (Equation 9.11; Wieder and Lang 1982), where, in this case, Y is the decimal fraction of mass remaining, T is time in years, and k is a constant. In CFLX, that expression is modified as

$$Y = \exp (k * T^p), \qquad (9.12)$$

where
   variables are as in Equation 9.11
   p is an additional constant (Kelly and Beauchamp 1987)

When $p = 1$, this form is identical to the classic expression (Equation 9.11); alternatively, rate of mass loss increases ($p > 1$) or decreases ($p < 1$) with time. Most decomposition data in the literature simply report k from Equation 9.11, but Equation 9.12 is used in CFLX to allow the functional form to conform more closely to some observations.

Litter Losses

In the aspen-birch version of CFLX, decomposition of each kind of input (herbs, foliar litterfall, and fine woody debris) is considered separately and their losses summed. In the black spruce version, all inputs are aggregated and treated by a single decomposition function for accumulating peat. In both cases, decomposition of preexisting forest floor or peat has unique functions. Herb-mass loss is relatively rapid ($k = -1.8$; Equation 9.11; Table 9.9). Rates of mass loss of overstory leaves decline with time ($k = -0.49$ and $p = 0.72$; Equation 9.12), consistent with an increase in the recalcitrance of the remaining material (Table 9.9). Decomposition of fine woody debris is even slower ($k = -0.208$, with $p = 1$; Equation 9.12).

   Measures of decomposition rates in peat are rare in the literature, and most are not easily adapted to either Equation 9.11 or 9.12. Data-based rates for accumulating peat were near those for fine woody debris ($k = -0.286$ and $p = 0.72$; Equation 9.12). For residual peat, low rates decreasing with depth were set (Table 9.10). In preliminary runs of CFLX, these low rates resulted in minimal C loss.

CWD Losses

The mass loss of standing dead trees (snags) is assumed to follow Equation 9.12 (Tables 9.9 and 9.10). When a snag reaches the ground (becomes a log), it continues to lose mass but the constants in Equation 9.12 change. The rate of mass loss of logs should be more rapid than that of snags because of log contact with the moist ground (Tables 9.9 and 9.10).

Initial Conditions

Initial conditions of the system must be specified, including the size of the major C pools and the stand age in years. Initial conditions may be at stand initiation following a harvest or fire or may be in mid to late rotation. The model initially was constructed on the basis of biomass because the literature contains more data on biomass and organic matter than on C. Outputs were modified by conversion from biomass or organic matter to C (biomass, including litterfall, C=48%, Alban and Perala 1990 and Raich et al. 1991; fine woody debris and residual forest floor, C=50%, Alban and Perala 1982; snags, C=50%, Alban and Pastor 1993; CWD, C=54%, Duvall and Grigal 1999; 0–25 cm and 25–50 cm peat, C=50%, 50–100 cm peat, C=49%, and 100–200 cm and 200–300 cm peat, C=48%, data from the MEF).

Notes/Caveats

There are important caveats associated with CFLX. First, it is simply an accounting tool; it does not attempt to mechanistically simulate forest stand development. Second, the fundamental drivers of CFLX are the temporal changes in BA and diameter described by the logistic function, which increases with time. As even-aged stands surpass maturity, evidence indicates that mortality increases and BA and biomass decrease, but those changes are not included in CFLX. It is designed to represent systems to the time that stands begin to decline. CFLX deals with even-aged stands and does not incorporate succession to another forest type.

Another caveat is associated with the uncertainty of the constants in the model. Each is derived from or based on one or a few studies. Changes in the parameters, through improved data or professional judgment, may better represent reality. For example, change in BA with time was based on midpoint ages from the empirical yield tables (e.g., midpoint from 10 to 20 years was 15 years). Although, for most age classes, this may be reasonable for the 0–10 year age class (midpoint = 5 years), a better assumption may have been the geometric mean or the final year of the age class. CFLX was designed to allow easy changes in the constants for virtually all functions.

Evaluation

The functions and constants used in CFLX were derived from the literature, albeit from a multitude of studies separated in time and space. After their derivation, the constants *were not* manipulated to achieve a better fit to any existing data. The selection of constants for rates of decomposition of deep peat required some preliminary runs of CFLX, but no other constants were manipulated. Iterative runs of CFLX, with adjustment of constants, are likely to better match observed data, but those iterations were not conducted. CFLX was evaluated by comparing its estimates with observations of C pools during early stand development (for aspen-birch) and with observations from mature stands (for both aspen-birch and spruce).

Initial Conditions

The aspen-birch version of CFLX was run using initial conditions from two studies that followed aspen stand development after whole-tree harvesting (Alban and Perala 1990; Silkworth and Grigal 1982). Average values for forest floor, CWD, and soil for the growing season following harvest were used as inputs. Neither study measured standing snags, but they were likely to be nearly absent after the harvest. CWD log data were available only from Alban and Perala (1990). Initial stand BA was set to $0\,m^2\,ha^{-1}$, and the simulation was carried for 75 years. In the case of black spruce, there were no measures of recently disturbed stands, so initial conditions were set arbitrarily with no trees or CWD and with a generic peat mass (~$550\,Mg\,ha^{-1}$ C to 100 cm).

Early Stand Development

To evaluate CFLX with respect to early stand development, the C pools from the model were compared to those reported by Alban and Perala (1990) and Silkworth (1980). Alban and Perala (1990) graphically presented changes in C pools of biomass, forest floor, CWD, and soil to 50 cm for three sites for up to 8 years following whole-tree harvest, and the data were extracted from those graphs. Silkworth (1980) reported biomass in trees, SHs, and herbs; forest-floor organic matter; and soil properties, including horizon depths, bulk densities, gravel content, and C and N concentrations for three sites at 2, 3, and 5 years after whole-tree harvest. From those data, C in biomass, forest floor, and soil to 50 cm was computed.

Estimates for years 2 through 8 from CFLX and the data from Alban and Perala (1990) and Silkworth (1980) were compared using ANOVA, where individual studies were considered factors, and year after harvest was a covariate. Results indicated that the size of C pools differed among studies, with the significance of differences in soil (prob. = 0.056) and CWD (prob. = 0.097) less than those for biomass (prob. = 0.001) and forest floor (prob. = 0.012). Year-to year differences were not significant for forest floor or for soil. Although C pools differed among studies, the estimates from CFLX were not uniformly higher or lower than observations (Figure 9.5). Baysian least significant difference (BLSD) (Smith 1978) indicated no differences between the estimates from CFLX and the observations from one or the other study. In other words, the estimates were within the range of the observations. Although modeled biomass C was at the high end of the data, a small change (decrease) in c

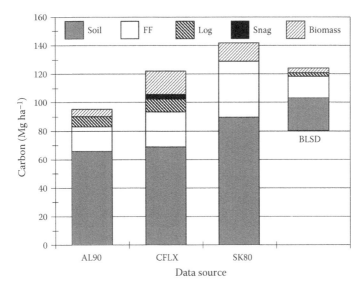

FIGURE 9.5
Comparison of ecosystem C estimates from model/spreadsheet for period of first 10 years following aspen harvest with literature measurements (AL90, Alban and Perala 1990; CFLX, model/spreadsheet; and SK80, Silkworth 1980). Based on Baysian least significant difference (BLSD), the estimates did not differ from the range of measurements.

in Equation 9.6 would lead to slower initial BA and biomass accretion and would more closely match the observations.

Mature Stands

The estimated C pools for mature aspen-birch stands from CFLX were compared to observations from a variety of detailed assessments of C pools reported by Alban et al. (1991), Ohmann et al. (1994), Alban and Perala (1990), Perala et al. (1995), Silkworth (1980), and data from the MEF. All data were from mature aspen stands, though in some cases, actual ages were not reported. Sample sizes varied widely, from 3 (Alban and Perala 1990) to 415 (MEF data). The mean and variation of each dataset were compared to the estimates from CFLX at 60 years with initial conditions as described previously.

For black spruce, the estimated C pools from CFLX were compared to data collected and/or reported by Grigal et al. (1985), Moore et al. (2002), Swanson (1988), Weishampel et al. (2009), the MEF, and in the case of peat, on five Histosols from the NRCS. Data were primarily from peatlands dominated by black spruce except those from Moore et al. (2002), which were not forested, and those from the NRCS, where vegetation type was not reported. Sample sizes varied from 1 (Weishampel et al. 2009) to 63 (Swanson 1988). The mean and variation of each dataset were compared with the estimates from CFLX at 110 years with initial conditions as described earlier.

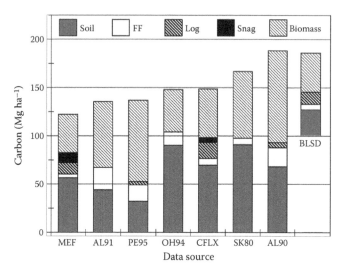

FIGURE 9.6

Comparison of ecosystem C estimates from model/spreadsheet for mature aspen stands with literature measurements (CFLX, model/spreadsheet; AL91, Alban et al. 1991; OH94, Ohmann et al. 1994; SK80, Silkworth 1980; MEF, inventory of Marcell Experimental Forest; AL90, Alban and Perala 1990; and PE95, Perala et al. 1995). Based on BLSD, the estimates did not differ from the range of measurements.

A one-way ANOVA was constructed using published means, variances, and numbers of observations. Data from an individual study were considered to be a treatment, including the estimates from CFLX (these latter data have no measure of variation). Due in part to the large number of within-treatment degrees of freedom for aspen-birch, differences in size of C pools among treatments (studies) were highly significant. However, there were no patterns in the ranking of any of the C pools by size; estimates from CFLX were not uniformly higher or lower than the observations (Figure 9.6), and BLSD indicated overlapping of means. For black spruce, differences among treatments were significant only for a few of the subpools (low SHs; herbs including low SHs, forbs, and sedges; overstory and peat to 100 cm). As was the case with aspen-birch, there were no patterns in the ranking of any of the C pools by size; CFLX estimates were not uniformly higher or lower than the observations (Figure 9.7), and where differences were significant, BLSD indicated overlapping of means. For both aspen-birch and black spruce, the estimates from CFLX were indistinguishable from means developed from detailed stand measurements.

Net Carbon Flux

The objective of the development of CFLX was to determine net C flux (also termed net ecosystem production [NEP] the net difference between C inputs

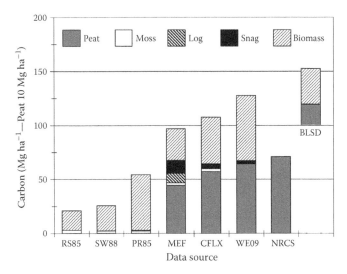

FIGURE 9.7

Comparison of ecosystem C estimates from model/spreadsheet for mature black spruce stands on peatlands with literature measurements (RS85, data for raised bogs from Grigal et al. 1985; SW88, Swanson 1988; PR85, data for perched bogs from Grigal et al. 1985; MEF, inventory of Marcell Experimental Forest; CFLX, model/spreadsheet; WE09, Weishampel et al. (2009); and NRCS, data from National Resource Conservation Service). Based on BLSD, the estimates did not differ from the range of measurements.

and losses from an ecosystem; Odum 1969). Annual NEP indicates whether an ecosystem is a source or sink to the atmosphere. In the case of aspen-birch, ecosystem C (including forest floor but excluding mineral soil) decreased annually for about 5 years following disturbance from harvest, increased, and peaked at about 1 Mg ha year^{-1} at 25–30 years after stand establishment, and declined and became negative at about 50 years. At 75 years, NEP was −0.4 Mg C ha year^{-1}. Aboveground, net primary production (ANPP) was about 2.3 Mg C ha year^{-1} at 25 years and 2.1 Mg C ha year^{-1} at 50 years. Other reports of ANPP for aspen-birch include 2.6 Mg C ha year^{-1} at MEF (Weishampel et al. 2009), 2.2 Mg ha year^{-1} in northeastern Minnesota (Reich et al. 2001), 3.5 Mg ha year^{-1} in boreal Canada (Gower et al. 1997), and 3.9 Mg ha year^{-1} (Burrows et al. 2003) and 5.1 Mg ha year^{-1} (Ruark and Bockheim 1988) in northern Wisconsin. To reiterate, CFLX was constructed to generically describe average stand behavior not to predict C fluxes at specific sites.

In the case of black spruce, NEP, including aboveground and belowground vegetation and peat to 100 cm, peaked at about 0.7 Mg ha year^{-1} at 30–40 years after stand establishment and declined to negative values at about 90 years, reaching −0.15 Mg C ha year^{-1} at 150 years. O'Connell et al. (2003) estimated NEP of −1.28 Mg ha year^{-1} for a 120 year old black spruce-*Sphagnum* forest in Saskatchewan. CFLX estimated ANPP of about 1.5 Mg C ha year^{-1} during a period of 30–50 years, excluding *Sphagnum* growth, or 3.3 Mg ha year^{-1} including *Sphagnum*, declining slightly to 3.1 Mg ha year^{-1} at 100 years. This can be compared to 2.8 Mg ha year^{-1} (including *Sphagnum*) at the MEF (Weishampel et al. 2009), 1.0 Mg ha year^{-1} (excluding *Sphagnum*) in northeastern Minnesota (Reich et al. 2001), and 3.4 Mg ha year^{-1} (including *Sphagnum*) in northern Minnesota (Grigal et al. 1985). These rates can be compared to other similar sites (all including *Sphagnum*), 1.5 Mg ha year^{-1} in Alberta (Szumigalski and Bayley 1996) and for Canadian boreal sites (Gower et al. 1997), and 1.8 Mg ha year^{-1} for a bog and muskeg in Manitoba (Reader and Stewart 1972). These wider ranges in ANPP for black spruce compared to aspen-birch are caused by wider variations in stand density, affecting both overstory and understory NPP.

Conclusions

Several conclusions can be drawn from this interlinked C inventory and model/spreadsheet. First, to emphasize what has been repeated by many, peatlands are important C pools in the northern landscape. The relative size of the C pool in forest vegetation is surprisingly uniform across the landscape, with apparent compensation by mixes of species yielding roughly similar masses among vegetation types. The mineral-soil C pool, at least in the morainic landscape of MEF, is also relatively uniform. Topographic variation does not play a strong role in creating major differences in C storage in mineral soil. As a result, prediction equations based on topographic variables have low explanatory power. However, the topographic variation that gives rise to peatlands has a profound effect on landscape C storage. In this area on the western edge of the mid-continental forests, both forest floor and peat have lower C than more northerly and easterly sites.

It also appears that there are nearly sufficient data in the literature with which to make reasonable predictions about landscape C storage without resorting to detailed measurements. Even a large suite of detailed measurements (Table 9.1) does not eliminate uncertainty (Table 9.7). Finding and accessing existing data can provide relatively inexpensive assessments of landscape C storage via categorical mapping and development of C estimates for categories.

There are many data on C flux in the literature, though often not under that rubric. The simple spreadsheet described here demonstrates that those data can be linked to provide realistic estimates of ecosystem C change with time. The same problem of finding and accessing relevant data also is true for estimates of C flux.

Two major points emerge from this work, carried out in a landscape with a poorly developed drainage network:

- C storage in vegetation and CWD did not differ appreciably among soils and differed little among mineral soils. The major difference in ecosystem C storage was related to differences between peatlands and other systems.
- Annual net C flux for both the most common upland (aspen-birch) and peatland (black spruce) forest types peaked at between 25 and 35 years after establishment, though C continued to accumulate with time.

The ramifications of these points with respect to potential climate change are that

- Climate-induced changes in peatland area will have profound effects on C storage.
- Manipulation of vegetation types by management will only marginally affect landscape C storage. Maintenance of maximum rates of C sequestration would best be achieved by management to maintain the forest in relatively young age classes.

Appendix 9.1

The data sources and rationale used to develop functional relationships describing C fluxes in aspen-birch and black spruce forest types at the MEF:

Basal Area versus Time

$$BA = a/\left(1+\exp(b*(T-c))\right), \tag{9.6a}$$

The data used to quantify the constants were from empirical yield tables for Minnesota collected by the USDA Forest Service Forest Inventory and Analysis (FIA) program (Hahn and Raile 1982). The tables provide the average BA and the number of observations by stand-age and site-quality class for 14 forest types. Inputs to CFLX for aspen were simple arithmetic averages of a, b, and c over six site classes (Table 9.9), inputs for black spruce, with only two site classes, were from all data (Table 9.10). Observed and predicted BA were strongly correlated; all r^2s were greater than 0.95 except for the highest aspen site class (28–30 m at 50 years).

Biomass versus Basal Area

$$BIOM = a*BA^{b}, \tag{9.7}$$

For aspen-birch, the constants were developed from a database of about 1000 forest stands (from Bell et al. 1996, Ohmann and Grigal 1985b, Swanson 1988, Wilson 1994, and the C inventory of MEF) spanning a range of vegetation types, including 447 aspen-birch stands (Table 9.9). The database for black spruce contained 117 spruce stands (from Grigal et al. 1985; Moore 1984; Swanson 1988; the MEF inventory, and tabulated data from a number of studies; Grigal and Brooks 1997; Table 9.10).

Diameter versus Time

$$DIAM = a/(1 + exp(b * (T - c))),   (9.6b)$$

For aspen-birch, a dataset from FIA plots measured between 2002 and 2006 in the aspen cover type in six northern Minnesota counties (M. Hatfield, USDA Forest Service, 2008, pers. commun.) was used. Because of minimal observations, stands with site quality <12 and >28 m at 50 years or with ages <10 and >100 years were removed, and constants were determined for each of the remaining six site-quality classes. The r^2s were greater than 0.90 for the four site classes with more than 100 observations and dropped to 0.76 (site class 28 m, n = 25) and 0.55 (site class 12 m, n = 39). The simple arithmetic average of a, b, and c over the six site classes was used (Table 9.9). There were no readily available age-diameter data for black spruce. Therefore, constants were estimated using both the results of the analysis with aspen-birch and several age-diameter data (Table 9.10).

Understory Mass

$$UBIOM = a * BIOM^b,   (9.8)$$

For aspen-birch, data were from the same database used for the continuous estimates of C storage (n = 453). Constants were determined for the linearized form of Equation 9.8 including a dummy variable determined by vegetation type; the solution was converted to the nonlinear form.

For black spruce, data were from a study of black spruce peatlands in northern Minnesota (Grigal et al. 1985). Tall woody SHs constitute a small proportion of biomass and were excluded. Because the herb stratum had low mass and high variability, Equation 9.8 was solved for low SHs and for the sum of low SHs and herbs, and herb mass was determined by difference.

Substrate

Mineral-soil C does not change based on both the lack of documentation of such change following aspen harvest and the difficulty of detecting measurable changes; belowground C inputs via root turnover and mortality are assumed to be in approximate equilibrium with C losses via decomposition. Following an extensive study, Alban and Perala (1990) stated that harvesting "...did not affect total soil carbon." An ANOVA of their table 4 indicated that differences related to time after harvest in forest floor plus soil C to 50 cm were not significant (prob. = 0.464). Similarly, analysis indicated that their mineral soil data and that from Silkworth (1980), who also sampled soils following aspen harvest, had virtually identical means (80.5 Mg ha^{-1}, prob. = 0.999) and that time after harvest again had no effect (prob. = 0.910). The pooled variance indicated that detection of a 5% change in soil C would require about 100 samples per stand (Freese 1962), demonstrating the difficulty of such detection.

In the case of the black spruce on peatland soils, inputs to peat via *Sphagnum* growth and vascular root turnover and mortality and their losses via decomposition are considered. In addition to the subpool of peat produced during the simulation, five subpools of preexisting peat, with only losses, are included (0–25, 25–50, 50–100, 100–200, and 200–300 cm). Because of the slow loss rates, especially in the lower layers (100–200 and 200–300 cm), their inclusion has little effect on net C flux.

Overstory and Shrub Litterfall Inputs

$$LIT = a * BA^b, \tag{9.9a}$$

Determination of a and b for aspen-birch was based on measured litterfall or foliage mass in a variety of aspen and other deciduous stands in Minnesota and Wisconsin, Alaska, and Canada (Alban et al. 1991; Alban and Perala 1982; Bernier et al. 2007; Lee et al. 2002; Pastor and Bockheim 1984; Reich et al. 2001; Ruark and Bockheim 1988; Steele et al. 1997; Van Cleve and Noonan 1975; Weishampel et al. 2009). Because of the wide geographic range, an additional explanatory variable, latitude (Lonsdale 1988) was used in the initial relationship,

$$LIT = a * BA^b * LAT^c \tag{9.9b}$$

where variables are as in Equation 9.9 and LAT is degrees (north) latitude. The solution was a significant improvement over the relationship without latitude ($r^2 = 0.46$ and 0.09, respectively; Table 9.9). In the case of black spruce, measured litterfall in stands in Minnesota, Wisconsin, Alaska, and Canada was used (Alban and Perala 1982; Reich et al. 2001; Steele et al. 1997; Van Cleve and Noonan 1975; Weishampel et al. 2009). Equation 9.9a had higher explanatory power for black spruce ($r^2 = 0.93$) than for aspen-birch (Tables 9.9 and 9.10). The annual contribution of overstory litterfall to forest floor or accumulating peat was estimated by substituting an appropriate latitude, which is 47.3° N for the MEF into Equation 9.9a (Tables 9.9 and 9.10).

In aspen-birch, foliage as 20% of aboveground tall shrub biomass from Equation 9.8 (Table 9.9) was annually added to forest floor (Grigal et al. 1976). Similarly, for black spruce, half of low shrub foliage as 30% of biomass from Equation 9.8 (Table 9.10) was added annually to accumulating peat (Grigal et al. 1985; Table 9.11). About half of tree seedling mass is in needles, and one-seventh (about 15%) of that was included as litterfall (Table 9.11). For both aspen-birch and black spruce, herb biomass (C) (from Equation 9.8) was also added to the substrate annually.

Fine Woody Debris Inputs

$$STEMS = a * exp(-b * A), \tag{9.10}$$

Estimates of fine-wood additions to aspen-birch forest floor by tall shrub mortality were based on shrub demography described by a negative exponential function where initial stem density (a) declines annually at a rate of b (Equation 9.10; Balogh and Grigal 1987). The pooled value of b did not differ significantly among shrub populations in closed upland aspen and conifer stands in northern Minnesota (b = −0.202, n = 34 stands; Balogh and Grigal 1987), implying that mortality rates are independent of shrub density and overstory and soil characteristics. Ideally, Equation 9.10 can be used to estimate annual shrub turnover as the reciprocal of the time required to reach 100% mortality, but, because it is exponential, there is no solution. It can be solved to 99% (time = 23 years and turnover = 0.04 year^{-1}) or 90% mortality (e.g., time = 11 years and turnover = 0.09 year^{-1}), with lesser mortality leading to faster turnover. A uniform b assumes a steady state; there are no significant changes in the age-class distribution. Although this may be the case for stands with closed canopies, it probably is not for recently disturbed stands. There, an initial burst of SHs is reduced by competition with the developing overstory, and mortality rates probably are higher than in stands with fully developed overstories. Because the pooled value of b in Equation 9.10 was from closed-canopy stands, and faster turnover is more reasonable for stands during canopy closure, time to 95% mortality was used to estimate shrub turnover rates over a spectrum of stand ages (Table 9.9). Dead stems (80% of tall shrub mass) were included as inputs of fine woody debris to the forest floor.

A study of demographics in black spruce stands in Minnesota indicated that about 40% of low SHs and 20% of tree seedlings turn over annually (Grigal et al. 1985), and those data were used to determine inputs of fine woody debris to peat (Table 9.10).

Moss and Root Inputs

Annual *Sphagnum* production in peatland black spruce forests of about 1.75 Mg C ha^{-1} year^{-1} was measured in Minnesota (Grigal 1985). That measure was consistent with literature values at similar latitudes in North America. Additional work, since then, including Chapin et al. (2004) and Moore et al. (2002), also is in agreement. Weishampel et al. (2009) measured *Sphagnum* productivity at MEF as only about a 15% of that value, but their study was carried out during an unusually dry growing season, which adversely affected *Sphagnum* growth.

Fine root turnover was estimated as 1.75 times litterfall (Table 9.11) based on data from a variety of boreal forests, including black spruce (Steele et al. 1997). Additions of roots to peat also include both overstory and shrub mortality. Woody roots constitute about 30% of tree and seedling and about 185% of low shrub aboveground biomass in Minnesota black spruce stands (Grigal et al. 1985). Those ratios were used to compute mortality-based additions (Table 9.11).

CWD Inputs

$$Y = \exp(-k * T) \tag{9.11}$$

In addition to self-thinning, mortality has been described by a simple rate per year, with estimated mortality in forests in the North Central States from 0.0146 to 0.0149 year^{-1} for aspen-birch stands and from 0.0070 to 0.0130 year^{-1} for upland spruce-fir (and no estimate for peatland black spruce—Harmon 1993). A sophisticated model of mortality for Lake States' trees also has been developed based on current diameter and past diameter growth (Buchman 1983) that predicted that smaller, slower growing trees have higher mortality rates than larger, faster growing trees. The self-thinning mortality rates in CFLX decline with tree size, spanning the values suggested by Harmon (1993) and tending to be lower than those tabulated by Buchman (1983).

There are limited data on the longevity of snags in north-central U.S. forests. Snags were monitored after the Little Sioux Fire in northeastern Minnesota (Ohmann and Grigal 1979; Slaughter et al. 1998), and the rate of an aspen-birch snag falling (k in Equation 9.11) was = −0.145 (n = 62), implying a half-life of 4.8 years, with 99% of snags down in about 21 years. That rate was used in the aspen-birch version of CFLX (Table 9.9). The rate of a black spruce or jack pine snag (as a surrogate for black spruce) falling was similar (k = −0.149, n = 154). Monitoring of snags during an intensive bog study (Grigal et al. 1985) showed that k = −0.063, with a half-life 11 years. The average of these two rates, k = −0.106, was used in the black spruce version CFLX (Table 9.10).

Litter Losses

Decomposition of herb leaves (*Aster* spp.) from northeastern Minnesota is relatively rapid (k = −1.8, Grigal and McColl 1977; −1.3, Ohmann and Grigal 1979, following Equation 9.11). The former figure, from a wider variety of sites, was used for herb litter (Table 9.9).

Mass loss of aspen leaves over 3 years, with additional cohorts added every year (data from Grigal and McColl 1977, yielded k = −0.49 and p = 0.72; Equation 9.12; Table 9.9). The rate of decomposition of fine woody debris was that from managed red pine stands (k = −0.208, with p = 1; Equation 9.12; Duvall and Grigal 1999). These stands were thinned continuously, and so the decomposing material was primarily branches and twigs. The annual rate of mass loss of residual forest floor is difficult to estimate, because most studies have determined rates of loss of fresh material, not of the multiple-aged residues in forest floor. The rate of loss of residual forest floor was considered to be that based on "turnover" of forest floor in aspen stands in northern Minnesota (k = −0.078, with p = 1; Equation 9.12; Alban and Perala 1982; Table 9.9).

Decomposition rates in peat are not easily adapted to Equation 9.11 or 9.12. Gorham et al. (2003) measured the age/mass relationship of the S2 bog on the MEF and reported a linear rate of peat accretion of 56.5 g m^{-2} year^{-1} over 9200 years. This is equivalent to an accretion rate of 0.65 mm year^{-1}, but there is no data indicating whether this rate also is linear with time. Near-surface peat has lower D_b than deeper peat, and a linear rate of mass accretion makes a linear rate of depth accretion unlikely. Farrish and Grigal (1985) measured annual rates of mass loss of both cellulose strips and of peat returned to its point of origin within the surface 100 cm of the S2 bog. The data showed significant linear reductions in rates of decomposition with depth within the upper 35 cm of peat (measured from the top of hummocks; $r^2 = 0.83$, n = 8, and prob. = 0.001), while, at greater depths, the slope was barely significant ($r^2 = 0.16$, n = 12, and prob. = 0.11). At those greater depths (ages), decomposition rate changed little. Both these studies provided useful background information, including a limit on cumulative mass losses through time (Gorham et al. 2003) and marked reductions in rates of mass loss with depth per time (Farrish and Grigal 1985). However, rates of decomposition for peat from Farrish and Grigal (1985) were much higher than indicated by data from Gorham et al. (2003). Removing the peat, placing it in litter bags, and replacing it apparently increased rates. The overall decomposition function for accumulating peat was based on data from Moore (1984), who monitored black spruce needles and *Ledum* leaves on two substrates (burned and unburned) for more than 2 years. Decomposition was well described by Equation 9.12 (k = −0.286 and p = 0.72, $r^2 = 0.76$; Table 9.10). Low rates, decreasing with depth, were set for residual peat (following Equation 9.11, 0–25 cm, k = −0.000500; 25–50 cm, k = −0.000100; 50–100 cm, k = −0.000050; 100–200 cm, k = −0.000010; and 200–300 cm, and k = −0.00000; Table 9.10).

CWD Losses

The mass loss of standing dead trees (snags) is assumed to follow Equation 9.12. Reports of decomposition rates of snags are rare. Duvall and Grigal (1999) measured density of snags in three decay classes, ranging from recently dead to a condition with most large branches missing and "unsound" wood. An empirical decomposition rate can be computed if the three classes of decay are assumed to represent time = 1 year (class 1), the half-life of standing snags (class 2), and the time when 90% of the snags have fallen (class 3), respectively. Following this approach, the result for aspen-birch snags was k = −0.037 and for "softwood" snags (not including red pine) was k = −0.0096 (with p assumed to be equal to 1; Tables 9.9 and 9.10). There are some data on log decay for aspen, none for black spruce, and a limited amount for jack pine (as a surrogate for black spruce), and most are one-time observations. The rate of decomposition of aspen-birch logs was based on data at 1 and 5 years from Miller (1983), at 14 and 17 years from Alban and Pastor (1993), at 50 years using Harmon's (1993) rate in aspen-birch forests, and at 100 years using the rate from Duvall and Grigal (1999). The resulting analysis yielded k = −0.0699 (Equation 9.11; Table 9.9). In the case of jack pine, data at 11 and 17 years from Alban and Pastor (1993), at 50 years using Harmon's (1993) rate in pine and spruce-fir forests, and at 100 years using the rate from Duvall and Grigal (1999) were used. The result yielded k = −0.0713, p = 0.774 (Equation 9.12; Table 9.10).

References

Abrahamson, W.G. 1979. Patterns of resource allocation in wildflower populations of fields and woods. *American Journal of Botany* 66:71–79.

Alban, D.H. and J. Pastor. 1993. Decomposition of aspen, spruce, and pine boles on two sites in Minnesota. *Canadian Journal of Forest Research* 23:1744–1749.

Alban, D.H. and D.A. Perala. 1990. Ecosystem carbon following aspen harvesting in the Upper Great Lakes. In *Aspen Symposium '89 Proceedings*, ed. R.D. Adams. General Technical Report NC-140. St. Paul, MN: USDA Forest Service, pp. 123–131.

Alban, D.H. and D.A. Perala. 1992. Carbon storage in Lake States aspen ecosystems. *Canadian Journal of Forest Research* 22:1107–1110.

Alban, D.H., D.A. Perala, M.F. Jurgensen, M.E. Ostry, and J.R. Probst. 1991. Aspen ecosystem properties in the upper Great Lakes. Research Paper NC-300. St. Paul, MN: USDA Forest Service.

Alemdag, I.S. 1983. Mass equations and merchantability factors for Ontario softwoods. Information Report PI-X-23. Canadian Forestry Service, Petawawa National Forestry Institute.

Alemdag, I.S. 1984. Total tree and merchantable stem biomass equations for Ontario hardwoods. Information Report PI-X-46. Canadian Forestry Service, Petawawa National Forestry Institute.

Arrouays, D., J. Daroussin, L. Kicin, and P. Hassika. 1998. Improving topsoil carbon storage prediction using a digital elevation model in temperate forest soils of France. *Soil Science* 163:103–108.

Ashmun, J.W., R.L. Brown, and L.F. Pitelka. 1985. Biomass allocation in *Aster acuminatus*: Variation within and among populations over 5 years. *Canadian Journal of Botany* 63:2035–2043.

Balogh, J.C. 1983. Tall shrubs in Minnesota. PhD dissertation. St. Paul, MN: University of Minnesota.

Balogh, J.C. and D.F. Grigal. 1987. Age-density distributions of tall shrubs in Minnesota. *Forest Science* 33:846–857.

Beauchamp, J.J. and J.S. Olson. 1973. Corrections for bias in regression estimates after logarithmic transformation. *Ecology* 54:1403–1407.

Beers, T.W., P.E. Dress, and L.C. Wensel. 1966. Aspect transformation in site productivity research. *Journal of Forestry* 64:691–692.

Bell, J.C., D.F. Grigal, P.C. Bates, and C.A. Butler. 1996. Spatial patterns in carbon storage in a Lake States' landscape. In *Proceedings, 1995 Meeting of the Northern Global Change Program*, ed. J. Hom, R. Birdsey, and K. O'Brian, March 14–16, 1995, Pittsburgh, PA, General Technical Report NE-214. Newtown Square, PA: USDA Forest Service, pp. 198–202.

Bell, J.C., D.F. Grigal, and P.C. Bates. 2000. A soil-terrain model for estimating spatial patterns of soil organic carbon. In *Terrain Analysis: Principles and Applications*, ed. J. Wilson and J.C. Gallant. New York: John Wiley & Sons, pp. 295–310.

Bernier, P.Y., M.B. Lavigne, E.H. Hogg, and J.S. Trofymow. 2007. Estimating branch production in trembling aspen, Douglas-fir, jack pine, black spruce, and balsam fir. *Canadian Journal of Forest Research* 37(6):1024–1033.

Berry, A.B. 1987. Plantation white spruce variable density volume and biomass yield tables to age 60 at the Petawawa National Forestry Institute. Information report PI-X-71. Petawawa National Forestry Institute, Canadian Forestry Service, Chalk River, Ontario.

Birdsey, R.A. and L.S. Heath. 1995. Carbon changes in U.S. forests. In Productivity of America's forests and climate change, ed. L.A. Joyce. General Technical Report RM-271. USDA Forest Service, pp. 56–70.

Birdsey, R.A. and G.M. Lewis. 2003. Carbon in U.S. forests and wood products, 1987–1997: State-by-state estimates. General Technical Report NE-310. Newtown Square, PA: USDA Forest Service.

Buchman, R.G. 1983. Survival predictions for major Lake States tree species. Research Paper NC-233. St. Paul, MN: USDA Forest Service.

Burrows, S.N., S.T. Gower, J.M. Norman, G. Diak, D.S. Mackay, D.E. Ahl, and M.K. Clayton. 2003. Spatial variability of net primary production for a forested landscape in northern Wisconsin. *Canadian Journal of Forest Research* 33:2007–2018.

Buttleman, C.G. 1982. Use of Rb/K ratio to define nutrient linkages between a perched bog and its surrounding upland. MS thesis. St. Paul, MN: University of Minnesota.

Chapin, C.T., S.D. Bridgham, and J. Pastor. 2004. pH and nutrient effects on above-ground net primary productivity in a Minnesota, USA bog and fen. *Wetlands* 24:186–201.

Chojnacky, D.C., R.A. Mickler, L.S. Heath, and C.W. Woodall. 2004. Estimates of down woody materials in eastern US forests. *Environmental Management* 33 (Suppl. 1):S44–S55.

Connolly-McCarthy, B.J. and D.F. Grigal. 1985. Biomass of shrub-dominated wetlands in Minnesota. *Forest Science* 31:1011–1017.

David, M.B. 1988. Use of loss-on-ignition to assess soil organic carbon in forest soils. *Communications in Soil Science and Plant Analysis* 19:1593–1599.

Davis, A.A., M.H. Stolt, and J.E. Compton. 2004. Spatial distribution of soil carbon in Southern New England hardwood forest landscapes. *Soil Science Society of America Journal* 68:895–903.

Duvall, M.D. 1997. Effects of timber harvesting on the distribution of coarse woody debris in red pine forests of the western Great Lakes region. MS thesis. Minneapolis, MN: University of Minnesota.

Duvall, M.D. and D.F. Grigal. 1999. Effects of timber harvesting on coarse woody debris in red pine forests across the Great Lakes states, U.S.A. *Canadian Journal of Forest Research* 29:1926–1934.

Farrish, K.W. and D.F. Grigal. 1985. Mass loss in a forested bog: relation to hummock and hollow microrelief. *Canadian Journal of Soil Science* 65:375–378.

Federer, C.A., D.E. Turcotte, and C.T. Smith. 1993. The organic fractionbulk density relationship and the expression of nutrient content in forest soils. *Canadian Journal of Forest Research* 23:1026–1032.

Freese, F. 1962. Elementary forest sampling. In *Agriculture Handbook 232*. Washington, DC: U.S. Department of Agriculture.

Galbraith, J.M., P.J.A. Kleinman, and R.B. Bryant. 2003. Sources of uncertainty affecting soil organic carbon estimates in northern New York. *Soil Science Society of America Journal* 67:1206–1212.

Gore, A., E.A. Johnson, and H. Lo. 1985. Estimating the time a dead tree has been on the ground. *Ecology* 66:1981–1983.

Gorham, E. 1991. Northern peatlands: Role in the carbon cycle and probable responses to climatic warming. *Ecological Applications* 1:182–195.

Gorham, E., J.A. Janssens, and P.H. Glaser. 2003. Rates of peat accumulation during the postglacial period in 32 sites from Alaska to Newfoundland, with special emphasis on northern Minnesota. *Canadian Journal of Botany* 81(5): 429–438.

Gosselink, J.G., R. Hatton, and C.S. Hopkinson. 1984. Relationship of organic carbon and mineral content to bulk density in Louisiana marsh soils. *Soil Science* 137:177–180.

Gower, S.T., J. Vogel, T.K. Stow, J.M. Norman, C.J. Kucharik, and S.J. Steele. 1997. Carbon distribution and aboveground net primary production of upland and lowland boreal forests in Saskatchewan and Manitoba. *Journal of Geophysical Research* 102:29–41.

Grigal, D.F. 1985. *Sphagnum* production in forested bogs of northern Minnesota. *Canadian Journal of Botany* 63:1204–1207.

Grigal, D.F. 2009. Ecosystem C storage on the Marcell Experimental Forest, Minnesota. Background paper for *Marcell Experimental Forest 50th Anniversary Symposium*. http://www.nrs.fs.fed.us/ef/marcell/pubs/proceedings/, accessed August 13, 2009.

Grigal, D.F. and P.C. Bates. 1992. Forest soils. A technical paper for a generic environmental impact statement on timber harvesting and forest management in Minnesota. Tarrytown, NY: Jaakko Pöyry Consulting, Inc.

Grigal, D.F. and K.N. Brooks. 1997. Impact of forest management on undrained northern forest peatlands. In *Northern Forested Wetlands: Ecology and Management*, ed. C.C. Trettin, M.F. Jurgensen, D.F. Grigal, M.R. Gale, and J. Jeglum. Boca Raton, FL: CRC/Lewis Publishers, pp. 379–396.

Grigal, D.F., S.L. Brovold, W.S. Nord, and L.F. Ohmann. 1989. Bulk density of surface soils and peat in the north central United States. *Canadian Journal of Soil Science* 69:895–900.

Grigal, D.F., C.G. Buttleman, and L.K. Kernik. 1985. Biomass and productivity of the woody strata of forested bogs in northern Minnesota. *Canadian Journal of Botany* 63:2416–2424.

Grigal, D.F., L.M. Chamberlain, H.R. Finney, D.V. Wroblewski, and E.R. Gross. 1974. Soils of the Cedar Creek Natural History Area. Minnesota Agricultural Experiment Station Miscellaneous Report 123.

Grigal, D.F. and P.S. Homann. 1994. Nitrogen mineralization, groundwater dynamics, and forest growth on a Minnesota outwash landscape. *Biogeochemistry* 27:171–185.

Grigal, D.F. and L.K. Kernik. 1984. Generality of black spruce biomass estimation equations. *Canadian Journal of Forest Research* 14:468–470.

Grigal, D.F. and J.G. McColl. 1975. Litterfall following wildfire in virgin forests of northeastern Minnesota. *Canadian Journal of Forest Research* 5:655–661.

Grigal, D.F. and J.G. McColl. 1977. Litter decomposition following forest fire in north-eastern Minnesota. *Journal of Applied Ecology* 14:531–538.

Grigal, D.F. and W.S. Nord. 1983. Inventory of heavy metals in Minnesota peatlands. Report to Minnesota Department of Natural Resources, Division of Minerals, St. Paul, MN.

Grigal, D.F. and L.F. Ohmann. 1992. Carbon storage in upland forests of the Lake States. *Soil Science Society of America Journal* 56:935–943.

Grigal, D.F., L.F. Ohmann, and R.B. Brander. 1976. Seasonal dynamics of tall shrubs in northeastern Minnesota: Biomass and nutrient element changes. *Forest Science* 22:195–208.

Gross, K.L., T. Berner, E. Marschall, and C. Tomcko. 1983. Patterns of resource allocation among five herbaceous perennials. *Bulletin of the Torrey Botanical Club* 110:345–352.

Guo, Y., R. Amundson, P. Gong, and Q. Yu. 2006. Quantity and spatial variability of soil carbon in the conterminous United States. *Soil Science Society of America Journal* 70:590–600.

Hahn, J.T. and G.K. Raile. 1982. Empirical yield tables for Minnesota. General Technical Report NC-71. St. Paul, MN: USDA Forest Service.

Harmon, M. 1993. Woody debris budgets for selected forest types. In *The Forest Sector Carbon Budget of the United States: Carbon Pools and Flux Under Alternative Policy Options*. Corvallis, OR: US Environmental Protection Agency, Environmental Research Laboratory, pp. 151–178.

Homann, P.S., P. Sollins, H.N. Chappell, and A.G. Stangenberger. 1995. Soil organic carbon in a mountainous, forested region: Relation to site characteristics. *Soil Science Society of America Journal* 59:1468–1475.

Johnson, M.G. and J.S. Kern. 2003. Quantifying the organic carbon held in forested soils of the United States and Puerto Rico. In *The Potential of U.S. Forest Soils to Sequester Carbon and Mitigate the Greenhouse Effect*, ed. J.M. Kimble, L.S. Heath, R.A. Birdsey, and R. Lal. Boca Raton, FL: CRC Press, pp. 47–72.

Johnston, M.H., P.S. Homann, J.K. Engstrom, and D.F. Grigal. 1996. Changes in ecosystem carbon storage over 40 years on an old-field/forest landscape in east-central Minnesota. *Forest Ecology and Management* 83:17–26.

Kelly, J.M. and J.J. Beauchamp. 1987. Mass loss and nutrient changes in decomposing upland oak and mesic mixed-hardwood leaf litter. *Soil Science Society of America Journal* 51:1616–1622.

Kern, J.S. 1994. Spatial patterns of soil organic carbon in the contiguous United States. *Soil Science Society of America Journal* 58:439–455.

Kolka, R.K. 1993. Cation release rates from weathering of five Upper Great Lakes forest soils. MS thesis. St. Paul, MN: University of Minnesota.

Kulmatiski, A., D.J. Vogt, T.G. Siccama, J. Tilley, K. Kolesinskas, T.W. Wickwire, and B.C. Larson. 2004. Landscape determinants of soil carbon and nitrogen storage in southern New England. *Soil Science Society of America Journal* 68:2014–2022.

Lee, J., I.K. Morrison, J.-D Leblanc, M.T. Dumas, and D.T. Cameron. 2002. Carbon sequestration in trees and regrowth vegetation as affected by clearcut and partial cut harvesting in a second-growth boreal mixedwood. *Forest Ecology and Management* 169(1–2):83–101.

Lonsdale, W.M. 1988. Predicting the amount of litterfall in forests of the world. *Annals of Botany* 61(3):319–324.

Malterer, T.J., E.S. Verry, and J. Erjavec. 1992. Fiber content and degree of decomposition in peats: Review of national methods. *Soil Science Society of America Journal* 56:1200–1211.

Methven, I.R. 1983. Tree biomass equations for young plantation-grown red pine (*Pinus resinosa*) in the maritime lowlands ecoregion [New Brunswick]. Information report M-X-147. Fredericton, New Brunswick: Maritimes Forest Research Centre, Canadian Forestry Service.

Miller, W.E. 1983. Decomposition rates of aspen bole and branch litter. *Forest Science* 29:351–356.

Moore, T.R. 1984. Litter decomposition in a subarctic spruce-lichen woodland, Eastern Canada. *Ecology* 65(1):299–308.

Moore, T.R., J.L. Bubier, S.E. Frolking, P.M. Lafleur, and N.T. Roulet. 2002. Plant biomass and CO2 exchange in an ombrotrophic bog. *Journal of Ecology* 90:25–36.

Mueller, T.G. and F.J. Pierce. 2003. Soil carbon maps: enhancing spatial estimates with simple terrain attributes at multiple scales. *Soil Science Society of America Journal* 67:258–267.

Nichols, D.S. and D.H. Boelter. 1984. Fiber size distribution, bulk density, and ash content of peats in Minnesota, Wisconsin, and Michigan. *Soil Science Society of America Journal* 48:1320–1328.

Nihlgård, B. 1972. Plant biomass, primary production and distribution of chemical elements in a beech and planted spruce forest in south Sweden. *Oikos* 23:69–81.

Nyberg, P.R. 1987. *Soil Survey of Itasca County, Minnesota*. Washington, DC: USDA Soil Conservation Service.

O'Connell, K.E.B., S.T. Gower, and J.M. Norman. 2003. Net ecosystem production of two contrasting boreal black spruce forest communities. *Ecosystems* 6:248–260.

Odum, E.G. 1969. The strategy of ecosystem development. *Science* 164:262–270.

Ohmann, L.F. and D.F. Grigal. 1979. Early revegetation and nutrient dynamics following the 1971 Little Sioux forest fire in northeastern Minnesota. *Forest Science Monograph 21*.

Ohmann, L.F. and D.F. Grigal. 1985a. Biomass distribution of unmanaged upland forests in Minnesota. *Forest Ecology and Management* 13:205–222.

Ohmann, L.F. and D.F. Grigal. 1985b. Plant species biomass estimates for 13 upland plant community types of northeastern Minnesota. Resource Bulletin NC-88. St. Paul, MN: USDA Forest Service.

Ohmann, L.F. and D.F. Grigal. 1991. Properties of soils and tree wood tissue across a Lake States sulfate deposition gradient. Resource Bulletin NC-130. St. Paul, MN: USDA Forest Service.

Ohmann, L.F., D.F. Grigal, S.R. Shifley, and W.E. Berguson. 1994. Vegetative characteristics of five forest types across a Lake States sulfate deposition gradient. Resource Bulletin NC-154. St. Paul, MN: USDA Forest Service.

Paavilainen, E. 1980. Effect of fertilization on plant biomass and nutrient cycle on a drained dwarf shrub pine swamp. *Communicationes Instituti Forestalis Fenniae* 98:1–71.

Pastor, J. and J.G. Bockheim. 1984. Distribution and cycling of nutrients in an aspen-mixed-hardwood-Spodosol ecosystem in northern Wisconsin. *Ecology* 65(2):339–353.

Perala, D.A. 1972. Stand equations for estimating aerial biomass, net productivity, and stem survival of young aspen suckers on good sites. *Canadian Journal of Forest Research* 3:288–292.

Perala, D.A. and D.H. Alban. 1982. Biomass, nutrient distribution and litterfall in *Populus*, *Pinus* and *Picea* stands on two different soils in Minnesota. *Plant and Soil* 64:177–192.

Perala, D.A. and D.H. Alban. 1994. Allometric biomass estimators for aspen-dominated ecosystems in the Upper Great Lakes. Research Paper NC-314. St. Paul, MN: USDA Forest Service.

Perala, D.A., J.L. Rollinger, and D.M. Wilson. 1995. Comparison between soil and biomass carbon in adjacent hardwood and red pine forests. *World Resource Review* 7(2):231–243.

Raich, J.W., E.B. Rastetter, J.M. Melillo, D.W. Kicklighter, P.A. Steudler, B.J. Peterson, A.L. Grace, and C.J. Vorosmarty. 1991. Potential net primary productivity in South America: Application of a global model. *Ecological Applications* 1:399–429.

Reader, R.J. and J.M. Stewart. 1972. The relationship between net primary production and accumulation for a peatland in southeastern Manitoba. *Ecology* 53:1024–1037.

Reich, P.B., P. Bakken, D. Carlson, L.E. Frelich, S.K. Friedman, and D.F. Grigal. 2001. Influence of logging, fire, and forest type on biodiversity and productivity in southern boreal forests. *Ecology* 82:2731–2748.

Remezov, N. and P.S. Pogrebnak. 1969. *Forest Soil Science*. Translated from Russian. Jerusalem: Israel Program for Scientific Translations. Distributed by H.A. Humphrey, London.

Rosenfield, G.H. and K. Fitzpatrick-Lins. 1986. A coefficient of agreement as a measure of thematic classification accuracy. *Photogrammetric Engineering and Remote Sensing* 52:223–227.

Ruark, G.A. and J.G. Bockheim. 1988. Biomass, net primary production, and nutrient distribution for an age sequence of *Populus tremuloides* ecosystems. *Canadian Journal of Forest Research* 18:435–443.

Santantonio, D., R.K. Hermann, and W.S. Overton. 1977. Root biomass studies in forest ecosystems. *Pedobiologia* 17:1–31.

Silkworth, D.R. 1980. Leaching losses of nutrients following whole-tree harvesting of aspen. MS thesis. St. Paul, MN: University of Minnesota.

Silkworth, D.R. and D.F. Grigal. 1982. Determining and evaluating nutrient losses following whole-tree harvesting of aspen. *Soil Science Society of America Journal* 46:626–631.

Simmons, J.A., I.J. Fernandez, R.D. Briggs, and M.T. Delaney. 1996. Forest floor carbon pools and fluxes along a regional climate gradient in Maine, USA. *Forest Ecology and Management* 84:81–95.

Slaughter, K.W., D.F. Grigal, and L.F. Ohmann. 1998. Carbon storage in southern boreal forests following fire. *Scandinavian Journal of Forest Research* 13:119–127.

Smith, C.W. 1978. Bayes least significant difference: A review and comparison. *Agronomy Journal* 70:123–127.

Smith, J.E. and L.S. Heath. 2002. A model of forest floor carbon mass for United States forest types. Research Paper NE-722. Newtown Square, PA: USDA Forest Service.

Smith, J.E., L.S. Heath, K.E. Skog, and R.A. Birdsey. 2006. Methods for calculating forest ecosystem and harvested carbon with standard estimates for forest types of the United States. General Technical Report NE-343. Newtown Square, PA: USDA Forest Service.

Soil Survey Staff. 1997. National characterization data, Soil Survey Laboratory, National Soil Survey Center. Lincoln, NE: USDA Natural Resources Conservation Service.

Steele, S.J., S.T. Gower, J.G. Vogel, and J.M. Norman. 1997. Root mass, net primary production and turnover in aspen, jack pine and black spruce forests in Saskatchewan and Manitoba, Canada. *Tree Physiology* 17(8–9):577–587.

Stutter, M.I., D.G. Lumsdon, M.F. Billett, D. Low, and L.K. Deeks. 2009. Spatial variability in properties affecting organic horizon carbon storage in upland soils. *Soil Science Society of America Journal* 73:1724–1732.

Suhartoyo, H. 1991. Analysis of vegetation of the Cedar Creek Natural History Area using methods of classification and ordination. MS thesis. Minneapolis, MN: University of Minnesota.

Swanson, D.K. 1988. Properties of peatlands in relation to environmental factors in Minnesota. PhD dissertation. St. Paul, MN: University of Minnesota.

Swanson, D.K. and D.F. Grigal. 1991. Biomass, structure and trophic environment of peatland vegetation in Minnesota. *Wetlands* 11:279–302.

SYSTAT, Inc. 2007. Intelligent software, version 12.0. Evanston, IL: SYSTAT, Inc.

Szumigalski, A.R. and S.E. Bayley. 1996. Net above-ground primary production along a bog-rich fen gradient in central Alberta, Canada. *Wetlands* 16:467–476.

Thompson, J.A. and R.K. Kolka. 2005. Soil carbon storage estimation in a forested watershed using quantitative soil-landscape modeling. *Soil Science Society of America Journal* 69:1086–1093.

Tremblay, S., R. Ouimet, and D. Houle. 2002. Prediction of organic carbon content in upland forest soils of Quebec, Canada. *Canadian Journal of Forest Research* 32:903–914.

Van Cleve, K. and L.L. Noonan. 1975. Litter fall and nutrient cycling in the forest floor of birch and aspen stands in interior Alaska. *Canadian Journal of Forest Research* 5:626–639.

Van Wagner, C.E. 1968. The line intersect method in forest fuel sampling. *Forest Science* 14:20–26.

Vejre, H., I. Callesen, L. Vesterdal, and K. Raulund-Rasmussen. 2003. Carbon and nitrogen in Danish forest soilscontents and distribution determined by soil order. *Soil Science Society of America Journal* 67:335–343.

Weishampel, P., R.K. Kolka, and J.Y. King. 2009. Carbon pools and productivity in a 1-km^2 heterogeneous forest and peatland mosaic in Minnesota, USA. *Forest Ecology and Management* 257:747–754.

Whittaker, R.H. and P.L. Marks. 1975. Methods of assessing terrestrial productivity. In *Primary Productivity of the Biosphere*, ed. H. Lieth and R.H. Whittaker. New York: Springer-Verlag, pp. 55–118.

Wieder, R.K. and G.E. Lang. 1982. A critique of the analytical methods used in examining decomposition data obtained from litter bags. *Ecology* 63(6):1636–1642.

Wilson, D.M. 1994. The differential effect of conifer plantations and adjacent deciduous forests on soil calcium in the Lake States. MS thesis. St. Paul, MN: University of Minnesota.

10

Carbon Emissions from Peatlands

Nancy B. Dise, Narasinha J. Shurpali, Peter Weishampel, Shashi B.
Verma, Elon S. Verry, Eville Gorham, Patrick M. Crill, Robert C. Harriss,
Cheryl A. Kelley, Joseph B. Yavitt, Kurt A. Smemo, Randall K. Kolka,
Kelly Smith, Joon Kim, Robert J. Clement, Timothy J. Arkebauer,
Karen B. Bartlett, David P. Billesbach, Scott D. Bridgham, Art E. Elling,
Patricia A. Flebbe, Jennifer Y. King, Christopher S. Martens, Daniel
I. Sebacher, Christopher J. Williams, and R. Kelman Wieder

CONTENTS

Introduction

Peatlands are one of the biosphere's most important reservoirs of recent atmospheric carbon (C). Covering only 3% of Earth's land area, they store some 15%–30% of the world's soil carbon as peat, and have acted as net sinks of atmospheric carbon dioxide (CO_2) for millennia (Gorham 1991; Limpens et al. 2008). Waterlogged conditions slow decomposition, and slow rates of subsurface flow allow the partly decayed organic matter to accumulate in place. But the same processes of anaerobic decomposition that allow C to accumulate also produce methane (CH_4), a strong greenhouse gas. Natural wetlands account for 20%–40% of global CH_4 emissions, with northern peatlands contributing some 30% of this source (Wuebbles and Hayhoe 2002; IPCC 2007). Over the time span of centuries, peatlands exert a net cooling effect on the global radiation balance, because the effect of removing long-lived atmospheric CO_2 ultimately surpasses that of releasing short-lived CH_4 (Frolking and Roulet 2007). However, should peatlands begin to degrade on a large scale, this stored C could be released, reducing—or even reversing—their climate cooling effect.

With their water and nutrient supplies primarily from precipitation, and cool temperatures limiting decomposition, peatlands are highly vulnerable to changes in climate (Gignac and Vitt 1990). Most peatlands are located in the boreal and subarctic northern hemisphere, where the climate is warming faster than anywhere else on Earth (IPCC 2007). Peatlands are also affected by numerous other environmental factors that are likely to change in the future, including precipitation amount and frequency, atmospheric deposition of reactive nitrogen (N) and sulfur (S), atmospheric CO_2 levels, extreme weather, and fire (Gorham 1991; Dise 2009a). Understanding and predicting the net response of peatlands to these interacting drivers presents a considerable challenge to the scientific community.

Over the last 30 years, the Marcell Experimental Forest (MEF) has played a significant role in the quantification and understanding of peatland C emissions. In this chapter, we review the investigations at the MEF on CH_4 and CO_2 emission from peatlands. We begin with CH_4, first describing the measurements of CH_4 emission at the MEF and how they are extrapolated to regional scales. We then turn to research on understanding the controls on CH_4 emission, including statistical analyses to identify the most important potential drivers, as well as field and laboratory manipulations of these drivers. We use the same approach to investigate CO_2 fluxes and drivers at the MEF, first by considering measures of net ecosystem CO_2 exchange, and then by separating out the components of respiration and photosynthesis. Next we open the peat "black box," describing studies on the activity and dynamics of the microbes that produce CH_4 and CO_2. Finally, we use simple models developed from this research to predict changes in the emission of C compounds in a future, warmer world.

Methane Emissions Measurements at the MEF: Assessing the Contribution of Northern Peatlands to the Global Budget of CH_4

Background

As with CO_2, the structure of the CH_4 molecule allows it to adsorb and emit infrared radiation, making it an atmospheric greenhouse gas. Methane has a global warming potential of 25 times that of CO_2, and atmospheric concentrations of CH_4 are increasing (IPCC 2007). As awareness of global warming began to increase in both the scientific and public arenas in the 1970s, quantifying the role of peatlands in the atmospheric CH_4 budget became a priority.

Methane is produced biologically by strictly anaerobic microorganisms belonging to the Domain Archaea. There are two major pathways for microbial methanogenesis in freshwater systems: autotrophic reduction of CO_2 using hydrogen as an energy source, and fermentation of simple C1 and C2 carbon compounds (methanol, methylamines, acetate) to CH_4 and CO_2 (Boone et al. 1993). The former process generally dominates in more recalcitrant peat, while the acetate fermentation pathway predominates where there is a higher level of labile organic C, such as from root exudates (Hornibrook et al. 1997; Lai 2009).

Methanogenesis occurs when concentrations of oxygen (O_2), nitrate (NO_3^-), iron ($Fe(III)$), and sulfate (SO_4^{2-}) are low (Kamal and Varma 2008), since these compounds change the redox state of the environment or stimulate more efficient competitors (aerobic, denitrifying, iron-reducing, and sulfate-reducing bacteria, respectively). Finally, CH_4 itself is a substrate for methanotrophic bacteria, which use oxygen (and other compounds) to oxidize CH_4 to CO_2. Both processes of CH_4 production and oxidation may occur simultaneously in peatlands, with production dominating in anoxic zones at or below the level of the water table, and oxidation in aerobic peat in contact with CH_4 (for instance, diffusing from below) (Kettunen et al. 2000).

Methane is emitted to the atmosphere by diffusion through peat, ebullition (bubbling), or transport through emergent vegetation. Vascular plants, particularly those with aerenchyma, such as sedges, can greatly increase the rate of CH_4 emission by acting as active or passive conduits and allowing CH_4 to bypass the zone of CH_4 oxidation in the upper peat (Sebacher et al. 1985; King et al. 1998; Joabsson et al. 1999). Plants can also increase CH_4 production via the release of root exudates and the production of labile litter (Jones 1998).

The ideal terrestrial habitat for producing and emitting CH_4, therefore, is an ecosystem in which oxygen levels are low, a suitable C source is available, aerenchymous vegetation is present, and there is little or no connection to sources of dissolved oxygen or nutrient ions in groundwater, throughflow, or seawater. Many peatlands are such habitats.

Initial studies of CH_4 dynamics at the MEF primarily consisted of flux measurements and environmental observations in a variety of peatland habitats (Figure 10.1). Research was structured to quantify the magnitude of emissions, and identify and model broad relationships to controlling factors. Subsequent intensive studies were conducted at a single peatland to understand temporal dynamics and controls on emission at finer scales.

FIGURE 10.1
Methane flux measurements by Nancy Dise at Junction Fen, 1990, using one of the same chambers as employed by Gill et al. (1988). (Photo courtesy of S. Verry, USDA Forest Service, Grand Rapids, MN.)

Methane Emission across Peatland Habitats

First Peatland CH₄ Measurements in the United States

Robert Harriss, Eville Gorham, and colleagues initiated at the MEF the first measurements of CH_4 emission from northern peatlands in the United States. In August 1983, they measured CH_4 fluxes from bogs in S1, S2, and S4 and from the fen in S3, in all taking 21 measurements (Appendix 10.1; Chapter 2). The MEF investigations were augmented by 25 measurements from four other wetlands in northern Minnesota. Methane flux was measured with a dynamic flow-through aluminum chamber, in which gas accumulating in the chamber was continuously circulated via a closed-loop hose to an infra-red gas analyzer on site (Sebacher and Harriss 1982).

Harriss and colleagues observed both low and high rates of CH_4 emission, with high-emission events (100–1000 mg CH_4 m^{-2} day^{-1}) the most frequent. They hypothesized that the lower fluxes reflect a combination of low rates of respiration and largely molecular diffusion, whereas the high values represent scattered "hot spots" of decaying, recently dead organic matter, or CH_4 bubbles. At the MEF, CH_4 emission was highest from bogs at S1 and S4, next highest from the S2 bog (with a lower water table than S1 or S4), and very low from the fen at S3, which they hypothesized was influenced by active input of flowing, aerated groundwater. Overall, CH_4 fluxes from the MEF and the other Minnesota wetlands were higher than those reported from other peatlands (at the time only Sweden and England), and were similar to, or higher than, fluxes measured from rice paddies and cypress swamps.

Published in *Nature* (Harriss et al. 1985) this landmark work showed that northern peatlands were potentially highly important in the global budget of CH_4. The authors pointed to the need for similar studies in peatlands in Canada and the (then) Soviet Union and, in a provocative note, ended with the speculation that the development of boreal peatlands over the last ~11,000 years may have contributed to a long-term increase in atmospheric CH_4.

Corroboration and Expansion

From the study at the MEF and surroundings, Harriss et al. (1985) proposed that peatlands could be a major source of atmospheric CH_4. But, was the CH_4 emission measured at the MEF and surroundings in 1983 representative of a typical summer and the wider peatland environment? This important question was addressed in the summer of 1986 by Patrick Crill and colleagues, who conducted a longer and more intensive survey of CH_4 fluxes at the MEF.

Crill and team measured seasonal CH_4 fluxes from the same sites as Harriss et al. (1985): a hollow and hummock in S2, two hollows in the open bog area of S4, and one site each in S1 and S3. Measurements were also made from Junction Fen just outside of the MEF (see Appendix 10.1) and from several other bogs and fens in the wider region. Porewater was collected at various depths and analyzed for concentrations of dissolved CH_4, and transects of CH_4 flux and porewater CH_4 concentration were conducted across bogs S2 and S4.

Crill and colleagues measured CH_4 emission from late April to the end of June 1986 (a generally warmer and wetter summer than the 25-year average). They used the same type of flow-through chamber as Harriss et al. (1985), or a closed-chamber technique, in which a chamber is placed over a fixed collar set in the peat (Figure 10.1). Gas accumulates in the chamber over time, and samples are periodically removed from the chamber headspace with a syringe and analyzed in the laboratory. There was no difference between these methods.

Averaging 203 mg CH_4 m^{-2} day^{-1}, CH_4 emission rates measured at the MEF in 1986 were not significantly different from the 152 mg CH_4 m^{-2} day^{-1} measured in 1983 (Crill et al. 1988). As in the previous survey, CH_4 fluxes were greater from the open bogs than from forested bogs, and greater from hollows than from hummocks (perhaps due to warmer temperatures and higher water tables in both open bogs and hollows). Porewater profiles of CH_4 showed the highest concentration just below the surface, at depths of 5–15 cm. Methane concentrations in the profiles also increased over time during the study period.

The increase in both emission and porewater concentration of CH_4 over the summer suggested that temperature was an important control on CH_4 production and fluxes. However, a 16 h study in S4, in which fluxes were measured every 2 h from 04:00 until 20:00, showed no distinct daily pattern in emission despite a 20°C range in air temperature. This indicated that the temperature of deeper peat, rather than surface peat or air temperature, controlled CH_4 emission from this site (temperature fluctuations were damped out by 10 cm). It also supported the validity of CH_4 fluxes measured during the day as being representative of emission at any time in the 24 h period. In addition, the lower CH_4 emission from hummocks versus hollows, and forested bogs versus open bogs, emphasized the importance of aerated surface peat for controlling CH_4 fluxes by providing conditions suitable for CH_4 oxidation.

Relating fluxes to chemical conditions in the peat also demonstrated the importance of C quality, as more decomposed peat (in both S2 and in the Red Lake peatlands) was associated with higher CH_4 emission. Higher CH_4 fluxes were also measured from sites with porewater pH > 4.5. The influence of groundwater was highlighted by work in the fen at S3, the only site with significantly different CH_4 emission from the earlier study. Harriss et al. (1985) found very low CH_4 emission from this site (only 1–2 mg CH_4 m^{-2} day^{-1}), whereas levels measured during the Crill survey were much higher (mean of 142 mg CH_4 m^{-2} day^{-1}). Crill and colleagues hypothesized that groundwater or throughflow can play a dual role in CH_4 emission: relatively fast-moving subsurface water (the situation in 1983) brings in oxygen and removes dissolved gases produced in the peat; however if the flow is slow and oxygen is depleted (as in 1986), CH_4 emission can be enhanced through the favorable acid-base balance and higher nutrient levels of groundwater.

Perhaps most importantly, the extensive study of Crill and colleagues showed that the high CH_4 fluxes measured at the MEF by Harriss et al. (1985)

were not an exception. Northern peatlands indeed had the potential to be major sources of global atmospheric CH_4.

Full-Year Measurements

In the autumn of 1988, Nancy Dise began a 2-year study to determine the seasonal and annual emission of CH_4 from a variety of peatland habitats at the MEF, and to investigate potential controls on emission. The growing seasons of 1988–1990 were warmer than the 30-year average, and mean precipitation declined from well above average in late 1988 to below average in 1990. To maximize continuity with the previous studies, Dise also studied the S2 bog (a 33 cm high hummock, hollow, and lagg), S4 (two hollows in the open part of the bog), and Junction Fen (slight hummock). The S2 hummock site and one of the S4 hollow sites were identical to those measured by Crill et al. (1988). The lagg in S2 is a distinct ecosystem: an area approximately 15 m wide at the upland–peatland intersection, where subsurface flow from the surrounding upland forest meets the edge of the domed bog (Appendix 10.1; Chapters 2, 4, and 7).

Methane emission was measured at weekly to monthly intervals from September 1988 through September 1990. Other studies included monthly porewater profile samples analyzed for CH_4 and dissolved ions, CH_4 gas flux transects across both S2 and S4, additional replicate sites in S2 and S4, flux measurements at other sites within the MEF and in the surrounding region, and collection of gas over 24 or more hours to determine the long-term linearity of CH_4 concentration changes over time in the chambers. Dise also conducted water table manipulation experiments in the S2 bog, and Junction Fen. These experiments are described later in the chapter.

Dise found that CH_4 emission increased in spring to a peak in mid-July, then decreased gradually through autumn and winter to the lowest level in late March (Dise 1991, 1993; Figure 10.2a). As in the previous studies, fluxes ranged over three orders of magnitude, from <1 to over 1000 mg CH_4 m^{-2} day^{-1}. However, Dise showed that this range was not only evident *across* sites, but *within* some peatlands (in this case, the S4 bog) over the course of a year. She also identified both spatial hot spots of emission (also shown by Harriss et al. 1985) along survey transects, and temporal "pulses" of emission in summer (as reported concurrently in other peatlands, e.g., Mattson and Likens 1990; Moore et al. 1990). Pulses of CH_4 were most pronounced from the hydrologically most variable sites S4 bog and S2 lagg. Methane emission in 1990 was lower than in 1989, primarily due to the absence of these pulses. Although it cannot be determined whether pulses were missed between measurements, 1990 was a dry year (precipitation about 25% below 1989 levels, which were near average), and it would be expected that CH_4 emissions would be correspondingly lower.

Across sites, CH_4 emission increased as peatland wetness, summer peat temperature, density of graminoids (particularly sedges), and porewater pH increased (with a threshold at pH 4.0–4.5), and as the level of peat decomposition at the mean water table decreased. The latter observation was opposite

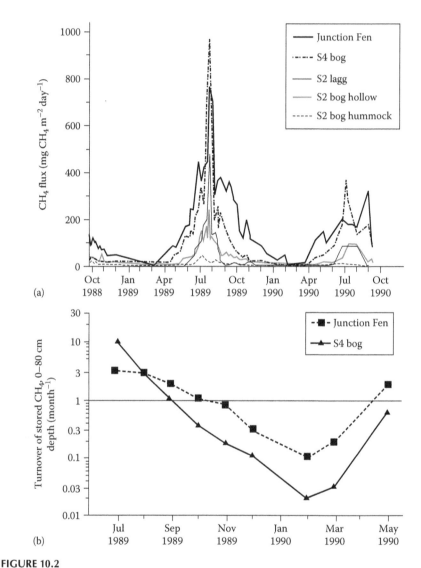

FIGURE 10.2

(a) Methane emission from peatlands in and near the MEF, September 1988–1990. (b) Monthly
CH_4 turnover rate at 0–80 cm depth, Junction Fen and S4 bog. (Modified from Dise, N.B., *Global
Biogeochem. Cycles*, 7, 123, 1993. With permission. Copyright 1993 American Geophysical Union.)

to that reported by Crill et al. (1988) from different areas in the same peatland. This discrepancy may be resolved by considering the broad differences in nutrients, hydrology, and vegetation composition across peatlands,
in contrast to the finer scales of variability in these factors within a peatland. Junction Fen has a higher proportion of productive vascular plants that
enhance CH_4 production and emission, as well as produce less decomposed
peat, than more ombrotrophic bogs such as S2. The rank order of annual CH_4

emission was: Junction Fen ≈ S2 lagg (summer only) > S4 bog > hollow in S2 bog ≈ S2 lagg (rest of year) > hummock in S2 bog.

Only the S2 lagg changed its relative order of CH_4 emission over the year, showing low emission in winter, spring, and autumn (similar to S2) and very high emission in summer (similar to Junction Fen). The flux chamber on this site, serendipitously, enclosed a large individual of *Calla palustris* (water arum), which was not visible when sites were selected in the autumn, but which emerged in early May. Methane emission closely paralleled the growth stage of the plant, and declined sharply with its senescence at the end of the summer (Dise 1993). These observations suggest that *Calla* provided a conduit for CH_4 transport, similar to that identified for other vascular peatland plants (Sebacher et al. 1985; Joabsson et al. 1999).

Dise (1993) also derived a simple measure to gain insight into the seasonal dynamics of CH_4 production and release. Comparing monthly CH_4 emission per square meter of peatland (g CH_4 m^{-2} month^{-1}) with the calculated amount of CH_4 stored in a hypothetical block of peat in the bog, with dimensions $1\,m^2$ in area by $0.8\,m$ deep (in g CH_4, by multiplying porewater CH_4 concentration by soil bulk density), she showed that in June through August the amount of CH_4 stored in this volume needed to turn over at least 10 times per month to account for the amount of CH_4 emitted, whereas the CH_4 emitted in February was only 1–10% of the reservoir (Figure 10.2b). This suggests a rapid net production of CH_4 in the summer months, and a near (or total) cessation of production in winter, with CH_4 flux dominated by diffusion of stored gas. Although the lowest CH_4 fluxes were consistently recorded in late March, the turnover calculations indicated that methanogenesis activity was already increasing by that time (Figure 10.2b).

One of the major surprises of the Dise study was significant CH_4 emission over the winter (Dise 1992). Methane emission from November 1988 through March 1989 (the year when most winter measurements were made) ranged from 4% to 21% of annual emission, with significant CH_4 emission over two winters despite these winters being very different: a deep snowpack with a thin surface freeze in 1988/1989, and a shallow snowpack with more than 15 cm of frozen peat in 1989/1990. Significant emissions were measured both when the snowpack was partially removed (allowing chambers to sit on the peat surface) and when the chamber was placed on the top of the snowpack. Using the same method over non-peatland soil, no CH_4 fluxes were measured. The ratio of March/July emission proved an excellent predictor of the winter fraction of annual flux ($r^2 = 0.99$), suggesting a way to estimate the contribution of winter CH_4 emission when direct measurement is not possible (Dise 1992).

Combining the Dise study with work by Harriss, Crill, and colleagues provides a picture of the multi-year "emission profiles" of the different peatland habitats of the MEF (Figure 10.3). Although fluxes generally seem higher in the 1986 survey, these differences are statistically significant only for the hummock in the S2 bog (too few data for comparison of the lagg fluxes or 1983 survey).

Methane Emission Dynamics within a Peatland

Taken together, the surveys of Harriss, Crill, Dise, and colleagues at the MEF became at the time the most detailed multi-annual record of CH_4 emission from a range of peatland ecosystems. But, although the peatlands at the MEF were analyzed over several years, each was measured only at a limited number of locations, and from a small area. How representative are these of the emission across the entire peatland? How frequent are pulses in CH_4 emission over time and space, and are these pulses associated with distinct

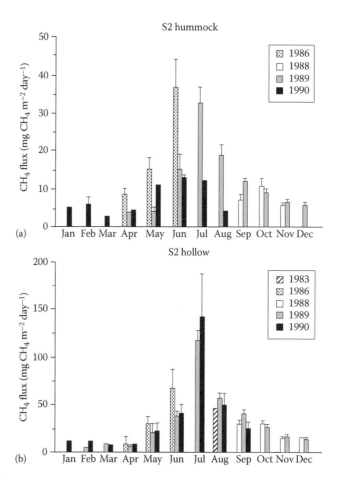

FIGURE 10.3

Monthly average CH_4 fluxes at peatlands at the MEF during different study years. Bars show one standard error of the mean and asterisks indicate fluxes measured as zero. Data from (a) bog S2 hummock, (b) bog S2 hollow, (c) the S2 lagg, (d) bog S4 and (e) Junction Fen (e). (Modified from Dise, N.B., *Global Biogeochem. Cycles*, 7, 123, 1993. With permission. Copyright 1993 American Geophysical Union.)

(continued)

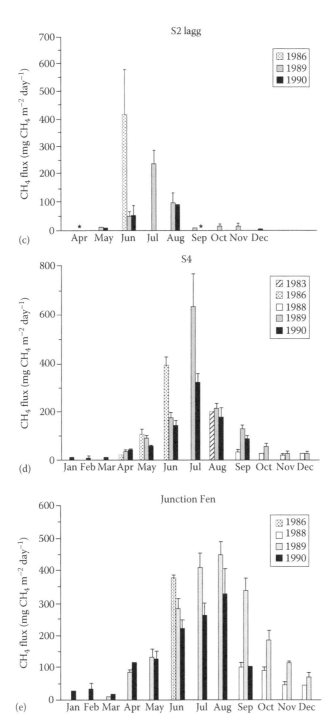

FIGURE 10.3 (continued)

environmental circumstances? These questions were addressed in two major studies conducted at the Bog Lake fen (BLF) in the early 1990s, with a systematic analysis of the spatial and temporal variation of CH_4 emission measured from chambers, and an intensive study of CH_4 and CO_2 emission at the landscape scale as measured by tower-based eddy covariance techniques.

Bog Lake Fen Chamber Studies

Bog Lake fen is an open, nutrient-poor fen (Appendix 10.1; Chapter 2). Methane emission was measured from an array of collars at BLF from July 1990 through October 1992 in a study by Kelly Smith with Sandy Verry (Smith 1993). Compared to the 30 year May–October average, the 1990 and 1991 sessions were drier and warmer, and the 1992 season was wetter and slightly cooler.

Smith and Verry sampled 16 collars over the 27-month period, with 8 collars each arranged north and south of a tower at which CH_4 and CO_2 fluxes from the same surface were measured using eddy covariance techniques. Within each of these rows, four hummocks and four hollows were measured. Sampling took place one to two times per week during the ice-free season, and approximately monthly in winter.

Smith and Verry found for BLF, as Dise (1993) had for S4, that CH_4 flux rates could range from zero to more than $1000\,mg\,CH_4\,m^{-2}\,day^{-1}$. Both spatial variabilitiy and temporal variability were highest when CH_4 emission was highest, in summer. They found that six replicates were necessary to reduce the 90% confidence interval on CH_4 emission to within 50% of the mean in summer; declining to three replicates in winter. This provided a (rather uncomfortable) approximation of the uncertainty inherent in the previous studies of Harriss, Crill, and Dise, who typically measured from one to three collars at a location.

The average CH_4 flux at BLF over the 1990–1992 measurement period, at $107\,mg\,CH_4\,m^{-2}\,day^{-1}$, was lower than that at the seemingly similar Junction Fen measured in 1988–1990 ($180\,mg\,CH_4\,m^{-2}\,day^{-1}$; Dise 1993), perhaps in part due to more favorable climatic conditions during the earlier study. The mean rate of CH_4 emission also declined over each of the three summers from 1990 through 1992. As found previously, CH_4 flux was related to topography (hummocks showed lower fluxes than hollows), and to seasonal temperature (fluxes peaked in summer). Vegetation also played a clear role: other factors being equal, there was a positive relationship between the number of sedges enclosed in a collar and the CH_4 emission.

The more intensive sampling and better statistical power of the Smith and Verry study allowed them to determine that CH_4 emission from hollows was *significantly* higher than from hummocks only slightly less than half of the time, and primarily in summer. Smith and Verry also found that the peat in hollows was more N-rich than those in hummocks (Smith 1993). This may reflect different vegetation composition between hollows and hummocks, which in turn could affect substrate quality for methanogenesis (Bubier and Moore 1994; Valentine et al. 1994; Lai 2009).

In addition to reporting, as others had previously, pulses of CH_4 during warm, wet periods in summer, Smith and Verry found that these events were immediately followed by days of very low CH_4 emission (Smith 1993). These pulses occurred several times in 1991 but were not observed in 1992. In a subsequent analysis of this work, Sandy Verry (unpublished data) found that the recurring pattern of high pulses followed by low emission was essentially independent of the broad controlling factors of temperature, hydrology, and vegetation. Verry hypothesized that, as found in previous studies, this pattern was likely related to the passing of low-pressure and high-pressure systems which enhance or retard gas flow from the peat, respectively (Mattson and Likens 1990).

Bog Lake Fen Eddy Covariance CH_4 Studies

Smith and Verry's study was concurrent with an intensive study of both CH_4 and CO_2 emission at the landscape scale at BLF conducted by Shashi Verma, N.J. Shurpali, and a team of researchers. This was the first field study at the MEF in which both CO_2 and CH_4 were measured simultaneously.

Verma and colleagues used a tower-based eddy covariance technique to estimate fluxes of CO_2 and CH_4 from the surface of the peatland. Eddy covariance determines fluxes by calculating the covariance between the concentration of a compound in the air and the vertical wind speed (Clement et al. 1995). The tower-based technique not only allows integrated measurements of C exchange at hectare to km^2 scales (depending on sensor heights and local meteorology), it also provides high temporal resolution. The technique thus allows the detection of pulses of gas and other short-term features that may be missed with chamber-based methods. Gas sensors were mounted at 2.5 m, and measurements were made of supporting variables such as wind speed and direction, humidity, photosynthetically active radiation, net radiation, and air temperature. Peat temperature, water table position, and CH_4 concentration in porewater were also measured.

Daily mean CH_4 fluxes in BLF in 1991 measured by eddy covariance gradually increased from about 30 mg CH_4 m^{-2} day^{-1} in May to 125–160 mg CH_4 m^{-2} day^{-1} from mid-July to mid-August, when peat temperatures were high and the water table was close to the surface (Figure 10.4a). Subsequently, fluxes declined rapidly with falling peat temperatures to the end of the measurement season in October. The average CH_4 emission over the 5-month period (145 days) was 105 mg CH_4 m^{-2} day^{-1}. Accounting for expected CH_4 emission during the rest of the year (based on data available at the time, including the full-year studies of Dise 1991), the estimated range for annual mean CH_4 emission of BLF was 44–53 mg CH_4 m^{-2} day^{-1} (Shurpali et al. 1993).

The eddy covariance method detected episodic pulses of CH_4 from BLF in summer 1991, similar in strength, and sometimes in timing, to those detected with the BLF chamber measurements of Smith and Verry, but with a much finer temporal resolution (Figure 10.4a). This allowed CH_4 pulses to be linked more

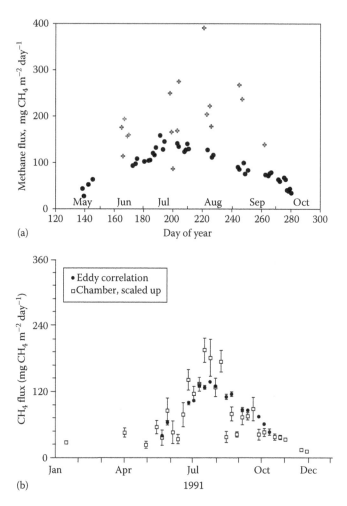

(a)

(b)

FIGURE 10.4

Seasonal distribution of daytime CH$_4$ flux at Bog Lake fen during 1991, eddy covariance measurements (a). Background level of emission is denoted by solid circles, episodic emission is denoted by plus signs. (Modified from Shurpali, N.J. et al., *J. Geophys. Res. D Atmos.*, 98, 20649, 1993. With permission. Copyright 1993 American Geophysical Union.) Comparison between 1991 CH$_4$ emission at Bog Lake fen measured by eddy covariance (filled circles) and using chambers (open squares), the latter adjusted for topography (b). Points are weekly means. Error bars are ±1 standard error of the mean. (From Clement, R.J. et al., *J. Geophys. Res. D Atmos.*, 100, 21047, 1995. With permission. Copyright 1995 American Geophysical Union.)

firmly to specific environmental conditions than had been previously possible (Shurpali et al. 1993). Methane pulses were associated with declining water table and atmospheric pressure, and a buildup of CH$_4$ gas in peat porewater. Dissolved CH$_4$ was often depleted in porewaters following pulses (Shurpali et al. 1993), similar to observations by Smith and Verry that CH$_4$ fluxes following pulses were low, even if conditions were conducive to production and ebullition

(Smith 1993). Also similar to the chamber studies at BLF, no large pulses of CH_4 were detected by eddy covariance in 1992. Disregarding the pulses, CH_4 emission in the 1992 season (May–October) was about 10% higher than in 1991.

Using the midpoint of the range of estimated annual emissions (described above), the mean CH_4 emission from BLF for January–December 1991 calculated by eddy covariance is 49 mg CH_4 m^{-2} day^{-1}. We can compare this to the chamber-derived estimates by taking the average of two annual integrations including this period from Smith and Verry: August 1990–1991 and May 1991–1992. This gives a mean emission of 91 mg CH_4 m^{-2} day^{-1} (Smith 1993), or nearly twice as high as the tower-derived value.

In an explicit comparison of these two methods, Robert Clement, Shashi Verma, and Sandy Verry found the same absolute differences in CH_4 emission in 1991 (43 mg CH_4 m^{-2} day^{-1}), with most weeks (8 out of 15) significantly higher from chambers and only one week lower (Clement et al. 1995). However, on average, the chambers were coincidentally located on lower ground than the mean height of the peatland as a whole. When chamber measurements were adjusted to the topography using a relationship developed between CH_4 emission and surface height, this difference was reduced to 12 mg CH_4 m^{-2} day^{-1}, with most weeks (10 of 15) not significantly different between chamber- and tower-derived estimates (Clement et al. 1995) (Figure 10.4b). In addition, during those weeks where estimates were different, the systematic bias toward high fluxes from chambers disappeared. This indicates that well-designed chamber-measurement programs can provide a good estimate of the gas emission from a landscape by capturing the range of variability within dominant habitat types and topographies. Several subsequent studies support this conclusion (e.g., Frolking et al. 1998; Christensen et al. 2000; Dinsmore et al. 2009).

Scaling Up CH_4 Emissions

Gas emission measurements from chambers or towers can be extrapolated over time in several ways. If measurements are made at frequent intervals and over a full year (e.g., Dise 1993; Smith 1993) then annual emissions can be calculated by integrating individual measurements. With less frequent measurements, or a study restricted to a shorter period (e.g., Crill et al. 1988; Shurpali et al. 1993), annual emissions can be estimated by multiplying measured values by an assumed season, or by using relationships between CH_4 emission and physical parameters for which annual data exist, such as air temperature, peat temperature, precipitation amount, or depth to water table. Scaling up annual estimates of CH_4 flux from chambers or towers to a whole peatland or wider region requires taking into account site-specific factors such as the topography and type of peatland.

Although there is a wide range in calculated annual CH_4 emission from the different peat habitats at the MEF, using the mean annual values of 4–66 g CH_4 m^{-2} year^{-1} for the sites measured by Dise (1993), and conservatively assuming a 75% proportion of forested bogs (which generally emit less CH_4 due to cooler and drier conditions) over open bogs or fens, gives an estimate of about 25 g CH_4 m^{-2} year^{-1} for the mean emission rate of CH_4 from peatlands at the MEF. Scaling this value over the calculated area of the Great Lake peatlands of Minnesota, Wisconsin, and Michigan (60 × 10^3 km^2; Kivinen and Pakarinen 1980) gives a projected annual emission of 1.5 Tg CH_4 (Dise 1993) (1 Tg = 10^{12} g or one million metric tons). This is about one-half of the CH_4 emission calculated for all peatlands in Canada, with nearly 20 times the area (3.2 Tg, Bridgham et al. 2006), and about the same level of CH_4 emission as has recently been calculated for all natural wetlands in China (1.76 Tg, Ding and Cai 2007).

The first CH_4 measurements at the MEF by Robert Harriss in 1983 have been followed by many more across the northern hemisphere. These now show that the original estimates of the contribution of peatlands to the global CH_4 budget, based largely on measurements at the MEF, were an overestimate by about an order of magnitude. This is because the MEF, and other (but not all) peatlands in the temperate–boreal transition zone, emit some 10–30 times more CH_4 per unit of surface area than their boreal and subarctic counterparts (e.g., Lai 2009).

To date, we do not fully understand this phenomenon. Temperature, favorable local (and perhaps microscale) hydrology, and high plant productivity all likely play important roles. Temperature at the MEF and similar temperate peatlands may be cold enough to limit rates of aerobic decomposition, but—with no permafrost—warm enough to allow active populations of anaerobic microbes to persist throughout the year, allowing high rates of CH_4 production when conditions are favorable (Yavitt et al. 1997). The regional hydrology also supports peatlands that are generally wetter than those in boreal Canada (P. Crill, 2009, personal communication), as well as fens such as Junction or BLF that have no outlet, which limits flushing rates and allows accumulation of organic sedge compounds (S. Verry, 2009, personal communication). In addition, the glacial topography at the MEF, with thick drifts and tills, may allow some subsurface and groundwater flow that at times could supply nutrients and raise the porewater pH (S. Bridgham, 2010, personal communication; Siegel et al. 1995). Even low inputs of higher-pH water could significantly enhance CH_4 emission: methanogenesis in peatlands has a pH optimum around 6, several units above the usual pH of ombrotrophic bogs (Dunfield et al. 1993), and an increase in pH from only 4.0–4.6 was found in a New England peatland to be associated with a tenfold increase in CH_4 fluxes (Duvall and Goodwin 2000).

Controls on the Emission of CH_4 from Peatlands

Background

Simply measuring gas fluxes gives insight into the amount, location, and timing of CH_4 emissions, and provides data for global models. Background measurements alone, however, cannot explicitly identify the drivers of emission, nor how emission will change should those drivers change. Assessing the consequences of factors such as climate change on CH_4 emission requires an understanding of the processes that control emission. This section provides an overview of research on the controls on CH_4 emission undertaken at the MEF, some within studies already described earlier in this chapter.

From the processes described previously, CH_4 emission would be expected to increase as peat temperature increases and as the water table approaches the surface (increasing the anaerobic zone for methanogenesis and decreasing the aerobic zone where CH_4 oxidation takes place). Methane emission would also be enhanced in peat with more easily-degraded C, more sedges, higher porewater pH and (within limits) higher nutrient concentrations. Methane emission would be lower in cooler peatlands, those with a lower water table, more recalcitrant peat, or with relatively high concentrations of NO_3^-, Fe(III), or SO_4^{2-}.

Correlative statistical analyses, field manipulations, and laboratory experimentation can all be used to shed light on controls on C fluxes. These techniques are complementary; ranging from the most "natural" systems with the least level of experimental control, to highly controlled but artificial systems. If the results of different techniques agree, they can provide strong support to a hypothesis. Conversely, differing results can prompt reexamination of assumptions and experimental methods, leading to new hypotheses and, ultimately, insights.

Correlative Analyses

Correlative analyses use statistical techniques such as correlations and regressions to identify potential drivers on a process such as CH_4 emission. Simple empirical models use these relationships to predict the rate of emission. Such analyses postulate a relationship between a process and a driver (or drivers) such that, as the magnitude of the driver changes, the rate of the process changes in a way that can be represented mathematically. Thus, sporadic events such as pulses of CH_4 emission may be poorly characterized by correlative analyses. Although correlation does not prove causation, strong statistical relationships between processes and putative drivers can lend good support to the hypothesis that one leads to the other, as well as supporting experimental studies. We shall discuss such correlative analyses undertaken at the MEF, first, across peatlands, and then, within a single peatland.

Models of CH₄ Emission across Peatlands

In the 1986 summer survey, Crill et al. (1988) determined that the strongest measured environmental variable linked to CH_4 emission was the peat temperature (measured at 10 cm below the water table surface). Regressing the

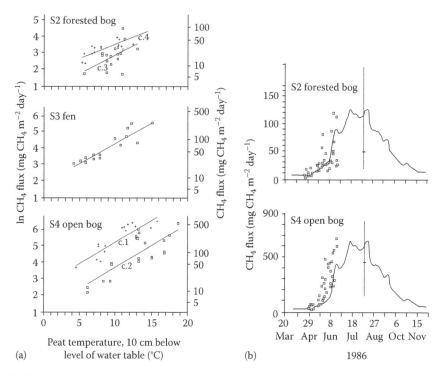

(a)

(b)

1986

FIGURE 10.5

(a) Natural logarithm of CH$_4$ emission versus temperature at 10 cm below the level of the water table for 1986 (a). C1 and C2 are from hollows in open area of the S4 bog; C3 is from a hummock in the forested bog of S2; C4 is from a hollow in S2. Methane fluxes from the S2 and S4 bogs, modeled for a full year using Equation 10.1 and temperature data at 30 cm depth in the nearby S1 bog (b). Data from 1986 are shown as open squares; vertical lines show the mean and range of flux measurements from August 1983 at the same sites (Harriss et al. 1985). (Modified from Crill, P.M. et al., *Global Biogeochem. Cycles*, 2, 371, 1988. With permission. Copyright 1988 American Geophysical Union.)

natural logarithm of CH$_4$ emission against the peat temperature at 10 cm below the water table led to good relationships for all five of the MEF sites examined (two collars in S2, one in S3 and one in S4; Figure 10.5a). Crill and colleagues found that an exponential relationship provided the best fit between CH$_4$ emission and hypothesized drivers of emission:

$$CH_4 \text{ flux} = \exp\left[-a\left(1/T\right)+b\right] \quad (n = 24; \; p < 0.05), \tag{10.1}$$

where

CH$_4$ flux is in units of mg CH$_4$ m^{-2} day^{-1}

T is temperature in K at 10 cm below the water table surface

a = 15,257.9 and 17,893.0, and b = 57.2 and 67.9, for S2 and S4, respectively

Equation 10.1 explained 74% and 66% of the variation in CH_4 emission from the S2 and S4 bogs, respectively. The relationship was then used to model the annual flux of CH_4 from S2 and S4 bogs, using a record of annual peat temperature from the nearby S1 bog. This model gave values of 12 and 58 g CH_4 m^{-2} year^{-1}, respectively (Figure 10.5b), which compare well to the estimates of Dise (1993) of 9 and 43 g CH_4 m^{-2} year^{-1} derived through integration of measured values over April 1989–1990.

Crill et al. (1988) argued that the strong relationship shown in Figure 10.5a demonstrated that temperature is the primary physical control on CH_4 emission from these peatlands. Catchment hydrology added a small, but significant, additional amount toward explaining the variability in CH_4 emission at a site: times of higher runoff relative to precipitation at both S2 and S4 were associated with lower CH_4 fluxes, presumably due to higher flow rates of oxygenated water and the removal of CH_4 in throughflow.

Dise (1991) and Dise et al. (1993) also showed that at is any single peatland, there is a strong significant relationship between the CH_4 emission on any day and the peat temperature on that day ($r^2 = 0.5$–0.9). The relationship between the CH_4 flux and water table position is much weaker, and there is no relationship if temperature is taken into account. However, if all sites are considered together, there is a clear segregation by hydrology: at any given temperature, drier peat habitats show lower fluxes than wetter sites (Figure 10.6a). Dise (Dise 1991; Dise et al. 1993) postulated that this reflects two different phenomena that act together to produce CH_4 emission characteristic of a peatland: a long-term effect of hydrology and peatland type on the overall composition and population balance of the microbial communities, and a short-term biochemical effect of temperature on microbial metabolic rates. In a study across a range of Canadian peatlands made around the same time, Roulet et al. (1992) also noted that the "persistent hydrological regime" was a critical factor in determining the magnitude of mean annual CH_4 flux. Subsequent studies have supported these observations (e.g., Kettunen et al. 1996).

Reflecting this, although CH_4 flux at individual survey sites correlated only to temperature, water table position became the most important correlate across peatland ecosystems, accounting for more than 60% of the variance in CH_4 flux. The best-fit model of CH_4 emission incorporated the peat temperature on the day emission was measured, but the water table averaged from the previous month (Dise 1991; Dise et al. 1993). A small, but significant additional correlate was the humification index measured on the Von Post scale:

$$Log\ (CH_4 + 1)\ flux = 0.214\ (MW) + 0.0765\ (T_{30}) - 0.154\ (VP)$$

$$+ 1.63 \quad (n = 141,\ p < 0.0001) \tag{10.2}$$

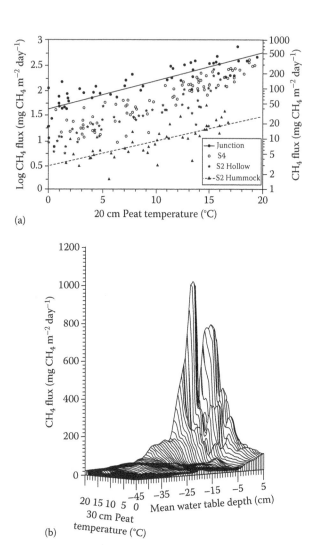

(a)

(b)

FIGURE 10.6

Log CH$_4$ flux versus peat temperature, 1988–1990 (a). Regression lines for Junction Fen and the hummock in the S2 bog are shown. Methane flux versus peat temperature (30 cm) and water table depth (monthly running average), 1988–1990 (b) and model results from Equation 10.2 (c). Kriging interpolation used to create surfaces. Peaks in surface curves are July pulses of CH$_4$ in the S4 bog and Junction Fen. Regression model (Equation 10.2) of the data shown in Figure 10.6b. Predicted versus measured geometric mean CH$_4$ flux, May to November 1989 (except where other dates are noted) (d). Models are similar to Equation 10.2; all validation data are independent of model development data. Sites are (1) the hummock in the S2 bog, (2) corral controls in the S2 bog, (3) the hollow in the S2 bog, (4) collar 2 in the S4 bog S4, (5) corrals in the S2 bog, (6) collar 1 in the S4 bog, (7) corrals (May to August 1990) in the S2 bog, (8) Junction Fen, and (9) control collars in Junction Fen (June to September 1989). (Modified from Dise et al., *Journal of Geophysical Research, D, Atmospheres*, 98 (D6), 10583, 1993. All equations in reference.)

(continued)

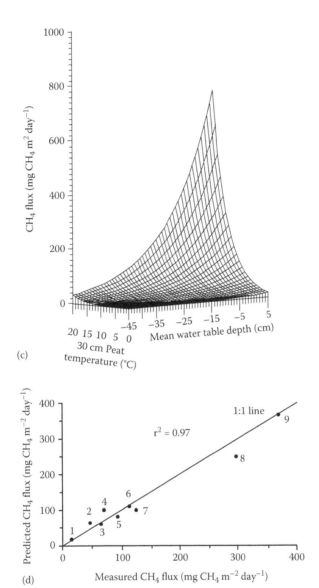

(c)

(d)

FIGURE 10.6 (continued)

where (in addition to CH_4 units as previously described),

MW is the monthly running mean water table in centimeters (0 = surface, negative values are below the surface)

T is temperature in degrees Celsius at 30 cm depth (both T and WT measured approximately once per week)

VP is the Von Post scale value at 30 cm, where higher values denote higher levels of humification (note a negative relationship)

The value of 1 is added to CH_4 emission to avoid negative log values for zero emission cases. Equation 10.2 explained 91% of the variation in the 2 year dataset of five peatland habitats studied by Dise (1993). Figure 10.6b shows the relationship between CH_4 emission, water table, and temperature from this study using kriging interpolation of the data; Figure 10.6c shows the corresponding modeled relationship described by Equation 10.2.

Equation 10.2 was tested with independent data from the MEF (Dise et al. 1993) using (1) a split data set (model developed on one subset of the data and validated on the other), (2) the control sites of the S2 bog and Junction Fen water table manipulation studies (described below), and (3) the 1986 measurements from Crill et al. (1988), as well as CH_4 flux data from boreal peatlands in Alaska (references in Dise et al. 1993). The model performed well in predicting CH_4 emission from the MEF peatlands (nine sets of data from different peatlands, experiments, or time periods) regardless of year ($r^2 = 0.7$–0.8), and very well at predicting the geometric mean emission from all sites at the MEF ($r^2 = 0.97$; Figure 10.6d). It was, however, poor at predicting CH_4 emission from the peatlands in Alaska ($r^2 = 0.15$).

Models of CH_4 Emission within a Peatland

Smith and Verry also used regression analyses to develop empirical relationships between CH_4 emission and peat temperature at BLF (Smith 1993). Data were divided by hydrology into two groups: water table above or below -2 cm. Models were developed on individual flux measurements, mean daily values of emission, and mean values averaged over 7–11 days ("weekly model"). Episodic pulses of CH_4, and emission during days when peat temperature was below freezing, were excluded from the model.

Smith and Verry also found, as was shown by Dise et al. (1993), that temperature was sufficient to describe CH_4 emission at the individual peatland BLF; water table explained little or no additional variability at this single site. All models fit the data well (r^2 typically >0.65) and model fits improved as data aggregation increased. The weekly model incorporated 48 measurements over the 27 month study:

$$\text{Log}_{10}(CH_4 \text{ flux}) = e^{(0.962)} - e^{[-0.000125 \times (T^{3.14})]}, \quad (n = 48, p < 0.05), \qquad (10.3)$$

where T is temperature in degrees Celsius measured at 10 cm depth (Figure 10.7). Equation 10.3 explained 86% of the variation in the data.

Shurpali, Verma and colleagues examined the CH_4 flux–peat temperature relationship at BLF over the same time period (1991–1992) as Smith and Verry, but using landscape-scale estimates from the eddy correlation towers. Relationships between CH_4 flux and peat temperature were examined for three classes of water table position: 15–25 cm below the surface, 0–15 cm below the surface, and above the surface. Within these categories, CH_4 emission was estimated with a Q_{10} model:

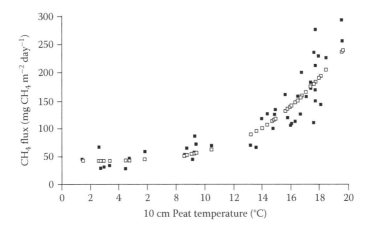

FIGURE 10.7
Weekly mean CH_4 emission versus soil temperature, Bog Lake fen, 1990–1992. Closed squares are measured data and open squares are modeled results (Equation 10.3). (Modified from Smith, K., Methane flux of a Minnesota peatland: Spatial and temporal variables and flux prediction from peat temperature and water table elevations, MS thesis, University of Minnesota, St. Paul, MN, 1993.)

$$CH_4 \, flux = (CH_{4(10C)}) \times Q_{10}^{(T-10)/10} \tag{10.4}$$

where

$(CH_{4(10C)})$ is the CH_4 emission at 10°C

T is temperature in degrees Celsius at 10 cm below the peat surface

Q_{10} is the Q_{10} coefficient for CH_4 emission at BLF

The Q_{10} coefficient comes from thermodynamics and describes the rate a process changes with a 10°C change in temperature; in ecosystem science Q_{10} is sometimes used in a less exact way to express net processes (in our example, CH_4 or CO_2 emission). Thus used, Q_{10} can serve as a standard currency for comparing the temperature response of a flux across different ecosystems.

Similar to the other investigators, Shurpali and colleagues found that, within given water table ranges, CH_4 emission increases exponentially as peat temperature increases (Table 10.1; Figure 10.8a). The model fit was better at higher and lower water tables ($r^2 = 0.64$ and 0.70, respectively; Table 10.1), suggesting that fluctuations in water table between 0 and 15 cm were important for CH_4 flux in BLF (reducing the simple relation with temperature), whereas fluctuations below or above this level were less important. Thus, this is probably the region where water table exerts the strongest influence on CH_4 production and consumption in the peatland.

Shurpali and colleagues also examined relationships between CH_4 flux and water table depth for three peat temperature classes: 7–10°C, 10–13°C, and 13–17°C. Within these classes, CH_4 flux rate was linearly related to water

TABLE 10.1

Q_{10} Model Fit between CH_4 Flux and Peat Temperature
(0.10 m Depth) for Three Different Water Table Classes
(Based on Equation 10.4), and Linear Regression
between CH_4 Flux and Water Table Depth for Three
Different Temperature Classes (Based on Equation 10.5)

CH_4 flux and peat temperature relationship

W class	$CH_{4(10°C)}$	Q_{10}	R^2	P
Low W	64.7	2.0	0.64	<0.0001
Medium W	74.1	2.1	0.34	0.04
High W	88.8	2.3	0.70	<0.0001

CH_4 flux and water table depth relationship

T class	A	b	R^2	P
Low T	108.9	75.4	0.54	0.001
Medium T	120.3	89.1	0.62	0.0009
High T	146.6	126.9	0.30	0.006

Notes: $CH_{4(10C)}$, CH_4 flux (mg CH_4 m^{-2} day^{-1}) at 10°C peat temperature; (a) and (b), slope and intercept of line (Equation
10.5). Standard errors of coefficients (as a proportion of
the mean values shown) are: 5–20% for $CH_{4(10C)}$, 15–40%
for Q_{10}, 23–32% for a, and 3–5% for b.

table depth, being low when the water table was lowest and increasing until
the water table was close to or just above the surface (Figure 10.8b). Both
regression intercepts and slopes of the equations increased with increasing
peat temperature. The overall relationship was

$$CH_4 \text{ flux} = a \times (W) + b, \qquad (10.5)$$

where water table (W) is in meters. The best model fit occurred at intermediate temperature, suggesting optimal acclimation of methanogens to this
temperature. It was not investigated if the mean water table of several weeks
rather than at the time of measurement provided a better fit to the data, as
found elsewhere (e.g., Dise et al. 1993; Kettunen et al. 1996).

Shurpali and Verma (1998) also found that CH_4 flux was linearly related to
the rate of photosynthesis when data from the entire growing season were
included. This would be expected if labile C from recent photosynthesis partially regulates the production of CH_4 (e.g., King and Reeburgh 2002; King et
al. 2002; Waddington et al. 1996;) or, alternatively, if net primary production
integrates a number of variables that act as primary controls on CH_4 emission, such as temperature, hydrology, vegetation type, and peat quality (e.g.,
Whiting and Chanton 1993). On a daily time scale, there was no clear relationship between CH_4 flux and photosynthesis rate (Shurpali et al. 1998). This

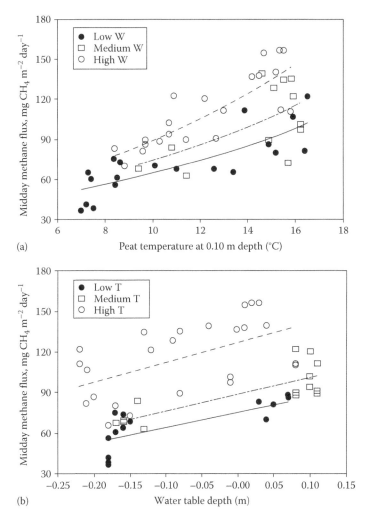

FIGURE 10.8
Methane flux at Bog Lake fen modeled with Equation 10.4 using peat temperatures (0.10 m depth) (a) and with Equation 10.5 using water table depths (b). W, water table depth; T, peat temperature; closed circles and solid curve or line, low W or T; open squares and dot-dashed curve or line, intermediate W or T; open circles and dashed curve or line, high W or T. Details are in Table 10.1.

is probably due to the different controls on these rates: CH_4 emission is primarily driven by belowground processes that would not respond as rapidly as photosynthesis to changing levels of light, temperature, and precipitation.

Since the research at the MEF described here, several studies have used regression or similar techniques to derive simple models of CH_4 emission using water table and peat temperature (e.g., Kettunen et al. 2000; Huttunen et al. 2003). These generally support the conclusions that CH_4 emission from peatlands can be modeled using temperature alone at a single site or where water table fluctuations are not large across sites, and by the recent mean water table together with temperature across several peatlands.

Field Manipulations to Identify Controls on CH_4 Emission

The correlative analyses indicate that temperature and hydrology are the primary controls on CH_4 emission, and provide insights into how these drivers act to determine fluxes. However, correlation does not prove causation, even when the models fit the data well. Furthermore, based on our understanding of the processes of CH_4 production, consumption, and emission, we have good reason to believe that these controls are modified, and may even be dominated, by other factors such as the vegetation composition (with certain vascular plants acting as conduits for emission and sources of labile C), the quality of the peat, and the chemistry of the porewater. This section describes field manipulations of the main variables in the correlative models to test the extent to which they are indeed drivers of CH_4 emission.

Water Table

The hypothesis that the recent mean water table has the strongest influence on peatland CH_4 emission across peatlands by Dise andcolleagues was tested at the MEF by Dise and colleagues (Dise 1991; Dise et al. 1993) with field experiments designed to manipulate the water level within the S2 bog and Junction Fen. Three "bog corrals," open-ended, square, sheet metal enclosures ($1.2 \times 1.2 \times 1.5$ m) were installed in each peatland in January 1989 by cutting the outline of the corral into the frozen peat with a chainsaw and hammering the corral vertically into the peat until only 30 cm was exposed above the surface. Lateral drainage was thus cut off and, because the lower peat is dense, vertical drainage was very slow. One collar was installed in a hollow in each corral. After 1 month of background CH_4 measurements, water level in the corrals in the S2 bog was raised to the surface in July 1989, and kept there through November. The experiment was repeated the following summer. In Junction Fen, the water level was lowered to 5 cm below the surface in late July 1989 by pumping water out of the corral. The experiment was continued for 5 weeks, until the corrals were overtopped by an extreme flooding event in early September.

Raising and maintaining the water table at the surface in the S2 bog led, after a lag of about 3 weeks, to significantly higher CH_4 fluxes (Figure 10.9a). Between August and November 1989, total CH_4 emission from corrals was 9.5 g CH_4 m^{-2}, 2.5 times higher than CH_4 flux from controls (3.8 g CH_4 m^{-2}), where water table averaged 6.1 cm below the peat surface. In 1990 (May through August), flux from the corrals was 15.7 g CH_4 m^{-2} versus 7.1 g m^{-2} from the controls (2.2 times higher), with the average water table difference at 10.4 cm. Although, as previously described, there was no correlation between

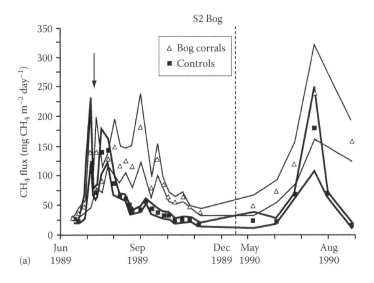

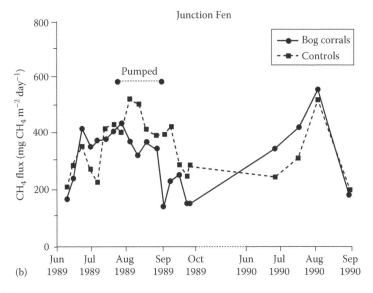

FIGURE 10.9

Methane emission from bog corrals and controls, June 1989 to August 1990. Each point in (a) is the mean of two to three measurements at the S2 bog. Boundary lines are ±1 standard error of the mean. Arrow indicates start of experiment in early July 1989. (Modified from Dise et al., *Journal of Geophysical Research, D, Atmospheres*, 98 (D6), 10583, 1993.) Data from Junction Fen shown in (b). The experiment started in late July and ended in late August 1989; the period is shown with line marked "pumped." Each point is the mean of two to three measurements. (Modified from Dise, N.B., Methane emission from peatlands in northern Minnesota, PhD dissertation, St. Paul, MN: University of Minnesota, 1991.)

CH$_4$ flux and water table position in the hollow site in the S2 bog, water table depth accounted for 34% of the variation in flux (in addition to 38% due to temperature) when the corrals and controls were considered together. This suggests that the part of the bog isolated by the corrals had in effect become a distinct "peatland within a peatland" (Dise et al. 1993).

Methane emission from Junction Fen, where the water table was lowered to 5 cm below the surface from July 25 through August 31, was also affected by the change in water table. Here, however, a change was observed sooner (by August 8) and, although only pumped for 5 weeks, the effect could be detected over an additional 8 weeks after pumping had ceased, until measurements ended in October (Figure 10.9b). This more pronounced response to the treatment may in part be an artifact of pumping, which may initially expel (unmeasured) stored porewater CH$_4$ through the relatively rapid drawdown, as well as enhancing the flow of oxygenated water through the subsurface peat. Total flux from the corrals between August 8 and September 28 was 16.8 g CH$_4$ m^{-2}, 28% less than that from the controls (23.3 g CH$_4$ m^{-2}). The experiment was not repeated the following year. A (nonsignificant) trend of higher CH$_4$ emission at the bog corral sites in 1990 (Figure 10.9b) may be related to the restricted lateral flow in the corrals, which would both reduce oxygen levels and allow sedge decomposition products to accumulate in the corrals.

The regression model developed from the study of Dise et al. (1993; Equation 10.2) was then used to test the field manipulation results with fully independent data. Only S2 had sufficient data suitable for this comparison. Since only 10 cm peat temperature was measured in the manipulation experiments, a new regression model was developed from the survey data using temperature at 10 cm rather than at 30 cm (Dise et al. 1993).

The model reproduced both the autumn decline in flux and the effect of the raised water table in enhancing CH$_4$ emission from the corral sites (Figure 10.10). It also successfully predicted the geometric mean fluxes from the controls (Figure 10.6d, point 2), as well as the increase in CH$_4$ emission from the corrals in both 1989 and 1990 (Figure 10.6d, points 5 and 7, respectively). In finer detail, the regression model predicted the control fluxes from the hollows in the S2 bog to be slightly higher than actually observed, and it predicted the increase in CH$_4$ emission in response to raising the water table to be less pronounced than actually observed (Figures 10.6d and 10.10). Both responses may be related, because the model combines several different peatlands, with the outcome of the model "smoothing" or integrating the responses from all sites. A similar smoothing result was found in a later study using a multiple regression approach to model CH$_4$ emission from peatlands in Finland (Kettunen et al. 2000).

Plant Transport

In 1992, Kelly Smith and Sandy Verry designed a simple field manipulation experiment at BLF to investigate the influence of plant transport on CH$_4$

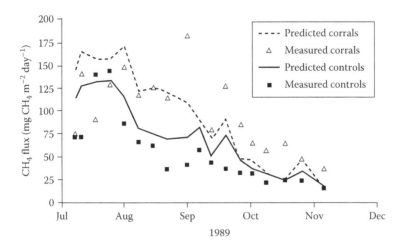

FIGURE 10.10
Test of regression model (similar to Equation 10.2) using the corral and control data from the
S2 bog, 1989. Equation in reference. (Modified from Dise et al., *Journal of Geophysical Research, D,*
Atmospheres, 98 (D6), 10583, 1993.)

emission: they clipped the emergent vascular plants (primarily *Rhynchospora*
alba and *Scheuchzeria palustris*) at collars in both the north- and south transects at BLF. Smith and Verry found that clipping vegetation decreased CH_4
flux by 64% compared to unclipped controls (Smith 1993). This experiment
was later conducted for boreal peatlands, with similar results (Waddington
et al. 1996). To explore whether this effect was primarily due to the vegetation acting as a passive conduit or through active transport, Smith and Verry
inserted straws to 5 cm below the peat surface at approximately the density
of the original stems. This manipulation increased CH_4 emission by 67%.

Although there are of course major differences between straws and living vegetation, we can conclude through the remarkable agreement between
the emission decline after clipping and the emission increase with the simulated vegetation that passive transport alone could potentially account for
increased CH_4 emission due to plants. It is not necessary to invoke active
transport through vascular tissue, or the continuous production of labile C
from root exudates. It is possible, however, that labile C leached from decomposing roots from the clipped vegetation may have stimulated methanogenesis and accounted for some of the increase in CH_4 flux.

Atmospheric Deposition of Sulfur and Nitrogen

As described previously, methanogenesis is suppressed in the presence of
oxidized inorganic compounds such as O_2, NO_3^-, Fe(III), and SO_4^{2-}, by raising redox levels, stimulating energetically more efficient bacteria, or both
processes. This leads to an intriguing question: are the relatively low levels

of $SO_4{}^{2-}$ or $NO_3{}^-$ in acid deposition sufficient to suppress CH_4 emission from peatlands?

In summer 1994, Nancy Dise and Sandy Verry set up a study in BLF to address this question. With very low background levels of acid deposition, the MEF was as an ideal location for the study. Over the growing season, SO_4^{2-} (as ammonium sulfate; $(NH_4)_2SO_4$) and N (as ammonium nitrate, NH_4NO_3) were added in low weekly doses to simulate the level of acid deposition of polluted areas of central Europe or Asia (about 30 kg S or N ha^{-1} year^{-1}). The ammonium salt was used to simulate a typical composition of acid deposition. Separate treatment plots were also harvested at peak season for biomass measurement, although disturbance in the field rendered the sulfate-treatment biomass samples unsuitable for analysis.

There was no significant difference in CH_4 emission between "designate" control and treatment plots measured over the 2 weeks before treatments began. One week after treatments began, CH_4 flux from the sulfate-amended sites was significantly suppressed in relation to controls, but there was no significant difference between N-amended plots and controls. Over the growing season, integrated total CH_4 emission from the sulfate-amended plots was 32% lower than for the controls (Dise and Verry 2001; Figure 10.11). This suppression was most important during the highest flux periods, so that the overall effect of the SO_4^{2-} treatment was to eliminate pulses of CH_4.

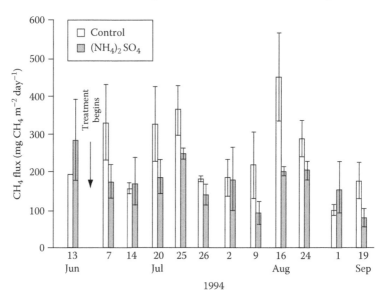

FIGURE 10.11
Emission of CH_4 from control versus treatment plots receiving weekly $(NH_4)_2SO_4$ additions, Bog Lake Fen. Adjacent columns represent measurements made on the same day; error bars show ±1 standard error of the mean. Treatment began on June 28, 1994 with nine weekly applications of 2.4 kg ha^{-1} NH_4^+–N and 2.7 kg ha^{-1} SO_4^{2-}–S. (Modified from Dise, N.B. and Verry, E.S., *Biogeochemistry*, 53, 143, 2001. With permission.)

There was no difference over the growing season between controls and treatments for N-amended plots, although there was a trend toward higher CH_4 emission and higher vascular plant biomass (Dise and Verry 2001). Together with the absence of an increase in nitrous oxide (N_2O) emission, this suggests that the main effect of the nitrogen addition (both nitrate, NO_3^-, and ammonium, NH_4^+) at BLF is to stimulate primary productivity, which can enhance CH_4 emission via plant transport and root exudates (Jones 1998; Joabsson et al. 1999). If, as would be expected, the input of NH_4^+ in the $(NH_4)_2SO_4$ plots would have also increased plant biomass, the suppression of CH_4 observed from the SO_4^{2-} treatment may be a conservative estimate, since some counterbalancing enhancement of CH_4 emission would be expected from an increase in vascular plant productivity.

Dise and Verry (2001) concluded that acid-rain levels of SO_4^{2-} could significantly suppress CH_4 emission from peatlands due to competition from SO_4^{2-}-reducing bacteria. This was the first time that such an effect was demonstrated in the field. In addition, the rapid response to the SO_4^{2-} addition, in contrast to the lagged response with the S2 water table manipulation (described previously), suggests that SO_4^{2-}-reducing bacteria (or at least enzymes) are already present and active in BLF. An intriguing hypothesis is that these organisms exist in a syntrophic relationship with methanogens through inter-species hydrogen transfer (Conrad et al. 1987) in the absence of SO_4^{2-}, and that this association shifts to a competition when the energetically more favorable SO_4^{2-} appears.

Carbon Dioxide Emissions, Budgets, and Controls

Background

Peatlands can be considered ecosystems that transfer, on millenium timescales, atmospheric CO_2 into peat (Lund et al. 2009). Although the aboveground net primary productivity of peatlands is lower than that of other terrestrial ecosystems, decomposition rates are proportionally lower still due to recalcitrant litter quality and anaerobic, cold soils. As a consequence, long-term C accumulation rates in peatlands are some 5–50 times larger than soil C accumulation rates for upland ecosystems (Schlesinger 1990; Frolking et al. 1998).

The two major C fluxes in undisturbed peatland ecosystems are CO_2 uptake through gross primary production (GPP), and CO_2 loss through the respiration of plants and microbes. Unlike CH_4, the rates of CO_2 production and consumption processes in peatlands are of similar magnitude, such that many peatlands can be either a significant source or a sink of CO_2 at different times and circumstances. The two processes of respiration and photosynthesis are linked, since a considerable portion of C fixed by plants through photosynthesis is sent belowground, where it can be used to support production as well as respiration of roots, mycorrhizal fungi, and heterotrophic rhizo-sphere microbes (Jones 1998). The difference between ecosystem respiration and GPP is the net ecosystem exchange (NEE), with positive values denoting a source of CO_2 to the atmosphere, and negative values a sink into the ecosystem. In practice, measurements of net CO_2 fluxes from enclosures or towers that integrate both soil and vegetation fluxes measure NEE.

Net ecosystem exchange and its components (GPP, respiration) clearly vary seasonally (with lowest rates in winter and highest in summer), but also from year to year at one site, and spatially across different sites (Lund et al. 2009). In an analysis of 12 peatland and tundra ecosystems, Lund et al. (2009) showed that the most important correlate to both GPP and respiration was the length of the growing season. Respiration rates were also significantly related to a number of environmental variables such as air temperature, growing degree days, normalized vegetation index, and vapor pressure deficit. Gross primary production, on the other hand, was more poorly correlated to environmental variables and more strongly related to vegetation metrics such as leaf area index.

Both components of NEE generally increase along the trophic sequence from bog to rich fen: bogs usually show lower rates of primary productivity, but also lower rates of respiration, than more productive wetlands (Bubier et al. 1998; Dise 2009b). In turn, these different habitats and plant communities may respond in different ways to changes in environmental conditions (Bubier et al. 2003; Humphreys et al. 2006; Ström and Christensen 2007). However, relatively strong feedbacks between rates of productivity and ecosystem respiration dampen the net response of peatland NEE to local-scale variability (Humphreys et al. 2006; Knorr et al. 2008; Limpens et al. 2008).

The MEF was an early site for the measurement and analysis of CO_2 fluxes from peatlands. The first studies focused on intensive tower-based landscape-scale measurements at a single site (BLF); more recent work has focused on extensive, plot-based measurements of different peatlands.

CO_2 Fluxes and Budgets within a Peatland

Net Ecosystem Production at BLF

Narasinha Shurpali, Shashi Verma, and colleagues used the same tower instrumentation as used for CH_4 (described previously) to measure C exchange from BLF in 1991–1992. The team reported their results in terms of net ecosystem production (NEP), or the C that is incorporated into the ecosystem (NEP = –NEE). Whether BLF was a net sink or a source of CO_2 depended on environmental and hydrological conditions during the growing season (Figure 10.12). Conditions favoring high CO_2 uptake rates (positive NEP) prevailed during measurement periods in June and early July in 1991 and July until mid-September in 1992. Generally, higher temperatures and vapor pressure deficits, together with significant drops in the water table

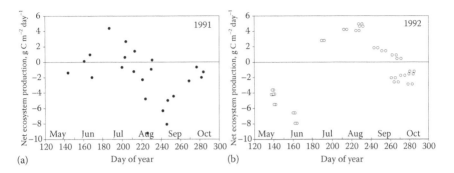

FIGURE 10.12
Net ecosystem production (NEP) in Bog Lake fen, May to October (a) 1991 and (b) 1992. (Modified from Shurpali, N.J. et al., *J. Geophys. Res. D Atmos.*, 100, 14319, 1995. With permission. Copyright 1995 American Geophysical Union.)

due to dry spells, occurred during August and September in 1991 and May and June in 1992. During such periods, vegetation responds to climatic and peat moisture stresses by restricting photosynthetic CO_2 uptake. At the same time, soil and plant respiration can be high since the decline in the water table exposes more peat to air. These combined conditions contributed to negative values of NEP.

Integration of the daily NEP indicated that BLF released about 71 g C m^{-2} (0.50 g C m^{-2} day^{-1}) over a 145 day period (mid-May to mid-October) in 1991. With conditions more favorable for high photosynthesis rates from the moss and vegetation canopy, and low soil and plant respiration rates, the ecosystem accumulated about 32 g C m^{-2} (0.22 g C m^{-2} day^{-1}) during a similar period in 1992 (Shurpali et al. 1995). Thus, the calculated net NEP over the 1991 and 1992 growing seasons was −39 g C m^{-2}, making BLF a net CO_2 source over this period, and possibly a stronger annual CO_2 source if winter respiration is accounted for.

Partitioning NEE into Photosynthesis and Respiration at Bog Lake Fen

Since the tower CO_2 flux measurements integrate both photosynthesis and respiration, these must be measured separately to determine their individual contributions. Leaf-level measurements of gas exchange in *Scheuchzeria palustris* (arrow grass) at BLF provided estimates of photosynthesis of the emergent vegetation. Shurpali et al. noted that this plant, together with *Sphagnum papillosum* and other vascular plants such as leatherleaf, *Chamaedaphne calyculata*, would be the dominant species contributing to canopy photosynthesis in BLF (Shurpali et al. 1995).

Shurpali et al. (1995) stated that on a short-term timescale, factors such as light intensity and duration, air and peat temperature, soil moisture, moss moisture content (water table used as a proxy), and atmospheric vapor pressure deficit would be important regulators of CO_2 uptake at BLF as in other peatlands. Thus, on days with moderate air temperature (20°C–28°C), water table close to the surface (1–8 cm below the surface) and low vapor pressure deficit (1.2–1.5 kPa), photosynthesis rates were high. During periods of high temperature (30°C–34°C), low water table (>20 cm below the surface) and high vapor pressure deficit (1.9–2.4 kPa), photosynthesis rates were much lower. By fitting a rectangular hyperbolic function to the data, the value of maximum canopy photosynthesis estimated from the light response curve under moderate conditions during the active season was 0.28 mg C m^{-2} s^{-1}; this value was reduced to 0.07 mg C m^{-2} s^{-1} under stressed conditions (Shurpali et al. 1995; Figure 10.13).

Root and microbial respiration are the main sources of soil CO_2 emission. At the cellular level, these are controlled by enzyme activity which, in the short term, is strongly regulated by temperature. Soil surface CO_2 fluxes (combining soil biota respiration and vegetation root respiration) were measured at BLF from June through October 1991 by Joon Kim and Shashi Verma. Gas flux was measured with a 0.75 L dark chamber connected to an infrared gas exchange system.

Kim and Verma found that CO_2 emission from hummocks was consistently higher than from hollows (9.8 and 5.4 g CO_2 m^{-2} day^{-1}, respectively, or 2.7 and 1.5 g C m^{-2} day^{-1}; Kim and Verma, 1992). Carbon dioxide fluxes from the soil

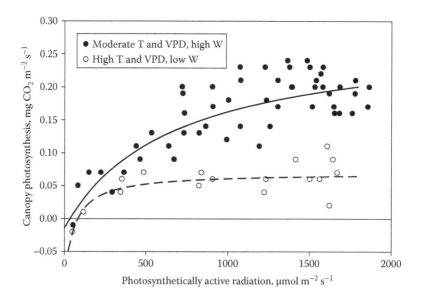

FIGURE 10.13
Relationship between canopy photosynthesis and photosynthetically active radiation under different climatic and hydrologic states. Data were fitted with a rectangular hyperbolic relationship. Closed circles represent moderate conditions (cool, wet); open circles represent stress conditions (hot, dry); and T is temperature, V is vapor pressure deficit, and W is water table depth. (Modified from Shurpali, N.J. et al., *J. Geophys. Res. D Atmos.*, 100, 14319, 1995. With permission. Equation in reference. Copyright 1995 American Geophysical Union.)

were low when the water table was close to the surface and peat temperature was low; fluxes were high when the water table was below the surface and peat temperature was high. Kim and Verma (1992) fitted a Q_{10} model to the measurements, which explained about 81% of the variation in the data:

$$CO_{2\ flux} = 6.97 \times [(0.002 + 0.025\ W)/0.012] \times (Q_{10})^{[(T-10)/10]} \qquad (10.6)$$

where

$CO_{2\ flux}$ is in g CO_2 m^{-2} day^{-1}
W is water table in meters
T is temperature in degrees Celsius
Q_{10} is the Q_{10} value for CO_2 fluxes measured over the study

Kim and Verma (1992) estimated a Q_{10} of about 3.7 for soil respiration (CO_2 efflux from soils) from both hollows and hummocks at BLF. With a similar temperature response, most of the difference in soil respiration can be explained by the differences in water table depth between the two microhabitats.

We can combine the above measurements of NEE, respiration, and photosynthesis to shed light upon the relative rates of these processes at BLF. Assuming an NEE of 0.5 g C m^{-2} day^{-1} emitted from mid-May to mid-October 1991 in BLF (Shurpali et al. 1995), and an average soil respiration CO_2 flux (assuming the same density of hummocks and hollows) of 2.1 g C m^{-2} day^{-1} over roughly the same period (Kim and Verma 1992) requires a mean value of photosynthesis of 1.6 g C m^{-2} day^{-1}. This is 2.5 h per day of photosynthesis at the mean rate between the maximum "active" and "stressed" conditions of 0.175 mg C m^{-2} s^{-1} (or 0.63 g C m^{-2} h^{-1}) calculated by Shurpali et al. (1995). Although this is a rather crude calculation, it suggests that photosynthesis rates are generally considerably lower than optimal at the MEF.

Soil CO_2 Measurements and Controls across Peatlands

Soil CO_2 Fluxes across Peatlands

From 2005 to 2006, Peter Weishampel and colleagues conducted an extensive study of peatland soil respiration and CO_2 efflux from the MEF (Weishampel et al. 2009). This work was initiated jointly between the Northern Global Change Research Program of the USDA Forest Service and NASA's Carbon Cycle Science program as part of a larger project to explore the use of intermediate landscape-scale sampling of C pools and fluxes to increase our understanding of C cycling (Bradford et al. 2008; Weishampel et al. 2009).

Soil respiration was measured approximately three times per week from April through October in 2005 and 2006 from the South Unit. Both summers of 2005 and 2006 were drier than average. Carbon dioxide emission was measured from 17 peatland plots, including 13 in open peatlands (poor

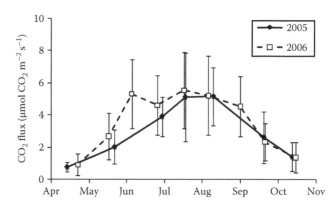

FIGURE 10.14
Carbon dioxide fluxes from 17 peatland soils measured at the MEF from April to October during 2005 and 2006.

fens to intermediate fens), two alder-dominated peatlands, and two conifer-dominated peatlands. A dynamic chamber system was used, with chambers 25 cm in diameter and 10 cm in height, together with an infrared gas analyzer. Peat temperature (10 cm depth) and water table elevation were recorded for each plot.

Mean soil CO_2 fluxes in 2005 and 2006 were under $1\,\mu mol\ CO_2\ m^{-2}\ s^{-1}$ (1.0 g C m^{-2} day^{-1}) in the initial spring sampling period in April (Figure 10.14). Fluxes peaked at approximately $5\,\mu mol\ CO_2\ m^{-2}\ s^{-1}$ (5.2 g C m^{-2} day^{-1}) during late July and early August, gradually declining to early spring levels by late October. Variability in fluxes tended to be highest in the mid-summer sampling periods. On average, peatland fluxes at the MEF during the growing seasons of 2005 and 2006 were 88% and 68%, respectively, of estimated global mean values (Roehm 2005). Carbon dioxide–temperature relationships and data on long-term winter soil temperature from the MEF were used to estimate that fluxes during winter (November to March) contribute approximately 13% to annual soil CO_2 emission.

Modeling Soil Respiration across Peatlands

Using regression analyses, Weishampel and colleagues analyzed the role of temperature and water table elevation in mediating CO_2 emissions from the MEF peatlands. Mean seasonal variability in CO_2 flux was closely related to mean seasonal variability in peat temperature. In 2005 and 2006, soil temperature at 10 cm explained 52% and 40%, respectively, of the variability in flux in an exponential relationship (Figure 10.15a, 2005 data shown). Variation in water table elevation of peatland plots was also significant, accounting for 24% and 8% of the variation in the natural log of peatland CO_2 flux in 2005 and 2006, respectively (Figure 10.15b, 2005 data shown). Generally, soil

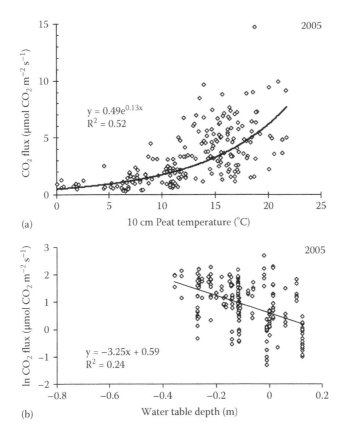

FIGURE 10.15
Relationships between CO_2 flux and peat temperature at 10 cm depth, (a) and ln CO_2 flux and water table elevation, from April to October during 2005 (b).

CO_2 fluxes were low at times of high water level; as water table declined, emissions increased.

Taken together, soil temperature and water table elevation were both significant in predicting CO_2 flux with stepwise multiple regression analyses for each year separately and for both years combined. The combined equation is

$$\text{Ln } (CO_2 \text{ flux}) = 0.115 \times T - 2.580 \times W + 0.089 \times TW \quad (p < 0.001) \tag{10.7}$$

where
CO_2 is in units of μmol CO_2 m^{-2} s^{-1}
T is temperature in degrees Celsius at 10 cm depth
W is the water table depth in meters relative to the soil surface
TW is an interactive term between temperature and water table

Equation 10.7 explained 52% of the variation in the data.

Weishampel and colleagues calculated the average daily fluxes of CO_2 from the MEF peatlands for the 2005 and 2006 growing seasons as 3.3 and 4.0 g C m^{-2}, respectively. These values are higher than the 2.7 g C m^{-2} day^{-1} (hummocks) and 1.5 g C m^{-2} day^{-1} (hollows) in 1991 calculated by Kim and Verma (1992) for BLF. This may reflect different weather conditions in different years, but may also indicate lower-than-average soil respiration rates at BLF due to a relatively high meanwater table.

Taking a mean value of 3.7 g C m^{-2} day^{-1} from the Weishampel survey, and comparing it to a calculated CH_4 emission of 0.051 g C m^{-2} day^{-1} from scaling up the Dise (1993) annual estimates of 25 g CH_4 m^{-2} year^{-1} (described previously), we can conclude that about 1.4% of total soil respiration C flux from the MEF peatlands to the atmosphere is emitted as CH_4.

Microbial Dynamics in Carbon Emissions

Most of the field investigations at the MEF have focused on net processes. Methane emission, for example, is the net effect of both production and oxidation of CH_4. However, understanding how carbon gas fluxes may change in the future in wetlands requires a thorough understanding of the controls on gas production and consumption. At the microbial level, these include species and functional groups, population sizes, activities, controls, substrates, and energy sources for metabolism. Several research teams at the MEF have studied such dynamics for the microbial populations that produce both CO_2 and CH_4.

Rates of Microbial Activity

In the early 1990s, Joseph Yavitt and colleagues used the MEF peatlands as part of an investigation of the controls on CH_4 and CO_2 production in peatland ecosystems (Yavitt et al. 1997). They collected samples in 12 peatlands (eight open bogs and four forested peatlands) across North America from Alberta, Canada, to Pennsylvania, spanning broad gradients in mean annual temperature (MAT), precipitation (MAP), and plant species composition. The S4 bog at the MEF was intermediate in the temperature range and toward the low end of the precipitation range of the open bogs. Peat soil was incubated in the laboratory at 2°C, 12°C, and 22°C, under oxic and anoxic conditions, and with and without added glucose.

Yavitt et al. (1997) found that all of the open bogs showed much higher CH_4 production, and higher CH_4 consumption, than the forested bogs. The S4 bog and a bog in New Hampshire, both at intermediate MAT, showed the highest rates of CH_4 production and the lowest rates of CH_4 consumption of the open bogs. In addition to temperature, the rate of CH_4 production was correlated with the level of certain phenolic compounds in the peat that are decomposition products of sedges (Williams et al. 1998).

The four coldest peatlands, including the S4 bog, also showed the lowest increase in microbial reaction rates in response to glucose, and the highest proportions of holocellulose in peat. Holocellulose is a major energy resource for decomposers, and is generally used first over more recalcitrant compounds such as lignin (Berg and McClaugherty 2007). The dominance of more labile C compounds in the colder sites, and the relatively slow response to glucose addition, suggest a stronger climatic, rather than substrate, limit on decomposition. This may partly explain the high CH_4 emission rates from the MEF peatlands compared to other peat-forming regions.

Yavitt and colleagues continued investigations on the controls on CH_4 and CO_2 production through reciprocal peat transplant experiments across three sites: a peatland in Ontario, Canada, one in West Virginia, and BLF (Yavitt et al. 2000). Peat samples from the surface (5 cm) and subsurface (40 cm) were transplanted across the three sites and left *in situ* for 4–25 months. Upon retrieval, the samples were incubated in the laboratory at 12°C and 22°C under both anoxic and oxic conditions as previously.

Similar to S4, BLF showed high levels of holocellulose (suggesting primarily climatic limits on decomposition rates), and the authors found a direct relationship between the CH_4 production rate and the ratios of phenolic compounds characteristic of *Carex* sedges. They suggested from this evidence that fresh sedges, where present, drive much of the CH_4 production in peatlands.

Yavitt et al. (2000) found that peat originating in BLF generally had higher rates of CH_4 production than peat samples from the other two sites. But, intriguingly, peat incubated within BLF had the highest rates of CH_4 production, regardless of origin. The transplanted peat samples experienced new climatic conditions in addition to new patterns in hydrologic fluxes, chemistry, and presumably, microbial communities. Yavitt et al. (2000) concluded that such site conditions, apparently optimized at the MEF, had an important influence on C mineralization. Rates of C mineralization to CO_2, in contrast, varied less with incubation site, and were primarily associated with peat type. The authors hypothesized that the organic matter composition of the peat plays a much more important role for CO_2 production and emission rates than it does for CH_4 production.

Methane production accounted for about 1.25% of the total molar C production of the three peatlands, with the rest as CO_2. This is similar to the 1.4% of respiration C emitted as CH_4 estimated by comparing the Dise (1993) with Weishampel studies across a range of peatlands at the MEF (described previously). It is low compared to measurements in other northern peatlands, which in some cases exceeded 30% of total C production as CH_4 (e.g., Yavitt and Lang 1990). These differences may in part be related to the overall productivity of the peatland: in a laboratory comparison of potential anaerobic C mineralization rates, Bridgham et al.

(1998) found the proportion of anaerobic C released as CH_4 approximately increased along an ombrotrophic–minerotrophic gradient, from very low levels (0.5%) in bogs to 10%–12% in intermediate fens and meadows, with BLF at about 2%.

Population Dynamics of Microbes

Stable Isotopes in the Field

Further insight into the processes driving CH_4 production came from the stable isotope work led by Cheryl Kelley during 1989–1990, coinciding with the field study of Dise (Kelley et al. 1992). The $\delta^{13}C$ value expresses the $^{13}C/^{12}C$ ratio in a substance relative to the standard. For CH_4 emission, this is influenced by several factors, primarily CH_4 oxidation rates, and the substrates used in methanogenesis. Since organisms preferentially assimilate CH_4 containing the lighter ^{12}C, CH_4 oxidation leads to a shift in the composition of emitted CH_4 toward a heavier (^{13}C-enriched) $\delta^{13}C$ value. In addition, acetate dissimilation is thought to produce CH_4 with a heavier $\delta^{13}C$ value than CO_2 reduction (Kelley et al. 1992; Hornibrook et al. 1997). Thus, heavier emitted CH_4 would indicate increased rates of CH_4 oxidation, a shift toward acetate fermentation in favor of CO_2 reduction in the CH_4 production process, or a combination of the two.

Kelley sampled from a hollow at S2 and in Junction Fen (at the same locations where CH_4 fluxes were measured by Dise) between June 1989 and April 1990. Methane emitted from the S2 bog was consistently lighter (^{13}C depleted) than from Junction Fen (Figure 10.16). She and colleagues hypothesized that the lighter $\delta^{13}C$ values in the S2 bog reflected differences in the main pathways of CH_4 production, with CO_2 reduction a more important pathway at

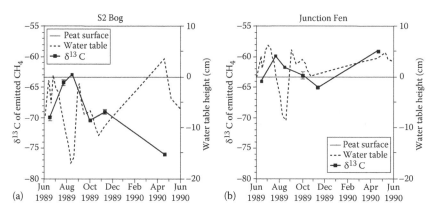

FIGURE 10.16
Isotopic composition of emitted CH_4, and water table level from June 1989 to June 1990 at the S2 bog (a) and Junction Fen (b). (Modified from Kelley, C.A. et al., *Global Biogeochem. Cycles*, 6, 263, 1992.)

S2, and acetate fermentation dominating at Junction Fen (Kelley et al. 1992). Carbon dioxide reduction generally dominates during the degradation of older organic material, whereas acetate dissimilation is the preferred pathway in the degradation of fresh organic matter (Hornibrook et al. 1997). Junction Fen is dominated by *Carex* sedges, which, in addition to acting as conduits for CH_4 release (Joabsson et al. 1999), can excrete sugars and low molecular weight organic acids as root exudates, precursors to acetate (Jones 1998). With no outlet (Appendix 10.1), these materials can accumulate in Junction Fen. The S2 bog peat is denser and more recalcitrant than Junction Fen peat, and recent decomposition products may flush out of the S2 watershed in streamflow.

Both wetland sites shifted toward heavier ^{13}C-enriched CH_4 during the warm summer months (Figure 10.16). The seasonal shift in $\delta^{13}C$ values was most pronounced for the S2 bog, and was strongly correlated to the level of the water table at S2, which was generally below the peat surface: as water table declined over the summer, $\delta^{13}C$ increased (Figure 10.16a). Kelley et al. (1992) hypothesized that this shift primarily reflects enhanced CH_4 oxidation during the drier summer months. Water table also declined at Junction Fen in August (Figure 10.16b), but was generally much higher, and the shift in the isotope ratios at Junction was hypothesized to be primarily due to a higher percentage of CH_4 produced by acetate dissimilation during the *Carex* growing season.

Anaerobic Oxidation of CH_4

Our estimates of CH_4 production and oxidation should sum to measured CH_4 emission, but often there is a wide disparity between the measured fluxes and the values we would expect based on our understanding of the controlling processes. This suggests that more complicated dynamics are involved in CH_4 production, consumption, and emission than are currently understood. Understanding the controls on CH_4 oxidation is particularly important for anticipating and modeling the impacts on CH_4 emission of hydrologic changes in peatlands, especially drought, that are predicted to accompany global warming.

One interesting new research area concerns the potential for anaerobic microbial oxidation and removal of CH_4, potentially occurring simultaneously with CH_4 production. It is well known that a significant proportion of CH_4 is oxidized by aerobic methanotrophic bacteria in peatlands, but the occurrence and importance of anaerobic CH_4 oxidation (involving non-oxygen electron acceptors such as NO_3^-, Fe(III), or SO_4^{2-}, and assumed to be widely occurring in marine wetlands) is not known. Indeed, no anaerobic CH_4 oxidizer has yet been isolated or grown in pure culture.

Kurt Smemo and Joseph Yavitt used peat cores from both BLF and Junction Fen to investigate anaerobic oxidation of methane (AOM) in freshwater peatlands. The MEF sites were part of a series of fen sites from Minnesota,

New York, and Sweden used to investigate this phenomenon. In the laboratory, Smemo and Yavitt (2007) showed with three independent methods that AOM occurred across all sites, with relatively low levels at the MEF peatlands, and the highest levels in the most minerotrophic fen in New York.

The experiments showed that AOM occurs simultaneously with methanogenesis, and fractionates C isotopes toward more ^{13}C-enriched CH_4. AOM is likely more important in minerotrophic peatlands because of the higher availability of inorganic electron acceptors. It is also possible that more *Carex*-dominated peatlands would support higher levels of AOM if organic matter decay products were important in the process. There remains much to be learned about this process, indeed it is possible that some of the isotopic changes observed in Junction Fen by Kelley et al. (1992) showing a shift toward heavier δ^{13}C could be due to AOM.

An interesting result from this research was that in the surface peat at both BLF and Junction Fen, Fe(III) addition *enhanced* methanogenesis, rather than reducing rates as expected (due to either Fe(III) acting as an electron acceptor for AOM or simple redox changes favoring Fe-reduction over methanogenesis). This suggests that iron may be a limiting trace nutrient for methanogenesis in Fe-poor peatlands such as those at the MEF (Basiliko and Yavitt 2001).

Climate Change and Peatland Carbon Emission

Having described research on the measurements, upscaling, controls, and processes involved in C emission from peatlands at the MEF, we turn to applying this knowledge to estimating perhaps the most important question: How will CO_2 and CH_4 production, removal, and emission change in a warmer future world? To gain insight toward the answer we first turn to Q_{10} estimates and then to applications of simple empirical models, both described previously in the chapter.

Response of Carbon Emission to Temperature Change

Q_{10} Values for Carbon Emission and Production Rates

Since they express the result of a temperature change on a process or flux, Q_{10} rates can be extrapolated to predict a response to climate change. These must be used with caution, however, since few if any ecosystem-scale processes respond solely to temperature changes. Respiration rates (both CH_4 and CO_2), for example, are driven by changes in soil moisture, microbial community dynamics, vegetation composition, and other factors that will mitigate or amplify the temperature response. Nevertheless, Q_{10} rates can provide some valuable insights into the potential for change, and in comparing the potential rates of different processes.

Values of Q_{10} for CH_4 emission for March to November 1989 for the five peat sites studied by Dise (1993) averaged 5.1, with the lowest (2.9) in the S2 hummock and the highest (7.7) in the S2 lagg. Q_{10} values increased with increasing mean water table, mean peat temperature, macrophyte cover, pH, and nutrient levels.

In BLF, a mean Q_{10} value of 2.1 was estimated by Shurpali and colleagues for CH_4 emission from late May through October (1991 and 1992) using eddy covariance (Table 10.1). Part of the reason for the higher Q_{10} rates in the Dise study may be that the Dise Q_{10} values incorporated early spring measurements, when there was a rapid increase in CH_4 emission with temperature. A corresponding early spring increase might have been missed by the eddy covariance flux measurements, which began in mid-May. However, an experimental Q_{10} of 2 was also estimated in the laboratory by Yavitt et al. (2000) for CH_4 production in BLF peat.

Rate constants for CH_4 consumption in BLF from Yavitt et al. (2000) were insensitive to the measurement temperature. This suggests that the Q_{10} value for CH_4 emission primarily reflects the response to temperature change of CH_4 production, rather than the combined response of CH_4 production and CH_4 oxidation. For a range of temperate and subarctic peat slurries incubated at 25°C, Dunfield et al. (1993) reported Q_{10} values of 1.4–2.1 for CH_4 consumption and 5.3–16 for CH_4 production. This supports the hypothesis that temperature is a less important driver for CH_4 oxidation than for CH_4 production.

For CO_2, Q_{10} values determined by Weishampel for a range of peatland habitats in the MEF from April through October (measured in 2005 and 2006) averaged 3.3, similar to the value of 3.7 determined in 1991 for BLF by Kim and Verma (1992). In laboratory incubations, Yavitt et al. (2000) found somewhat lower rates for CO_2 production for BLF, about 1.4, with no seasonal difference.

If we assume that the Dise and Weishampel studies sampled across representative habitats, comparing Q_{10} values between CH_4 (Q_{10} of 5.1) and CO_2 (Q_{10} of 3.3) suggests that CH_4 emission will respond more strongly to changes in peat temperature than CO_2 fluxes. In addition, Yavitt et al. (2000) reported lower Q_{10} values for CH_4 production in peat collected in mid-summer than earlier or later in the season. This led them to conclude that the response of CH_4 emission to summer warming will be less than to warming at other times of the year, and that peatlands located in colder climates, when warmed, will show larger increases in C mineralized to CH_4 than more temperate peatlands. Since climate models generally predict that rates of warming will be greatest at higher latitudes and, in many areas, highest in the non-summer months (IPCC 2007), the conclusion from Q_{10} studies that CH_4 emission responds to changing temperature more strongly than CO_2, with the response most pronounced in spring and autumn, are particularly relevant.

Model Predictions

For peatlands, Q_{10} values provide good relative estimates of how CH_4 emission will change with changing temperature alone, but not on the relationship between flux and other factors such as hydrology or substrate quality. Changes in CH_4 or CO_2 emission as a function of multiple drivers can, however, be estimated with some of the empirical models described previously.

Simulating a 3°C increase in April–October peat temperature at 30 cm peat depth (similar to 5°C increase at 10 cm), from 10°C to 13°C, with mean water table remaining at the surface, equation 10.2 from Dise et al. (1993) predicts a 72% increase in CH_4 emission (89 to 154 mg CH_4 m^{-2} day^{-1}). If this temperature increase is accompanied by a water table decline of 5 cm, however, CH_4 emission increases by only 26% (89 to 112). If both water table (−5 cm to surface) and temperature increase, CH_4 emission greatly increases, by 140% (65 to 154 mg CH_4 m^{-2} day^{-1}).

Using the model of Weishampel (Equation 10.7), which was also developed across a range of the MEF peatlands, and the same starting conditions, a 5°C peat temperature increase (at a peat depth of 10 cm) would result in a 78% increase in soil CO_2 emission (3.2 to 5.6 g C m^{-2} day^{-1}). If both water table (−5 to 0) and temperature (10°C to 15°C) increase, soil CO_2 emission increases by 61% (3.4 to 5.6 g C m^{-2} day^{-1}). A decline in water table (surface to −5 cm) together with an increase in temperature leads to an 89% increase in soil CO_2 emission (3.4 to 6.0 g C m^{-2} day^{-1}).

From these calculations, one would predict that both soil CO_2 and CH_4 emissions will change by a similar proportion (70–80%) in response to a mean annual 10 cm depth peat temperature change of about 5°C. However, coupled with a change in water table, the amplitude of both synergistic and antagonistic effects is larger for CH_4 than for CO_2. In other words, the equations generated from the field studies at the MEF predict that CH_4 emission will exhibit more extreme responses to combined changes in water table and temperature than soil CO_2 fluxes. Although the CO_2 equation applies to soil respiration only (no comparable equations were developed for NEE), a recent review of peatland C fluxes found that peatland CH_4 emission is up to several orders of magnitude more variable in time and space than that of peatland NEE (Limpens et al. 2008), supporting the MEF model predictions.

Summary and Conclusions

In the early 1980s, the MEF became the first peatland in the United States, and one of the first in the world, where CH_4 emissions were measured in the field (Harriss et al. 1985). This pioneering work provided critical evidence to support the importance of peatlands as a major global CH_4 source. Among other firsts, work at the MEF:

- Produced the first empirical model of CH_4 emission from peatlands (Crill et al. 1988)
- Produced the first reported CH_4 emission from peatlands in winter (Dise 1992)
- Pioneered simultaneous measurements of CH_4 and CO_2 emission by eddy covariance techniques with laser spectroscopy (Shurpali et al. 1993)
- Produced the first field manipulations of water table (Dise et al. 1993) and atmospheric chemistry (Dise and Verry 2001) for investigating controls on CH_4 emission
- Served as a keystone field site for process-based networks to study controls on CH_4 and CO_2 production (Kelley et al. 1992; Yavitt et al. 1997, 2000; Smemo and Yavitt 2007)
- Served as a keystone field site for the USDA–NASA network of land-scape-scale research on C pools and fluxes (Weishampel et al. 2009)

Field studies of C emission at the MEF have taken the form of extensive and intensive measurements of CH_4 and CO_2, using both field chambers and towers. They have provided broad overviews onto the controlling processes of CH_4 and CO_2 emission, and allowed habitat-based upscaling to estimate the contribution of northern bogs and fens to the global CH_4 budget. Subsequent campaigns have shown that CH_4 emission from the MEF is high in comparison to many northern peatlands in the sub-boreal and boreal region. This may reflect a combination of climatic and local factors at the MEF that optimize the production and emission of CH_4.

Long-term measurements have been supported by analyses on the controls on CO_2 and CH_4 production. These range from laboratory experiments on CO_2 and CH_4 production, detailed investigations of novel pathways of anaerobic CH_4 oxidation, reciprocal transplants to separate climatic from edaphic factors in C emission rates, soil respiration and CO_2 efflux studies, and field manipulation experiments on water table, deposition chemistry, and emergent vegetation to understand processes of CH_4 emission.

Tower-based measurements of peatland CO_2 exchange such as those pioneered at the MEF have evolved into global networks, allowing international comparisons of high-intensity, integrated fluxes at a landscape scale (e.g., Lund et al. 2010). Finally, forming (even today) one of the world's most complete datasets on CH_4 emission in relation to environmental variables, research and data from the MEF have been used as the basis for several large-scale extrapolations, including the development of coupled biogeochemical–physical models of global CH_4 emission (Walter and Heimann 2000) and a global assessment of the combined influences of sulfur pollution and climate change on past, current, and future CH_4 emission from natural wetlands (Gauci et al. 2004). Building upon this legacy, the contributions of the MEF to addressing some of the most important questions concerning peatland C emission and sequestration will no doubt continue well into the future. Indeed, the global recognition of the central role of peatlands in climate regulation make C emissions research at the MEF more important now than at any time in its long history.

Appendix 10.1

Brief description of the major peatland study sites involved in C emissions research at the MEF (from Dise 1993 and Dise and Verry 2001). More extensive descriptions of the peatlands and associated watersheds may be found in the reference publications, in the other the MEF publications listed in this chapter, in Boelter and Verry (1977), and in Chapter 2.

S2 Bog

The coolest and driest of the peatlands studied, S2 is a very slightly domed, ombrotrophic bog that is 3 ha in area and completely forested with black spruce (*Picea mariana*). Porewater pH (0–40 cm) measured over 1988–1990 averaged 3.9 (range 3.7–5.3). Understory vegetation is dominated by *Sphagnum angustifolium*, *Sphagnum magellanicum* (on drier hummocks) and *Ledum groenlandicum* (labrador tea), with *Chamaedaphne calyculata* (leatherleaf), *Eriophorum spissum* (cottongrass), *Sarracenia purpurea* (pitcher plant), and *Smilacina trifolia* (three-leafed false Solomon's seal) occurring in abundance.

Runoff from the domed bog portion of S2 mixes with that from the surrounding upland to form a fen lagg around the edge of the bog (Chapters 4 and 7). The lagg, approximately 15 m wide at the study site, is a distinct ecosystem type. It is vegetated with *Alnus rugosa* (speckled alder) over a diverse ground vegetation including *Sphagnum angustifolium*, *Sphagnum centrale*, *Sphagnum teres*, *Calla palustris* (water arum), *Lycopus uniflorus* (water horehound), *Equisetum fluviatile* (horsetail), and *Viola* spp. (violet). Porewater pH (10–30 cm) measured during 1988–1990 was the highest of the study sites, averaging 5.0 (4.5–6.0).

S4 Bog

The bog is a partially open 8 ha peatland with a small central pond. Vegetation in the open peatland area, where flux measurements were taken, is dominated by *Chamaedaphne calyculata* over *Sphagnum capillifolium*, with *Carex oligosperma*, *Eriophorum virginicum*, and *Rhynchospara alba* (beak rush) the most prevalent sedges. *Larix laricina* (tamarack) and small *P. mariana* are scattered; *Sarracenia purpurea* (pitcher plant) and *Menyanthes trifoliata* (buckbean) are also common. Porewater pH (0–80 cm) averaged 4.6 over 1988–1990, with a range of 3.9–6.0. The occasional high-pH events measured suggest that S4 is occasionally influenced by throughflow from the surrounding uplands. It therefore acts hydrologically as a transitional poor fen, at least some of the time.

Junction Fen

Junction Fen is an open poor fen strongly dominated by *Carex oligosperma* with some *Scheuchzeria palustris* (arrowgrass) and *Vaccinium oxycoccus* (cranberry) over *Sphagnum angustifolium, Sphagnum capillifolium,* and *Sphagnum fuscum.* Average porewater pH, 4.5, is similar to S4, but with a narrower range (4.1–5.6).

Bog Lake Fen

The Bog Lake Fen (BLF) is an open nutrient-poor fen receiving some minor subsurface flow from the surrounding land and groundwater. With no outlet, the water table is consistently at or near the surface. Peat porewater pH is similar to Junction, ranging from 4.0 to 5.2. *Sphagnum papillosum* occurs on 60% of the peatland, forming the major species on carpets and many of the hummocks. Common emergent plants are *R. alba, S. palustris,* and *Andromeda glaucophylla* (bog rosemary). *Sphagnum angustifolium* occurs near the peatland margin in concert with *C. calyculata, Kalmia polifolia* (bog laurel), *Iris versicolor* (blue flag iris), and *Glyceria* spp. (manna grass).

References

Basiliko, N. and J.B. Yavitt. 2001. Influence of Ni, Co, Fe and Na additions on methane production in *Sphagnum*-dominated Northern American peatlands. *Biogeochemistry* 52:133–153.

Berg, B. and C. McClaugherty. 2007. *Plant Litter: Decomposition, Humus Formation, Carbon Sequestration.* New York: Springer.

Boelter, D.H. and E.S. Verry. 1977. Peatland and water in the northern lake states. General Technical Report NC-31. St. Paul, MN: USDA Forest Service.

Boone, D.R., W.B. Whitman, and P. Rouvière. 1993. Diversity and taxonomy of methanogens. In *Methanogenesis,* ed. J.G. Ferry. New York: Chapman & Hall, pp. 35–80.

Bradford, J.B., P. Weishampel, M.-L. Smith, R. Kolka, D.Y. Hollinger, R.A. Birdsey, S. Ollinger, and M.G. Ryan. 2008. Landscape scale carbon sampling strategy–lessons learned. In *Field Measurements for Forest Carbon Monitoring,* ed. C.M. Hoover. New York: Springer, pp. 227–238.

Bridgham, S., J.P. Megonigal, J.K. Keller, N. Bliss, and C. Trettin. 2006. The carbon balance of North American wetlands. *Wetlands* 26:889–916.

Bridgham, S.D., K. Updegraff, and J. Pastor. 1998. Carbon, nitrogen, and phosphorus mineralization in northern wetlands. *Ecology* 79(5):1545–1561

Bubier, J.L., G. Bhatia, T.R. Moore, N.T. Roulet, and P.M. Lafleur. 2003. Spatial and temporal variability in growing-season net ecosystem carbon dioxide exchange at a large peatland in Ontario, Canada. *Ecosystems* 6:353–367.

Bubier, J.L., P.M. Crill, T.R. Moore, K. Savage, and R.K. Varner. 1998. Seasonal patterns and controls on net ecosystem CO_2 exchange in a boreal peatland complex. *Global Biogeochemical Cycles* 12:703–714.

Bubier, J.L. and T.R. Moore. 1994. An ecological perspective on methane emissions from northern wetlands. *Trends in Ecology and Evolution* 9:460–464.

Christensen, T.R., T. Friborg, M. Sommerkorn, J. Kaplan, L. Illeris, H. Soegaard, C. Nordstroem, and S. Jonasson. 2000. Trace gas exchange in a High-Arctic Valley 1. Variations in CO_2 and CH_4 flux between tundra vegetation types. *Global Biogeochemical Cycles* 14(3):701–713.

Clement, R.J., S.B. Verma, and E.S. Verry. 1995. Relating chamber measurements to eddy correlation measurements of methane flux. *Journal of Geophysical Research, D, Atmospheres* 100(D10):21047–21056.

Conrad, R., F.S. Lupton, and J.G. Zeikus. 1987. Hydrogen metabolism and sulfate-dependent inhibition of methanogenesis in a eutrophic lake sediment (Lake Mendota). *FEMS Microbial Ecology* 45:107–115.

Crill, P.M., K.B. Bartlett, R.C. Harriss, E. Gorham, E.S. Verry, D.I. Sebacher, L. Madzar, and W. Sanner. 1988. Methane flux from Minnesota peatlands. *Global Biogeochemical Cycles* 2:371–384.

Ding, W.-X. and Z.-C. Cai. 2007. Methane emission from natural wetlands in China: Summary of years 1995–2004 studies. *Pedosphere* 17:475–486.

Dinsmore, K.J., U.M. Skiba, M.F. Billett, R.M. Reees, and J. Drewer. 2009. Spatial and temporal variability in CH_4 and N_2O fluxes from a Scottish ombrotrophic peatland: Implications for modeling and up-scaling. *Soil Biology and Biochemistry* 41:1315–1323.

Dise, N.B. 1991. Methane emission from peatlands in northern Minnesota. PhD dissertation. St. Paul, MN: University of Minnesota.

Dise, N.B. 1992. Winter fluxes of methane from Minnesota peatlands. *Biogeochemistry* 17:71–83.

Dise, N.B. 1993. Methane emission from Minnesota peatlands: Spatial and seasonal variability. *Global Biogeochemical Cycles* 7:123–142.

Dise, N.B. 2009a. Peatland response to global change. *Science* 326:810–811.

Dise, N.B. 2009b. The critical role of carbon in wetlands. In *The Wetlands Handbook*, ed. E. Maltby and T. Barker. London: Blackwell Publishers, pp. 249–265.

Dise, N.B., E. Gorham, and E.S. Verry. 1993. Environmental factors controlling methane emissions from peatlands in northern Minnesota. *Journal of Geophysical Research, D, Atmospheres* 98(D6):10583–10594.

Dise, N.B. and E.S. Verry. 2001. Suppression of peatland methane emission by cumulative sulfate deposition in simulated acid rain. *Biogeochemistry* 53(2):143–160.

Dunfield, P., R. Knowles, R. Dumont, and T.R. Moore. 1993. Methane production and consumption in temperate and subarctic peat soils: Response to temperature and pH. *Soil Biology and Biochemistry* 25:321–326.

Duvall, B. and S. Goodwin. 2000. Methane production and release from two New England peatlands. *International Microbiology* 3:89–95.

Frolking, S. and N.T. Roulet. 2007. Holocene radiative forcing impact of northern peatland carbon accumulation and methane emissions. *Global Change Biology* 13:1079–1088.

Frolking, S.E., J.L. Bubier, T.R. Moore, J.T. Ball, L.M. Bellisario, J.A. Bhardwaj, J.P. Carrol, P.M. Crill, P.M. Lafleur, J.H. McCaughey, N.T. Roulet, A.E. Suyker, S.R. Verma, M. Waddington, and G.T. Whiting 1998. Relationship between ecosystem productivity and photosynthetically active radiation for northern peatlands. *Global Biogeochemical Cycles* 12:115–126.

Gauci, V., E. Matthews, N.B. Dise, B. Walter, R. Koch, G. Granberg, and M. Vile. 2004. Sulfur pollution suppression of the wetland methane source in the 20th and 21st centuries. *Proceedings of the National Academy of Sciences of the USA* 101:12583–12587.

Gignac, D. and D.H. Vitt. 1990. Habitat limitations of *Sphagnum* along climatic, chemical and physical gradients in mires of western Canada. *The Bryologist* 93:7–22.

Gorham, E. 1991. Northern peatlands: Role in the carbon cycle and probable responses to climatic warming. *Ecological Applications* 1:182–195.

Harriss, R.C., E. Gorham, D.I. Sebacher, K.B. Bartlett, and P.A. Flebbe. 1985. Methane flux from northern peatlands. *Nature* 315:652–654.

Hornibrook, E.R.C., F.J. Longstaffe, and W.S. Fyfe. 1997. Spatial distribution of microbial methane production pathways in temperate zone wetland soils: Stable carbon and hydrogen isotope evidence. *Geochimica et Cosmochimica Acta* 61: 745–753.

Humphreys, E.R., P.M. Lafleur, L.B. Flanagan, N. Hedstrom, K.H. Syed, A.J. Glenn, and R Granger. 2006. Summer carbon dioxide and water vapor fluxes across a range of northern peatlands. *Journal of Geophysical Research, G, Biogeosciences* 111:G04011.

Huttunen, J.T., H. Nykänen, J. Turunen, and P.J. Martikainen. 2003. Methane emissions from natural peatlands in the northern boreal zone in Finland, Fennoscandia. *Atmospheric Environment* 37:147–151.

IPCC. 2007. *Intergovernmental Panel on Climate Change, 2007: The Physical Science Basis*. Contribution of Working Group I to the Fourth Assessment Report of the Intergovernmental Panel on Climate Change, ed. S. Solomon, D. Qin, M. Manning, Z. Chen, M. Marquis, K. B Avery, M. Tignor, and H.L.Miller. New York: Cambridge University Press.

Joabsson, A., T.R. Christensen, and B. Wallen. 1999. Vascular plant controls on methane emissions from northern peat-forming wetlands. *Trends in Ecology and Evolution* 14:385–388.

Jones, D.L. 1998. Organic acids in the rhizosphere: A critical review. *Plant and Soil* 205:25–44.

Kamal, S. and A. Varma. 2008. Peatland microbiology. In *Microbiology of Extreme Soils*, ed. P. Dion and C.S. Nautiya. Soil Biology 13. Berlin: Springer-Verlag, pp. 177–203.

Kelley, C.A., N.B. Dise, and C.S. Martens. 1992. Temporal variations in the stable carbon isotopic composition of methane emitted from Minnesota peatlands. *Global Biogeochemical Cycles* 6:263–269.

Kettunen, A., V. Kaitala, J. Alm, J. Silvolva, H. Nykänen, and P.J. Martikainen. 1996. Cross-correlation analysis of the dynamics of methane emissions from a boreal peatland. *Global Biogeochemical Cycles* 10:457–471.

Kettunen, A., V. Kaitala, J. Alm, J. Silvolva, H. Nykänen, and P.J. Martikainen. 2000. Predicting variations in methane emissions from boreal peatlands through regression models. *Boreal Environment Research* 5:115–131.

Kim, J. and S.B. Verma. 1992. Soil surface CO_2 flux in a Minnesota peatland. *Biogeochemistry* 18:37–51.

King, J.Y. and W.S. Reeburgh. 2002. A pulse-labeling experiment to determine the contribution of recent plant photosynthates to net CH_4 emissions in arctic wet sedge tundra. *Soil Biology and Biochemistry* 34:173–180.

King, J.Y., W.S. Reeburgh, and S.K. Regli. 1998. Methane emission and transport by arctic sedges in Alaska: Results of a vegetation removal experiment. *Journal of Geophysical Research, D, Atmospheres* 103:29083–29092.

King, J.Y., W.S. Reeburgh, K.K. Thieler, G.W. Kling, W.M. Loya, L.C. Johnson, and K.J. Nadelhoffer. 2002. Pulse-labeling studies of carbon cycling in Arctic tundra ecosystems: The contribution of photosynthates to methane emission. *Global Biogeochemical Cycles* 16:1062–1069.

Kivinen, E. and P. Pakarinen. 1980. Peatland areas and the proportion of virgin peatlands in different countries. In *Proceedings of the 6th International Peat Congress*, Duluth, MN.

Knorr, K.-H., M.R. Oosterwoud, and C. Blodau. 2008. Experimental drought alters rates of soil respiration and methanogenesis but not carbon exchange in soil of a temperate fen. *Soil Biology and Biochemistry* 40:1781–1791.

Lai, D.Y.F. 2009. Methane dynamics in northern peatlands: A review. *Pedosphere* 19:409–421.

Limpens, J., F. Berendse, C. Blodau, J.G. Canadell, C. Freeman, J. Holden, N.T. Roulet, H. Ryden, and G. Schaepman-Strub. 2008. Peatlands and the carbon cycle: From local processes to global implications – A synthesis. *Biogeosciences* 5:1475–1491.

Lund, M., P.M. Lafleur, N.T. Roulet, A. Lindroth, T.R. Christensen, M. Aurela, B.H. Chojnick, L.B. Flanagan, E. R. Humphreys, T. Laurila, W.C. Oechel, J. Olejnik, J. Rinne, P. Schubert, and M.B. Nilsson. 2010. Variability in exchange of CO_2 across 12 northern peatland and tundra sites. *Global Change Biology* 16:2436–2448.

Mattson, M.D. and G.E. Likens. 1990. Air pressure and methane fluxes. *Nature* 347:713–714.

Moore, T.R., N.T. Roulet, and R. Knowles. 1990. Spatial and temporal variations of methane flux from subarctic northern boreal fens. *Global Biogeochemical Cycles* 4:29–46.

Roehm, C.L. 2005. Respiration in wetland ecosystems. In *Respiration in Aquatic Ecosystems*, ed. del Giorgio, P. and P. Williams. New York: Oxford University Press, pp. 83–103.

Roulet N.T., R. Ash, and T.R. Moore. 1992. Low boreal wetlands as a source of atmospheric methane. *Journal of Geophysical Research* 97(D4):3739–3749.

Schlesinger, W.H. 1990. Evidence from chronosequence studies for a low carbon-storage potential of soils. *Nature* 348:232–234.

Sebacher, D.E. and R.C. Harriss. 1982. A system for measuring methane fluxes from inland and coastal wetland environments. *Journal of Environmental Quality* 11:34–37.

Sebacher, D.L, R.C. Harriss, and K.B. Bartlett. 1985. Methane emissions to the atmosphere through aquatic plants. *Journal of Environmental Quality* 14:40–46.

Shurpali, N.J. and S.B. Verma. 1998. Micrometeorological measurements of methane flux in a Minnesota peatland during two growing seasons. *Biogeochemistry* 40(1):1–15.

Shurpali, N.J., S.B. Verma, R.J. Clement, and D.P. Billesbach. 1993. Seasonal distribution of methane flux in a Minnesota peatland measured by eddy correlation. *Journal of Geophysical Research, D, Atmospheres* 98(D11):20649–20655.

Shurpali, N.J., S.B. Verma, J. Kim, and T.J. Arkebauer. 1995. Carbon dioxide exchange in a peatland ecosystem. *Journal of Geophysical Research, D, Atmospheres* 100(D7):14319–14426.

Siegel, D.I., A.S. Reeve, P.H. Glaser, and E.A. Romanowicz. 1995. Climate-driven flushing of pore water in peatlands. *Nature* 374:531–533.

Smemo, K.A. 2003. The biogeochemistry of methane in northern peatland ecosystems: A potential role for anaerobic methane oxidation. PhD dissertation. Ithaca, NY: Cornell University.

Smemo, K.A. and J.B. Yavitt. 2007. Evidence for anaerobic CH_4 oxidation in freshwater peatlands. *Geomicrobiology Journal* 24(7):583–597.

Smith, K. 1993. Methane flux of a Minnesota peatland: Spatial and temporal variables and flux prediction from peat temperature and water table elevations. MS thesis. St. Paul, MN: University of Minnesota.

Ström, L. and T.R. Christensen. 2007. Below-ground carbon turnover and greenhouse gas exchanges in a sub-arctic wetland. *Soil Biology and Biochemistry* 39:1689–1698.

Valentine, D.W., E.A. Holland, and D.S. Schimel. 1994. Ecosystem and physiological controls over methane production in northern wetlands. *Journal of Geophysical Research, D, Atmospheres* 99(D1):1563–1571.

Verma, S.B., F.G. Ullman, D. Billesbach, R.J. Clement, J. Kim, and E.S. Verry. 1992. Eddy correlation measurements of methane flux in a northern peatland ecosystem. *Boundary-Layer Meteorology* 58:289–304.

Waddington, J.M., N.T. Roulet, and R.V. Swanson. 1996. Water table control of CH_4 emission enhancement by vascular plants in boreal peatlands. *Journal of Geophysical Research D, Atmospheres* 101(D1):22775–22785.

Walter, B.P. and M. Heimann. 2000. A process-based, climate-sensitive model to derive methane emissions from natural wetlands: Application to five wetland sites, sensitivity to model parameters, and climate. *Global Biogeochemical Cycles* 14:745–765.

Weishampel, P., R.K. Kolka, and J.Y. King. 2009. Carbon pools and productivity in a 1-km^2 heterogeneous forest and peatland mosaic in Minnesota, USA. *Forest Ecology and Management* 257:747–754.

Whiting, G.J. and J.P. Chanton. 1993. Primary production control of methane emission from wetlands. *Nature* 364:794–795.

Williams, C.J., J.B. Yavitt, R.K. Wieder, and N.L. Cleavitt. 1998. Cupric oxide products of northern peat and peat-forming plants. *Canadian Journal of Botany* 76:51–62.

Wuebbles, D.J. and K. Hayhoe. 2002. Atmospheric methane and global change. *Earth-Science Reviews* 57:177–210.

Yavitt, J.B. and G.E. Lang. 1990. Methane production in contrasting peatland sites: Responses to substrate quality and to sulfate reduction. *Geomicrobiology Journal* 8:27–46.

Yavitt, J.B., C.J. Williams, and R.K. Wieder. 1997. Production of methane and carbon dioxide in peatland ecosystems across North America: Effects of temperature, aeration and organic chemistry of peat. *Geomicrobiology Journal* 14:299–316.

Yavitt, J.B., C.J. Williams, and R.K. Wieder. 2000. Controls on microbial production of methane and carbon dioxide in three *Sphagnum* dominated peatland ecosystems as revealed by a reciprocal field peat transplant experiment. *Geomicrobiology Journal* 17(1):61–88.

11

Mercury Cycling in Peatland Watersheds

Randall K. Kolka, Carl P.J. Mitchell, Jeffrey D. Jeremiason,
Neal A. Hines, David F. Grigal, Daniel R. Engstrom,
Jill K. Coleman-Wasik, Edward A. Nater, Edward B. Swain,
Bruce A. Monson, Jacob A. Fleck, Brian Johnson, James E. Almendinger,
Brian A. Branfireun, Patrick L. Brezonik, and James B. Cotner

CONTENTS

Introduction

Mercury (Hg) is of great environmental concern due to its transformation into the toxic methylmercury (MeHg) form that bioaccumulates within the food chain and causes health concerns for both humans and wildlife (U.S. Environmental Protection Agency 2002). Mercury can affect neurological development in fetuses and young children. In adults, exposure to Hg can lead to the deterioration of the nervous system, decreased sensory abilities, and a lack of muscle control (Ratcliffe et al. 1996). Methylmercury in fish poses a severe health risk for fish-eating animals such as otter, mink, bald eagle, kingfisher, osprey, and the common loon (Walcek et al. 2003), and even for fish themselves (Sandheinrich and Miller 2006).

Studies in Minnesota, Wisconsin, and Sweden indicate that 18%–25% of the atmospheric Hg deposited on terrestrial basins ultimately reaches associated lakes; most of the deposited Hg is initially retained by the terrestrial landscape (Swain et al. 1992; Lee et al. 2000). In most terrestrial landscapes in northern latitudes, peatlands play a key role in determining the fate of Hg (St. Louis et al. 1994; Driscoll et al. 1995; Kolka et al. 1999a). Peatlands are sources of dissolved and particulate organic carbon (OC) that have sulfur (S) groups that bind Hg (Skyllberg et al. 2000), leading to transport Hg to surface waters (Mierle and Ingram 1991; Kolka et al. 1999a, 2001). Peatlands are also areas of high-MeHg production and contribute to MeHg loading of downstream aquatic ecosystems (Branfireun and Roulet 2002; Branfireun et al. 2005). These studies demonstrate the importance of determining the loadings of Hg from terrestrial components of watersheds to aquatic systems. Additional research is needed to better understand the fate and transport of Hg in terrestrial environments (Wiener et al. 2006).

In the mid-1990s, studies at the USDA Forest Service's Marcell Experimental Forest (MEF) began to measure the transport of Hg through the terrestrial environment (Kolka et al. 1999a,b, 2001). These studies assessed the mechanisms responsible for the deposition of Hg (including that in litterfall) and the uptake of Hg by trees (Fleck et al. 1999). Because of the organic soil peatlands present in the watershed and the documented role of OC as a carrier of hydrologically transported Hg (Aastrup et al. 1991; Johansson and Iverfeldt 1994; Driscoll et al. 1995; Ravichandran 2004), total Hg (THg) relationships with dissolved organic carbon (DOC) and particulate organic carbon (POC) (OC > 0.45 µm) were assessed. More recently, scientists have investigated in-lake cycling of Hg, the impact of sulfate deposition on the production of MeHg, the influence of the snowmelt period on annual Hg transport, and the presence of MeHg "hotspots" in peatland landscapes. In this chapter, we review the progression of Hg research conducted on the MEF and its significance with respect to the larger scientific questions related to environmental Hg.

Total Mercury Mass Balance in Upland–Peatland Watersheds

The first major effort to study Hg at the MEF was aimed at understanding both the inputs and outputs of THg and the influence of DOC and POC on Hg transport at the watershed scale. See Chapter 2 for description of study watersheds. A conceptual model was developed of the major hydrologic pathways that transport Hg (or other constituents) through the terrestrial system (Figure 11.1). Studies of each component in the mass balance made it possible to determine the ecosystem level THg inputs and outputs and the sinks and sources within upland–peatland watersheds at the MEF (Grigal et al. 2000; Kolka et al. 2001).

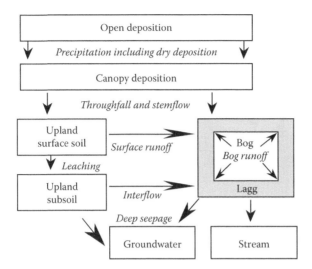

FIGURE 11.1
Conceptual hydrological model used to assess the transport of Hg through upland/peatland forested watersheds. (Modified from Kolka, R.K. et al., *Soil Sci. Soc. Am. J.*, 65, 897, 2001.)

Mercury in Atmospheric Deposition

The majority of atmospheric THg deposition in a typical-forested watershed occurs on terrestrial landscapes due to (1) the greater proportion of land relative to surface water and (2) the higher rate of dry deposition associated with tree canopies (Kolka et al. 1999b). As a result, a study was initiated to measure the atmospheric deposition of THg and DOC in throughfall and stemflow in both the upland and peatland portions of the S2 watershed (Kolka et al. 1999b). Throughfall collectors were installed across a range of canopy densities and canopy types, while stemflow collectors were installed across a range of tree species and diameters. Although a minor hydrologic flux (~2%), THg in stemflow accounted for 8% of the open wet-only deposition for THg (Kolka et al. 1999b). Deposition of THg in throughfall is a function of both canopy cover and canopy type (Figure 11.2), providing strong evidence that dry deposition is an important contributor to the total Hg load to terrestrial environments. Deposition of DOC in throughfall is related only to canopy cover, indicating that DOC fluxes appear to be a function of the leaching of the vegetative material and that dry deposition apparently exerts little influence on DOC concentrations (Figure 11.3). Because of greater apparent leaf and branch surface area (Oliver and Larson 1996), conifer canopies collect greater dry deposition of Hg than deciduous canopies, leading to greater Hg accumulation and deposition to the forest floor (Figure 11.2). Significantly, more THg (2×) and DOC (7×) deposition occurs under a forest canopy than

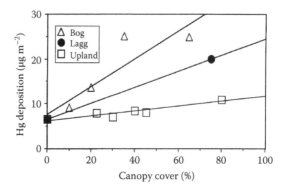

FIGURE 11.2

Relationship between canopy cover and annual Hg deposition for groups of collectors based on canopy type and density in the watershed (slope = 0.31 for bog, r^2 = 0.82; slope = 0.06 for upland, r^2 = 0.87; slope = 0.18 for lagg). The bog canopy consists of conifers (mainly black spruce), the lagg canopy is a combination of conifer and deciduous species, and the upland canopy is primarily deciduous (mainly aspen and paper birch). (From Kolka, R.K. et al., *Water Air Soil Pollut.,* 113(1), 273, 1999b.)

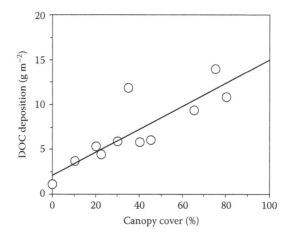

FIGURE 11.3

Relationship between canopy cover and annual DOC deposition in throughfall for all collectors, DOC = 2.3 + 0.13 (Canopy%), R^2 = 0.72. (From Kolka, R.K. et al., *Water Air Soil Pollut.,* 113(1), 273, 1999b.)

in a nearby opening (Figure 11.4). Throughfall deposition of THg at MEF (130 mg ha^{-1} year^{-1}) is similar to that measured at Walker Branch, TN (140 mg ha^{-1} year^{-1}; Lindberg 1996), which is considerably higher than that measured at the relatively remote Experimental Lakes Area (ELA) in Ontario, Canada (80 mg ha^{-1} year^{-1}; St. Louis et al. 2001), but less than that from areas in northern Sweden (150 mg ha^{-1} year^{-1}) and considerably less than the more polluted southern Sweden (240 mg ha^{-1} year^{-1}; Lee et al. 2000). From this work, it is

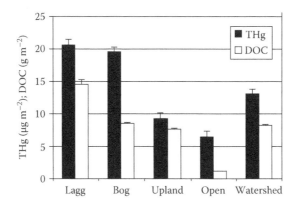

FIGURE 11.4
Annual throughfall + stemflow deposition of THg and DOC at the MEF. (Modified from Kolka, R.K. et al., *Water Air Soil Pollut.*, 113(1), 273, 1999b.)

clear that any attempt to quantify Hg inputs to forested systems must consider both wet and dry deposition as well as the influence of forest types on deposition (Kolka et al. 1999b).

Mercury in Litterfall and Tree Uptake

Inputs of THg from litterfall are of similar magnitude to inputs from atmospheric deposition (Grigal et al. 2000). In a study on the MEF, litterfall was collected in the fall of 1995 (August 8–November 15) within the two main forest types (upland aspen/birch, and peatland black spruce) in the S2 watershed. Although leaf litterfall in the upland (38.3 ng $g^{-1} \pm 1$ standard error of 1.4 ng g^{-1}) had significantly higher Hg concentrations than the leaf (needle) litterfall in the peatland (29.7 ng $g^{-1} \pm 2.4$), differences in the mass of deposition and the Hg concentrations of nonleaf material in litterfall led to similar areal deposition of THg among uplands (12.5 µg $m^{-2} \pm 0.85$) and peatlands (11.7 µg $m^{-2} \pm 1.36$; Grigal et al. 2000). Litterfall THg deposition at MEF (123 mg ha^{-1} $year^{-1}$) is similar to that at ELA in Ontario (120 mg ha^{-1} $year^{-1}$; St. Louis et al. 2001), but much less than that from areas in northern Sweden (180 mg ha^{-1} $year^{-1}$) and southern Sweden (230 mg ha^{-1} $year^{-1}$; Lee et al. 2000) and Walker Branch in Tennessee (300 ha^{-1} $year^{-1}$; Lindberg 1996).

An assessment of tree uptake of Hg found no relationships between Hg in soil or forest floor and that in woody tissue or foliar concentrations (Fleck et al. 1999). The lack of relationships between the plant tissues and the soil indicated that Hg, in plant tissue, is derived directly from the atmosphere and not the soil.

Hydrologic Pathways of Mercury

To better understand the importance of various hydrological pathways by which both THg and OC (including DOC and POC) are transported, measurements were made in each component of the hydrologic cycle in a typical upland/bog watershed (i.e., the S2 watershed) and modeled with the Peatland Hydrologic Impact Model (Guertin et al. 1987; Kolka et al. 2001; Chapter 15; Figure 11.1). In addition to measuring concentrations of THg and OC, water flow from one hydrologic component to the next was estimated. Because water flows were dominated by the peatland (bog in this case), most of the THg and DOC fluxes originated from the bog (Figure 11.5). Fluxes from the upland are important in the spring during snowmelt and in the fall following the cessation of evapotranspiration. Upland and bog waters both flow to the lagg (Figure 11.1), which ultimately coalesces into the stream that exits the watershed (Chapter 7). THg fluxes to the lagg are approximately twice those that exit the S2 watershed, while DOC fluxes are comparable to those of the stream (Figure 11.6). These results suggest that the bulk of the THg losses from the lagg can be attributed to

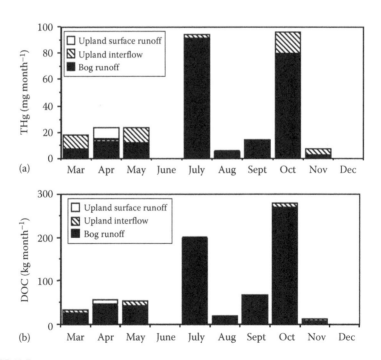

FIGURE 11.5

Hydrologic fluxes of (a) THg and (b) DOC from upland (surface runoff and interflow) and bog components of the S2 watershed at the MEF in 1995. Annual precipitation in 1995 was 890 mm and the long-term mean was 775 mm. (Modified from Kolka, R.K. et al., *Soil Sci. Soc. Am. J.*, 65, 897, 2001.)

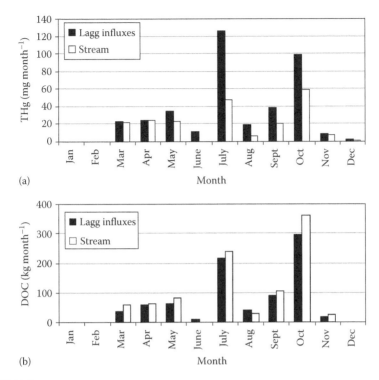

FIGURE 11.6
Hydrologic fluxes of (a) THg and (b) DOC to the lagg compared to the fluxes in the stream exiting the S2 watershed in 1995. (Modified from Kolka, R.K. et al., *Soil Sci. Soc. Am. J.*, 65, 897, 2001.)

either accumulation in lagg soils or volatilization between our sampling points and the stream outlet. The soils in the lagg are organic and likely have a high ability to complex THg. If we assume that soil accumulation was the only loss of Hg, the lagg soil would accumulate about 35 µg THg m^{-2} year^{-1}, within the range of that reported for bog soils (Benoit et al. 1994). If volatilization is considered, the only Hg loss, the mean rate from the lagg is 6.8 ng m^{-2} h^{-1}, which is near the upper end of the range reported by Kim et al. (1995) but lower than the geometric mean of 11 ng m^{-2} h^{-1} reported in a review of studies by Grigal (2002). Both processes (soil accumulation and volatilization) likely contribute to THg losses in the lagg.

Stream Fluxes of Mercury

To better understand watershed fluxes of THg and OC, outlet waters were sampled from other experimental watersheds at the MEF to compare with results from the S2 watershed. Streams exiting the other MEF watersheds

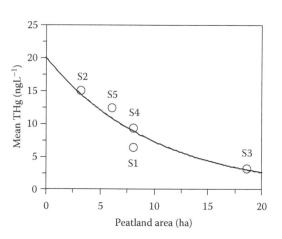

FIGURE 11.7
Relationship between mean flow-weighted THg concentration and peatland area for watersheds at the MEF, $HgT = 20.0 \times 10^{(-0.044 \times peatland\ area)}$, $r^2 = 0.94$, $p = 0.01$. (Modified from Kolka, R.K. et al., *J. Environ. Qual.*, 28(3), 766, 1999a.)

originate from a range of peatland types including oligotrophic bogs similar to S2 to a mesotrophic fen (S3). Watershed flow-weighted concentrations of Hg are related to the area of peatland (Figure 11.7), demonstrating that larger peatlands had lower THg concentrations. As peatland size increased, there was an increase in overall flow from the watershed and a greater connection with regional groundwater sources that are lower in THg concentration than runoff from perched water tables such as found in the S2 bog. Regression models relating THg flux to landscape variables for five watersheds indicated that upland and peatland areas combined explained approximately 89% of the variation in THg stream flux from MEF watersheds (Kolka et al. 1999a). Through further regression analysis, it was estimated that 67%–98% of stream THg is derived from peatlands while 2%–33% is derived from upland sources. Similar regressions indicated that 92%–99% of DOC and 49%–96% of POC originate from the peatland component of the watersheds (Kolka et al. 1999a). Streamflow yields of THg range from 0.70 to 2.82 $\mu g\ m^{-2}\ year^{-1}$ at MEF, which brackets the mean yield of 1.7 $\mu g\ m^{-2}\ year^{-1}$ reported by Grigal (2002) and is similar to comparable wetland ecosystems at ELA in Ontario (range for export from wetland watersheds = 0.7–2.1 $\mu g\ m^{-2}\ year^{-1}$; St. Louis et al. 1996), Finland (1.0–1.8 $\mu g\ m^{-2}\ year^{-1}$; Porvari and Verta 2003), and Sweden (1.0–3.4 $\mu g\ m^{-2}\ year^{-1}$; Lee et al. 2000).

An important feature of watershed export at MEF is that short-term events can play a large role in annual fluxes. Earlier studies on the hydrologic fluxes of THg (Kolka et al. 1999a) did not isolate the importance of event-based stream fluxes; however, more recent studies indicate that 26%–39% of the annual THg flux (depending on the watershed) and 22%–23% of the annual MeHg flux occur during the short snowmelt period (~12 days; Mitchell et al. 2008a; Table 11.1). Both upland and peatland fluxes were important during snowmelt periods, whereas peatland fluxes of both THg and MeHg dominated watershed fluxes later in the growing season. Studies in northern

TABLE 11.1

Upland, Peatland, and Total Watershed Fluxes of THg and MeHg during the Studied Snowmelt Period Compared to Total Watershed Annual Fluxes

	S2 Watershed		S6 Watershed	
	THg Flux (mg)	MeHg Flux (mg)	THg Flux (mg)	MeHg Flux (mg)
Snowmelt upland runoff flux	24	0.14	8.5	0.054
Snowmelt peatland runoff flux	35	0.52	13	1.0
Total snowmelt watershed flux	59	0.66	21	1.1
2005 annual	230	3.0	54	4.7

Source: Mitchell, C.P.J. et al., Biogeochemistry, 90, 225, 2008a.

Sweden in comparable wetland watersheds also showed that snowmelt is an important time for Hg transport, accounting for 34% of THg and 12% of annual MeHg fluxes (Bishop et al. 1995). From the event-based data, it is apparent that the temporal distribution and frequency of sampling can substantially increase our understanding of the sources of Hg from terrestrial to aquatic ecosystems.

Total Mercury Relationships with Organic Carbon

For waters sampled at MEF in the 1990s, we analyzed unfiltered samples, primarily because clean filtering techniques were still untested. Although many studies have identified a strong relationship between Hg species and DOC concentrations (e.g., Driscoll et al. 1995), it was found that POC (>0.45 μm) explained more of the variation in unfiltered THg at the MEF (Kolka et al. 1999a, 2001). Relationships between THg and POC among terrestrial waters in S2 included an r^2 of 0.52 for upland surface runoff, r^2 of 0.65 for upland interflow (subsurface runoff), r^2 of 0.84 for bog runoff, and r^2 of 0.49–0.76 for streamflow (Kolka et al. 1999a). Few relationships are evident between THg and DOC as correlation coefficients did not exceed 0.40. If our THg samples had been filtered, correlations with DOC may have been higher. When all stream data were merged, POC was clearly an important carrier of THg (Figure 11.8). Other studies have found a close relationship between particulate THg and POC (e.g., Shanley et al. 2002) and showed that POC is an important carrier of Hg during high-flow events and in larger, more disturbed watersheds (Grigal 2002). A new technique using specific ultraviolet absorbance at 254 nm to determine the amount and type of DOC is successfully predicting THg in streams (Dittman et al. 2009).

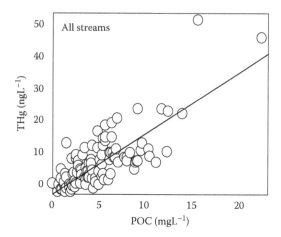

FIGURE 11.8
Relationship between THg and POC concentrations for watersheds (S1–S5) at the MEF, THg = 0.99 + 1.86 (POC), $r^2 = 0.60$, $p = 0.0001$. (Modified from Kolka, R.K. et al., *J. Environ. Qual.*, 28(3), 766, 1999a.)

Total Mercury Mass Balance

Mass balance studies at the MEF indicate that the terrestrial portion of a watershed is a sink for THg (Grigal et al. 2000; Kolka et al. 2001). The mass balance for the S2 watershed in 1995 showed that the forest canopy captured significant amounts of atmospheric THg as measured in throughfall and stemflow, while the watershed soils are a major sink. The overall THg accumulation is about 4.3 µg THg m^{-2} year^{-1}, if volatilization back to the atmosphere is not considered (Kolka et al. 2001). When litterfall inputs are included, the S2 watershed becomes a larger sink for THg, approximately 22.5 µg m^{-2} year^{-1} (Table 11.2; Grigal et al. 2000). St. Louis et al. (1996) found that similar watersheds in Ontario are sinks for THg and range from 1.9 to 3.4 µg THg m^{-2} year^{-1} (excluding litterfall inputs), similar in magnitude to the S2 watershed, excluding litterfall. Research in Acadia National Park in Maine indicated that watersheds are a sink for about 9.0 µg THg m^{-2} year^{-1} not including litterfall inputs (Nelson et al. 2007). Presumably, the Acadia watersheds represent a larger sink for THg than those in central North America, because atmospheric Hg inputs are higher on the East Coast. Similarly, a

TABLE 11.2

Total Hg Mass Balance for the S2 Watershed at the MEF

Component	Entire Watershed (µg m^{-2} Year^{-1})
Throughfall and stemflow inputs	13.0 (0.35)
Litter input	12.3 (0.73)
Hydrologic output	2.81 (0.28)
Input–output	22.5 (0.8)

Note: Standard errors in parentheses; Grigal et al. (2000).

watershed in the Adirondack Mountains is a sink of THg of 12.9 μg THg m^{-2} year^{-1}, when litterfall is not included (Selvendiran et al. 2009). When litterfall is included as an additional input to the mass balance, watershed sinks range as high as 44 μg THg m^{-2} year^{-1} in Sweden (Lee et al. 2000) and as low as 6.9 μg THg m^{-2} year^{-1} in Norway (Larssen et al. 2008). Although the work at the MEF and from other regions indicates that watershed soils are important long-term storage compartments for Hg (Grigal 2003), the mass of Hg transported from these small watersheds remains an important contributor to downstream ecosystems.

Lake Cycling of Mercury

Beginning in 2000, a series of studies were conducted investigating Hg dynamics of a small seepage lake with no defined inlets or outlets (Spring Lake) on the MEF. These studies found that sediment accumulation rates were 21.4 μg THg m^{-2} year^{-1} and 0.20 μg MeHg m^{-2} year^{-1}, respectively, from 1990 to 2000 (Hines et al. 2004). The rate of lake sediment THg accumulation is nearly identical to the calculated watershed sink rate of 22 μg m^{-2} year^{-1} at the MEF (Grigal et al. 2000), indicating that watershed soils and lake sediment accumulate THg at similar rates in these ecosystems. Accumulation rates of both THg and MeHg are higher at the sediment surface than at depth (Figure 11.9), approximately 4 and 10 times preindustrial rates for THg and MeHg, respectively (Hines et al. 2004). Patterns of accumulation rates are

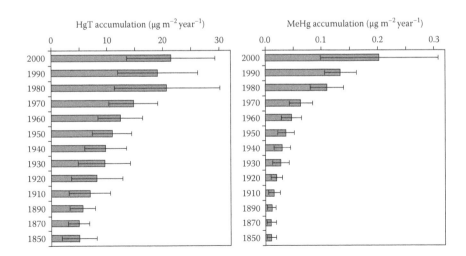

FIGURE 11.9
THg (HgT) and MeHg accumulation rates since preindustrial times in Spring Lake on the MEF. (From Hines, N.A. et al., *Environ. Sci. Technol.*, 38(24), 6610, 2004. With permission.)

similar for lakes in northeastern Minnesota (Engstrom et al. 2007). However, it is important to note that the apparent increase in MeHg accumulation near the surface may be a consequence of the demethylation of MeHg following deposition and burial. Sediment pore water concentrations (5 and 15 cm depth) of MeHg were low in the spring and peaked in late summer and roughly correlated with peaks in sulfate reduction (Hines et al. 2004). Because of temporal gradients in pore water concentrations, there was an apparent diffusive flux of MeHg from sediment to lake water. However, there was also advective transport of MeHg to deeper sediment within the sediment column that occurred in late summer and early fall leading the lake sediment to be an overall sink of MeHg (Hines et al. 2004).

Additional research on Spring Lake assessed the role of photochemistry on Hg dynamics. Evasional losses of Hg^0 during midsummer were 5–6 pmol $m^{-2}\ h^{-1}$ as a result of photoreduction, which tended to be twice the rate of photooxidation (Hines and Brezonik 2004a,b) and similar to rates observed at the ELA in Ontario (Sellers et al. 2001). For Spring Lake, wet deposition was the largest THg input, while outputs from the lake were dominated by burial in the sediment (67% of outputs) and evasion of Hg^0 from the lake surface (26% of outputs; Hines and Brezonik 2007). For MeHg, wet deposition only accounted for 9% of inputs while runoff and sediment pore water contributions only accounted for an additional 7% of inputs, indicating that the bulk of the MeHg production was occurring in the water column of the lake (Hines and Brezonik 2007). Photolysis of MeHg was the dominant sink (output), removing approximately three times the entire lake mass of MeHg annually. During open water season, residence times of THg and MeHg in the lake were 61 and 48 days, respectively, while the residence time of Hg^0 in the photic zone was on the order of hours (Hines and Brezonik 2007). Finally, sampling of seston, zooplankton, and fish indicated that seston accounted for the highest mass of Hg in lake biota, much higher (10–100×) than the higher trophic levels. The bioconcentration factor of Hg (dry mass concentration in biota: concentration in water) for northern pike, the top fish predator, was 1.2×10^7 (Hines and Brezonik 2007), which is in the range for northern pike in lakes on nearby Isle Royale National Park in Lake Superior (range, 7.4×10^6–2.3×10^7) (Gorski et al. 2003).

Factors Affecting Methylmercury Production

Following the characterization of the pools and fluxes of THg in upland–peatland watersheds at the MEF, much of the research has focused on the factors that lead to methylation of Hg. Previous laboratory experiments (Gilmour et al. 1992) and small-scale field mesocosm research (Branfireun et al. 1999, 2001) established a link between sulfate inputs and Hg methylation in S-limited sediments. In 2001, an ecosystem-scale manipulative experiment began in the S6 wetland to determine the effects of sulfate deposition on both peatland and watershed-level MeHg production. The wetland was divided into an upgradient control and a downgradient experimental treatment. Samples were collected along transects that extended north to south across the treated and untreated halves of the peatland. From 2001 to 2006, four times the annual ambient sulfate deposition was applied via a sprinkler system to the experimental treatment. Shallow sampling wells were located throughout the lagg and bog zones of S6 both in the treated and untreated halves of the peatland. Sampling occurred 1 day prior to each addition and periodically following the addition (e.g., 1, 3, 7, and 14 days after sulfate application). In 2006, the experimental treatment was further divided, and sulfate addition was halted to the upgradient one-third of the treatment to simulate a recovery phase after excess sulfate deposition. Sulfate applications ceased after 2008.

Relationships developed via the paired watershed approach between MeHg fluxes from the S6 treatment watershed and the S7 reference watershed allowed comparisons at the watershed scale. The first year of enhanced sulfate deposition resulted in a 2.4-time increase in MeHg flux from the watershed (Figure 11.10; Jeremiason et al. 2006). Following the spring sulfate addition, peat pore water concentrations of MeHg in the experimental treatment peaked at three times those in the control half of the S6 peatland (Jeremiason et al. 2006) and remained elevated relative to the control

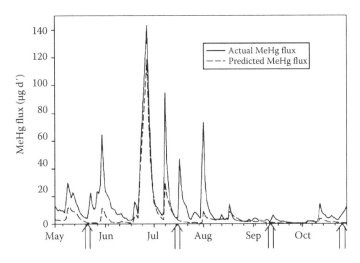

FIGURE 11.10
Actual and predicted fluxes of MeHg from the S6 watershed for 2002. The predicted flux is that which would have occurred in the absence of sulfate addition and is based on a correlation of 2001 MeHg fluxes from S6 with those from a nearby reference wetland (S7). Arrows indicate dates of experimental sulfate applications. The pretreatment r^2 was 0.77. (From Jeremiason, J.D. et al., *Environ. Sci. Technol.*, 40(12), 3800, 2006. With permission)

throughout the remainder of the year. It is clear from the data that increased sulfate deposition led to increased in situ methylation and increased MeHg flux from the watershed. Other researchers have demonstrated that sulfate-reducing bacteria (SRB) can methylate Hg and that the methylation process is stimulated in the presence of sulfate (Gilmour et al. 1998; Benoit et al. 1999). The current working hypothesis for the S6 sulfate-addition study is that SRB populations in this S-limited system increase MeHg production when they are stimulated by a dose of sulfate; they continue to methylate Hg at an elevated rate so long as sulfate is available. The focus of the sulfate addition work changed in 2005 with an assessment of chronic effects of sulfate addition, wetland recovery processes, and the overarching influence of climatic and hydrologic processes on Hg and S cycling (Coleman-Wasik 2008). Preliminary results indicate that a wetland affected by elevated sulfate deposition may take years to return to baseline conditions and that climatic and hydrologic variability affects the recovery process.

Following the early results from the sulfate addition experiment, investigations into the controls on MeHg production continued. More recent mesocosm studies have assessed the influence of both labile carbon (C) and sulfate additions on MeHg production. Mesocosms were installed in the Bog Lake peatland site, and a factorial experiment assessed the influence of three levels of sulfate and two levels of five C sources including glucose, acetate, lactate, deciduous litter leachate, and conifer litter leachate on MeHg production (Mitchell et al. 2008b). There was little response to the labile C additions, indicating that C was not limiting methylating bacterial metabolism in the peatland. Responses to 4 and 10 times annual sulfate deposition were strong but not different among the two levels, indicating that sulfate was limiting, but also that the fourfold treatment was adequate to alleviate the microbial limitation. However, when labile C and sulfate were combined, there was generally an increase in MeHg compared to the response of sulfate alone, indicating a potential for colimitation of organic C and sulfate (Figure 11.11). The C additions with the 10-fold sulfate addition did not increase MeHg production above those with fourfold sulfate plus C additions, indicating that sulfate limitations were alleviated, even with additional labile C, at fourfold annual deposition (Mitchell et al. 2008b).

Additional studies have assessed important methylation locations or hotspots in these upland–peatland landscapes. To identify location of hotspots, porewater was collected from the S2 and S6 watersheds at the MEF as well as two watersheds at the ELA in Ontario in landscape positions just upslope from the upland peatland margin and then into the lagg and bog zones of the peatland (Mitchell et al. 2008c, 2009). Methylation hotspots were based on the distribution of MeHg:THg with a location designated as a methylation hotspot when 22% of the THg was present as MeHg (i.e., the 90th percentile of all measurements). Results indicated that the lagg zone (Figure 11.1), both spatially and temporally, was the location that typically met the hotspot criterion (Figure 11.12; Mitchell et al. 2008c). It appears that methylation is

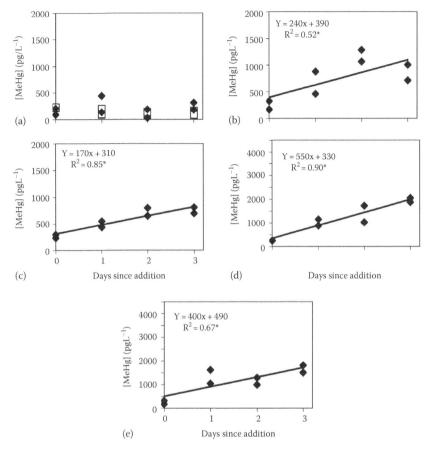

FIGURE 11.11
Response of MeHg to C and sulfate additions (From Mitchell, C.P.J. et al., *Appl. Geochem.*, 23(3), 503, 2008b). No response was seen when adding C as glucose (a). Responses were similar when adding 4× (b) and 10× (c) annual sulfate deposition, and methylation was higher when combining C as glucose plus 4× (d) and 10× (e) sulfate additions. There were no differences between the 4× and 10× sulfate plus C additions.

enhanced in the lagg zone when upland-derived soil water, which is relatively rich in sulfate, nutrients, and labile DOC mix with bog water, which is relatively low in nutrients and has DOC that is more recalcitrant to microbial degradation. Upland waters may be too well oxygenated for sulfate reduction to occur, while bog waters lack the labile DOC and sulfate needed for methylation. When the waters mix in the lagg zone, conditions are ideal for methylation because sufficient sulfate and labile DOC present under saturated conditions allows sulfate reduction to occur.

Further investigation of the landscape influences on methylation indicates that upland topographic configuration plays an important role in the location of MeHg hotspots within the lagg. Upland positions that focus subsurface

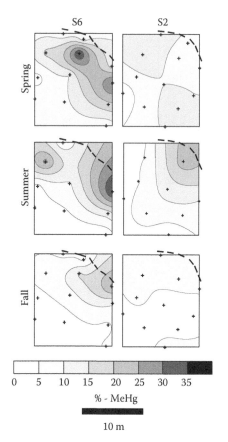

FIGURE 11.12
Contour plots of %MeHg in high-resolution sampling grids illustrate seasonal variability in spatial pattern of %MeHg in pore water at the upland–peatland interfaces in the S6 and S2 watersheds. The dashed line in the upper right corner of each plot represents the upland–peatland interface, corresponding to a change in topography and a change from organic to mineral soils. Upland hillslope runoff is from the upper right toward the lower left of each plot. Each+symbol represents a pore water sampling point (From Mitchell, C.P.J. et al., *Environ. Sci. Technol.*, 42(4), 1010, 2008c).

flow to the lagg (concave configuration) generally exhibit higher MeHg, sulfate, and DOC concentrations than upland landscape positions that lead to more diffuse flow to the lagg (Figure 11.13). Focused flow from the uplands transported greater quantities of labile DOC and sulfate to the lagg zone, likely promoting MeHg production. Lagg areas that received lower, diffuse, upland flow receive primarily low-sulfate, recalcitrant-DOC waters from the bog, resulting in significantly less methylation.

Conclusion

The extensive hydrologic and biogeochemical monitoring across varied upland–peatland ecosystems has created unique opportunities to address important Hg research questions at the MEF. Research in the northern ecosystem at the MEF has shown that forest canopies enhance Hg deposition

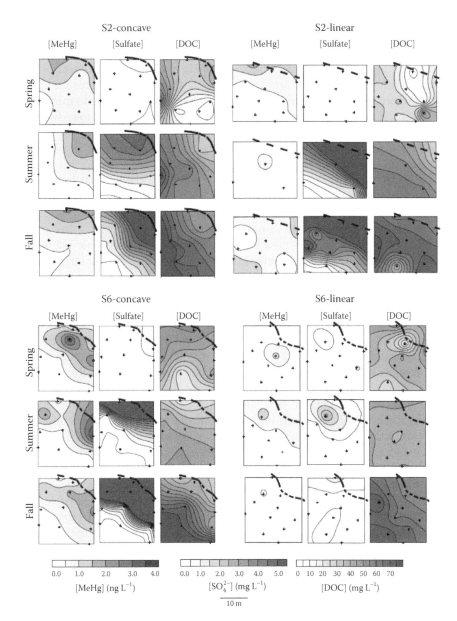

FIGURE 11.13
Spatial patterns of pore water MeHg, sulfate, and DOC concentration at the upland–peatland interface adjacent to concave or linear upland hillsopes in the S2 and S6 watersheds. The dashed line in each represents the upland interface. Fluxes from the upland are across the interface from the upper right of each plot toward the lower left (From Mitchell, C.P.J. et al., *Water Resour. Res.*, 45:W02406, 2009).

and has documented the important hydrologic pathways through which Hg is transported to surface waters. Because of enhanced atmospheric deposition, storage in peat soils, and efficient methylation, the ecosystem represented by the MEF is sensitive and vulnerable to global atmospheric Hg contamination. Watershed mass balances of Hg at the MEF are revealing the perturbations that mobilize and methylate Hg that already is present in the environment. Such knowledge can help direct human activities to minimize the negative impact of Hg pollution, even as governments work to reduce the amount of Hg released into the atmosphere.

After Hg reaches aquatic systems, additional processes such as photoreduction and demethylation can occur and affect whether lakes are a source or a sink of atmospheric Hg. Studies at the MEF indicate that both THg and MeHg accumulate significantly in lake sediment, although most of the MeHg is eventually demethylated. In Spring Lake, most of the MeHg in the lake is not transported from connected terrestrial systems or from atmospheric deposition but is formed by the methylation of inorganic Hg in the water column or surface sediment. Photoreduction is by far the largest sink (loss) of MeHg from the lake system, with uptake by biota a minor sink. Although uptake by biota is minor in the overall mass balance, it is a critical link between abiotic Hg sources and bioaccumulation in the food web.

Identifying the processes that lead to the methylation of Hg is critical to efforts aimed at mitigating the effects of global Hg contamination. Recent work indicates that both increased sulfate and labile C lead to higher methylation rates. Within watersheds, MeHg production is highest at the upland–peatland interface or lagg, which is considered a hotspot for MeHg production because of available Hg, labile C, and sulfate, and the anaerobic conditions present there. Understanding both the controls on MeHg production and locations in the landscape where methylation occurs can foster management approaches and land-use decisions that will lessen the contribution of MeHg to aquatic systems where it can accumulate in the biota.

On the basis of well-cited past studies and novel current Hg studies, the MEF is poised to address research aimed at new questions related to Hg. Future research in MEF will investigate the chemical and hydrologic controls on Hg methylation, particularly those relating to organic matter composition controls on Hg dynamics. The use of Hg isotopes has considerable promise to elucidate processes that lead to methylation. A new study will use Hg isotopes to determine the effects of forest harvesting on Hg transport and MeHg production in upland soils. Preliminary investigations have measured Hg in terrestrial invertebrates. Future research could build on this work to assess the trophic cycling of Hg and OC in ecosystem food chains. Climate change has considerable potential to exacerbate Hg contamination in northern ecosystems through enhanced bacterial growth and metabolism, sulfate cycling, or the release of stored Hg from peat. We anticipate a number of opportunities related to the impact of climate change on Hg cycling. Past and current research clearly demonstrates that the MEF is an ideal field laboratory to investigate these important questions about Hg in the environment.

References

Aastrup, M., J. Johnson, E. Bringmark, I. Bringmark, and A. Iverfeldt. 1991. Occurrence and transport of mercury within a small catchment area. *Water, Air, and Soil Pollution* 56:155–167.

Benoit, J.M., W.F. Fitzgerald, and A.W.H. Damman. 1994. Historical atmospheric mercury deposition in the mid-continental U.S. as recorded in an ombrotrophic peat bog. In: *Mercury Pollution: Integration and Synthesis*, ed. C.J. Watras and J.W. Huckabee. Chelsea, MI: Lewis Publishers, pp. 187–202.

Benoit, J.M., C.C. Gilmour, R.P. Mason, and A. Heyes. 1999. Sulfide controls on mercury speciation and bioavailability in sediment pore waters. *Environmental Science & Technology* 33:951–957.

Bishop, K., Y.-H. Lee, C, Pettersson, and B. Allard. 1995. Methylmercury output from the Svartberget catchment in northern Sweden during spring flood. *Water, Air, and Soil Pollution* 80:445–454.

Branfireun, B.A. and N.T. Roulet. 2002. Controls on the fate and transport of methylmercury in a boreal headwater catchment, northwestern Ontario, Canada. *Hydrology and Earth System Sciences* 6:785–794.

Branfireun, B.A., N.T. Roulet, C.A. Kelly, and J.W.M. Rudd. 1999. In situ sulphate stimulation of mercury methylation in a boreal peatland: Toward a link between acid rain and methylmercury contamination in remote environments. *Global Biogeochemical Cycles* 13:743–750.

Branfireun, B.A., K. Bishop, N.T. Roulet, G. Granberg, and M. Nilsson. 2001. Mercury cycling in boreal ecosystems: The long-term effect of acid rain constituents on peatland pore water methylmercury concentrations. *Geophysical Research Letter* 28:1227–1230.

Branfireun, B.A., D.P. Krabbenhoft, H. Hintelmann, R.J. Hunt, J.P. Hurley, and J.W.M. Rudd. 2005. Speciation and transport of newly deposited mercury in a boreal forest wetland: A stable mercury isotope approach. *Water Resources Research* 41:W01016.

Coleman-Wasik, J.K. 2008. Chronic effects of atmospheric sulfate deposition on mercury methylation in a boreal wetland: Replication of a global experiment. MS thesis. St. Paul, MN: University of Minnesota.

Dittman, J.A., J.B. Shanley, D.T. Driscoll, G.R. Aiken, A.T. Chalmers, and J.E. Towse. 2009. Ultraviolet absorbance as a proxy for total dissolved mercury in streams. *Environmental Pollution* 157:1953–1956.

Driscoll, C.T., V. Blette, C. Yan, C.L. Schofield, R. Munson, and J. Holsapple. 1995. The role of dissolved organic carbon in the chemistry and bioavailability of mercury in remote Adirondack lakes. *Water, Air, and Soil Pollution* 80:499–508.

Engstom, D.R., S.J. Balogh, and E.B. Swain. 2007. History of mercury inputs to Minnesota lakes: Influences of watershed disturbance and localized atmospheric deposition. *Limnology and Oceanography* 52:2467–2483.

Fleck, J.A., D.F. Grigal, and E.A. Nater. 1999. Mercury uptake by trees: An observational experiment. *Water, Air, and Soil Pollution* 115(1–4):513–523.

Gilmour, C., E. Henry, and R. Mitchell. 1992. Sulfate stimulation of mercury methylation in freshwater sediments. *Environmental Science & Technology* 26:2281–2287.

Gilmour, C., G. Riedel, M. Ederington, J. Bell, J. Benoit, G. Gill, and M. Stordal. 1998. Methylmercury concentrations and production rates across a trophic gradient in the northern Everglades. *Biogeochemistry* 40:327–345.

Gorski, P.R., L.B. Cleckner, J.P. Hurley, M.E. Sierszen, and D.E. Armstrong. 2003. Factors affecting enhanced mercury bioaccumulation in inland lakes of Isle Royale National Park, USA. *Science of the Total Environment* 304:327–348.

Grigal, D.F. 2002. Inputs and outputs of mercury from terrestrial watersheds: A review. *Environmental Reviews* 10:1–39.

Grigal, D.F. 2003. Mercury sequestration in forests and peatlands. *Journal of Environmental Quality* 32:393–405.

Grigal, D.F., R.K. Kolka, J.A. Fleck, and E.A. Nater. 2000. Mercury budget of an upland-peatland watershed. *Biogeochemistry* 50:95–109.

Guertin, D.P., P.K. Barten, and K.N. Brooks. 1987. The peatland hydrologic impact model: Development and testing. *Nordic Hydrology* 18:79–100.

Hines, N.A. and P.L. Brezonik. 2004a. Mercury dynamics in a small northern Minnesota lake: Water to air exchange and photoreactions of mercury. *Marine Chemistry* 90(1–4):137–149.

Hines, N.A. and P.L. Brezonik. 2004b. Mercury and methylmercury in Spring Lake, Minnesota: A mass balance approach comparing redox transformations, MeHg photodegradation, sediment loading, and watershed processes. *Materials and Geoenvironment* 51(2):1055–1059.

Hines, N.A. and P.L. Brezonik. 2007. Mercury inputs and outputs at a small lake in northern Minnesota. *Biogeochemistry* 84(3): 265–284.

Hines, N.A., P.L. Brezonik, and D.R. Engstrom. 2004. Sediment and porewater profiles and fluxes of mercury and methylmercury in a small seepage lake in northern Minnesota. *Environmental Science & Technology* 38(24):6610–6617.

Jeremiason, J.D., D.R. Engstrom, E.B. Swain, E.A. Nater, B.M. Johnson, J.E. Almendinger, B.A. Monson, and R.K. Kolka. 2006. Sulfate addition increases methylmercury production in an experimental wetland. *Environmental Science & Technology* 40(12):3800–3806.

Johansson, K. and A. Iverfeldt. 1994. The relationship between mercury content in soil and the transport of mercury from small catchments in Sweden. In: *Proceedings on International Conference on Mercury as a Global Pollutant*, ed. C.J. Watras and J.W. Huckabee, May 31–June 4, 1992, Monterey, CA. Palo Alto, CA: Electric Power Research Institute, pp. 323–328.

Kim, K.-H., S.E. Lindberg, and T.P. Meyers. 1995. Micrometeorological measurements of mercury vapor fluxes over background forest soils in eastern Tennessee. *Atmospheric Environment* 29:267–282.

Kolka, R.K., D.F. Grigal, E.A. Nater, and E.S. Verry. 1999a. Mercury and organic carbon relationships in streams draining forested upland/peatland watersheds. *Journal of Environmental Quality* 28(3):766–775.

Kolka, R.K., E.A. Nater, D.F. Grigal, and E.S. Verry. 1999b. Atmospheric inputs of mercury and organic carbon into a forested upland/bog watershed. *Water, Air, and Soil Pollution* 113(1):273–294.

Kolka, R.K., D.F. Grigal, E.A. Nater, and E.S. Verry. 2001. Hydrologic cycling of mercury and organic carbon in a forested upland-bog watershed. *Soil Science Society of America Journal* 65:897–905.

Larssen, T., H.A. de Wit, M. Wiker, and K. Halse. 2008. Mercury budget of a small forested boreal catchment in southeast Norway. *Science of the Total Environment* 404:290–296.

Lee, Y.H., K.H. Bishop, and J. Munthe. 2000. Do concepts about catchment cycling of methylmercury and mercury in boreal catchments stand the test of time? Six years of atmospheric inputs and runoff export at Svartberget, northern Sweden. *Science of the Total Environment* 260:11–20.

Lindberg, S.E. 1996. Forests and the global biogeochemical cycle of mercury: The importance of understanding air/vegetation exchange processes. In: *Global and Regional Mercury Cycles: Sources, Fluxes and Mass Balances*, ed. W. Baeyens, R. Ebinghaus, and O. Vasiliev. NATO ASI Series. Dordrecht, the Netherlands: Kluwer Academic Publishers, Vol. 21, pp. 359–380.

Mierle, G. and R. Ingram. 1991. The role of humic substances in the mobilization of mercury from watersheds. *Water, Air, and Soil Pollution* 56:349–357.

Mitchell, C.P.J., B.A. Branfireun, and R.K. Kolka. 2008a. Total mercury and methylmercury dynamics in upland-peatland watersheds during snowmelt. *Biogeochemistry* 90:225–241.

Mitchell, C.P.J., B.A. Branfireun, and R.K. Kolka. 2008b. Assessing sulfate and carbon controls on net methylmercury production in peatlands: An in situ mesocosm approach. *Applied Geochemistry* 23(3):503–518.

Mitchell, C.P.J., B.A. Branfireun, and R.K. Kolka. 2008c. Spatial characteristics of net methylmercury production hot spots in peatlands. *Environmental Science & Technology* 42(4), 1010–1016.

Mitchell, C.P.J., B.A. Branfireun, and R.K. Kolka. 2009. Methylmercury dynamics at the upland-peatland interface: Topographic and hydrogeochemical controls. *Water Resources Research* 45:W02406.

Nelson, S.J., K.B. Johnson, J.S. Kahl, T.A. Haines, and I.J. Fernandez. 2007. Mass balances of mercury and nitrogen in burned and unburned forested watersheds at Acadia National Park, Maine, USA. *Environmental Monitoring and Assessment* 126:69–80.

Oliver, D.C. and B.C. Larson. 1996. *Forest Stand Dynamics*. New York: John Wiley & Sons.

Porvari, P. and M. Verta. 2003. Total and methyl mercury concentrations and fluxes from small boreal forest catchments in Finland. *Environmental Pollution* 123:181–191.

Ratcliffe, H.E., G.M. Swanson, and L.J. Fischer. 1996. Human exposure to mercury: a critical assessment of the evidence of adverse health effects. *Journal of Toxicology and Environmental Health* 49:221–270.

Ravichandran, M. 2004. Interactions between mercury and dissolved organic matter—A review. *Chemosphere* 55:319–331.

Sandheinrich, M.B. and K.M. Miller. 2006. Effects of dietary methylmercury on reproductive behavior of fathead minnows (*Pimephales promelas*). *Environmental Toxicology and Chemistry* 25:3053–3057.

Sellers, P., C.A. Kelly, and J.W.M. Rudd. 2001. Fluxes of methyl mercury to the water column of a drainage lake: The relative importance of internal and external sources. *Limnology and Oceanography* 46(3):623–631.

Selvendiran, P., C.T. Driscoll, M.R. Montesdeoca, H.-D. Choi, and T.M. Holsen. 2009. Mercury dynamics and transport in two Adirondack lakes. *Limnology and Oceanography* 54(2):413–427.

Shanley, J.B., P.F. Schuster, M.M. Reddy, D.A. Roth, H.E. Taylor, and G.R. Aiken. 2002. Mercury on the move during snowmelt in Vermont. *EOS Transactions* 83(5):45–48.

Skyllberg, U., K. Xia, P.R. Bloom, E.A. Nater, and W.F. Bleam. 2000. Binding of mercury(II) to reduced sulfur in soil organic matter along upland-peat soil transects. *Journal of Environmental Quality* 29(3):855–865.

St. Louis, V.L., J.W.M. Rudd, C.A. Kelly, K.G. Beaty, N.S. Bloom, and R.J. Flett. 1994. Importance of wetlands as sources of methyl mercury to boreal forest ecosystems. *Canadian Journal Fisheries and Aquatic Science* 51:1065–1076.

St. Louis, V.L., J.W.M. Rudd, C.A. Kelly, K.G. Beaty, R.J. Flett, and N.T. Roulet. 1996. Production and loss of methylmercury and loss of total mercury from boreal forest catchments containing different types of wetlands. *Environmental Science & Technology* 30:2719–2729.

St. Louis, V.L., J.W.M. Rudd, C.A. Kelly, B.D. Hall, K.R. Rolfhus, K.J. Scott, S.E. Lindberg, and W. Dong. 2001. Importance of the forest canopy to fluxes of methyl mercury and total mercury to boreal ecosystems. *Environmental Science & Technology* 35:3089–3098.

Swain, E.B., D.R. Engstrom, M.E. Brigham, T.A. Henning, and P.L. Brezonik. 1992. Increasing rates of atmospheric mercury deposition in midcontinental North America. *Science* 257:784–787.

U.S. Environmental Protection Agency. 2002. Workshop on the fate, transport, and transformation of mercury in aquatic and terrestrial environments. EPA-625/R-02/005. Washington, DC.

Walcek, C., S. De Santis, and T. Gentile. 2003. Preparation of mercury emissions inventory for eastern North America. *Environmental Pollution* 123:375–381.

Wiener, J.G., B.C. Knights, M.B. Sandheinrich, J.D. Jeremiason, M.R. Brigham, D.R. Engstrom, L.G. Woodruff, W.F. Cannon, and S.J. Balogh. 2006. Mercury in soils, lakes, and fish in Voyageurs National Park (Minnesota): Importance of atmospheric deposition and ecosystem factors. *Environmental Science & Technology* 40:6261–6268.

12

Forest Management Practices and Silviculture

Donald A. Perala and Elon S. Verry

CONTENTS

Introduction

This chapter is an overview of forest management and silviculture practices, and lessons learned, on the Marcell Experimental Forest (MEF). The forests there are a mosaic of natural regeneration and conifer plantations. Verry (1969) described forest-plant communities in detail for the study watersheds (S1 through S6) on the MEF. The remaining area is described in standard USDA Forest Service classification of cover types.

Interest in forest management was a driving factor for early studies on the MEF: long-term aspen regrowth on the calibrated watersheds, conversion of aspen to conifers, and the effect of harvest techniques on forest soil properties and forest productivity. Grigal and Brooks (1997) reviewed how forest management can affect undrained peatlands in North America. This chapter further explores the links of traditional forest management to current interest in forest habitat typing and intensive management whereby carbon can be sequestered in soils by leaving thinnings and tops or removed as biofuels.

History of Forest Management Prior to Establishment of the MEF

The forests of northern Minnesota were logged during the late 1800s and early 1900s, especially for white pine. Forests in the MEF were cut between 1865 and 1897 by either the Lorene Day family, Mike McAlpine, and Kirkpatrick, or the Itasca Lumber Company. Perhaps all three were involved as reentry into previous cutovers to remove overlooked pockets of timber was common (Hawkinson and Jewett 2003). Forest fires fueled by logging slash often devastated whatever reproduction remained. Much of the area in and around the MEF burned in widespread fires of 1917. By the late 1920s, much of these cutovers were "too poor to support agriculture and too grim to attract tourists" (Sommer 2008).

On October 29, 1929, the stock market crash forced many older school children to forgo their education and work at odd jobs to help support their family. This dire situation led in part to the creation of the Civilian Conservation Corps (CCC) on April 5, 1933, whereby young men aged 18–25 years old could enroll in CCC camps across the United States. There, they worked on land and water conservation projects for $1 per day, room, board, and healthcare. Nearly, 150 camps were located in Minnesota alone (Sommer 2008).

Red Pine Plantations on the MEF

The CCC crews planted red pine to form several adjacent stands comprising about 52 ha within the South Unit of the MEF (Figure 12.1). These crews almost certainly were stationed at Day Lake (Camp F-34), about 13 km southwest of the MEF. Company 1724 served there from 1934 until 1941 when the camp closed.

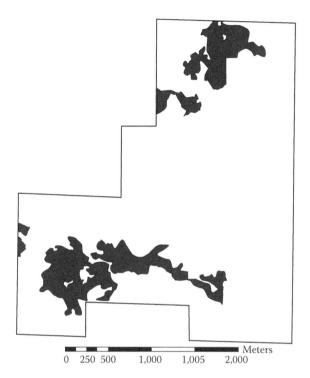

FIGURE 12.1
Red and jack pine stands (105 ha) on the South Unit of the MEF.

Thinning the MEF Red Pine Plantations

By 1958, 2 years before the MEF was established, these CCC plantations were 17–24 years old; some were overly dense and in need of thinning. Concurrently, thinning schedules to maximize yields for given product objectives were being developed by Buckman (1962). The first application of these schedules on the MEF occurred in the summer of 1958 (Roger Bay, emeritus, USDA Forest Service, 2009, personal communication) near Bog Lake. At that time, logging machines were not well matched for use in such dense young stands, and markets for small-diameter products were limited. However, advances within the timber industry and in mechanized harvesters propelled the intensive management of red pine to the elevated state we see today. Red pine is now commonly thinned at intervals of about 10 years (Figure 12.2) as recommended by Buckman (1962) and Buckman et al. (2006).

FIGURE 12.2
Thinned (three times) CCC red pine plantation on the South Unit. Established circa 1935.

Wetland Strip-Cut Harvest to Promote Black Spruce Regeneration

Guidelines for regenerating black spruce (*Picea mariana* (Mill.) B.S.P.) are well established for extensive peatlands (Johnston 1977), but application on small, isolated, lake-filled peatlands had not been documented. Verry and Elling (1978) studied natural seeding in two stands on an 8.1 ha, nonbrushy black spruce bog (S1) on the MEF beginning in 1968 (Table 12.1).

In January and February 1969, these stands were partially clearcut in 30.5 m-wide strips (east-west oriented), leaving 45.7 m-wide uncut strips (Figure 12.3). Almost all slash was piled in windrows. Thus, microenvironmental gradients across these strips could be studied in relation to dispersal of spruce seed, seedling establishment, and general vegetation recovery following harvest.

TABLE 12.1

Black Spruce Stand Characteristics on the S1
Peatland Prior (1968) to Strip Cutting

Stand Characteristic	Older Stand	Younger Stand
Age in 1968	73	62
SI at 50 years (m)	3.7	2.7
Basal area ($m^2\ ha^{-1}$)	8.26	6.43
Volume ($m^3\ ha^{-1}$)	439	313

FIGURE 12.3
The S1 bog with clearcut strips. (Photo courtesy of Verry, USDA Forest Service, Grand Rapids, MN.)

In the summer of 1971, Brown (1973) surveyed vegetation on the S1 strip cuts in the same study area. He found that ericaceous shrubs decreased in coverage but increased in frequency after clearcutting, while sedge increased five times in dry weight production in clearcuts compared to sedges beneath the uncut canopies.

In November 1972, Verry and Elling (1978) inventoried wind-caused mortality in the uncut strips (Figure 12.4) and combined their data with that of Heinselman (1957) to develop a model with remarkable predictive ability. The model (Equation 12.1) predicts wind-caused mortality in strip-cut peatland black spruce using three stand factors: (1) length of exposed edge (both sides), (2) area of the residual strip, and (3) site index (SI).

$$X = \frac{1}{2}\left[\frac{\text{Length of exposed edge (m}^2)}{\text{Area (m}^2)}\right] \times \left[\text{Site Index (m)}\right]^2, r^2 = 0.93, \quad (12.1)$$

The remaining uncut strips were harvested in January 1974 (Verry and Elling 1978). All slash from the 1974 cutting was progressively piled and burned.

FIGURE 12.4
Wind damage to black spruce on clearcut edge, an example from a black spruce stand on the Big Fall Experimental Forest, Big Falls, Minnesota. (Photo courtesy of unknown photographer, USDA Forest Service, Grand Rapids, MN.)

In November 1975 (seven growing seasons after harvesting), spruce reproduction was sampled in the first 30.5 m-wide clearcut strips (they had been sampled in August 1971, after three growing seasons). In November 1976 (three growing seasons after harvest), reproduction in the 45.7 m-wide strips was sampled. Natural seeding in the first strip cuts was adequate (4400 ha^{-1}; 80% milacre stocking) but inadequate (1600 ha^{-1} with 40% stocking) in the 45.7 m-wide strips when few mature trees with seed remained. Progressive strip cuts every other year were recommended.

Forest Harvest and Aspen Suppression with Cattle Grazing to Convert Upland Aspen to Softwoods

In the western United States, cattle and sheep commonly graze aspen forests. If grazing is monitored judicially, there is little harm to ecosystems, but excessive grazing can be harmful, particularly within regenerating aspen stands (Debyle 1985). In the early twentieth century in northern Minnesota, homesteading farmers commonly grazed the forestlands. This practice waned gradually in favor of clearing forestland for pasture.

To examine the efficiency of cattle grazing in preparing harvested aspen stands for conversion to conifers in lieu of herbicides, a series of experiments was established on the MEF. From January to June 1980, mature aspen on the upland of S6 (6.9 ha) along with two other uplands on the east side of Forest Road 3143 were harvested. A Drott feller/buncher and rubber-tired skidders delivered trees to landings where tops and limbs were removed. The S6 upland was fenced in June 1980. During the summers from 1980 to 1982, cattle provided by the University of Minnesota Extension Service were pastured at an intensity of 2.5 cow/calf per ha (Figure 12.5). About 75 A.U.M. (1 animal unit month = 2.5 cow/calf unit grazing for 1 month per ha) of grazing on 8.9 fenced ha accrued over the three summers.

At the time of study planning, increasing commercial demand for red meat and the prospect of exploiting aspen forage for beef production were additional reasons why cattle grazing was considered. The grazing approach was thought to be viable, because the measured nitrogen (N) content of aspen leaves used to calculate total N (or "protein") mass for the uplands area was identical to that of alfalfa. Cattle require 0.75 kg of protein per day. About 1 month of growth after clearcutting (or at the start of the growing season plus ongoing summer growth) was expected to provide enough aspen biomass to make grazing an alternative food source for cattle.

In the summer of 1980, the north lobe to the east of the road was sprayed aerially with Weedone 170 (a mixture of 2-4 D and 2-4-5 T). In the spring of 1981, this area was planted with 4-0 red pine and white spruce seedlings and further released from competing vegetation in 1983 with Esterone 99. This

FIGURE 12.5
Cattle grazing the S6 watershed (left) pasture conditions suitable for planting conifers contrast with the aspen outside the fence. (Photo courtesy of unknown photographer, USDA Forest Service, Grand Rapids, MN.)

provided an area in which to compare and contrast these two site preparation treatments. Coincidentally, the south lobe east of the road was treated similarly except that the area was planted with 6 month-old container stock of genetically selected white spruce and eastern European larch.

Three years of grazing controlled the aspen suckers and reduced the soil surface to pasturelike conditions (Figure 12.5). The cattle lost 9 kg during the first month they were introduced but gained 11 kg the following month. Overall, they gained a satisfactory 0.28 kg per day. In subsequent years, the cattle lost an average of 7.5 kg during the first month of grazing, and cattle gained 0.23 kg d^{-1} during the second month for an average gain of 9.3 kg after acclimation to the aspen diet.

From 1980 through 1983, the cost of site preparation by grazing or herbicides was similar (ca. $250 ha^{-1}), and both provided good vegetation control. However, in the fourth year after planting the pasture (1987), willow from windblown seed increased substantially. On August 3 and 4, 1987, the conifer seedlings were released from this competition by spraying with Garlon 4. Survival and growth of the conifers (Figure 12.6) was satisfactory in 1987 (Table 12.2).

To determine the effects of harvesting and cattle grazing on soil compaction, soil-bulk density was measured in 1980, 1981, 1982, 1984, and 1987 to compare the clearcut and grazed areas to uncut and ungrazed soils. Bulk densities on soils that were logged in winter and not grazed were similar to control soils where trees were not harvested (Figure 12.7). Soil-bulk densities were higher at the skid landing and on the skid trail but were similar between grazed and ungrazed landing soils and between landing and skid trail soils. Soil-bulk densities were similar between logged and grazed

FIGURE 12.6
European larch and "super" white spruce in the herbicided conversion area. (Photo courtesy of unknown photographer, USDA Forest Service, Grand Rapids, MN.)

TABLE 12.2

Survival and Growth of the Conifers on and Adjacent to the S6 Watershed

Treatment	Species	Age (Year)	Height (cm)	Survival (%)
Grazed	Red pine	7	48	55
	White spruce	7	48	81
Herbicide	Red pine	10	56	88
	White spruce	10	104	78
Genetic trial; herbicide	Super white spruce	6	145	85
	European larch	6	203	68

areas and logged but not grazed areas. Error bars (±1 standard deviation) are shown for the uncut control area (Figure 12.7). The error bars for the other areas are as large or larger (up to 270%), indicating that the overall effects of harvesting and cattle grazing on soil-bulk density were heterogeneous and generally minimal.

Development of Allometric Relationships

In December 1981, Grigal and Kernik (1984a) felled a sample of 24 black spruce trees representing the range of diameters in perched bog sites on the MEF. Using usual subsampling techniques, the trees were further separated into bole bark, bole wood, cones, foliage, and live and dead branches. Nine trees were sampled in the summer of 1981 to determine stump and

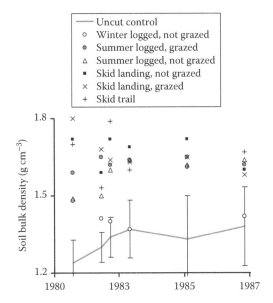

FIGURE 12.7
Changes in soil-bulk density after logging and grazing on the S6 watershed. Standard deviation error bars are shown for the control (uncut) data; error bars for other data are about twice
as large.

root weights greater than 3mm in diameter. Subsamples of each of these
components were ovendried to determine dry weight–fresh weight ratios for
estimates of component dry weight.

Relationships between diameter at breast height (dbh) and mass of various
components were established by fitting the allometric model

$$Y = a \times D^b,$$ (12.2)

where
Y is mass in kilogram
D is dbh in centimeter

Total above ground biomass conformed well to this model and some of the
more variable components (cones and foliage) less so.

Grigal and Kernik (1984b) thereafter compared their equation predictions
for total aboveground biomass and foliage biomass with predictions from
nine published equations for black spruce across its range in North America
(Figure 12.8). Their equation for total aboveground biomass conformed well
to the other equations, indicating wide generality and thus justifying its use
regionally. However, estimates for foliage biomass differed greatly, apparently due to variation in stand stocking density. Including a term for relative
stocking also may be useful in generalizing foliage prediction.

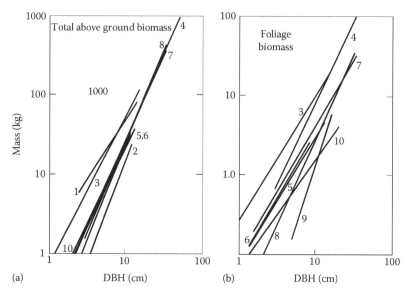

FIGURE 12.8

Comparison of 10 equations for total aboveground biomass (a) or foliage (b) equations. (From Grigal, D.F. and Kernik, L.K., *Can. J. For. Res.*, 14(3), 468, 1984a; Grigal, D.F. and Kernik, L.K., Biomass estimation for black spruce (*Picea mariana* (Mill.) B.S.P.) trees, Research note 290, University of Minnesota Department of Forestry, St. Paul, MN, 1984b.)

North American Long-Term Soil Productivity Study Site

Modern logging machinery is heavy and can compact forest soils. Removal or displacement of soil organic matter by machine operation is also common. This could result in increased soil-bulk density with less pore space for root aeration and water absorption. Slower tree growth and increased water runoff follow.

Soil scientists had been studying this problem sporadically until the late 1980s when a coordinated effort was undertaken. In 1989, the Forest Service initiated the North American Long Term Productivity (LTSP) study to provide "…a network of installations… across a broad range of forest ecosystems throughout the United States and Canada" (USDA Forest Service 2001).

In 1990, an LTSP study site was established on the MEF by David H. Alban (Alban et al. 1994). This site is one of four established by Alban in the Lake States; all are part an LTSP study to better understand how soils are affected by forest harvesting. Indeed, the MEF served as a pilot site for establishing a common protocol in designing LTSP experiments across North America.

The MEF study was installed in a well-stocked, 70 year-old quaking and bigtooth aspen stand with a SI of 20.7 m at age 50. Aspen accounted for 88% of the aboveground biomass versus 1% for shrubs and herbs; the soil is Cutaway loamy sand. Eight 30 by 40 m treatment plots were established. In February 1991, six plots were total tree harvested (TTH) with a feller–buncher weighing 13,000 kg. In April 1991, the forest floor was hand-raked to simulate forest floor removal (FFR) from three of the plots (Figure 12.9). In May, four of these plots were compacted (CPT) by four or five passes of an 8100 kg rubber-tired pneumatic roller pulled by a D-6 Caterpillar tractor (Figure 12.9). Of the harvested plots, two were compacted (TTH + CPT), two were compacted with the forest floor removed (TTH + CPT + FFR), one had only the forest floor removed (TTH + FFR), and one had neither treatment (TTH).

Alban et al. (1994) reported that 2 year-old aspen coppice regeneration was the most vigorous with TTH alone. Total biomass, including shrubs and herbs, showed the same trend. Number of species in the ground flora increased considerably 2 years after TTH, especially after TTH + CPT + FFR. The rate of rainfall infiltration was greatly reduced by TTH + CPT after 2 years.

Stone and Elioff (1998) reported that TTH + CPT significantly increased bulk density and strength of the surface 30 cm of soil and that neither has recovered 5 years after treatment. Total vegetation biomass on TTH plots was more than two times greater than for any other treatment.

Page-Dumrose et al. (2006) reported how soil-bulk density and strength change as indicators of soil compaction 1 and 5 years after timber harvesting and other site treatment on 12 LTSP sites, including MEF. Severe soil-compaction treatments produced nearly root-limiting bulk densities for all soil textures studied and were noticeable to a depth of 30 cm. After 5 years, bulk density recovered in coarse soils down to 10 cm but less so from 10 to 30 cm. Fine-textured soils recovered little.

FIGURE 12.9
Forest-floor litter layer raked off the LTSP plots by hand (left); compacting the soil using a "wobble wheel." (Photo courtesy of unknown photographer, USDA Forest Service, Grand Rapids, MN.)

Powers et al. (2005) summarized the impacts of organic-matter removal and soil compaction over the first 10 years for the 26 oldest installations in the North American network of LTSP sites, including the MEF. Total removal of surface organic matter led to declines in concentrations of soil carbon to a depth of 20 cm and less nutrient availability, due primarily to the loss of the forest floor. Stored soil carbon seemed unchanged but is confounded by changed soil-bulk density and decomposition of roots left by harvesting. Biomass harvesting did not influence forest growth. Effects of soil compaction depended on initial bulk density. Soils with densities greater than 1.4 Mg m^{-3} resisted compaction. Density recovery was slow, particularly in frigid temperature regimes. Forest productivity declined on compacted clay soils, increased on compacted sands, and generally was unaffected where an understory was absent.

After 15 years, soil strength and soil-bulk density still were considerably greater in compacted plots at the MEF, and plots with compaction or with the forest floor removed averaged 31% less total biomass (unpublished data). This result indicates that recovery to pretreatment condition will not occur in the near future.

Demonstration of Cut-to-Length Technology

In June 1998, the MEF hosted more than 300 people interested in learning how to minimize logging damage to soil and forest productivity (light-on-the-land-logging) (Verry 1998). As part of an equipment demonstration, they watched Timberjack cut-to-length (CTL) processors and forwarders (Figure 12.10) of two sizes harvesting trees on a variety of sites, including forests on fragile peatlands. These highly computerized machines grab a tree,

FIGURE 12.10
Cut-to-length feller/buncher (left) and forewarder demonstrated in an MEF aspen stand. (Photo courtesy of unknown photographer, USDA Forest Service, Grand Rapids, MN.)

sever it at the stump, and automatically shear off the branches and cut the bole into optimum programmed lengths for transport by a forwarder. One of the questions about these machines was whether they would leave the soils with optimum pore space to maintain forest productivity. Soils, water table, and stand condition were evaluated before and after logging aspen, jack pine, and black spruce for five stand conditions, and under several silvicultural systems. The demonstration was mostly successful in leaving the harvested sites in good condition. The single failure was an increase of 10%–20% in organic soil-bulk density and organic soil rutting caused by a loaded forwarder on peat soil; even a half load caused the forwarder to rut deeply on the third pass. As a result, CTL harvesting on undrained and/or unfrozen organic soils is not recommended. On the upland sites, machine travel on slash seemed to protect mineral soils better than travel with no slash. This can be accomplished by logging from the back to the front of the sale.

Growth and Yield of Aspen Regeneration after Clearcutting Watershed S4

Experiments within aspen ecosystems on the MEF have contributed greatly to our understanding of aspen stand development, growth, and yield. The S4 watershed (Figure 12.11) was clearcut harvested during the late autumn of 1970 (1.28 ha), in the winter of 1970–1971 (10.18 ha), in the summer of 1971 (5.38 ha), and in late summer and fall of 1971 through January 1972 (11.61 ha). Aspen coppice regenerated the cutovers quickly, vigorously, and completely.

FIGURE 12.11
The S4 watershed in late winter 1972 after clearcutting the last aspen stand. (Photo courtesy of Verry, USDA Forest Service, Grand Rapids, MN.)

A hailstorm in the summer of 1973 caused great damage and mortality that, in turn, initiated new coppice that was mostly ephemeral.

The entire upland, except for 0.6 ha reserved for fertilizer rate trials, was fertilized by air in 1978 at 336 kg ha^{-1} of N supplied as ammonium–nitrate. By age 38, SI had increased as much as 2.4 m from that of the original stand (22.6 m). Increased SI was hypothesized to be a response to the N fertilization and the decay of large amounts of broadcast-scattered logging slash. Diameters and heights of regenerated trees and shrubs were measured every 2 years through 2002 and in 2008. From these, biomass and merchantable timber could be calculated using equations from Perala and Alban (1993) and Wenger (1984).

The development of these regenerating stands over 38 years is demonstrated in Figure 12.13a and b. The near total dominance of aspen in biomass per hectare is well illustrated in Figure 12.12a. At age 38, only the "other" category approached 8% of the aspen biomass. Figure 12.12b shows that aspen sucker densities offer overwhelming competition to other regeneration for the first 3 or 4 years. Hazel then surpassed aspen in numbers, but, by then, aspen dominants were 3.7 m tall, and their canopies captured most of incoming solar energy. By age 38, aspen stem densities were a small fraction of the total.

Figure 12.13c and d show aspen development by season of harvest. In late fall (1970) and late winter (March 1971), logging produced the most 1 year-old coppice stems, whereas logging during summer to fall produced the most 2 year-old stems. Late fall and winter logging is followed by a full summer to initiate and grow coppice regeneration. However, summer or fall logging disturbs the litter layer and competing ground vegetation to allow more soil warming and thus more abundant suckering. A snowpack may limit such disturbance, delaying and slowing the production of coppice. There may be little practical advantage with harvest season. Figure 12.13c shows stocking differences among time of harvest are confounded in several years.

Figure 12.13d shows that logging in late fall or winter produced the most aspen biomass for the first 10–12 years. After that time, the development of aspen biomass is even more confounded by harvest season than stem numbers, indicating that aspen can be logged in any season on soils that are well drained or frozen to minimize compaction. Figure 12.13d further shows two waves of accelerated self-thinning at age 14 (1985) and 30 (2001). This common phenomenon was little appreciated until pointed out by Graham et al. (1963).

Aspen Self-Thinning Tested

This allows an opportunity to independently test the predictive usefulness of aspen growth and self-thinning models (Perala et al. 1996). The first task is to estimate, for each of the four harvest timings, SI (S, meters at 50 years) according to Lundgren and Dolid (1970):

$$S = \frac{H}{1.48 \times [1 - e^{-0.0214 \times A}]^{0.9377}} , \qquad (12.3)$$

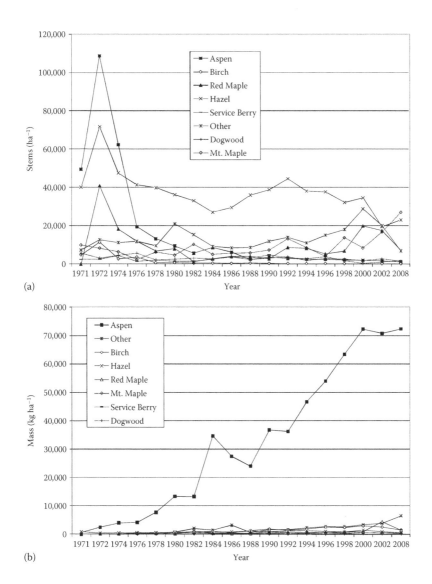

(a)

(b)

FIGURE 12.12

Regeneration, growth, and survival of aspen and associated species after clearcutting the S4 upland. Species density (a) and biomass (b)

(continued)

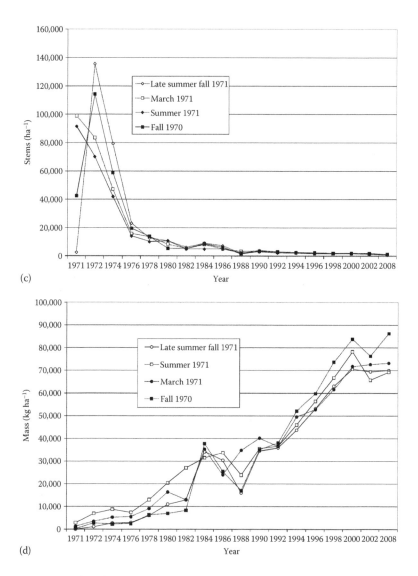

(c)

(d)

FIGURE 12.12 (continued)
aspen stem production and survival (c), and aspen biomass accretion and loss (d) by season of harvest.

where
  H is the dominant height in meters
  A the stand age in years

Equation 12.4 now can be used to predict quadratic mean diameter (D, cm) from given A and S (we used the age 37 or 38 S value) and from long-term (30 years) local mean July temperature (J, Celsius; 18.7°C at the MEF):

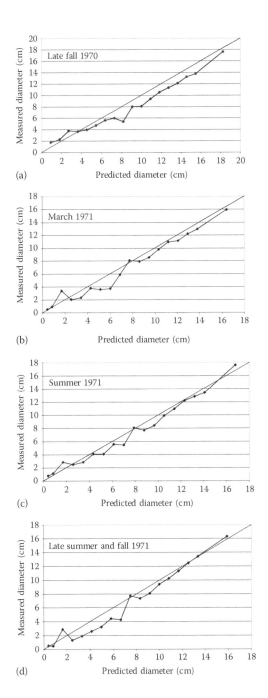

FIGURE 12.13
Measured aspen quadratic mean dbh plotted over predicted values by season of harvest; the diagonal indicates parity. Reductions in dbh occur with periodic thinning episodes, but diameter growth tends to recover to predicted values with time.

TABLE 12.3

Measured Site Index for the Upland Harvest Study on the S4 Watershed (Original Stand 54 Years Old and Regenerated Stand 38 Years Old), Diameter, Density, and Basal Area of the Regenerated Stand (Age 38) and Predicted Values Using Perala's (1996) Self-Thinning Equations

Logging Date	SI 1968 (m)	SI 2008 (m)	Measured or Predicted Diameter (cm)	Diameter (cm)	Number (ha⁻¹)	Basal Area (m² ha⁻¹)
Late fall 1970	22.6	25.9	Meas.	17.6	766	18.6
			Pred.	18.3	836	22.0
March 1971	21.6	24.4	Meas.	16.0	1058	21.3
			Pred.	16.4	994	21.0
Summer 1971	22.6	25.0	Meas.	17.6	1203	29.3
			Pred.	16.8	959	21.2
Fall 1971	22.6	23.5	Meas.	16.3	988	20.6
			Pred.	15.9	1048	20.8

$$D = 0.04895 \times A^{1.009} \times 1.036^{S} \times 1.071^{J}, \tag{12.4}$$

Next, normal stocking, N (stems ha⁻¹), is predicted from:

$$N = 4,088,000 \times (1 - 0.7022^{D}) \times 0.8213^{J} \times D^{-1.657}, \tag{12.5}$$

where D is the predicted value from Equation 12.4.

The results are summarized as follows (predicted values in parentheses) for these stands as they near commercial rotation age. Note the close agreement between predicted and measured values (Table 12.3).

Figure 12.13 shows how well the predicted values of D compare with measured values over the life of the stands. In general, predicted values were overestimates during the first 3 or 4 years but tended toward convergence with maturity. Part of this bias would be mitigated by taking into account the lesser SI values in 1968. How to proceed with such analysis is beyond the scope of this chapter. For example, how long does it take until the full growth response to fertilizer and decaying coarse woody debris is fully expressed? The results of the variable rate fertilizing (Figure 12.14) suggest that SI may have improved in the first 2 years after broadcasting 336 kg ha⁻¹ of N. Note that both 168 and 504 kg ha⁻¹ produced lesser growth gains. Others have shown growth gains of as much as 177% (e.g., review by Berguson and Perala, 1988). Note also that nitrate concentrations in soil-water extracts and in stream water (Chapter 14) were greatly diminished 1.5 years after fertilizing. This could indicate that much of the N already had

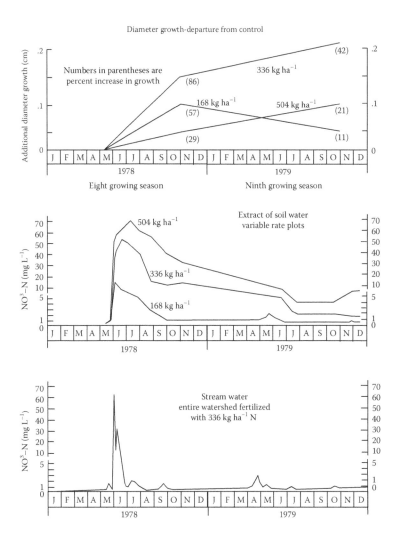

FIGURE 12.14
Response of aspen growth and water chemistry to 168, 336, and 450 kg ha^{-1} of N fertilizer application on plots in the S4 watershed.

been taken up by vegetation and is contributing to N-cycling in the upland forest ecosystem.

The self-thinning model greatly underpredicted stem density for the first 5–10 years (Figure 12.15). This is expected, because the model estimates stocking for "normal" stands that have attained maximum canopy, usually by age 10. The estimates improve after that but there is much variability, especially with the self-thinning waves mentioned. The estimates for stands approaching maturity become more stable.

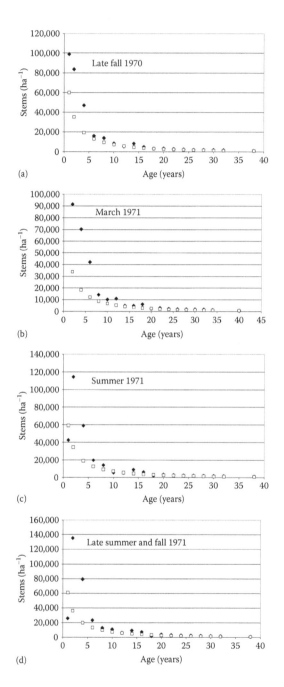

FIGURE 12.15
Measured (diamonds) and predicted aspen stem density over age for the upland harvest study on the S4 watershed. Initial high densities collapsed to predicted densities at about age 18. While summer and fall (1971) harvest had the highest initial densities, time of harvest did not affect density after age 18.

New Directions in Forest Management

Aimo K. Cajander was Prime Minister of Finland during the Winter War from November 1939 to March 1940. Thirteen years earlier, as a professor of forestry at the University of Helsinki, Cajander (1926) theorized that forest productivity is best predicted by understory vegetation. The combination of forestry professor and high political office also occurred in the United States. In 1890, Gifford Pinchot, a professor of forestry, founded both the Yale University School of Forestry and the Society of American Foresters. Gifford's role as one of American's first professional foresters was foreshadowed by his father James, whose land speculation and lumbering business brought great family wealth but also regret for the damage his business had done to the land. James subsequently made conservation a priority and placed his capable son at Yale by endowing the School of Forestry at the University. Grey Towers, the family estate at Milford, PA, became a "nursery" for the American forestry movement. Pinchot became the first Chief of the Forest Service in 1905 and espoused a national vision of American forestry by recruiting professionally trained foresters throughout the nation. Pinchot promoted the efficient use and renewal of the nation's forests resources for the greatest good.

In recent decades, the Forest Service presented a continental vision of land classification for both land and water ecological units. The aquatic classification is based on large river basins and nested watersheds within basins (Maxwell et al. 1995). The terrestrial classification is based on geology, climate, soil taxa, and potential vegetation (Keys et al. 1995). At its lowest level, the National Hierarchical Framework of Ecological Units defines the Ecological Land Type Phase (ELTP) as a combination of landform (slope position, slope, aspect, etc.) and soil type. In many cases, an ELTP and a habitat type based on total vegetation structure are the same. In other cases, one ELTP may contain more than one habitat type. The merger of land form and soil-type classification with vegetation habitat-type classification has come to maturity in the Lake States some 75 years after Cajander (1926, 1949) used habitat type and landform (esker, coastal, mire, etc.) to classify forests in Finland.

Bakuzis (1959) and Bakuzis and Kurmis (1978) arrayed forest types (trees, shrubs, and herbaceous associations) in Minnesota on a four factor grid of synecological coordinates (moisture, nutrients, light, and heat): factors of landform, soils, and climate. Combinations of these factors on a scale of 1 to 5 predicted the plant association requirements for optimal growth and survival. Grigal and Arneman (1970) quantified relationships among vegetation and soil classification units in northeastern Minnesota. Habitat typing (potential tree and herbaceous vegetation) was first developed in the Upper Peninsula of Michigan by Coffman et al. (1980). Subsequently, Kotar (1986) and Kotar et al. (1988) published habitat types for several counties in

Michigan and Wisconsin. A rush of publications on habitat-type publications followed.

Almendinger and Hanson (1998) listed 18 native community types (habitat types) for the Northern Lake Plains subsection and the Chippewa National Forests in Itasca and Cass Counties in Minnesota. Kotar et al. (1999) found that habitat types explained much of the variation in forest productivity on Forest Service forest-inventory plots in Wisconsin. In the same year, the Boise-Cascade Corp. (International Fall, MN) published habitat types as an ecosystem diversity matrix for forest land in Koochiching County in northern Minnesota and Ontario, Canada (Kernohan et al. 1999). In 2000, the Blandin Paper Co. (Grand Rapids, Minnesota) and Boise Cascade cosponsored the publication of "A Field Guide to Forest Habitat Type Classification for North Central Minnesota" (Kotar and Burger 2000). Bakuzis' moisture and nutrient synecologic coordinates were correlated with 10 dominant and codominant habitat types, and Cajander's concept of using forest understory to predict forest productivity was expanded to predict the successional pathways of forest trees despite natural or logging disturbances (Kotar 1986). Forest management tools had mixed the amalgam of potential tree and existing herbaceous vegetation, soil moisture, nutrient status, and soil-mapping units to hardened steel, founded on the belief and field expression that optimum forest productivity is based on the natural plant associations adapted to the landforms, soils, and climate of the region.

As Pinchot envisioned, the efficient use and renewal of our forest resources should provide the greatest good for the greatest number of people in the long run. The greatest good is provided when the optimal growth of all tree species on a habitat type is achieved at the least expense of time and money. Forest management broadened from single-species site management supplying a single mill, to the concept of forest-community vision (a community-based approach to forest management reflected in a variety of products and of forest-land uses). In a forest-community vision, all trees adapted to a site reach optimum growth, and a mix of trees is supplied to a mix of mills using one or several species.

Knowledge of habitat types enlarges the palette of forest management choices for both early- and late-succession canopy covers. Cheryl Adams with the Blandin Paper Company, a division of UPM Kymmene, in Grand Rapids, Minnesota, has invested over a decade in learning new management approaches to supply a mixture of aspen, spruce, and balsam fir to their mill and other species to other mills while optimizing forest productivity on all their forest land. Three habitat types and management options are briefly reviewed to illustrate how the concept of efficient use with the least expenditure for the greatest productivity is applied in north central Minnesota. The ATiCa habitat type (*Acer saccharum-Tilia Americana/Cauliphyllum thalictroides* or sugar maple-basswood/blue cohosh) is found on mesic, nutrient-rich sites: hilly to steep slopes of stagnation moraines with well-drained silt loams over sandy loams or sandy loam caps over clay loam till. The naming convention lists the two dominant trees at the late-succession stage and the dominant herbaceous plant at any succession stage. Identifying the dominant herbaceous plant at any stage avoids site typing by the current canopy cover. This also avoids limiting management choices associated only with a canopy-cover type.

The ATiCa-habitat type is managed for northern hardwoods (sugar maple and basswood with inclusions of red maple and ironwood) in the late-succession stage. Where there are mature aspen in the early succession stage, this type can be managed for aspen with a combination of clearcutting, precommercial, and commercial thinning with a final harvest at age 40. Only sites with an aspen SI of 21.3 m at age 50 or higher are suitable for this management option. Precommercial thinning at age 7 or 8 allows the efficient use of brush saws and avoids the introduction of hypoxylon canker found increasingly in sucker stands after age 10 as self-thinning damage opens tree wounds. Thinning around dominant suckers in a 2.1 m circle (1.1 m radius) leaves 2000–2225 trees ha^{-1} with the intervening suckers laid on the ground. Commercial thinning at age 25 yields clear wood when half of the trees are removed. Final harvest maximizes the production of clear wood at age 40. One person with a power brush saw can thin 0.8 ha per day. Typically, six-person crews are used. Commercial thinning yields product useable wood but costs about $40 ha^{-1}.

The AbFnAu habitat type (*Abies-Fraxinus nigra/Asarum canadense* or the balsam fir-black ash/wild ginger) is found on wet mesic to mesic, medium-nutrient sites: slightly rolling lake plains, till plains, and outwash plains that are somewhat poorly to poorly drained. This is mixed-wood type and is common in northern Minnesota and wide areas in central and western Canada. Balsam fir and black ash along with red maple are the late-succession cover type, but white spruce is a potential cover type in this habitat, as is white pine. As with many Lake State forest stands, mature aspen may be the current dominant species. Where aspen is age 55 or older, one prescription for mixed-wood management is to clearcut the aspen and plant white spruce in the same year as the clearcut. On the basis of a decade of planting trials, planting white spruce at 740–990 ha^{-1} is recommended. White pine may substitute for 15% of the spruce where dense slash piles provide a planting site that is less susceptible to deer browsing. The site is examined at age 3 and if total trees (aspen suckers and white spruce or white pine) exceed 5000 ha^{-1}, the aspen is thinned at age 4 with a brush saw downing a 1.1 m swath around each conifer and leaving a total tree count of 2000–2225 trees ha^{-1}. At age 45, aspen basal area is thinned to 15 m^2 ha^{-1}. Heavier aspen thinning may encourage competition from hazel and heavy balsam fir regeneration that is subject to dieback. White spruce is planted again after aspen thinning if needed. The aspen overstory is removed at age 50. The stand is subsequently regenerated by commercial thinning. Balsam fir, white spruce, white pine, patch-derived aspen, and black ash are removed periodically. However, barring wide spread fire and wind throw, the site retains a mature forest with a mixed wood canopy.

The AbPiV habitat type (*Abies-Picea glauca/Vaccinium angustifolium* or the balsam fir-white spruce/blueberry) is found on dry to mesic, nutrient-poor sites: rolling, sandy, well drained. Any mixture of balsam fir, white spruce, aspen, paper birch, and jack/ red/white pine is common. Light scarification favors aspen and balsam fir reproduction while heavy scarification favors pine. On National Forests, managers tend to favor pine species and remove balsam fir.

These examples illustrate how habitat type drives management choices based on landowner objectives. In many harvests, CTL equipment retains slash on site. This practice has been recommended as the most effective using harvester finesse and many-wheeled forwarders. Stands with large trees require the power of conventional shearers and cable skidders, but the return of slash broadcast on the site has been recommended. In aspen clearcuts and all harvests, cutting progresses from the back to the front of the sale (main access road) to minimize soil disturbance as soils in cleared or thinned areas become wet after initial entry. Protecting the soil structure ensures the continuity of habitat-type structure and potential productivity. The destruction of soil structure by compaction and rutting reduces the productivity of crop trees by 10%–35% (Stone and Elioff 1988). Major compaction on skid trails can nearly be eliminated with the use of slash mats prior to skidding.

In the following section, we discuss the use of long-term data on aspen growth and yield from the MEF, how much biomass is returned to the soil for various aspen-thinning scenarios, and how many nutrients are returned along with the thinning slash. The development of mixed-wood and aspen-thinning scenarios adds more tools to allow a wide mix of harvest species while maintaining a mature forest canopy. Additional innovations in forest management are also possible using habitat types as a guide.

Forest-management choices are watershed-management choices. Maintaining mature forest canopies near streams (riparian areas) ensures protection of streambanks from equipment crushing and erosion and provides continued stream amenities of shade and long-term inputs of large wood. All-season forest roads with stream-wide culverts or bridges ensure passage of fish during critical spawning runs and light-on-the-land access for many forest uses.

Watersheds with too much open land or young forest land (<16 years old) produce runoff at accelerated rates without erosion of the land surface, but these accelerated runoff rates are high enough to extend gullies and to erode and reshape stream channels when open and young forest land accumulates to 60% of watersheds 0.4–4 ha in size. The accelerated stream channel flow is caused by soil compaction that reduces infiltration and by the melting of snowpacks in open areas or young forest stands. Differences in snowmelt in young forests are related to adolescent crown height. Even aspen thinning directed at a 40 year rotation can produce a forest canopy with high crowns (wide crowns on the upper third of the tree bole) in 10 years compared to self-thinning in aspen that typically produces a high-crown forest in 16 years or more and retains high stem counts. Watershed experiments at the MEF can measure the effect of new forest-management scenarios on total water yield and, more importantly, for dynamically stable streams, the rate of channel forming flows at the bankfull stage that occur with a frequency of about every 1.5 years. Evaluations of thinning and mixed-wood impacts on total water yield, peak flows, and channel-forming flows are prime candidates for watershed research.

Biomass and Nutrient Additions to Soils from Thinnings, Logging Slash, and Fertilizer

We again use the growth-and-yield model of Perala et al. (1996) to demonstrate how management choices affect carbon and nutrient budgets. Specifically, we apply Equation 12.6 to aspen growth data from the already-featured S4 watershed study to predict the amount of organic matter that is allocated to the felled thinnings. Equation 12.6 assumes each thinning is taken

$$D_2 = D_1 \times \left(\frac{N_2}{N_1} \right)^{-0.1967} , \qquad (12.6)$$

in ascending order from a modeled-ranked dbh distribution until the final target of residual stem density is attained. Thus, the most robust trees are left to grow freely until asymptotically approaching maximum leaf area.

$$D_2 = D_1 \times \left(\frac{N_2}{N_1} \right)^{-0.1967} , \qquad (12.6)$$

The thinning prescriptions developed for the AbFnAu (age 4) and ATiCa (age 8) habitat types are used as examples here because they meet management goals for producing clear-wood products. Nutrient values are derived from Table 2 of Perala and Alban (1982) describing the distribution of organic matter and nutrients in a 40 year-old aspen stand growing on a Warba very fine sandy loam soil; they are applied here to the thinnings (Table 12.4). How well these nutrient concentrations represent the soils of S4 (a Nashwauk sandy loam over clay loam), and particularly the nutrient status of young stands is unknown.

TABLE 12.4

Original S4 Aspen Stand (54 Years Old) and Aspen Stand Assessed for Nutrient Content at 40 Years Old

Aspen Stand Condition	Age (Year)	Quadratic Mean Diameter (cm)	Density (Stems ha⁻¹)	Oven Dry Weight Bole and Branches (kg ha⁻¹)	N (kg ha⁻¹)	P (kg ha⁻¹)	K (kg ha⁻¹)	Ca (kg ha⁻¹)	Mg (kg ha⁻¹)
S4 prior to harvest 1971	54	21.1	632	320,000	570	82	396	1,852	133
Logging slash left on site	54			173,000	308	44	214	1,001	72
Fertilizer added to regen.	1				336	0	0	0	0
Total (fertilizer + slash)					644	44	214	1,001	72
% of original 1971 stand					113	54	54	54	54
Perala and Alban 1972	40	18.0	1,344	148,800	265	38	184	861	62
Regenerating S4 stand									
Prior to thinning	4	3.8	25,210	9,610	17.1	2.5	11.9	55.6	4.0
After thinning	4	6.8	500	1,540	2.7	0.4	1.9	8.9	0.6
In the thinning	4		24,710	8,070	14.4	2.1	10.0	46.7	3.4
% of Perala and Alban					5	5	5	5	5
Regenerating S4 stand									
Prior to thinning	8	4.0	5,298	14,143	25.2	3.6	17.5	81.8	5.9
After thinning	8	4.8	850	8,068	14.4	2.1	10.0	46.7	3.4
In the thinning	8		4,448	6,075	10.8	1.6	7.5	35.2	2.5
% of Perala and Alban					4	4	4	4	4

Source: Perala, D.A. and Alban, D.H., *Plant Soil*, 64, 177, 1982.

Notes: Two thinning scenarios, using the regenerating aspen sucker stand on S4 at age 4 (recommended on habitat type ATiCa), are shown. Stand age, density, weight, and nutrient content are shown for each along with the amount of N fertilizer applied to the regenerating S4 stand.

The age 4 scenario would contribute 8070 kg ha^{-1} of organic matter in thinnings to the soil, and the age 8 scenario would contribute 6075 kg ha^{-1} if only the crop trees remain (Table 12.4). This is counterintuitive, but the 4 year-old stand had nearly five times greater stem density while mean stand dbh is nearly equal to the 8 year old stand. This demonstrates the weakness of Equation 12.6 to predict D values in stands younger than age 10, because they have not yet attained maximum leaf area. If the thinnings are confined to the 0.9–1.1 m radius circles as specified, these estimates must be reduced by 45% or 25%, respectively.

Nutrients in 4 and 8 year-old thinnings amount to only 4%–5% of the 40 year-old Perala and Alban stand. Whether this is a significant return to the nutrient cycle is beyond the scope of this chapter. The early aspen thinning accelerates the growth of residual trees and produces clear wood with significantly less disease impact (C. Adams, 2009, personal communication).

Table 12.4 also estimates nutrient amounts in the original S4 stand (origin about 1917) harvested in 1971, the amount of logging slash left on site, and the amount of N added in fertilizer at age 1. It is important to recognize that the regenerating aspen stand on S4 (at age 38) had increased in SI by 11% (Table 12.3). A higher SI stand will also support greater basal areas, and so the actual increase in wood production may be 15%–20%. Logging slash remaining from the 1971 harvest returned 54% of the organic matter and N, phosphorus (P), potassium (K), calcium (Ca), and magnesium (Mg) found in the mature stand. The application of ammonium-nitrate fertilizer to the age 1 aspen stand added 59% of the N in the original stand (113% total addition; Table 12.4).

Guidelines for harvesting biomass in Minnesota (MFRC 2007) suggest that conventional harvesting leaves about 40% of the original aboveground bole and branch wood and that one-third of this should remain if the remainder is harvested as biomass. The 1971 conventional harvesting on the S4 watershed left 54% of the original stand on the ground (173,000 kg ha^{-1}, Table 12.4). Removing two-thirds (115,333 kg ha^{-1}) of the biomass would also remove 205, 29, 142, 667, and 48 kg ha^{-1} of N, P, K, Ca, and Mg, respectively. The regenerating aspen stand on S4 received 308 kg ha^{-1} of slash N and 336 kg ha^{-1} of N as ammonium–nitrate fertilizer. Presumably, these substantial additions of N to the regenerating stand both caused the 11% increase in SI after 38 years. The MFRC guidelines suggest that nutrient additions from the decay of organic matter over 50 years can balance nutrient losses if bole wood and two-thirds of the logging slash are removed. However, this single-item evaluation does not incorporate 50 years of soil nutrient leaching nor account for possible changes in SI during the subsequent rotation. The long-term estimates of biomass and nutrients at S4 show that large returns of organic matter and nutrients to the soil after harvest can significantly increase SI; it may be that large removals of logging slash in biomass also can reduce SI. Long-term research is needed to incorporate all nutrient-cycling pathways (e.g., leaching, litter fall return, mineral weathering, and changes in SI from depauperate organic matter return or soil compaction).

References

Alban, D.H., G.E. Host, J.D. Elioff, and D.A. Shadis. 1994. Soil and vegetation response to soil compaction and forest floor removal after aspen harvesting. Research paper NC-315. St. Paul, MN: USDA Forest Service.

Almendinger, J. and D. Hansen. 1988. *Ecological Land Classification Handbook for the Northern Minnesota Drift & Lake Plains and the Chippewa National Forest.* Minneapolis, MN: Minnesota Department of Natural Resources.

Bakuzis, E.V. 1959. Synecological coordinates in forest classification and in reproduction. PhD thesis. Ann Arbor, MI: University of Minnesota. Microfilm and Xerox Publ. Univ. Microfilms.

Bakuzis, E.V. and V. Kurmis. 1978. Provisional list of synecological coordinates and selected ecographs of forest and other plant species in Minnesota. Staff Series Paper No. 5. St. Paul, MN: Department of Forest Resources, University of Minnesota.

Berguson, W.E. and D.A. Perala. 1988. Aspen fertilization and thinning research results and future potential. In: *Minnesota's Timber Supply: Perspectives and Analysis,* eds., A.R. Ek and H.M. Hoganson. Grand Rapids, MN: University of Minnesota College of Natural Resources and Agricultural Experiment Station, pp. 176–183.

Brown, J.M. 1973. Effect on overstory removal on production of shrubs and sedge in a northern Minnesota bog. *Journal of the Minnesota Academy of Science* 38(2–3):96–97.

Buckman, R.E. 1962. Growth and yield of red pine in Minnesota. Tech. Bull. 1272. St. Paul, MN: USDA Forest Service.

Buckman, R.E., B.B. Bishaw, T.J. Hanson, and F. Benford. 2006. Growth and yield of red pine in the Lake States. General technical report NC-271. St. Paul, MN: USDA Forest Service.

Cajander, A.K. 1926. The theory of forest types. *Acta Forestalia Fennica* 29(3):1–108.

Cajander, A.K. 1949. Forest types and their significance. *Acta Forestalia Fennica* 56:1–71.

Coffman, M.S., E. Alyank, and J. Ferris. 1980. *Field Guide to Habitat Type Classification System for Upper Peninsula of Michigan.* Houghton, MI: School of Forestry and Wood Products. Michigan Technological University.

Debyle, N.V. 1985. Animal impacts. In: *Aspen: Ecology and Management in the Western United States.* General Technical Report RM 119. Fort Collins, CO: Rocky Mountain Forest and Range Experiment Station, USDA Forest Service.

Graham, S.A., R.P. Harrison Jr., and C.E. Westell Jr. 1963. *Aspens: Phoenix Trees of the Great Lakes Region.* Ann Arbor, MI: University of Michigan Press.

Grigal, D.F. and H.F. Arneman. 1970. Quantitative relationships among vegetation and soil classifications from northeastern Minnesota. *Canadian Journal of Botany* 48(3):555–566.

Grigal, D.F. and K.N. Brooks. 1997. Forest management impacts on undrained peatlands in North America. In: *Northern Forested Wetlands: Ecology and Management,* ed. C.C. Trettin, M.F. Jurgensen, D.F. Grigal, M.R. Gale, and J.K. Jeglum. Boca Raton, FL: CRC/Lewis Publishers, pp. 379–396.

Grigal, D.F. and L.K. Kernik. 1984a. Generality of black spruce biomass estimation equations. *Canadian Journal of Forest Research* 14(3):468–470.

Grigal, D.F. and L.K. Kernik. 1984b. Biomass estimation for black spruce (*Picea mariana* (Mill.) B.S.P.) trees. Research note 290. St. Paul, MN: University of Minnesota Department of Forestry.

Hawkinson, S. and W. Jewett. 2003. *Timber Connections: The Joyce Lumber Story*. Grand Rapids, MN: Bluewaters Press.

Heinselman, M.L. 1957. Wind-caused mortality in Minnesota black spruce in relation to cutting methods and stand conditions. In: *Proceedings of Society of American Foresters*, pp. 74–77.

Johnston, W.F. 1977. Manager's handbook for black spruce in the north central states. General Technical Report GRT-NC-264. St. Paul, MN: USDA Forest Service, North Central Forest Experiment Station.

Kernohan, B.J., J. Kotar, K. Dunning, and J.B. Hauler. 1999. Ecosystem diversity matrix for the northern Minnesota and Ontario peatlands landscapes. Unpublished technical manual on file at Boise Cascade Corp. International Falls, MN.

Keys, J.E. Jr., C.A. Carpenter, S. Hooks, F. Koenig, W.H. McNab, W.W. Russell, and M.L. Smith. 1995. Ecological units of the eastern United States: First approximation. USDA Forest Service.

Kotar, J.E. 1986. Soil-habitat type relationships in Michigan and Wisconsin. *Journal of Soil and Water Conservation* 41(5):348–350.

Kotar, J. and T.L. Burger. 2000. Field guide to forest habitat type classification for north central Minnesota: Northern Minnesota and Ontario peatlands section and eastern portion of the northern Minnesota Drift and Lake Plain section. Madison, WI: Terra Silva Consultants.

Kotar, J., A. Kovach, and C.T. Locey. 1988. *Field Guide to Forest Habitat Types of Northern Wisconsin*. Madison, WI: Department of Forestry, University of Wisconsin and Wisconsin DNR.

Kotar, J., J. Kovach, and G. Brand. 1999. Analysis of the 1996 Wisconsin forest statistics by habitat type. General Technical Report NC-207. St. Paul, MN: USDA Forest Service, North Central Research Station.

Lundgren, A.L. and W.A. Dolid. 1970. Biological growth functions describe published site index curves for Lake States timber species. Research paper NC-36. St. Paul, MN: USDA Forest Service.

Maxwell, J.R., C.J. Edwards, M.E. Jensen, S.J. Paustian, H. Parrott, and D.M. Hill. 1995. A hierarchical framework of aquatic ecological units in North America (nearctic zone). General technical report NC-176. St. Paul, MN: USDA Forest Service, North Central Forest Experiment Station.

Minnesota Forest Resource Council. 2007. Biomass harvesting guidelines on forest management sites. St. Paul, MN: Minnesota Forest Resource Council.

Page-Dumroese, D.S., M.F. Jurgensen, A.E. Tiarks, F. Ponder, F.G. Sanchez, R.L. Fleming, M. Kranabetter, R.F. Powers, D.M. Stone, J.D. Elioff, and D.A. Scott. 2006. Soil physical property changes at the North American Long-Term Soil Productivity study sites: 1 and 5 years after compaction. *Canadian Journal of Forest Research* 36(3):551–564.

Perala, D.A. and D.H. Alban. 1982. Biomass, nutrient distribution and litterfall in *Populus*, *Pinus*, and *Picea* stands on two different soils in Minnesota. *Plant and Soil* 64:177–192.

Perala, D.A. and D.H. Alban. 1993. Allometric biomass estimators for aspen-dominated ecosystems in the upper Great Lakes. Research Paper NC-314. St. Paul, MN: USDA Forest Service, North Central Forest Experiment Station.

Perala, D.A., G.E. Host, J.K. Jordan, and C.J. Ceiszewski. 1996. A multiproduct growth and yield model for the circumboreal aspens. *Northern Journal of Applied Forestry* 13(4):164–170.

Powers, R.F., S.D. Andrew, F.G. Sanchez, R.A. Voldseth, D.S. Page-Dumroese, J.D. Elioff, and D.M. Stone. 2005. The North American long-term soil productivity experiment: Findings from the first decade of research. *Forest Ecology and Management* 220(1–3):31–50.

Sommer, B.W. 2008. *Hard Work and a Good Deal: The Civilian Conservation Corps in Minnesota*. St. Paul, MN: Minnesota Historical Society Press.

Stone, D.L. and J.D. Elioff. 1998. Soil properties and aspen development five years after compaction and forest floor removal. *Canadian Journal of Soil Science* 78:51–58.

U.S. Department of Agriculture, Forest Service. 2001. Long-term soil productivity research: A program in sustaining forest ecosystems. Pamphlet FS-714, 8 pp. Available online at http://www.fs.fed.us/reseach/fgdc/pdf/soilprod.pdf

Verry, E.S. 1969. 1968 vegetation survey of the Marcell Experimental Watersheds. Report GR-W2–61. Grand Rapids, MN: USDA Forest Service.

Verry, E.S. 1998. Cut-to-length timber demonstration draws over 300. NC News. St. Paul, MN: USDA Forest Service.

Verry, E.S. and A.E. Elling. 1978. Two years necessary for successful natural seeding in nonbrushy black spruce bogs. Research Note NC-229. St. Paul, MN: USDA Forest Service.

Wenger, K.E. 1984. *Forestry Handbook*. 2nd edn. New York: John Wiley & Sons.

13

Hydrological Responses to Changes in Forest Cover on Uplands and Peatlands

Stephen D. Sebestyen, Elon S. Verry, and Kenneth N. Brooks

CONTENTS

Introduction

Long-term data are used to quantify how ecosystem disturbances such as vegetation management, insect defoliation, wildfires, and extreme meteorological events affect hydrological processes in forested watersheds (Hibbert 1967; Swank et al. 1988; Likens and Bormann 1995; Lugo et al. 2006). The long-term, paired-watershed approach has been used at many sites to measure the effects of vegetation manipulations (e.g., harvesting and cover-type conversions) on stream-water yield. When data from multiple studies and diverse ecosystems are compared, worldwide results show that (1) increases of water yield are detectable when 20% or more of forest basal area is harvested and (2) the magnitudes of responses vary among sites due to differences in the percentage of the watershed that was affected by afforestation and reforestation (Hibbert 1967; Bosch and Hewlett 1982; Hornbeck et al. 1993; Stednick 1996; Brown et al. 2005).

Paired-watershed studies in forests of the United States are widely dispersed, span climate extremes, and are typically located in steep, mountainous headwaters. To fill a gap in the USDA Forest Service network of research watersheds, the Marcell Experimental Forest (MEF) was established during the 1960s in an environmental setting in north-central Minnesota that is distinct from the steep mountainous terrains of other experimental forests (Bay 1962). The MEF has little topographic relief, no dominant aspect, abundant peatlands, and a climate that is continental with long, cold winters (Chapter 2). Relative to steep upland watersheds, the magnitude and timing of peak streamflow are lagged and attenuated due storage in wetlands such that runoff responses to rainfall or snowmelt events are longer in duration and peak streamflow is lower (Bay 1968; Verry 1997). These characteristics are common in boreal ecosystems that span northern latitudes including headwaters in postglacial landscapes of the northern Lake States of Minnesota, Wisconsin, and Michigan. Studies at the MEF were initiated to investigate the unique physical, silvicultural, and hydrological aspects of lowland ecosystems where uplands drain to peatlands, streams, and lakes (Chapter 1).

Peatland Watersheds at the MEF

The landscape at the MEF is a mosaic of lakes and peatlands interspersed among low-elevation hills that were deposited as glacial moraines (Wright 1972). Sandy glacial outwash overlies Precambrian greenstone bedrock, is up to 50 m deep, and forms a regional groundwater aquifer across north-central Minnesota. A thin till layer tops much of the sandy outwash. The surficial soils on uplands are sandy loams above clay loams that are permeable but retard the vertical flow of water. In low-lying areas underlain by the clay loam aquitard, shallow postglacial lakes slowly filled with peat (Histosols) as partially decomposed organic matter accumulated under saturated and anaerobic conditions (Boelter and Verry 1977).

The six research watersheds at the MEF have uplands that route upland runoff waters to central peatlands. Each watershed is 12%–33% peatland by area and drains to series of downgradient wetlands, lakes, and streams via a short intermittent or perennial stream (Bay 1968). All the peatlands in the research watersheds receive water inputs from precipitation and runoff from the surrounding upland mineral soils whether perched above or embedded in the regional groundwater aquifer.

Fens are minerotrophic peat-forming wetlands with inputs of water from regional aquifers. The peatland in the S3 watershed is a fen, and the stream draining S3 flows perennially unlike the S1, S2, S4, S5, and S6 research watersheds, which are perched several meters above unsaturated sands and the regional aquifer.

Because a confining till layer separates perched watersheds from the regional aquifer and groundwater inputs, water in perched watersheds originates solely from precipitation falling on the peatland and surrounding upland soils (Chapter 7). In these hydrogeologic settings, ombrotrophic peat-forming wetlands, such as those in the S1, S2, S4, S5, and S6 watersheds, are bogs (Chapter 4). Clay layers route runoff water from the overlying silty loams via lateral flowpaths into lagg zones that surround raised dome bogs (Timmons et al. 1977; Verry and Timmons 1982; Tracy 1997). Streams drain the lagg zones, and streamflow usually stops during dry summers and autumns. During winter, streamflow has stopped every year on record (Chapter 2). Despite low-hydraulic conductivities, up to 40% of annual precipitation inputs to the perched peatlands may recharge the regional groundwater aquifer via deep seepage through the clay aquitard (Nichols and Verry 2001), primarily through clay loam tills under the lagg zones of bogs (Chapters 4 and 7).

Conifer forests were prevalent throughout northern Minnesota before European settlement. Cutting of uplands forest was widespread during the late 1800s and early 1900s. After harvesting and subsequent fire suppression, much of the former white (*Pinus strobus*), red (*Pinus resinosa*), and jack (*Pinus banksiana*) pine forest is now a primary succession forest of trembling aspen (*Populus tremuloides*), bigtooth aspen (*Populus grandentata*), paper birch (*Betula papyrifera*), and balsam fir (*Abies balsamea*) with other mixed northern hardwoods. Stand types, stem densities, and basal areas were surveyed for uplands and peatlands in each watershed during 1968 (Verry 1969) as summarized in Table 13.1.

Aspen-dominated stands and pure aspen stands grow prolifically on uplands soils throughout the region. Aspen is typically harvested by whole-tree clearcutting during winter on frozen soils and during summer on dry soils to minimize soil disturbance. Aspen is merchantable after 40–60 years of growth, and forests are actively managed throughout the region for the

TABLE 13.1

Stand Age, Stem Density, and Basal Area in Uplands and Peatlands at the Time of the 1968 Vegetation Survey

Watershed	Stand Type	Stand Ages	Stem Density (Stems ha⁻¹)	Basal Area (m² ha⁻¹)
Uplands				
S1	Two aspen stands	44 and 52	645	22.0
S2	Aspen with red maple and paper birch	50	480	23.2
S3	One aspen and two jack pine stands	51, 52, and 64	—	37.6
S4	Aspen	49	645	21.8
S5	Two aspen, aspen-birch, two spruce-fir, red and white pine, mixed hardwood, and cedar stands	48 (aspen) and older (other stands)	—	13.5–49.8
S6	Aspen	56	356	29.6
Peatlands				
S1	Black spruce	62 and 73	3520	7.0
S2	Black spruce	99	1759	13.3
S3	Black spruce	100	685	14.0
S4	Black spruce	49	188	15.2
S5	Black spruce	100	1055	13.3
S6	Black spruce and tamarack	64	—	12.4

Source: Verry, E.S., 1968 vegetation survey of the Marcell Experimental Watersheds, Report GR-W2-61, USDA Forest Service, Grand Rapids, MN, 1969.

pulp industry (Bates et al. 1989) or to restore conifer cover to aspen-covered lands. To convert to a conifer forest in the absence of wildfires, excessive growth of aspen and other competitors in clearcuts must be managed, usually by controlled burns, herbicides, or mechanical treatment.

Black spruce (*Picea mariana*), eastern tamarack (*Larix laricina*), and northern white cedar (*Thuja occidentalis*) grow in the overstory of forested peatlands. Trees in bog forests often are even-aged, probably as a result of regeneration after wildfires or other catastrophic disturbances such as straight-line windstorms and tornados (Clark 1990).

Hydrological Measurements

The six research watersheds were instrumented during the 1960s with outlet streamflow gages, peatland water table wells, and precipitation gages to study the effects of forest management on water yield. The watersheds, instrumentation, and measurements are detailed in Chapter 2. Daily water table levels were measured at central wells in each peatland. Streamflow was initially measured at each watershed, and stage-discharge relationships or regressions between peatland water table level and stream outflow were used to calculate stream discharge. A water year begins on the first of March and ends on the last day of February of the following calendar year. This water year reflects the annual water cycle of Minnesota bog watersheds where intermittent flow starts during spring snowmelt and typically stops during winter. Average annual water yield ratios were calculated for each water year by dividing stream-water yield by precipitation and averaging annual values.

Daily precipitation is measured in forest clearings in the S2 and S5 watersheds (Chapter 2). Annual precipitation is calculated for water years that begin on November 1 and end on October 31, which reflects the accumulation of snow starting in November that does not contribute to streamflow until snowmelt the following spring.

Comparisons of Responses at Experimental and Control Watersheds

Responses of annual stream-water yield from harvest treatments at the S1, S3, S4, and S6 watersheds are compared to the S2 and S5 control watersheds with unmanipulated upland and peatland forests. We consider water yields to be significantly different from preharvest conditions during postharvest years and decades that exceeded the 95% confidence intervals about the regressions, or if regressions slopes significantly differed (ANCOVA where $p < 0.05$). Peak daily water yields were compared between harvested and control watersheds to assess changes in peak flow response to forest harvesting. Peak flow responses were compared during snowmelt and rainfall-runoff events (June–October). This approach, though not a generalizable test of peak flow responses due to stochasticity of long-return-interval storm events, may show responses to the specific watershed experiments.

The 9.7 ha S2 watershed has a 3.2 ha peatland. During water years from 1961 to 2008, annual precipitation at the meteorological station (South) in the S2 watershed averaged 76.9 ± 11.1 cm year^{-1} (mean ± 1 standard deviation). Mean stream-water yield from water years 1961 to 2008 was 16.7 ± 5.5 cm year^{-1}, and the mean runoff ratio was $21\% \pm 6\%$. When inventoried in 1967 before any harvest, mean basal area of the aspen dominated stand was 23.2 m^2 ha^{-1}, and the stand was 50 years old (Verry 1969). The basal area of black spruce in the S2 peatland was 13.3 m^2 ha^{-1}, and the average age was 99 years during 1968.

The S5 watershed is 52.6 ha. Upland soils and five small satellite wetlands drain into a 6.1 ha central peatland. Mean stream-water yield from 1962 to 2005 was 10.9 ± 4.2 cm year^{-1}, the mean runoff ratio was $13\% \pm 4\%$, and mean precipitation was 78.6 ± 11.2 cm year^{-1}. The stream-water yield from the S5 watershed is less than at the S2 watershed, because 5 ha of the S5 upland soil is deep sand and that area may recharge directly to the regional groundwater table. In 1968, the S5 uplands had stands of aspen (two age classes), spruce-fir, red and white pine, mixed hardwoods, and white cedar. Mean basal area was $23.6\,m^2\,ha^{-1}$, and the average age was 100 years (Verry 1969).

Hydrologic Responses to Upland Harvests

Studies at the MEF were designed to determine the effects of upland-only harvests and peatland-only harvests. This information was needed to maintain forest productivity, understand water-yield responses to forest management, and maintain water quality in lowland watersheds with uplands that drain to peatlands. Even when peatland vegetation is not disturbed by upland tree harvesting, watershed-scale hydrologic responses are influenced by the routing of water through peatlands on the way to the watershed outlet. The effects of upland tree harvesting on the hydrology of undrained peatlands had not been quantified before manipulative watershed studies were initiated at the MEF in the 1970s.

Bog water table elevations are controlled by lateral outflow via streams as well as vertical fluxes to the underlying regional groundwater system via deep seepage and the atmosphere via evaporation and transpiration (Chapter 7). If increased water flows to peatlands from hillslopes after upland clearcutting and vegetation change, bog water levels may fluctuate more in response to changes of water inputs from upland source areas. In particular, maximum bog water levels may increase if lateral inflow of upland runoff water from hillslopes increases after clearcutting. However, if high water levels are in contact with the shallow root systems of bog vegetation, evapotranspiration may increase resulting in a negative feedback that may mask any changes of annual water yield.

Harvest and Regeneration of an Upland Aspen Forest

The 25.9 ha upland of the S4 watershed was harvested from 1970 to 1972. Hydrologic responses to the S4 upland clearcut previously were reported for the first year after the clearcutting (Verry 1972), updated later to include the first 9 years after the clearcut (Verry et al. 1983), and updated again to include the first 15 years after the clearcutting (Verry 1987). Herein, we interpret hydrological responses based on 35 years of post clearcutting data that spanned the regrowth of the aspen forest to the stage of full aspen stocking.

The S4 watershed is 34.0 ha. The central 8.1 ha is a black spruce peatland that surrounds a 0.4 ha open bog and a small pond. Mean stream-water yield before clearcutting was $19.4 \pm 5.1\,cm\,year^{-1}$ (± 1 standard deviation), and the ratio of precipitation to stream runoff was $25\% \pm 5\%$. Prior to clearcutting, the upland forest was predominantly a 52 year-old mature aspen stand. Mean basal area of the S4 and S5 uplands was similar prior to clearcutting (Table 13.1). About 2% of the upland forest was a mixed cover of 57 year-old paper birch and aspen. At the time of cutting, basal area was $21.8\,m^2\,ha^{-1}$, and aspen biomass was 240,000 kg ha^{-1}.

All merchantable timber taller than 3.0 m was cut and removed from the S4 uplands (Figure 12.12). The first 1.3 ha of forest were clearcut during December 1970 on an area northeast of the peatland. Clearcutting west of the peatland continued through winter and about half of the upland area, or 34% of the total watershed area, was clearcut before snowmelt in 1971. The remaining upland forest was clearcut from September 1971 to January 1972. By September 1971, 70% of the annual stream-water yield already had occurred. The progression of the clearcut provided an opportunity to assess the effects of partial (water year 1971) and total (post-1971) upland clearcutting on stream hydrological responses (Verry et al. 1983; Verry 1987). All non-commercial trees larger than 8.9 cm in diameter at breast height (dbh) were felled and left in place. Because the black spruce forest on the central peatland of the S4 watershed was not harvested, the completed aspen clearcutting included 71% of total watershed area. Disturbance from road and major skid trails affected less than 2% of the watershed area (Verry et al. 1983). Logging slash was left in place, and the tangled branches finally settled to the forest floor about 6 years after clearcutting (Verry 1987).

Aboveground tree biomass was surveyed during 1968, 1971, every 2 years from 1972 to 2002, and again during 2008. Aspen regrowth from root suckering was rapid. Aspen tree density was 101,000 stems per ha by August 1971, and the trees were about 2 m tall in clearcut areas (Verry et al. 1983). By the spring of 1979, aspen were about 6 m tall. The biomass of overstory trees greater than 8.9 cm dbh exceeded 90% of the preharvest biomass during 2000 and was similar during 2008.

Changes in Annual Water Yield

Preharvest streamflow data were collected during water years at S4 and S5 from 1962 to 1970. To determine effects of the uplands clearcut and model, a "no clearcutting" scenario, predictions of annual stream-water yield at the S4 watershed were calculated for water years from 1971 to 2006 using a regression equation relating preharvest stream-water yield between S4 and the S5 control watershed:

$$Q_{S4\text{-predicted}} = -46.0 + 36.7 \times Q_{S5}{}^{0.25}, \quad \left(R^2 = 0.99, \ p < 0.0001\right), \qquad (13.1)$$

where Q is the annual stream-water yield at the S4 or S5 watershed (units of centimeter per year). Stream-water yield during 1970 was included in the precalibration dataset, because less than 5% of the watershed area was harvested during December 1970, most of the annual water yield occurred during snowmelt during the preceding spring, streamflow stopped on 28 December, and streamflow during the December harvesting was less than 1% of the annual stream-water yield. Change in annual stream-water yield was calculated for each water year by subtracting the predicted from the observed annual water yield:

$$\Delta Q_{S4} = Q_{S4\text{-observed}} - Q_{S4\text{-predicted}}, \tag{13.2}$$

A positive ΔQ_{S4} indicates an increase of stream-water yield relative to the preharvest period. Hydrological responses during the clearcut and early regrowth period from water years 1971 to 1979 and the postclearcut decades from water years 1980 to 1989 and 1990 to 2005 were compared to the pre-clearcut stage from water years 1962 to 1970.

Prior to the harvest, the difference between measured and predicted annual water yield ranged from −4% to +5% with a mean difference of 0%±3% (Figure 13.1). During the first postharvest water year when 34% of the watershed area was clearcut, annual stream-water yield increased by 25% relative to S5 (Verry 1972). Beginning during the second water year (1972) when 71% of the watershed area had been clearcut, annual water yield was 38% higher relative to the preharvest period and remained higher until water year 1982 (Figures 13.1 and 13.2). During water years from 1971 to 1979, the mean increase in stream-water yield was 22%±10%. The change in yield (ΔQ_{S4}) ranged from −10% to +18% and averaged +0%±8% during water years from 1980 to 1989. The ΔQ_{S4} ranged from −22% to +16% and averaged −7%±11% during water years from 1990 to 2005. The recent decrease in annual water yield occurred when biomass had regrown to the level of the preharvest forest, and the site index exceeded the original stand by 2 m (Chapter 12).

Changes in Peak Flow to Snowmelt and Rainfall

If clearcutting affected precipitation inputs by changing snow accumulation and ablation, peak streamflow responses during snowmelt would also be expected to change. During the first postharvest snowmelt (April 1971), when the uplands were partially clearcut at the S4 watershed, the magnitude of snowmelt peak streamflow was 35% less than predicted from streamflow at the S5 control watershed (Verry 1972). Snowmelt began 2 days earlier relative to the S5 control watershed and streamflow peaked twice in contrast to the single peak at S5. The timing of snowmelt likely changed due to earlier melting of snow along the exposed road and runoff of that snowmelt upstream of the S4N weir. During April 1972, when the uplands were completely clearcut, the snowmelt runoff peak more than doubled the predicted peak and streamflow peaked 4 days earlier than that at the control (Verry 1972).

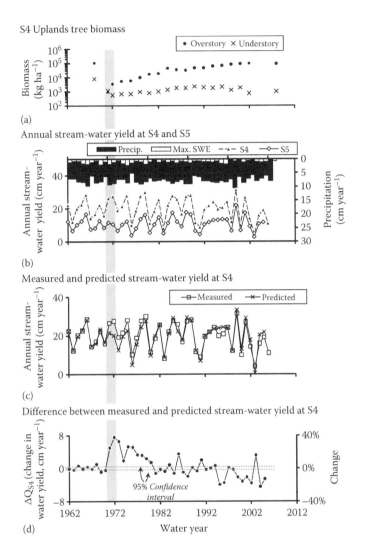

FIGURE 13.1
Uplands tree biomass (a); annual precipitation, maximum snow water equivalent (SWE), and annual stream-water yield from the S4 and S5 watersheds (b); observed and regression model predictions of annual streamflow at S4 (c); and difference in water yield between observed and predicted streamflow (d). Overstory biomass includes trees more than 8.9 cm diameter at breast height, and understory biomass includes woody vegetation less than that dbh. The shading shows the timing of clearcutting from winter 1970 and spring 1972.

When the first two snowmelt responses were considered within the 1971–1979 postharvest period, there was no detectable change in peak stormflow magnitude during snowmelt (Figure 13.3a), suggesting a short-term net effect on snow interception, sublimation, and melt. The differences between the first snowmelt with a complete clearcut and subsequent years highlight

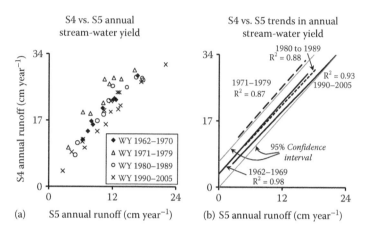

FIGURE 13.2
Relationships of stream-water yield between the clearcut S4 and the S5 control watersheds. Annual stream-water yield changed in the water years after harvest (1971–1979) relative to preharvest water years (1962–1970) and decades after harvest (1980s, 1990s, and 2000s). Annual data (a), and trends (b) by response period.

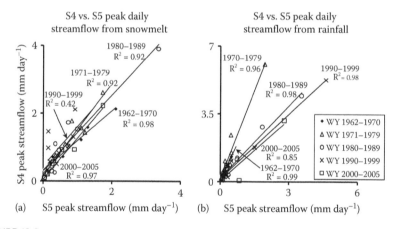

FIGURE 13.3
Peak daily stream-water yield from snowmelt (a) and rainfall (b) during each water year, relationships between the clearcut S4 and the S5 control watersheds during the preclearcut (1962–1970), clearcut (1971–1979), and postclearcut periods (1980 and after).

different responses from a cleared condition (e.g., pasture or shrub land) to a regenerating aspen forest (see Chapter 7).

Clearcutting should affect rainfall–runoff responses during the growing season due to differences in interception, evaporation, and transpiration between mature and regenerating aspen forests. Verry et al. (1983) reported that streamflow peak size doubled during the first 3–5 years after clearcutting (Figure 13.3b). During the 1980s, rainfall–runoff responses reversed when forest biomass increased above preharvest levels, and the relative magnitude of peak rainfall–runoff continued to decrease from 1990 to 2005. The timing of storm flow in response to rainfall events did not change (Lu 1994).

Changes in Bog Water Levels

Mean water levels in the S4 bog increased relative to the S5 control watershed during the 1970s after the uplands were clearcut and then decreased during the 1990s when the overstory basal area was about the same as the preharvest period (Figure 13.4). The increase in bog water level reflects increased inflow of upland runoff water from hillslopes after clearcutting.

Land Management Implications of the S4 Upland Clearcutting Experiment

Upland clearcutting at the S4 watershed affected hydrological responses when areas exceeding 34% of the watershed were clearcut. Syntheses of past studies suggest a threshold of about 20% basal area harvesting before effects on stream-water yield are observed (Hibbert 1967; Stednick 1996; Verry 2004).

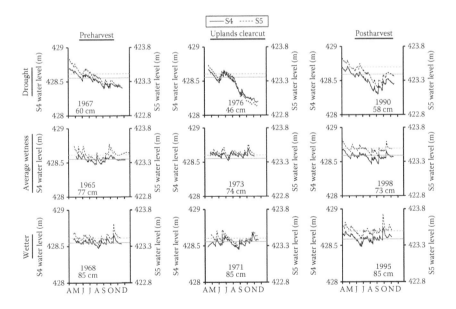

FIGURE 13.4
Daily water levels in the S4 bog for preharvest, clearcut, and early regrowth, and third decade of regrowth stages relative to the S5 control watershed. Water levels are shown during drought, average, and above average wetness conditions. The annual precipitation input is shown below the water year. The horizontal lines show the mean bog water level for S4 (solid) and S5 (hashed) during the preharvest water years from 1964 to 1969, during the immediate postharvest water years during the 1970s, and the 1990s when aspen biomass was increasing above preharvest amounts.

Verry (2004) showed a response curve that included data from the S4 partial and total upland clearcuts. Interpretation of these data indicated little or no increase in stream peak flows until 50% of an aspen-dominated landscape area was cleared. In larger watersheds where 2%–3% of the landscape may be clearcut in a given year such as the managed forests of northern Minnesota, effects of aspen harvesting on streamflow peaks in large rivers would not be detectable, and increases of peak streamflow may not be measurable with distance downstream of small clearcut watersheds.

Upland Forest Conversion from Hardwoods to Conifers

The upland forest at the S6 watershed was clearcut during 1980 to study effects of harvesting and aspen forest conversion to conifers (Figures 12.6, 12.7, and 13.5). Alternatives to the widespread use of herbicides were being considered at the time, and the S6 experiment was used to evaluate cattle grazing as a forestry tool for site preparation (see Chapter 12). The results, which have not previously been published, show how watershed hydrology responded and provided an example of a management approach to convert forest cover. The 8.9 ha S6 watershed has a 2.0 ha peatland. Prior to harvest, mean annual water yield was 15.1 ± 10.5 cm year^{-1}, and the mean ratio of annual stream runoff to precipitation was $19\% \pm 11\%$. Aspen was 84% of the overstory while red maple, paper birch, and red oak were each 7% or less of the overstory. The mature aspen stand at the S6 watershed had the lowest preharvest basal area of any research watershed (Table 13.1). The nearly equal mix of black spruce and tamarack cover on the bog was 76 years old, when the uplands were clearcut, and the trees were 12–15 m tall.

The first 2.1 ha of upland forest were commercially clearcut during March 1980. Road restrictions during snowmelt prevented further logging until

FIGURE 13.5
The S6 upland clearcut during 1981, facing east. (Photo courtesy of A.E. Elling, USDA Forest Service, Grand Rapids, MN.)

the remaining 7.1 ha of upland trees were clearcut during June. A Drott feller–buncher was used to shear trees during the whole-tree harvest. The total logged area exceeded the S6 uplands area because the clearcutting extended beyond the watershed boundaries to include an edge buffer. Some logging slash was piled and burned on site but most was removed by full-tree skidding. The aspen clearcut removed trees on 77% of the watershed area because black spruce, tamarack, and other vegetation were not harvested from the peatland.

The watershed was fenced in June 1980 and 12–17 cows, steers, or calves grazed the uplands from September to October 1980, June to September 1981, and June to July 1982 to suppress aspen regeneration by sprouting from root suckering (Chapter 12). Cattle were removed from the S6 watershed when all forage was consumed or trampled. The uplands were relatively free of woody vegetation during the 3 years before conifers were planted (Figure 12.6). The grazing effectively suppressed aspen regeneration and was a viable alternative to the use of herbicide applications. At $247 per ha, the cost was the same as for herbicide application.

After grazing, red pine seedlings were planted on more than 70% of the clearcut on the western end, and white spruce was planted in the remaining area during May 1983. Both species were 4 year-old seedlings at the time of planting, and 2200 seedlings per ha were planted. By 1987, the average height of conifer trees was 43 cm. Survival of red pine was 55%, and survival of white spruce was 81%. Because the growth of willows (*Salix* sp.), paper birch, and hazel shaded conifer seedlings, the herbicide Garlon 4 was sprayed on uplands vegetation on August 1987 after it was decided that additional grazing might damage seedlings.

Changes in Annual Water Yield

Preharvest streamflow data were collected from 1964 to 1979 at the S6 watershed. As a result of the change in the stream gage structure from a flume to a V-notch weir during 1976, water transiently backed into the peatland resulting in higher bog water levels after 1976 (Chapter 2). Consequently, only 4 years of pretreatment data from water years 1976 to 1979 were used to calculate expected water yield, because the pre- and post-1976 data were not comparable. Expected annual stream-water yield was calculated from a regression of preharvest data:

$$Q_{S6\text{-}predicted} = 4.6 - 0.2 \times P + 1.5 \times Q_{S2}, \left(R^2 = 1.0, \ p < .0075\right), \quad (13.3)$$

where
  Q in units of centimeter per year is annual stream-water yield at the harvested S6 watershed or S2 control watershed
  P in units of centimeter is total precipitation during the calendar year

Expected water yield represents the yield if there was no effect of clearcutting and forest conversion at S6. Change in annual stream-water yield was calculated for each water year by subtracting the predicted from the observed annual yield:

$$\Delta Q_{S6} = Q_{S6\text{-observed}} - Q_{S6\text{-predicted}}, \tag{13.4}$$

A positive ΔQ_{S6} indicates an increase of stream-water yield relative to the preharvest period. Data from the clearcutting, grazing, and initial conifer growth stage (water years from 1980 to 1989) and conifer growth stages (water years from 1990 to 1999 and 2000 to 2008) were compared to the preharvest stage during water years from 1976 to 1979. Despite the short duration, the calibration period included extremes from the driest year (1976) and a year in the top 10th percentile of wet years (1979). Nonetheless, the statistical power was low given the short precalibration period. The results for the S6 experiment must be considered within this limitation.

During clearcutting, grazing, and the initial years of conifer growth from 1980 to 1989, annual stream-water yield at the S6 watershed increased relative to the S2 control watershed (Figures 13.6 and 13.7). The difference between observed and predicted annual stream-water yield prior to clearcutting ranged from −1% to +1% with a mean difference of 0% ± 1% (Figure 13.6c). During the water year of 1980 when the uplands were clearcut and first grazed, stream-water yield increased by 59% above the amount expected if the uplands had not been clearcut. The largest increase of 78% occurred during 1982, the third and final year of aspen suppression by grazing. The change in annual stream-water yield remained positive until water year 1999 and averaged +39% ± 25% relative to the predicted S6 water yield from 1980 to 1989 (Figure 13.6b). During water years from 1990 to 1999, the mean change in stream-water yield was +17% ± 31%. The mean change in annual stream-water yield from 2000 to 2006 was −28% ± 28% with a maximum change of 69% less yield than expected if the uplands had not been converted to conifers. The recent decrease in annual stream-water yield relative to a mature aspen forest occurred with conifer canopy closure.

The decreased stream-water yield from 2000 to 2006 indicates that the maturing conifer forest yielded less water to the stream than the original aspen forest (Figure 13.7). Prior to clearcutting, stream-water yield was slightly higher at the S2 control watershed than at S6, but the regression slope of S6 on S2 was not different from 1.0 (ANCOVA test, $p > 0.05$). Water yield increased during and after the clearcut and grazing period with no change in slope of the regression of stream-water yield at the S6 and S2 watersheds. When substantial conifer biomass had accrued during the 1990s and conifers began to dominate the overstory, the water yield response changed. Wet years yielded less water during the 1990s and 2000s than wet years prior to the 1990s, as shown by slopes that were different before the 1990s (ANCOVA

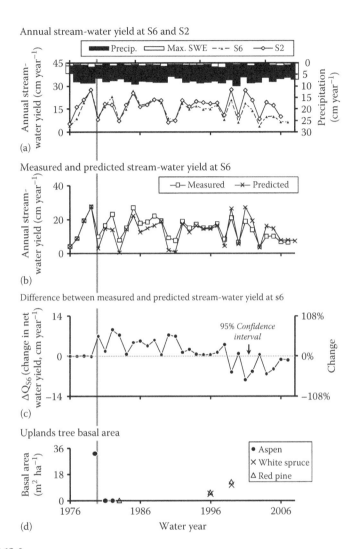

FIGURE 13.6
Annual precipitation, maximum SWE, and annual stream-water yield from the S6 and S2 watersheds during each water year (a), measured and regression model predictions of annual streamflow at S6 (b), the difference in water yield between measured and predicted streamflow (c), and uplands tree basal area (d). The shaded line shows the timing of clearcutting during water year 1981.

test, p < 0.0001; Figure 13.7). From 2000 to 2006, water yield at S6 decreased even more than in the 1990s.

Streamflow is intermittent at both S2 and S6. Although water yield decreased at S6 during the first decade of the 2000s, the relative number of days per year with streamflow at S6 increased progressively from the preharvest through the postharvest decades as the conifer forest accrued

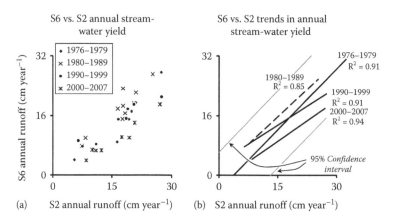

FIGURE 13.7

Stream-water yield relationships between the clearcut S6 and the S2 control watersheds. Annual stream-water yield changed in the water years after the uplands aspen harvest (1980–1989) relative to preharvest water years (1976–1979) and decades after conversion to a conifer forest (1990–1999 and 2000–2006). Annual data (a), and trends (b) by response period. To account for wide confidence intervals, high uncertainty, and low statistical power due to the short calibration period at S6 (only 4 years), differences in slope were detected by ANCOVA.

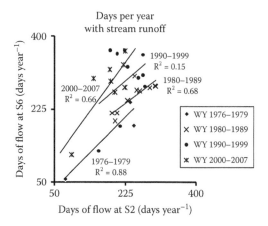

FIGURE 13.8

Days per year with streamflow at S6 and S2. The duration of streamflow increased progressively at S6 relative to S2 from the preclearcut stage (1976–1979) through the harvest and grazing (1980s) and conifer afforestation (1990s and 2000s) stages.

biomass (Figure 13.8). Changes to the duration of streamflow at S6 are consistent with increased magnitude of low flow during summer (Figure 13.9). Measurements of sap flux for trees on S2 and S6 were initiated in 2009 to determine whether water use differs between upland conifer and aspen forests.

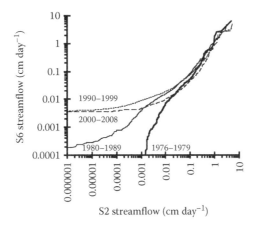

FIGURE 13.9
Daily streamflow at S6 versus S2. Low flow increased at S6 during the 1980s after the clearcut into the 1990s and 2000s. The new flow regime stabilized during the 1990s and 2000s.

Changes in Peak Flow to Rainfall and Snowmelt

Peak streamflow responses to both snowmelt and rainfall differed between the S6 and S4 upland clearcut studies. Peak streamflow during snowmelt was higher at the S6 than the S2 watershed before the clearcutting (Figure 13.10a). The snowmelt peak response was similar prior to clearcutting and during water years from 1990 to 1999. During the harvest and grazing stage

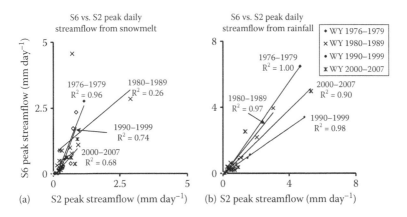

FIGURE 13.10
Peak daily stream-water yield from rainfall (a) and snowmelt (b) in each water year, relationships between the clearcut S6 and the S2 control watersheds during the preclearcut (1976–1979), clearcut (1980s), and postclearcut periods (1990s and 2000s).

(1980–1989) and during water years from 2000 to 2006, the magnitudes of peak flow during snowmelt were about the same at S2 and S6.

Peak streamflow in response to rainfall increased marginally during the decade after clearcutting and progressively decreased during following decades as annual stream-water yield decreased (Figure 13.10b). At S6, peak streamflow response to rainfall during the clearcutting and grazing period decreased relative to the 1976–1979 preclearcutting period (Figure 13.10a). Peak streamflow response decreased further during water years from 1990 to 1999. Peak streamflow response to rainfall increased from the 1990s to the 2000s, but peak streamflow at S6 was still low relative to the preharvest stage.

Interception of rain and snow water by conifer canopies exceeds that of hardwood species (Helvey 1967; Zinke 1967; Verry 1976). Consequently, more intercepted rain may evaporate from conifer canopies, reduce precipitation inputs to upland soils, and reduce lateral subsurface runoff from hillslopes to peatlands. The decreased water yield at the S6 watershed following conversion of uplands to conifers is similar to the results at Coweeta Hydrologic Laboratory in North Carolina where a hardwood forest was converted to eastern white pine (*P. strobus*) and is consistent with increased canopy interception after conversion of hardwoods to conifers (Swank and Douglass 1974; Swank et al. 1988; Hornbeck et al. 1993). Precipitation inputs decrease due to increased interception on conifer canopies and sublimation of snow relative to open areas and leafless hardwood canopies. Transpiration when moisture is not limiting also is higher for conifer than hardwood canopies due largely to leaf area indices that are up to 10 times higher than hardwoods (Kozlowski 1943; Swank et al. 1988). Because most of the annual yield of stream water occurs in response to spring snowmelt, increased losses of winter precipitation with increased interception and sublimation in forests converted to conifers may influence peak streamflow during snowmelt events, which is consistent with the disproportionate decrease of water yield during wetter years, decreased peak streamflow during snowmelt, and a longer duration of streamflow.

Changes in Bog Water Levels

Water levels in the S6 bog increased relative to the S2 bog after the upland clearcutting, grazing, and forest conversion (Figure 13.11). Data on bog water levels are available since 1977. Mean water table elevations and water table fluctuations were similar between the 1977 and 1979 calibration and 1980s postclearcutting periods. In the 1990s and 2000s, mean water level decreased at the S2 control watershed and increased at the S6 watershed. Although not intuitive given the decrease in water yield that occurred over the past two decades, the increased water level was consistent with increased days of low flow and the increased magnitude of low flow relative to the S2 control watershed. The change in boundary layer conditions at a watershed

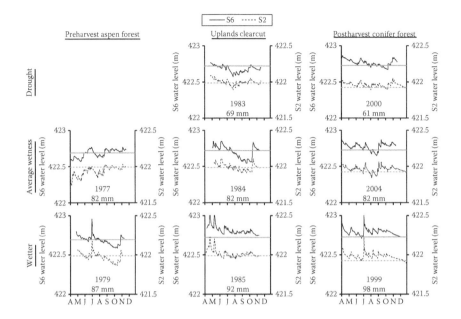

FIGURE 13.11

Daily bog water levels in the S6 bog during a year for preharvest aspen, clearcut, and postharvest conifer forest stages shown relative the control S2 watershed. Water levels are shown during drought, average, and above average wetness conditions except for the pretreatment stage when comparable dry conditions did not occur. The precipitation input is shown below the water year. The horizontal lines show the mean bog water level for S6 (solid) and S2 (hashed) during the preharvest water years from 1977 to 1979, during the postharvest water years from 1980 to 1989, and the accruing conifer forest stage (1990s and 2000s).

converted to an upland conifer canopy may also affect vapor fluxes from the bog conifers.

Land-Management Implications of the S6 Upland Clearcutting and Forest Conversion Experiment

The observed decrease in annual streamflow after establishment of a conifer forest on the S6 uplands suggests that annual stream-water yields and streamflow peaks probably were lower before European settlement and widespread forest harvesting during the late 1800s and early 1900s. Pines dominated presettlement northern forests that included mixed hardwoods. Currently, the most widespread forests are primary succession and have deciduous species with few or no conifers in the overstory canopy. Although the S6 conversion to a dense spruce and pine plantation represents an extreme version of a presettlement forest, annual peak streamflow at the 1.5 year recurrence interval would be smaller in magnitude for presettlement forests (Lu 1994).

Effects of Upland Forest Harvesting on Streamflow Peaks

The mixed upland/wetland watersheds at the MEF are typical of watersheds throughout the Upper Great Lakes region. Forest harvesting of the uplands increases or reduces stormflow peaks depending on the relative amount of the entire watershed harvested. Figure 13.12 shows the progression of peak-flow change from normal forests more than 15 years of age to watersheds with an increasing percentage of young forests (15 years and less) or open pasture and shrub land (Verry et al. 1983; Verry 2004).

Increases in peak flow for rainstorms are caused by logging on wet soils that compact the soil and reduces infiltration. Increases or reductions in peak flow for snowmelt are caused by the desynchronization of snowmelt in the watershed (Verry 2004). The frequency of the peak flows in Figure 13.13 is unknown, because each data point is based on actual measurements from MEF watershed studies (e.g., Verry et al. 1983), long-term data from the Mississippi and Minnesota river basins (Verry 1986), channel morphology studies in Wisconsin (Fitzpatrick et al. 1999), and from a modeling study at MEF (Lu 1994). Lu examined stormflow frequency by modeling peak-flow response using an 80 year precipitation record that was measured nearby at a meteorological station in Grand Rapids, MN. He found that the magnitude

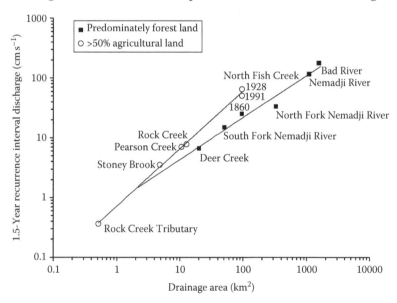

FIGURE 13.12

Change in annual peak streamflow from forested watersheds in the Lake States as a percentage of the basin in open or young forests (<16 years old). (Reproduced from Verry, E.S., Land fragmentation and impacts to streams and fish in the central and upper Midwest, in *Lessons for Watershed Research in the Future; A Century of Forest and Wildland Watershed Lessons*, Ice, G.G. and Stednick, J.D. (eds.), Society of American Foresters, Bethesda, MD, pp. 129–154, 2004.)

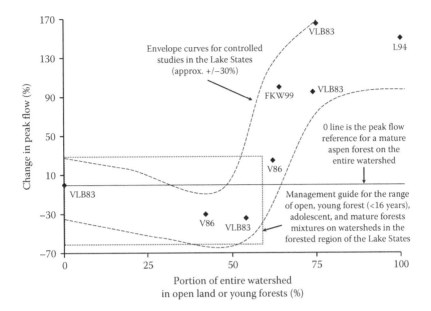

FIGURE 13.13
Change in the annual maximum daily snowmelt peak-flow frequency from a mature aspen watershed (solid line) to a recently harvested aspen watershed (dashed line) in Minnesota. Modified from Lu (1994). Lightly dashed lines are 95% confidence intervals. (Reproduced from Verry, E.S., Land fragmentation and impacts to streams and fish in the central and upper Midwest, in: *Lessons for Watershed Research in the Future; A Century of Forest and Wildland Watershed Lessons*, Ice, G.G. and Stednick, J.D. (eds.), Society of American Foresters, Bethesda, MD, pp. 129–154, 2004.)

of storms at a given frequency was increased by upland forest harvesting; storm frequencies ranged from 1 to 30 years (Figure 13.14).

Bankfull or channel-forming flow occurs on average with a 1.5 year frequency (Leopold et al. 1958). That is significant because bankfull flow maintains channel morphology and dimensions. When bankfull flow increases, the channel becomes wider, deeper, or both to accommodate the higher discharge at the 1.5 year frequency. The additional sediment caused by channel scour destabilizes the channel, resulting in a long period of adjustment to new channel dimensions. During this time, fine sediment in the channel can cover gravel used for spawning beds. Changes in bankfull discharge have been documented for clay basins along the south shore of Lake Superior based on long-term streamflow records of the U.S. Geological Survey (Figure 13.14).

The change in bankfull discharge at the MEF was verified with long-term US Geological Survey streamflow data for watersheds along the south shore of Lake Superior in Minnesota and about 200 km east in Wisconsin. In these basins on clay soils, the 1.5 year bankfull peak flow increased about 2.5-fold when predominately forested watersheds were converted to predominately agricultural, "open" lands (Verry 2004).

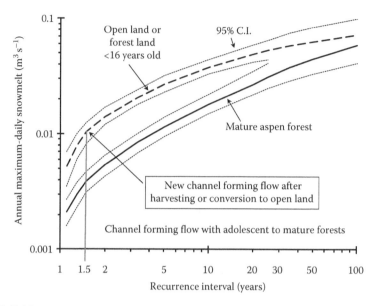

FIGURE 13.14

Effects of forest conversion from forests to agriculture for basins in the Southern Lake Superior Clay Belt on bankfull flow (about 1.5 year discharge response). Curves were fit visually. The North Fish Creek data represents a progression from forest cover (1860) to row crop and pasture agriculture in more than half of the basin (1928) to eventually reversion to forests (1991). (Modified from Verry, E.S., Land fragmentation and impacts to streams and fish in the central and upper Midwest, in *Lessons for Watershed Research in the Future; A Century of Forest and Wildland Watershed Lessons*, Ice, G.G. and Stednick, J.D. (eds.), Society of American Foresters, Bethesda, MD, pp. 129–154, 2004 with data on North Fish Creek from Fitzpatrick, F.A. et al., Effects of historical land-cover changes on flooding and sedimentation North Fish Creek, Wisconsin, Water Resources Investigation Report 99-4083, U.S. Geological Survey, U.S. Department of the Interior, Middleton, WI, 1999.)

Hydrologic Responses to Peatland Harvests

Although less common than harvests of upland forests in the north-central United States, black spruce may be commercially clearcut, stripcut, and patch cut from peatlands (Johnston 1977; Grigal and Brooks 1997). The effects of clearcutting on water yield and water tables could differ in relation to the unique hydrological and hydrochemical settings of bogs and fens. Although water levels in both bogs and fens are drawn down by evaporation and transpiration (Nichols and Brown 1980), the differences are more pronounced in bog watersheds where precipitation is the only source of inflowing water. Fens have a more stable hydrological regime due to connectivity with a regional aquifer and groundwater inputs (Bay 1967). The hydrological differences between fens and bogs may affect the restocking of peatland forests after clearcutting. If black spruce is harvested from peatlands, changes to the hydrological setting such as hydroperiod or maximum water levels and loss of a viable seed source from isolated peatlands like those at the MEF could inhibit the regrowth of black spruce.

Clearcutting is more common than stripcutting or patch cutting of black spruce (Grigal and Brooks 1997). Most peatland conifer stands cleared by fire, wind, or large cuts need 20–25 years to reach full stocking. Black spruce seeds remain viable for only 1 year, and seeding or seedling planting may be required to regenerate fully stocked black spruce stands where clearcuts remove the seedstock (Verry and Elling 1978). Strip and patch cuts within the 120 m dispersal distance of black spruce retain seedstock to regenerate fully stocked black spruce forests on clearcut areas. Once seedlings are established in cutover areas, strips and patches of remaining black spruce may be harvested to complete a commercial clearcut.

Bog Stripcutting and Clearcutting

The peatland in the S1 watershed was strip clearcut in 1969, and the uncut strips were clearcut in 1974 to remove the remaining black spruce. The study was designed to quantify hydrological responses to stripcutting and clearcutting as well as evaluate regeneration of black spruce on the peatland. Verry (1981) published results on the hydrological responses during the first 10 years after stripcutting. In addition, effects on micrometeorology (Brown 1972a,b) and black spruce regeneration (Verry and Elling 1978) have been reported for the S1 study.

Streamflow and bog water levels were measured from 1961 to 1980 at the 33.2 ha S1 watershed (Chapter 2). The 8.1 ha wetland area is surrounded by 25.1 ha of upland hardwood forest. Two aspen stands on the uplands were aged 44 and 52 years old and had a basal area of $22 \, m^2 \, ha^{-1}$ prior to black spruce stripcutting. The two black spruce stands on the bog were 62 and 73 years old and had a basal area of $7.0 \, m^2 \, ha^{-1}$ (Verry 1969; Brown 1973). Eight 30 m-wide strips of black spruce were cut during January and February 1969 when the bog surface was frozen and snow covered. The strips were cut on a 110° azimuth that was perpendicular to the axis of the peatland. Black spruce was removed from 43% of the peatland surface area or 8% of the entire watershed area (Figure 12.3). Wood was cut, piled, and then hauled from the bog using a crawler tractor and a skidder. The stripcuts were separated by 46 m-wide uncut strips that served as a seed stock for the harvested strips. A narrow strip of black spruce was cut along the eastern margin of the bog to move the timber among the strips to a yarding area in the uplands between the road and the bog. The upland vegetation was not disturbed except for the logging yarding area where trees were stored before being loaded on to trucks. Slash was piled in two rows within each strip to expose most of the bog surface for black spruce seeding from natural dispersal. The strips of black spruce remaining from the 1969 cuts were then clearcut between December 4, 1973 and January 17, 1974 to complete the commercial harvest. The logging slash from the 1974 cutting was progressively piled and burned (Verry and Elling 1978).

Short- and long-term changes in floristics were found. Increased growth of grasses and sedges was observed during 1970 and 1971. Brown (1973) measured the increased production of sedges and decreased total coverage of shrubs 3 years after the first stripcutting. By 1971, the cut strips were stocked with 4370 black spruce per ha (Verry and Elling 1978). Regrowing seedlings on the first stripcuts were not affected by the 1974 cutting, and stem density exceeded 5400 trees per ha by 1976 (Verry and Elling 1978). However, the second set of strips were understocked with only 1680 trees per ha by 1977. Verry and Elling (1978) concluded that seed stock from uncut strips needed to be available for at least 2 years to naturally regenerate adequately stocked black spruce stands on small isolated peatlands.

Net radiation was measured 1 m above the bog surface in the center of a stripcut and 3 m above the black spruce canopy in an uncut strip during 1969 and 1970 (Brown 1972a). Shortwave radiation was measured about 1 km to the west (Berglund and Mace 1972; Chapter 2). During the first summer after stripcutting, net radiation did not differ between the clearcut strip and adjacent black spruce strips (Brown 1972a). In the second summer, less solar radiation (14% less during June and 20% during September) was absorbed in the clearcut strip than above the black spruce. Brown (1972a) attributed these differences to decreased albedo and decreased conversion of solar radiation to sensible heat in cut strips relative to the intact black spruce canopy in the undisturbed strips. These differences showed that while more energy was available for evapotranspiration above the open bog surface in the stripcuts, more solar energy was converted to sensible heat by black spruce in an uncut strip. During 1970, windspeed was measured with three-cup anemometers that were located with the net radiometers (Brown 1972b). Windspeed and duration increased in the cut strip, which suggested increased evapotranspiration from surface vegetation due to the greater advection of energy and the possibility of enhanced wind-driven water flux (Brown 1972b). Subsurface temperatures at depths of 2.5, 30, 100, and 200 cm did not change in cut strips despite changes of energy fluxes, wind dynamics, and temperatures above the surface (Brown 1976).

Changes in Annual Water Yield and Bog Water Tables

As reported by Verry (1981), water yield did not change when only 8% of the entire watershed was devegetated due to stripcutting during 1969 or when 24% of the total area was free of overstory vegetation after clearcutting during 1974 (Figure 13.15). Despite slight differences, the slopes of the regression relationships between stream-water yields at S1 and S5 are not significantly different during the preharvest, stripcutting, and clearcutting stages.

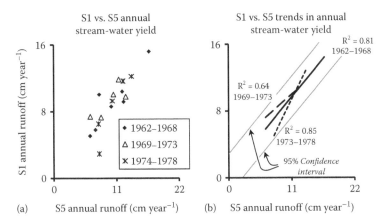

FIGURE 13.15

Stream-water yield relationships between the stripcut S1 and the S5 control watersheds. Annual stream-water yield did not change in response after the 1969 strip cut or the 1974 clearcut that removed the uncut strips of black spruce.

When Verry (1981) first interpreted results after 10 years, he concluded that the amplitude of bog water table fluctuation increased after clearcutting. The maximum annual bog water table elevation appeared to be higher and the minimum lower. Subsequent analysis of the data showed that bog water table responses were similar except during the severe drought of 1976 when the decline in water level at the S1 bog was more pronounced than that at the control bogs in the S5 (Figure 13.16) and S2 watersheds. Although the severe decline in water table at S1 coincided with the postclearcut stage, this pronounced decline was plausibly related more to drought severity and intersite variability, because the berm on the downgradient end of the S1 bog is more permeable than the more well-sealed outlets of the S5 and S2 control watersheds.

Forest Management and Hydrological Implications of the S1 Peatland Harvest Experiment

The ensemble of studies at the S1 watershed show that stream-water yield was not affected by stripcutting or clearcutting and that small isolated peatlands were hydrologically resilient to stripcutting and clearcutting when the uplands were not deforested. Studies at S1 showed changes in floristics (Brown 1973) and differences in evaporation and energy budgets between forested and cut strips (Brown 1972a,b). However, these differences did not affect the annual responses of stream-water yield. In contrast to the clearcutting of upland forests at the S4 and S6 watersheds, effects on annual stream-water yield were not detectable at the S1 watershed after stripcutting or

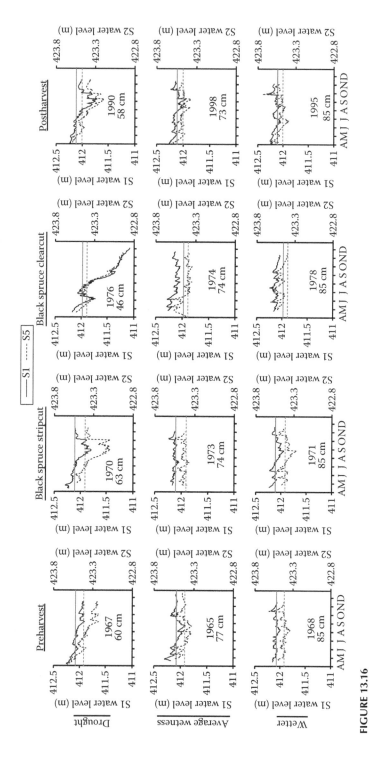

FIGURE 13.16

Daily water levels in the S1 bog during a year for preharvest, stripcut, clearcut, and postharvest stages relative the control S2 watershed. Water levels are shown during drought, average, and above average wetness conditions. The annual precipitation input is shown below the water year. The horizontal lines show mean bog water levels for S1 (solid) and S5 (hashed) during the preharvest water years from 1964 to 1968, the post-stripcut period from 1969 to 1973, the post-clearcut period from 1974 to 1978, and the post-harvest period from 1980 to 2008.

clearcutting. These differences were perhaps related to the small area (24%) that was cleared of black spruce in the S1 bog.

Fen Clearcut

Trees on the S3 peatland were clearcut to evaluate how the effects of peatland black spruce harvesting differed between bogs and fens. All trees taller than 3 m were cut between December 15, 1972 and January 24, 1973. Merchantable timber was removed with a rubber tired loader, skidder, and tractor. An area of alder was not economical to commercially clearcut, but the alder were cut for research purposes. The fen surface was prepared for planting of black spruce and white cedar seedlings in July1973 using a controlled burn of the slash (Knighton and Stiegler 1981).

Bay (1967, 1968) first described different hydrological responses of the S3 fen watershed relative to perched bog watersheds. Unlike the intermittent flow from the perched bogs, outflow from the S3 watershed is perennial. The size of the S3 watershed, 72 ha, is more than seven times larger than S2. Annual stream-water yield at S3, which averages 6777 ± 3994 cm, is more than two orders of magnitude larger, because annual stream-water yield from the S3 watershed is dominated by water inflow from the regional groundwater system (Bay 1967). Because so much of the S3 outflow water originated from groundwater inputs from beyond the watershed boundaries, effects of clearcutting on water yield were not expected and were not apparent in the paired-watershed data (Figure 13.17). Moreover, fluctuations in water yield and water table elevations at the S3 watershed often decoupled from the responses at the S2 control watershed. In the years following the fen clearcutting, annual stream-water yield increased for three consecutive years at the S2 watershed during which stream-water yield at the S3 watershed decreased for 2 years and then increased. From the late 1970s through the early 2000s, stream-water yield at the S3 watershed increased relative to the previous decades (Figure 13.17c). This increase was likely related to the statewide increase in groundwater levels that accompanied an increasing trend in rainfall during late autumn after the drought of the 1960s. These differences make clear that the S3 fen clearcut was not comparable to the S2 control watershed using the paired-watershed approach.

Summary

- The findings from paired-watershed studies at the MEF were consistent with a broader range of watershed studies. However, some findings reflect the unique hydrological responses of lowland watersheds with uplands that drain to peatland-fed streams.

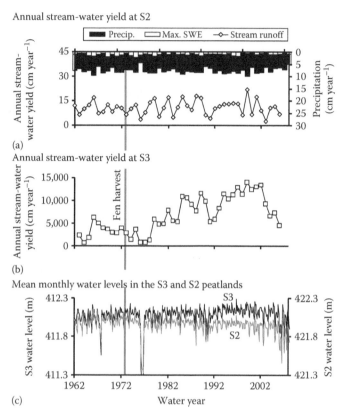

FIGURE 13.17
Annual precipitation, maximum SWE, and annual stream-water yield at S2 bog during each water year (a), annual streamflow at the S3 fen (b), and the mean monthly water levels in the S3 and S2 peatlands (c). The shaded vertical line marks the time of the 1963 fen clearcut.

- Annual water yield increased following clearcutting of upland aspen forests. Stream-water yields from watersheds with large central bogs were high for about a decade and then decreased as biomass increased relative to the preharvest stages. Water yields decreased more quickly at the S6 watershed where the aspen forest was converted to conifers. Peak streamflow in response to rainfall and snowmelt differed between the S4 and S6 clearcuts, and responses were consistent with changes in interception and evapotranspiration between aspen and conifer forests. Bog water tables increased transiently in response to both upland clearcuts probably as a result of increased subsurface runoff from surrounding hillslopes into the peatlands when the uplands had no or little tree cover to intercept and evapotranspire precipitation inputs.

- Stripcutting and clearcutting of black spruce in a bog affected floristics and micrometeorology, but effects on annual water yield and the water table fluctuation dynamics were not detectable, perhaps because of the small portion of the watershed area (24%) that was harvested. When a 37 year-long time series was considered, the magnitude of water table fluctuations did not change in response to bog clearcutting as interpreted originally after 10 years of postharvest data. Interpretation of the shorter data series was confounded by an unexpected water table response due to an extreme drought that occurred 2 years after total clearcutting of the S1 peatland. Our ability to reinterpret initial findings from the clearcut studies illustrates the importance of long-term records.

- The effects of regional groundwater inputs masked the effects of clearcutting, and the paired-watershed approach was not valid for deciphering annual water yield responses at the S3 watershed. Fen groundwater tables vary with regional groundwater aquifer and transboundary groundwater inflow to the S3 watershed that is driven by responses to decadal scale cycles of drought and lags that are associated with the time needed to recharge the spatially extensive aquifer.

References

Bates, P.C., C.R. Blinn, A.A. Alm, and D.A. Perala. 1989. Aspen stand development following harvest in the Lake States Region. *Northern Journal of Applied Forestry* 6:178–183.

Bay, R.R. 1962. Establishment report, Marcell Experimental Forest (Chippewa National Forest and adjacent private, Itasca county, and State of Minnesota lands). Grand Rapids, MN: USDA Forest Service.

Bay, R.R. 1967. Ground water and vegetation in two peat bogs in northern Minnesota. *Ecology* 48(2):308–310.

Bay, R.R. 1968. The hydrology of several peat deposits in northern Minnesota, USA. In: *Proceedings of the Third International Peat Congress*. Quebec, Canada: National Research Council of Canada, pp. 212–218.

Berglund, E.R. and A.C. Mace. 1972. Seasonal albedo variation of black spruce and sphagnum-sedge bog cover types. *Journal of Applied Meteorology* 11(5):806–812.

Boelter, D.H. and E.S. Verry. 1977. Peatland and water in the northern lake States. General Technical Report NC-31. St. Paul, MN: USDA Forest Service.

Bosch, J.M. and J.D. Hewlett. 1982. A review of catchment experiments to determine the effect of vegetation changes on water yield and evapotranspiration. *Journal of Hydrology* 55(1–4):3–23.

Brown, J.M. 1972a. Effect of clearcutting a black spruce bog on net radiation. *Forest Science* 18(4):273–277.

Brown, J.M. 1972b. The effect of overstory removal upon surface wind in a black spruce bog. Research Note NC-137. St. Paul, MN: USDA Forest Service.

Brown, J.M. 1973. Effect on overstory removal on production of shrubs and sedge in a northern Minnesota bog. *Journal of the Minnesota Academy of Science* 38(2–3):96–97.

Brown, J.M. 1976. Peat temperature regime of a Minnesota bog and the effect of canopy removal. *The Journal of Applied Ecology* 13(1):189–194.

Brown, A.E., L. Zhang, T.A. McMahon, A.W. Western, and R.A. Vertessy. 2005. A review of paired catchment studies for determining changes in water yield resulting from alterations in vegetation. *Journal of Hydrology* 310(1–4):28–61.

Clark, J.S. 1990. Fire and climate change during the last 750 year in northwestern Minnesota. *Ecological Monographs* 60(2):135–159.

Fitzpatrick, F.A., J.C. Knox, and H.E. Whitman. 1999. Effects of historical land-cover changes on flooding and sedimentation North Fish Creek, Wisconsin. Water Resources Investigation Report 99–4083. Middleton, WI: U.S. Geological Survey, U.S. Department of the Interior.

Grigal, D.F. and K.N. Brooks. 1997. Forest management impacts on undrained peatlands in North America. In: *Northern Forested Wetlands: Ecology and Management*, ed. C.C. Trettin, M.F. Jurgensen, D.F. Grigal, M.R. Gale, and J.K. Jeglum. Boca Raton, FL: CRC/Lewis Publishers, pp. 379–396.

Helvey, J.D. 1967. Interception by eastern white pine. *Water Resources Research* 3(3):723–729.

Hibbert, A.R. 1967. Forest treatment effects on water yield. In: *Forest Hydrology*, eds., W.E. Sopper and H.W. Lull. New York: Pergamon Press, pp. 527–543.

Hornbeck, J.W., M.B. Adams, E.S. Corbett, E.S. Verry, and J.A. Lynch. 1993. Long-term impacts of forest treatments on water yield: A summary for northeastern USA. *Journal of Hydrology* 150(2–4):323–344.

Johnston, W.F. 1977. Manager's handbook for black spruce in the north central states. General Technical Report NC-34. St. Paul, MN: USDA Forest Service.

Knighton, M.D. and J.H. Stiegler. 1981. Phosphorus release following clearcutting of a black spruce fen and a black spruce bog. In: *Proceedings of the Sixth International Peat Congress*, Duluth, MN, August 17–23, 1980. Eveleth, MN: International Peat Society, pp. 577–583.

Kozlowski, T.T. 1943. Transpiration rates of some forest tree species during the dormant season. *Plant Physiology* 18(2):252–260.

Leopold, L.B., M.G. Wolman, and J.P. Miller. 1958. *Fluvial Processes*. New York: Dover Publishers.

Likens, G.E. and F.H. Bormann. 1995. *Biogeochemistry of a Forested Ecosystem*. New York: Springer-Verlag.

Lu, S.-Y. 1994. Forest harvesting effects on streamflow and flood frequency in the northern Lake States. PhD dissertation. St. Paul, MN: University of Minnesota.

Lugo, A.E., F.J. Swanson, O.R. González, M.B. Adams, B.J. Palik, R.E. Thill, D.G. Brockway, C. Kern, R. Woodsmith, and R.C. Musselman. 2006. Long-term research at USDA Forest Service's experimental forests and ranges. *BioScience* 56(1):39–48.

Nichols, D.S. and J.M. Brown. 1980. Evaporation from a sphagnum moss surface. *Journal of Hydrology* 48(3–4):289–302.

Nichols, D.S. and E.S. Verry. 2001. Stream flow and ground water recharge from small forested watersheds in north central Minnesota. *Journal of Hydrology* 245(1–4):89–103.

Stednick, J.D. 1996. Monitoring the effects of timber harvest on annual water yield. *Journal of Hydrology* 176(1–4):79–95.

Swank, W.T. and J.E. Douglass. 1974. Streamflow greatly reduced by converting deciduous hardwood stands to pine. *Science* 185(4154):857–859.

Swank, W.T., L.W. Swift, and J.E. Douglass. 1988. Streamflow changes associated with forest cutting, species conversions, and natural disturbances. In: *Forest Hydrology and Ecology at Coweeta*, ed. W.T. Swank and D.A. Crossley. New York: Springer-Verlag, Vol. 66, pp. 297–312.

Timmons, D.R., E.S. Verry, R.E. Burwell, and R.F. Holt. 1977. Nutrient transport in surface runoff and interflow from an aspen-birch forest. *Journal of Environmental Quality* 6(2):188–192.

Tracy, D.R. 1997. Hydrologic linkages between uplands and peatlands. PhD dissertation. St. Paul, MN: University of Minnesota.

Verry, E.S. 1969. 1968 vegetation survey of the Marcell Experimental Watersheds. Report GR-W2–61. Grand Rapids, MN: USDA Forest Service.

Verry, E.S. 1972. Effect of an aspen clearcutting on water yield and quality in northern Minnesota. In: *Proceedings of a symposium on Watersheds in Transition*, eds., S.C. Csallany, T.G. McLaughlin, and W.D. Striffler, Fort Collins, CO, June 19–22, 1972. Urbana, IL: American Water Resources Association, pp. 276–284.

Verry, E.S. 1976. Estimating water yield differences between hardwood and pine forests. Research Paper NC-128. St. Paul, MN: USDA Forest Service.

Verry, E.S. 1981. Water table and streamflow changes after stripcutting and clearcutting an undrained black spruce bog. In: *Proceedings of the Sixth International Peat Congress*, Duluth, MN, August 17–23, 1980. Eveleth, MN: International Peat Society, pp. 493–498.

Verry, E.S. 1986. Forest harvesting and water: The Lake States experience. *Water Resources Bulletin* 22(6):1039–1047.

Verry, E.S. 1987. The effect of aspen harvest and growth on water yield in Minnesota. In: *Forest Hydrology and Watershed Management–Hydrologie forestière et aménagement des bassins hydrologiques*. Wallingford, UK: International Association of Hydrological Sciences, Vol. 167, pp. 553–562.

Verry, E.S. 1997. Hydrological processes of natural, northern forested wetlands. In: *Northern Forested Wetlands Ecology and Management*, ed. C.C. Trettin, M.F. Jurgensen, D.F. Grigal, M.R. Gale, and J.K. Jeglum. New York: CRC Lewis Publishers, pp. 163–188.

Verry, E.S. 2004. Land fragmentation and impacts to streams and fish in the central and upper Midwest. In: *Lessons for Watershed Research in the Future; A Century of Forest and Wildland Watershed Lessons*, ed. G.G. Ice and J.D. Stednick. Bethesda, MD: Society of American Foresters, pp. 129–154.

Verry, E.S. and A.E. Elling. 1978. Two years necessary for successful natural seeding in nonbrushy black spruce bogs. Research Note NC-229. St. Paul, MN: USDA Forest Service.

Verry, E.S., J.R. Lewis, and K.N. Brooks. 1983. Aspen clearcutting increases snowmelt and storm flow peaks in north central Minnesota. *Water Resources Bulletin* 19(1):59–67.

Verry, E.S. and D.R. Timmons. 1982. Waterborne nutrient flow through an upland-peatland watershed in Minnesota. *Ecology* 63(5):1456–1467.

Wright, H.E. 1972. Quaternary history of Minnesota. In: *Geology of Minnesota: A Centennial Volume*, ed. P.K. Sims and G.B. Morey. St. Paul, MN: Minnesota Geological Survey, pp. 515–592.

Zinke, P.J. 1967. Forest interception studies in the United States. In: *Forest Hydrology*, ed. W.E. Sopper and H.W. Lull. New York: Pergamon Press, pp. 137–159.

14

Effects of Watershed Experiments on Water Chemistry at the Marcell Experimental Forest

Stephen D. Sebestyen and Elon S. Verry

CONTENTS

Introduction

Watershed studies reveal important information on the effects of forest harvesting and ecosystem disturbances on water and solute yield. Since the first study by Bormann and Likens (1967), studies of nutrient cycling in steep mountainous watersheds have shown that changes to stream yields of nutrients and mineral weathering products vary with vegetation cover, climate, and forest management practices. Most studies show increases in concentrations of nitrate after clearcutting (Bormann et al. 1968; Swank 1988; McHale et al. 2008). Concentrations of other solutes such as sulfate and cations have increased at some watersheds after clearcutting (Bormann et al. 1968; Neal et al. 1992) while other watersheds have shown little or no response (Swank 1988; Martin and Harr 1989).

Although studies in lowland watersheds with large wetlands are less common, results from short-duration studies on peatland ecosystems have reported effects of forest management on concentrations of stream solutes (Verry 1972; Knighton and Stiegler 1981; Ahtiainen 1992; Rosén et al. 1996; Sørensen et al. 2009). The Marcell Experimental Forest (MEF) was established during the 1960s (Chapters 1 and 2) to study the hydrology and ecology of lowland watersheds where upland mineral soils drain to central peatlands (Boelter and Verry 1977). The effects of seven large-scale manipulations on water chemistry have been studied on the MEF watersheds and the data now span up to four decades. In this chapter, we review effects of watershed experiments on solute concentrations and yields in streams and present some findings for the first time.

Watershed Studies at the MEF

Each of six research watersheds at the MEF has a bog or a fen that drains to one or more streams. The ecology, hydrology, and chemistry of bogs differ from fens (Bay 1967; Boelter and Verry 1977). Fens are embedded in the regional aquifer and groundwater flow into fens is a source of water in addition to precipitation and upland runoff. The MEF bog watersheds are perched 5–10 m above the regional aquifer and groundwater from the aquifer does not flow into the bogs. Therefore, the only sources of water are precipitation to bog surfaces and inputs of water via lateral flowpaths through upland soils (Boelter and Verry 1977; Timmons et al. 1977; Chapter 7). Responses to forest harvesting on both types of peatlands have been studied at the MEF: the black spruce at the S1 bog were clearcut with a set of stripcuts in 1969 and the remaining mature trees were cut in 1974; and the forest on the S3 fen was clearcut during 1972.

Even when peatland forests are not harvested, stream-water yield and chemical responses may be influenced by cutting on adjacent uplands. Therefore, clearcutting of upland aspen forests was studied at two watersheds including an experiment in which the uplands regenerated in aspen cover at the S4 watershed and another experiment in which the forest was converted from aspen to conifer cover after clearcutting at the S6 watershed. Vegetation and hydrological responses to harvesting of bog, fen, or upland forests at the MEF are described in Chapters 12 and 13. Water chemistry responses are described in this chapter.

In conjunction with the forest harvest treatments, three additional large-scale experiments included watershed manipulations. The regenerating aspen forest at the S4 watershed was fertilized with ammonium nitrate during 1978 (Berguson and Perala 1988). Aspen sprouting on the S6 uplands was suppressed by cattle grazing during the first three growing seasons after harvest. In addition, ambient sulfate wet deposition was quadrupled on the S6 bog from 2001 to 2008 to determine effects on methylmercury production in a peatland (Chapter 11, Jeremiason et al. 2006).

Long-term measurements in the research watersheds are briefly described here with more details in Chapter 2. Water table elevations in peatlands are measured in each research watershed. Stream stage was measured and streamflow was calculated from stage-discharge relationships at each watershed for at least several years. Stream stage has been measured at permanent stream gages at the outlets of the S2, S4, S5, and S6 watersheds for the duration of the studies. Regressions between water levels at peatland wells and streamflow have been used to calculate monthly and annual water yields for the S1 and S3 watersheds during periods when stream stage was not measured.

Samples of stream water have been collected and analyzed for a variety of solute concentrations in unfiltered water samples. During the 1960s and 1970s, sampling frequency and analyses varied among studies and study sites. Routine biweekly sampling of stream or peatland waters began during 1975 (Chapter 2). Although some data have been reported for samples collected during the 1960s and 1970s (Verry 1972, 1975; Timmons et al. 1977; Knighton and Stiegler 1981; Verry and Timmons 1982), many values from the 1960s, 1970s, 1980s, and 1990s have not yet been entered as electronic data due to the lengthy process of discovering, entering, and checking values from hand-written records and computer printouts. In coming years, as we devote renewed attention to these legacy datasets, we hope that filling gaps as electronic data will enable us to more rigorously evaluate concentrations and yields of solutes in streams.

Nitrate (nitrate + nitrite), ammonium, total organic nitrogen (TON), total phosphorus (TP), and calcium concentrations are the longest and most complete data series. There are sufficient data for those solutes to calculate annual volume-weighted concentrations and solute yields for many years before, during, and after watershed experiments. For some studies, concentrations of other solutes were measured for shorter periods. Stream TON concentrations were calculated by subtracting the ammonium concentration from total Kjeldahl nitrogen, which was measured frequently between 1978 and 1997. Sample concentrations were interpolated linearly to estimate daily concentrations during the intervals between successive samples. Daily concentrations were multiplied by the daily stream-water yield to calculate stream loadings. Loadings were integrated to estimate annual solute yields for each water year. A water year begins on March 1 of the calendar year and ends on February 28 or 29 of the following year. This water year corresponds to the hydrology of perched bog watersheds of north-central Minnesota where streamflow begins during spring snowmelt and stops during the winter (Chapter 2). Annual volume-weighted concentrations were calculated by dividing annual solute yields by annual stream-water yields.

Water and solute yields at the harvested sites were compared to unmanipulated control watersheds (S2 or S5). Temporal patterns of annual volume-weighted concentrations and yields may vary between the control watersheds. To distinguish responses due to experiments from natural variation among different watersheds, we compare the time series of ratios of a particular manipulated watershed to a control watershed (e.g., S6:S2 for the S6 uplands clearcut study) to the ratios of the control watersheds (e.g., S5:S2). We refer to "treatment" ratios when the numerator is the annual concentration or yield data from a manipulated watershed divided by that of a control watershed and to "control" ratios when data from the second control watershed are divided by the control of the manipulated watershed.

Effects of Harvest and Regeneration of Upland Aspen Forests on Stream Chemistry

Aspen (*Populus sp.*), an early succession species, is managed for pulp production throughout northern Minnesota on a short-rotation basis (Bates et al. 1989). The S4 study was designed to determine the effects of upland aspen clearcutting on stream-water and solute yields as well as management strategies for aspen regeneration (Chapter 12). The 25.9 ha upland aspen forest at S4 was harvested with a whole-tree commercial clearcut during the winters of 1970–1971 and 1971–1972. No vegetation on the 8.1 ha central peatland was harvested. Half of the upland area or 34% of the total watershed area was clearcut before snowmelt in 1971. Seventy-one percent of total watershed area was clearcut when the uplands harvesting was completed during January 1972. The upland forest began to regenerate naturally from root sprouting of aspen during the first growing seasons after the two clearcutting periods. In 1978, as a study of forest management in aspen regrowth, ammonium nitrate fertilizer was applied by helicopter at a rate of 340 kg ha^{-1} to the S4 upland forest (Berguson and Perala 1988).

Stream-water yield after harvesting was compared to water yield that was predicted from the pretreatment calibration period during water years from 1962 to 1969 to evaluate the effects of uplands clearcutting on water yield (Chapter 13). Beginning in 1967, water samples were collected from the S5 control watershed and one of the two outlet stream gages (S4N, which is the north outlet) of the S4 watershed (Verry 1975). No samples were collected at S4N and S5 from 1983 to 1988.

Verry (1972) reported values for pH, specific conductivity, apparent color, organic nitrogen, nitrate, nitrite, total N, aluminum, chloride, iron, calcium, copper, sodium, magnesium, manganese, potassium, total phosphate, and zinc before the harvest and the first year after the harvest. After the partial clearcutting, streamflow was 35% lower during peak snowmelt (Verry 1972), and the annual water yield increased by 25% relative to the preharvest period (Chapter 13). Despite the hydrological changes, Verry (1972) showed that stream water chemistry at the partially clearcut S4 watershed did not change relative to the S5 control.

Until now, stream chemistry responses to the S4 clearcutting have not been updated since Verry (1972) published first-year results. From 1972 (second postharvest year) to 1979, stream-water yield increased by 9%–38% relative to the predicted S4 water yield during water years from 1962 to 2005 (Verry 1987; Chapter 13; Figure 14.1a). Unlike increased water yields that persisted for a decade after the harvest, stream nitrogen responses were of short duration. Annual volume-weighted concentration and yield ratios of nitrate + nitrite (Figure 14.1) and ammonium (Figure 14.2) were similar between the control and clearcut watershed until the upland fertilization during 1978. The control (S2:S5) concentration and yield ratios were higher and more variable than the treatment (S4:S5) ratios for nitrate + nitrate during water years before and after 1978. By contrast, the relationships were much different during 1978 when the treatment ratio of nitrate + nitrite concentrations was 96.7 times larger than the control ratio and the yield ratio was 245 times larger (Figure 14.1). The treatment ratio of ammonium concentrations was 6.0 times larger than the control ratio during the 1978 fertilization, which contrasted with the most other years when the ratios were similar (Figure 14.2). The treatment ratio of ammonium yields was 12.0 times larger than the control ratio during the 1978 fertilization. Although the control ratio was 6.6 times larger than the treatment ratio during 1990, yield ratios were similar that year because the water yield that had remained elevated during the first decade after the uplands clearcutting returned to preharvest levels during the 1980s and 1990s. For TON, no data before 1978 were available. We calculated annual volume-weighted concentrations and yields of TON for water years from 1978 to 1997 (Figure 14.3). The treatment ratios were the highest during 1978 and decreased over the next 2 years but remained higher than the levels of the controls through the 1990s.

The magnitudes and fluctuation patterns of calcium concentration ratios were similar during water years before and after clearcutting (Figure 14.4). After fertilization during 1978 through the 2000s, the S4:S5 (treatment) yield ratios were larger than the control ratios. The yield ratios have converged during the last 6 years of record corresponding to water years during the 2000s when the water yield had decreased by as much as 21% relative to the preharvest period (Chapter 13).

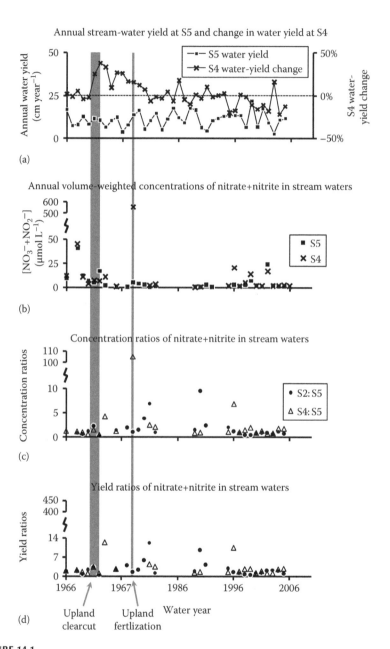

FIGURE 14.1
Annual water yields at the S5 control watershed and changes in stream-water yield at the
S4 watershed (a); annual volume-weighted concentrations of nitrate + nitrite for the S5 and S4
watersheds (b); annual stream nitrate + nitrite concentration ratios at S2 relative to S5 (S2:S5)
for the control watersheds and S4 relative to S5 (S4:S5) for the experiment (c); and annual stream
nitrate + nitrite yield ratios (d).

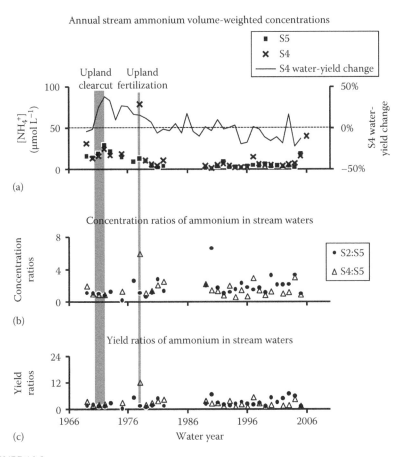

FIGURE 14.2
Changes in annual water yield at the S4 watershed and annual volume-weighted concentrations of ammonium for the S5 control and S4 watersheds (a); annual stream ammonium concentration ratios at S2 relative to S5 (S2:S5) and S4 relative to S5 (S4:S5) (b); and annual stream ammonium yield ratios (c).

Conversion of an Upland Forest from Hardwoods to Conifers

The 6.9 ha upland aspen forest at the S6 watershed was harvested with a whole-tree commercial clearcut from December 1980–June 1981. Most streamflow occurs during spring snowmelt; 22% of the uplands were clearcut before snowmelt and 78% clearcut after most of the annual stream-water yield had occurred. When completed in June, the clearcut was 78% of the watershed area because no vegetation on the 2.0 ha central peatland was harvested. The watershed was fenced during 1981 and grazed with cattle during the first three growing seasons to prepare the site for

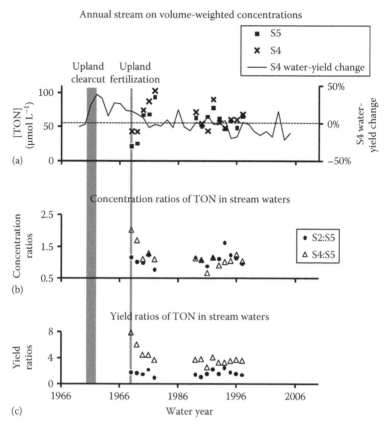

FIGURE 14.3
Changes in annual stream-water yield at the S4 watershed and annual volume-weighted concentrations of TON for the S5 control and S4 watersheds (a); annual stream TON concentration ratios at S2 relative to S5 (S2:S5) and S4 relative to S5 (S4:S5) (b); and annual stream TON yield ratios (c).

planting of red pine and white spruce seedlings during 1983. Grazing was evaluated to assess the cost effectiveness relative to the more common use of herbicides to control vegetation regrowth (Chapter 12). Grazing suppressed the natural regeneration of aspen sprouts, shrubs, and grasses until conifers were planted. During grazing, the uplands were relatively devoid of woody vegetation for 3 years until the conifers were planted during 1983. The canopy of the growing conifer forest closed during the late 1990s. About the time of canopy closure, an experiment was initiated to determine the effects of sulfate deposition on methylmercury production in peatlands. Sulfate was added to the downstream half of the peatland (11% of the total watershed area) as a sodium sulfate solution to increase the ambient wet deposition by a factor of four between the autumn of 2001 and autumn of 2008.

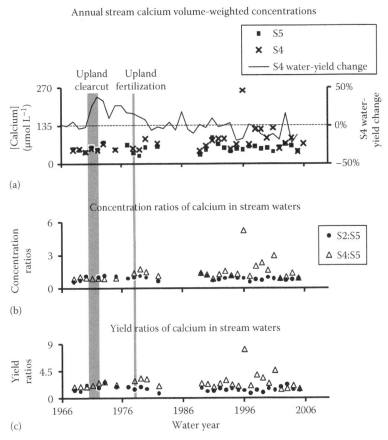

FIGURE 14.4
Changes in annual water yield at the S4 watersheds and annual volume-weighted concentrations of calcium for the S5 control and S4 watersheds (a); annual stream calcium concentration ratios at S2 relative to S5 (S2:S5) and S4 relative to S5 (S4:S5) (b); and annual stream calcium yield ratios (c).

Effects of Uplands Clearcutting and Forest Conversion on Stream Solutes

Stream-water yield after harvesting was compared to water yield predicted from the pretreatment relationship during the calibration period during water from 1976 to 1979 (Chapter 13). Increases in water yield during the first decade after clearcutting averaged 23% ± 16% (±1 standard deviation) and the largest increase of 63% occurred during 1982 (Figure 14.5a). The average water yield (+1% ± 28%) during the 1990s was similar to preharvest levels. Water yield during the 2000s decreased by an average of 39% ± 19% relative to preharvest levels.

We report the stream-chemistry responses to the clearcutting on the S6 uplands and the forest conversion for the first time. In contrast to the

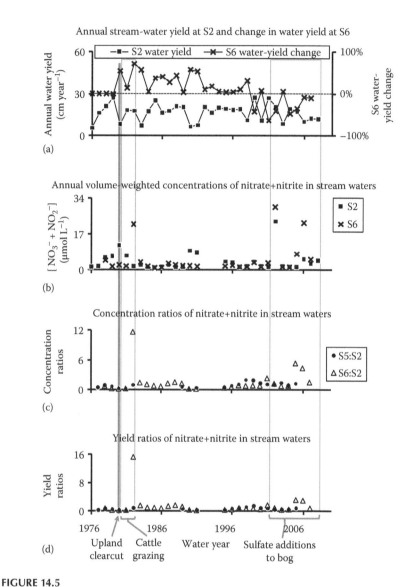

(a)

(b)

(c)

(d)

FIGURE 14.5
Annual water yields at the S2 control watershed and changes in stream-water yield at the S6 watershed (a); annual volume-weighted concentrations of nitrate + nitrite for the S2 control and S6 watersheds (b); annual stream calcium concentration ratios at S5 relative to S2 (S5:S2) and S6 relative to S2 (S6:S2) (c); annual stream nitrate + nitrite yield ratios (d).

immediate increase of the water yield that persisted for a decade after harvesting, responses of stream nitrogen species were offset and of short duration (Figure 14.5a). During and 1 year after the upland clearcutting when cattle grazed the watershed, the range and variation of concentrations for nitrate + nitrate (Figure 14.5b), ammonium (Figure 14.6a), and TON

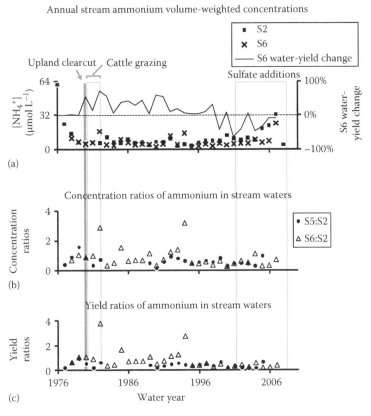

FIGURE 14.6
Changes in annual stream-water yield at the S6 watershed and annual volume-weighted concentrations of ammonium for the S2 control and S6 watersheds (a); annual stream ammonium concentration ratios at S5 relative to S2 (S5:S2) and S6 relative to S2 (S6:S2) (b); and annual stream ammonium yield ratios (c).

(Figure 14.7a) were similar to those for the control watershed. Nitrate + nitrite concentration and yield ratios increased by an order of magnitude during the second water year after the harvest (1982). Annual stream nitrate + nitrite ratios during other years were similar between the controls (S5:S2) and treatment (S6:S2). Stream nitrate + nitrite ratios also were higher after the sulfate addition during water years from 2005 to 2007.

The annual volume-weighted concentration of ammonium was much higher during 1976 than other years at S2 (data for S6 start during 1978). There was a severe summer drought and the lowest stream-water yields on record during 1976. Annual concentration and yield ratios for ammonium were similar before the clearcutting and again from the mid-1990s through 2004. The treatment ratios for annual concentrations and yields were markedly higher during 1982 and 1994 than controls ratios (2 and 14 years after the

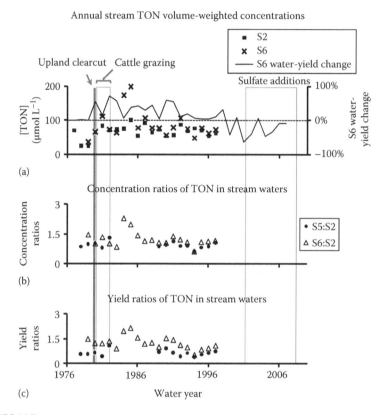

(a)

(b)

(c)

FIGURE 14.7
Changes in annual stream-water yield at the S6 watershed and annual volume-weighted concentrations of TON for the S2 control and S6 watersheds (a); annual stream TON concentration ratios at S5 relative to S2 (S5:S2) and S6 relative to S2 (S6:S2) (b); and annual stream TON yield ratios (c).

clearcutting, respectively). For example, the concentrations were 2.9 times larger and the yield ratios were 3.8 times higher during 1982.

During the fourth and fifth water years after the clearcutting (i.e., 1984 and 1985), the annual volume-weighted concentrations of TON at S6 increased relative to other years and were nearly double those of S2 (Figure 14.7a). From 1989 until 1997, concentrations at the control and experimental watersheds had similar magnitudes and interannual fluctuation patterns (Figure 14.7b).

Responses of TP (Figure 14.8) and calcium (Figure 14.9) did not show brief, large increases like concentrations and yields of nitrogen species. Annual volume-weighted concentrations of TP in stream waters were similar in magnitude and patterns of interannual variation at the S2 and S6 watersheds except the first 7 years after the harvest and several years during the sulfate-addition study. Annual concentration and yield ratios for TP were more variable and larger in magnitude for the treatment than the control watersheds (Figure 14.8). For both TP and calcium, the paucity

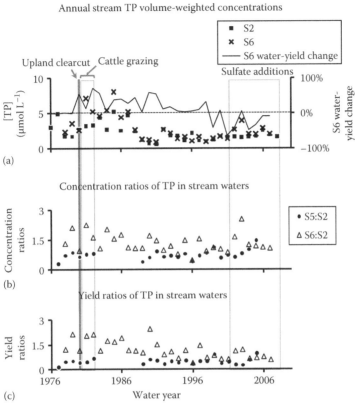

FIGURE 14.8
Changes in annual water yield at the S6 watershed and annual volume-weighted concentrations of TP for the S2 control and S6 watersheds (a); annual stream TP concentration ratios at S5 relative to S2 (S5:S2) and S6 relative to S2 (S6:S2) (b); and annual stream TP yield ratios (c).

of preharvest data prevents a meaningful comparison to the preharvest period. Nonetheless, a pattern does emerge for TP yields after clearcutting. Annual yield ratios have decreased since the upland clearcutting (linear regression having a $p = 0.0004$ and R^2 of 0.49) and not changed for the controls ($p = 0.34$, $R^2 = 0.05$). This convergence pattern reflects increases in stream-water yield at S6 during the 1980s that diminished as stream-water yields returned to preharvest levels during the 1990s and decreased further during the 2000s (Figure 14.8). Annual volume-weighted concentrations of calcium in stream waters at the S6 watershed increased relative to the S2 control and were higher throughout the 1980s (Figure 14.9a).

As noted earlier, the uplands were only partially clearcut before snowmelt and most runoff during water year 1980 occurred with snowmelt before the entire upland area was clearcut. This sequence of harvesting may have been a factor related to the similarity of stream solute concentrations and yields between the S6 and control watersheds during the year when clearcutting occurred.

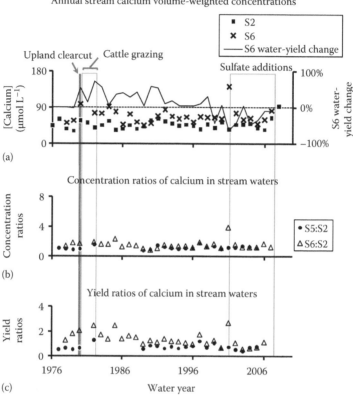

(a)

(b)

(c)

FIGURE 14.9
Annual water yield changes at S6 and annual volume-weighted concentrations of calcium for the S2 control and S6 watersheds (a); annual stream calcium concentration ratios at S5 relative to S2 (S5:S2) and S6 relative to S2 (S6:S2) (b); and annual stream calcium yield ratios (c).

Effect of Cattle Grazing on Stream Fecal Coliform Counts in Streams

The grazing study provided an opportunity to monitor the effects of cattle on stream-water quality during forest conversion. The uplands were grazed for 5–12 weeks each year from 1980 to 1982 to consume aspen regrowth. Cattle were not excluded but did not venture onto the soft organic soils of the peatland. The presence of more than one positive fecal coliform count (per 100 mL sample) per month exceeds the maximum contaminant level for drinking water (U.S. Environmental Protection Agency 2009). To determine whether manure that accumulated on bare upland soils was transported to streams during rainfall and snowmelt runoff events, fecal coliform colonies were counted starting September 1980 (after clearcutting but before the addition of cattle) and ending in 1986. Samples were collected at the S6 weir, about 60 m downstream of the S6 weir where the stream entered a road culvert at the fence boundary, and the S2 weir.

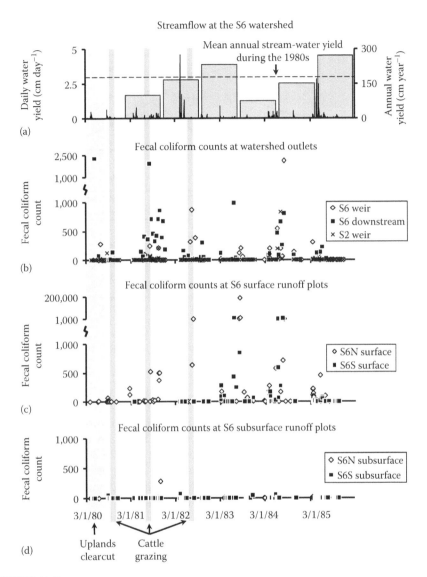

FIGURE 14.10
Streamflow (a) and fecal coliform counts (per 100 mL aliquot) in stream waters (b), upland surface runoff waters (c), and upland subsurface runoff waters (d) from the S2 and S6 watersheds.

Fecal coliform counts in stream water ranged from 0 to 834 at the S2 control watershed and from 0 to 2150 at the S6 weir from 1980 to 1986 (Figure 14.10a). Positive counts at S2 ranged from 5% to 53% of samples among years and the percentages at the S6 weir equaled or exceeded those at the S2 control during every year (Figure 14.11). The mean fecal coliform count of 13 ± 73

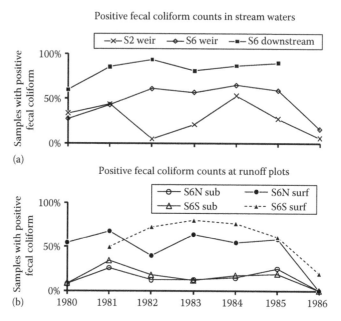

FIGURE 14.11
Positive fecal coliform counts in stream (a) and runoff plot (b) waters from the S2 and S6 watersheds.

(±1 standard deviation) at the S2 watershed was less than the mean of 48±142 at the S6 weir (t-test p = .028).

Fecal coliform counts frequently were higher downstream because cattle entered the channel downstream of the weir. The mean of downstream samples from September 1980 to November 1982 of 212±367 (±1 standard deviation) counts was four times the upstream mean of 53±153 (Figure 14.10b). In addition to cattle sources, unknown natural sources introduced fecal coliform to stream waters as shown by positive counts at the S2 control watershed and positive counts before cattle grazed the S6 uplands. One sample collected at the S6 downstream site before grazing had the highest stream-water count of 2300. Despite natural sources, the frequency of large counts increased after grazing. Larger counts persisted at the S6 weir and downstream for 2 years when levels and the frequency of positive counts returned to levels at the S2 control during 1986. Above-average streamflow during 1985 may have hydrologically flushed residual fecal coliform colonies from the S6 watershed after 2 years of below-average water yield during 1983 and 1984 that coincided with the first 2 years after cattle were removed.

Fecal coliform counts also were measured at surface and subsurface runoff plots on north-facing (site S6S) and south-facing (site S6N) hillslopes. Waters from these samplers represent two flowpaths that transport solutes from upland soils to the lagg zones that surround the bogs (Chapters 2 and 7). Lagg zones preferentially route water and solutes to the watershed outlet during snowmelt and rainfall runoff events when peatland water levels are the highest. Therefore, lagg zones that receive upland waters may route fecal coliform to the watershed outlet. After cattle first grazed the uplands during September 1980, the autumn was relatively dry and fecal coliform counts in surface (Figure 14.10c) and subsurface (Figure 14.10d) runoff were zero. In contrast to consistently low counts at the subsurface runoff plots with means of 2 ± 9 (S6S) and 2 ± 20 (S6N) from 1980 to 1986, counts in water samples from the surface runoff plots increased during grazing at S6N and after grazing at S6S. The highest count of 191,000 was measured in surface runoff from the S6S plot.

Data from the surface runoff plot, S6 weir, and downstream sites suggest that the downstream fecal coliform originated from manure in the stream channel during 1981 and 1982. When counts in surface runoff and S6 weir samples were high during 1982, 1983, and 1984, we hypothesize that some fecal coliform was transported from uplands via the lagg to the outlet stream.

Effect of Sulfate Additions on Stream Methylmercury

Sulfate is an atmospherically deposited pollutant that affects aquatic and peatland ecosystems (Gorham et al. 1984; Hedin et al. 1987) and is linked to the production of highly toxic methylmercury in anoxic organic soils (St. Louis et al. 1996; Branfireun et al. 1999). The sulfate addition at the S6 watershed was designed to assess: (1) the watershed-scale effects of sulfate deposition on methylmercury production in peatlands, and (2) the effects of decadal scale recovery from peak sulfate deposition during the 1970s on methylmercury production and outflow in peatland watersheds. Two important responses of stream chemistry were measured in direct response to the S6 sulfate additions between November 2001 and November 2008. During the first year after the sulfate addition, merthylmercury yields in stream water doubled after sulfate additions quadrupled the wet deposition of sulfate to half the bog surface area (Jeremiason et al. 2006). Sulfate yields in stream waters draining the S6 watershed increased relative to the S2 watershed during water years in the 2000s in contrast to the similar relationships between sulfate yields at the watersheds during the 1980s and 1990s (Figure 14.12).

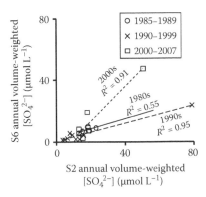

FIGURE 14.12
Annual volume-weighted sulfate concentrations at S6 versus S2.

Additional findings reveal rapid responses of net methylmercury production to sulfate additions: net methylmercury production increased when water levels were the highest, particularly after snowmelt; and net methylmercury production was elevated relative to controls during the first year of the recovery period after sulfate additions ended (Coleman-Wasik 2008). Importantly, these dramatic changes in the production of methylmercury occurred when only 11% of the entire watershed area was affected by the elevated inputs of sulfate.

Chemical Responses to Upland Forest Harvesting in Lowland Peatland Watersheds in Relation to Other Studies

Nitrate and ammonium are redox-sensitive dissolved species of nutrients that play a major role in regulating ecosystem productivity. At the MEF where peatlands are nutrient-poor, little inorganic nitrogen is transported through the peatlands to streams (Urban and Eisenreich 1988; Urban et al. 1988; Chapter 8). The responses of stream nitrogen chemistry to commercial harvests of upland forests at the MEF were of short duration and delayed relative to other upland watershed studies that include commercial clearcuts (Swank 1988; Martin and Harr 1989; Neal et al. 1992; Martin et al. 2000; McHale et al. 2008; Stednick 2008), experimental strip and block cuts (Hornbeck et al. 1987), and multiyear devegetation studies (Bormann et al. 1968; Kochenderfer and Aubertin 1975). However, these patterns at the MEF may be related to the partial clearcutting of the watersheds before snowmelts with completion of the upland clearcutting after most of the annual stream water-yield had already occurred. Climatic variation over the decade that separated the S6 and S4 studies at the MEF may explain some differences in stream nutrient responses to treatments. Nonetheless, results of the experiments show stream nutrient responses to the particular forest-management practices (i.e., 3 year deforestation and cattle grazing at S6 versus the immediate resprouting of aspen at S4).

Nitrate concentrations in stream waters typically increase from twofold to several orders of magnitude within 1 year of clearcutting and higher concentrations may persist from 3 to 20 years (Bormann et al. 1968; Swank 1988; Martin and Harr 1989; Neal et al. 1992; Martin et al. 2000; McHale et al. 2008). First-year increases in stream nitrate concentrations and yields were not detected during 1971 after 34% of the S4 watershed area was clearcut or during 1972 after 71% was clearcut. Annual stream nitrate concentrations and yields changed only during the seventh year after clearcutting when the regenerating forest on the S4 uplands was fertilized. The greater than 500-fold increase in stream nitrate concentrations with fertilization is among the largest reported responses to a forest-management practice. The S4 study results differed from the S6 study in that nitrate concentrations in stream waters increased by an order of magnitude 2 years after the clearcutting and in the third year of cattle grazing.

Comparisons with steep upland watersheds reveal some striking differences from lowland watersheds where oxidized species such as nitrate are chemically reduced in saturated and anoxic organic soils to produce ammonium. In addition, TON concentrations are high due to leaching of soluble organic matter and production of organic acids in peatlands. Nitrogen limitation also leads to near-complete biological assimilation of inorganic nitrogen in peatlands (Urban and Eisenreich 1988; Urban et al. 1988; Chapter 8). Although different from upland watershed studies, the timing response of stream nitrate increases at the MEF S6 clearcutting study was similar to another lowland watershed with wetlands and boreal tree species. Stream nitrate concentrations increased incrementally during the first three growing seasons after uplands clearcutting at a lowland Swedish watershed, 277 Balsjö (Löfgren et al. 2009). However, increases in concentration were at least an order of magnitude larger at 277 Balsjö. One pertinent difference may be drainage ditching of the 277 Balsjö peatland and the consequences of lowered water tables and the altered oxygen status of peatland soils relative to the MEF peatlands that have not been drained.

Concentrations and yields of ammonium in stream waters increased after upland clearcutting at S6 and at 277 Balsjö (Löfgren et al. 2009). The timing of increases in ammonium concentrations and yields at the S6 watershed was similar to nitrate increases but much smaller in magnitude. Ammonium concentrations increased threefold and yields increased fourfold during the third water year after the S6 clearcutting. Ammonium concentrations increased 6-fold and yields increased 23-fold after the S4 fertilization. The increased ammonium concentrations and yields from lowland watersheds with peatlands differed from studies at steep upland watersheds where little to no ammonium was present both before and after clearcuts (Bormann et al. 1968; Hornbeck et al. 1987; Swank 1988).

TP concentrations usually are driven by particulate transport in steep catchments. By contrast, the hummock and hollow microtopography of the MEF peatlands effectively retains most organic and inorganic particulates. Stream TP concentrations at the S6 watershed were higher than at the S2 control during most years from the second through seventh water years after harvest. At two other lowland watersheds with peatlands, TP concentrations showed little or no response during or after clearcutting until peatlands drainage (Ahtiainen 1992; Löfgren et al. 2009). Water-yield deficits drive the long-term declines in TP and calcium yields at watersheds on the MEF. Due to the forest conversion at the S6 watershed, water-yield deficits (relative to the preharvest aspen forests) were larger in magnitude than deficits at the S4 watershed where the upland aspen forest was clearcut a decade earlier and the forest was subsequently fertilized.

Hydrologic Responses to Peatland Harvests

The spatial variation of solute sources, sinks, and transformations between well-aerated upland soils and saturated anaerobic peatland soils affects how water and solutes are delivered to peatland laggs and streams. Solute sources and nutrient availability also differ between ombrotrophic bogs and minerotrophic fens that are embedded in regional groundwater aquifers. Bogs at the MEF are isolated from groundwater inputs due to hydrogeologic settings. The raised domes of ombrotrophic bogs maintain hydraulic gradients that drive water and solute fluxes to lagg zones that surround bogs in perched watersheds (Chapter 7). In contrast to bogs, groundwater flow transports chemically distinct groundwater into fens from regional aquifers (Bay 1967). Fen groundwater tables and streamflow from fens vary in response to interannual variation in groundwater inflow from aquifers. For these reasons, stream solute concentrations and yields may differ when forests on peatlands are clearcut rather than the uplands that drain to those peatlands. Two peatland forests were harvested at the MEF to: (1) determine effects of peatland clearcutting when the upland forests were left intact on water yields and solutes, and (2) determine whether effects on stream solutes differed between bog and fen watersheds. Peatland hydrology was not modified before or after the clearcutting by drainage, a common practice for increasing forest productivity on peatlands in northern Europe and Siberia (Vompersky 1976; Mitsch and Gosselink 1993).

Although solute concentrations have been measured at the S1 and S3 watersheds since the 1960s, there are insufficient data entered in electronic formats for rigorous evaluation of long-term responses to the two peatland harvest studies at the MEF. These analyses will be forthcoming. Here, we summarize and update findings of Knighton and Stiegler (1981). To quantify solute responses of bogs and fens to peatland forest harvesting, Knighton and Stiegler collected water samples from a small pool in each peatland near the stream outlet. They compared stream phosphorus yields to peatland clearcutting at the S3 fen and S1 bog watersheds from 1968 to 1978, compared chemical responses at the treatment watersheds to the S2 bog watershed, and speculated on processes that controlled differences in response between the two wetland types.

Bog Stripcut and Clearcut

Stripcutting retains a seed source to regenerate black spruce stands and is a best management practice to commercially harvest black spruce on peatlands in the Lake States region (Grigal and Brooks 1997). The black spruce forest on the 8.1 ha S1 bog was harvested during staggered whole-tree clearcutting in strips. Eight 30.5 m-wide strips were clearcut during the winter of 1969. The clearcut strips alternated with 45.7 m-wide strips of black spruce that served as the seed source to regenerate black spruce on the cut strips (Verry and Elling 1978). The mature trees that remained in the previously uncut strips were harvested during the winter from 1973 to 1974 after seedlings established on the previously cut strips (Verry and Elling 1978). The aspen and birch forest on the uplands was not harvested except for a small area that was used as a log yarding area.

Changes in water yields and bog water table levels were not detected after the 1969 or 1974 stripcuttings (Chapter 13). TP concentrations in bog waters showed no response to the 1969 stripcutting and then tripled during the year after the 1974 clearcutting (Figure 14.13). Elevated TP concentrations may have lasted longer but no samples were collected during 1975 and concentrations were similar to those at the S2 control by 1976. Clearcutting of the black spruce strip next to the S1 outlet where the samples were collected during 1974 also may be a factor that explains the brief concentration responses at the S1 watershed.

Knighton and Stiegler (1981) measured a large soluble phosphorus pool when surficial peat samples were incubated in a laboratory study (Figure 14.14a). They concluded that more phosphorus may have been transported from the S1 bog if the flow of water through the peat was greater. This finding implied that TP yields to streams may be elevated when wet years and higher flows follow bog clearcuttings.

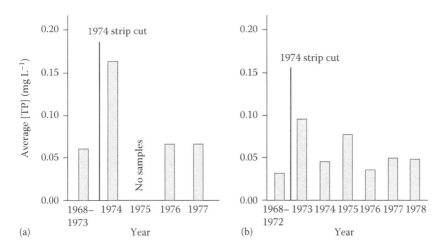

FIGURE 14.13
Average TP concentration of waters from the S1 bog (a) and S3 fen (b). (Modified from Knighton, M.D. and Stiegler, J.H., Phosphorus release following clearcutting of a black spruce fen and a black spruce bog, in *Proceedings of the Sixth International Peat Congress*, August 17–23, 1980, Duluth, MN, International Peat Society, Eveleth, MN, pp. 577–583, 1981.)

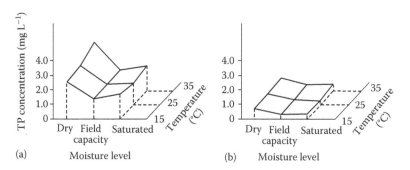

FIGURE 14.14
Concentration of TP in leachates from S1 bog (a) and S3 fen (b) peats. The leachates were extracted from peats that were incubated at different moisture and temperature levels. (Modified from Knighton, M.D. and Stiegler, J.H., Phosphorus release following clearcutting of a black spruce fen and a black spruce bog, in: *Proceedings of the Sixth International Peat Congress,* August 17–23, 1980, Duluth, MN, International Peat Society, Eveleth, MN, pp. 577–583.)

Knighton and Stiegler (1981) discussed how the increased magnitude of water table fluctuations, which Verry (1981) reported for S1 after the 1974 clearcutting, may increase phosphorus mineralization due to transient saturation and anoxia of surficial peats. We reject their hypothesis about controls on phosphorus mineralization after reassessing the entire 1961–2007 record of water table fluctuations. When the original data presented by Verry (1981) were reevaluated, we found that the amplitude of water table fluctuations did not change during or after clearcutting (Chapter 13) and therefore, could not have been a factor affecting phosphorus mineralization.

Fen Clearcut

The 18.6 ha forest on the S3 fen was commercially clearcut during the winter of 1972. The study was designed as site preparation for cedar restoration and to assess effects on stream chemistry. Black spruce and northern white cedar were seeded on the fen after clearcutting and slash burning. Black spruce, alder, and willow currently are the most abundant overstory species because survivorship of northern white cedar was poor in areas that were not protected from browsing by white-tailed deer (*Odocoileus virginianus*). The 34.0 ha upland aspen and birch forest was not harvested though a small portion northwest of the fen had previously been clearcut and replanted in red pine during the 1960s (Chapter 2).

Water yield via streamflow did not change after the peatland harvest. This finding is similar to that in the S1 study (Chapter 13). TP concentrations tripled relative to the pretreatment period from 1968 to 1972. Concentrations remained higher through 1978 when data collection ended. Knighton and Stiegler (1981) speculated that slash burning after clearcutting-induced increases in concentrations and yields after mineralization of phosphorus to soluble forms. The soluble phosphorus pool in surficial peats was less at S3 than at S1 (Figure 14.14b). In contrast to S1, phosphorus release to the stream may have been source (not transport) limited at the S3 fen. This finding suggests that stream TP concentrations are more sensitive to the biogeochemical processes that control phosphorus solubility in fen organic soils than interannual variation in water yields.

Summary

- The physical, chemical, and hydrological settings of lowland watersheds affect water and solute responses to forest harvesting. Because uplands route water through central peatlands that drain to streams, management strategies that affect forest cover and composition on uplands are coupled with the physical and biogeochemical processes that affect water movement and solute availability in peatlands.

- The increases in nitrate and ammonium concentrations in response to upland harvests at the MEF lasted no more than 1 year, whereas the water-yield increases lasted about a decade. The short duration nitrogen responses at the MEF contrast with changes of longer duration after experimental or natural vegetation changes at steep upland watersheds.

- The differences between nitrogen, TP, and calcium reflect fundamentally different biogeochemical processes that affect individual nutrient species when water yields changed after upland clearcuts in the watersheds of the MEF.

- Annual water yields did not change and TP concentrations increased after peatland harvesting on a bog and a fen at the MEF. The duration of increases in TP concentrations at the fen watershed exceeded that at the bog watershed.

- Internal peatland processes regulated the immobilization of upland inputs to laggs and the mobilization of solutes from organic soils after forest management or other forms of ecosystem change such as deposition of atmospheric pollutants. The sulfate addition study at S6 showed that peatlands have a disproportionate effect on the production and release of methylmercury, and that hotspots of biogeochemical transformations are important controls on solute fluxes from lowland watersheds.

References

Ahtiainen, M. 1992. The effects of forest clear-cutting and scarification on the water quality of small brooks. *Hydrobiologia* 243–244(1):465–473.

Bates, P.C., C.R. Blinn, A.A. Alm, and D.A. Perala. 1989. Aspen stand development following harvest in the Lake States Region. *Northern Journal of Applied Forestry* 6:178–183.

Bay, R.R. 1967. Ground water and vegetation in two peat bogs in northern Minnesota. *Ecology* 48(2):308–310.

Berguson, W.E. and D.A. Perala. 1988. Aspen fertilization and thinning research results and future potential. In: *Minnesota's Timber Supply: Perspectives and Analysis*, ed. A.R. Ek and H.M. Hoganson. Grand Rapids, MN: University of Minnesota, pp. 176–183.

Boelter, D.H. and E.S. Verry. 1977. Peatland and water in the northern lake States. General Technical Report NC-31. St. Paul, MN: USDA Forest Service.

Bormann, F.H. and G.E. Likens. 1967. Nutrient cycling. *Science* 155(3761):424–429.

Bormann, F.H., G.E. Likens, D.W. Fisher, and R.S. Pierce. 1968. Nutrient loss accelerated by clear-cutting of a forest ecosystem. *Science* 159(3817):882–884.

Branfireun, B.A., N.T. Roulet, C.A. Kelly, and J.W.M. Rudd. 1999. In situ sulphate stimulation of mercury methylation in a boreal peatland: Toward a link between acid rain and methylmercury contamination in remote environments. *Global Biogeochemical Cycles* 13(3):743–750.

Coleman-Wasik, J.K. 2008. Chronic effects of atmospheric sulfate deposition on mercury methylation in a boreal wetland: Replication of a global experiment. MS thesis. St. Paul, MN: University of Minnesota.

Gorham, E., S.E. Bayley, and D.W. Schindler. 1984. Ecological effects of acid deposition upon peatlands: A neglected field in "acid-rain" research. *Canadian Journal of Fisheries and Aquatic Sciences* 41(8):1256–1268.

Grigal, D.F. and K.N. Brooks. 1997. Forest management impacts on undrained peatlands in North America. In: *Northern Forested Wetlands: Ecology and Management*, ed. C.C. Trettin, M.F. Jurgensen, D.F. Grigal, and J.K. Jeglum. Boca Raton, FL: CRC/Lewis Publishers, pp. 379–396.

Hedin, L.O., G.E. Likens, and F.H. Bormann. 1987. Decrease in precipitation acidity resulting from decreased SO_4^{2-} concentration. *Nature* 325(6101):244–246.

Hornbeck, J.W., C.W. Martin, R.S. Pierce, F.H. Bormann, G.E. Likens, and J.S. Eaton. 1987. The northern hardwood forest ecosystem: Ten years of recovery from clearcutting. RP-NE-596. Broomall, PA: USDA Forest Service.

Jeremiason, J.D., D.R. Engstrom, E.B. Swain, E.A. Nater, B.M. Johnson, J.E. Almendinger, B.A. Monson, and R.K. Kolka. 2006. Sulfate addition increases methylmercury production in an experimental wetland. *Environmental Science and Technology* 40(12):3800–3806.

Knighton, M.D. and J.H. Stiegler. 1981. Phosphorus release following clearcutting of a black spruce fen and a black spruce bog. In: *Proceedings of the Sixth International Peat Congress*, August 17–23, 1980, Duluth, MN. Eveleth, MN: International Peat Society, pp. 577–583.

Kochenderfer, J.N. and G.M. Aubertin. 1975. Effects of management practices on water quality and quantity: Fernow Experimental Forest, W. Virginia. In: *Municipal*

Watershed Management Symposium Proceedings. Upper Darby, PA: USDA Forest Service, pp. 14–24.

Löfgren, S., R. Ring, C. von Brömssen, R. Sørensen, and L. Högbom. 2009. Short-term effects of clear-cutting on the water chemistry of two boreal streams in northern Sweden: A paired catchment study. *Ambio* 38(7):347–356.

Martin, C.W. and R.D. Harr. 1989. Logging of mature Douglas-fir in western Oregon has little effect on nutrient output budgets. *Canadian Journal of Forest Research* 19(1):35–43.

Martin, C.W., J.W. Hornbeck, G.E. Likens, and D.C. Buso. 2000. Impacts of intensive harvesting on hydrology and nutrient dynamics of northern hardwood forests. *Canadian Journal of Fisheries and Aquatic Sciences* 57(S2):19–29.

McHale, M.R., P.S. Murdoch, D.A. Burns, and B.P. Baldigo. 2008. Effects of forest harvesting on ecosystem health in the headwaters of the New York City water supply, Catskill Mountains. New York Scientific Investigations Report 2008–5057: US Geological Survey.

Mitsch, W.J. and J.G. Gosselink. 1993. *Wetlands*. New York: Van Nostrand Reinhold.

Neal, C., R. Fisher, C.J. Smith, S. Hill, M. Neal, T. Conway, G.P. Ryland, and J.A. Jeffrey. 1992. The effects of tree harvesting on stream-water quality at an acidic and acid-sensitive spruce forested area: Plynlimon, mid-Wales. *Journal of Hydrology* 135(1–4):305–319.

Rosén, K., J.-A. Aronson, and H.M. Eriksson. 1996. Effects of clear-cutting on streamwater quality in forest catchments in central Sweden. *Forest Ecology and Management* 83(3):237–244.

Sørensen, R., M. Meili, L. Lambertsson, C. von Brömssen, and K.H. Bishop. 2009. The effects of forest harvest operations on mercury and methylmercury in two boreal streams: Relatively small changes in the first two years prior to site preparation. *Ambio* 38(7):364–372.

St. Louis, V.L., J.W.M. Rudd, C.A. Kelly, K.G. Beaty, R.J. Flett, and N. Roulet. 1996. Production and loss of methylmercury and loss of total mercury from boreal forest catchments containing different types of wetlands. *Environmental Science and Technology* 30(9):2719–2729.

Stednick, J.D. 2008. Long-term water quality changes following timber harvesting. In: *Hydrological and Biological Responses to Forest Practices*, ed., J.D. Stednick. New York: Springer, Vol. 199, pp. 157–170.

Swank, W.T. 1988. Stream chemistry responses to disturbance. In: *Forest Hydrology and Ecology at Coweeta*, eds., W.T. Swank and D.A. Crossley. New York: Springer-Verlag, Vol. 66, pp. 339–357.

Timmons, D.R., E.S. Verry, R.E. Burwell, and R.F. Holt. 1977. Nutrient transport in surface runoff and interflow from an aspen-birch forest. *Journal of Environmental Quality* 6(2):188–192.

Urban, N.R. and S.J. Eisenreich. 1988. Nitrogen cycling in a forested Minnesota bog. *Canadian Journal of Botany* 66(3):435–449.

Urban, N.R., S.J. Eisenreich, and S.E. Bayley. 1988. The relative importance of denitrification and nitrate assimilation in midcontinental bogs. *Limnology and Oceanography* 33(6 part 2):1611–1617.

U.S. Environmental Protection Agency. 2009. National primary drinking water regulations. 816-F-09-004. Washington, DC: U.S. Environmental Protection Agency.

Verry, E.S. 1972. Effect of an aspen clearcutting on water yield and quality in northern Minnesota. In: *Proceedings of a symposium on Watersheds in Transition*, eds., S.C. Csallany, T.G. McLaughlin, and W.D. Striffler. Fort Collins, CO, June 19–22, 1972. Urbana, IL: American Water Resources Association, pp. 276–284.

Verry, E.S. 1975. Streamflow chemistry and nutrient yields from upland-peatland watersheds in Minnesota. *Ecology* 56(5):1149–1157.

Verry, E.S. 1981. Water table and streamflow changes after stripcutting and clearcutting an undrained black spruce bog. In: *Proceedings of the Sixth International Peat Congress*, August 17–23, 1980, Duluth, Minnesota. Eveleth, Minnesota: International Peat Society, pp. 493–498.

Verry, E.S. 1987. The effect of aspen harvest and growth on water yield in Minnesota. In: *Forest Hydrology and Watershed Management—Hydrologie Forestière et Aménagement des Bassins Hydrologiques*. Wallingford, UK: International Society of Hydrological Sciences, Vol. 167, pp. 553–562.

Verry, E.S. and A.E. Elling. 1978. Two years necessary for successful natural seeding in nonbrushy black spruce bogs. Research Note NC-229. St. Paul, MN: USDA Forest Service.

Verry, E.S. and D.R. Timmons. 1982. Waterborne nutrient flow through an upland-peatland watershed in Minnesota. *Ecology* 63(5):1456–1467.

Vompersky, S.E. 1976. Biological foundations of forest drainage efficiency. Washington, DC: USDA Soil Conservation Service and the National Science Foundation.

15

Multiple Resource and Hydrologic Models for Peatland–Upland Forests of the Northern Lake States

Kenneth N. Brooks, Peter F. Ffolliott, D. Phillip Guertin, Shiang-Yue Lu, John L. Neiber, Steve R. Predmore, and Paul K. Barten

CONTENTS

Introduction

Integrated assessments of hydrologic and biological responses of watersheds to forest and peat management in the northern Lake States have been made possible by 50 years of long-term monitoring, laboratory experiments, and linked field-plot research at the Marcell Experimental Forest (MEF). The iterative process—field plots, laboratory and watershed-scale experiments, and model calibration and verification—have facilitated model development. Research at the MEF has generated a continuous, quality-assured monitoring database that along with linked research has been used to develop empirical and dynamic modeling approaches for a variety of applied management considerations, for example, assessing environmental impacts from peat mining and forest harvesting. The MEF's long-term data sets and predictive models also have been used to evaluate potential impacts of climate change on wetlands in Minnesota. This chapter includes an overview of models that have originated or evolved from research and monitoring activities on the MEF.

Multiple-Resources Modeling

Predicting, displaying, and evaluating tradeoffs among the biophysical and socioeconomic effects of forest management became a focus of planning activities by the USDA Forest Service in the 1970s. This focus was aimed at facilitating more effective forest-, rangeland-, and watershed-management practices by improving methods for formulating viable alternatives, simulating likely future outcomes of each, and assembling comparative displays that effectively communicate the tradeoffs among alternatives. A shortcoming of planning policies at the time was that they were too narrow in scope. However, most landscapes in the National Forest System were and continue to be managed for ecosystem-based multiple goals and benefits. Harvesting timber, ensuring a sustainable water supply, and creating or enhancing wildlife habitat and recreational opportunities represent the types of objectives considered by managers during the planning process. Practices designed to sustain timber production, improve water supplies, and provide recreational opportunities have a variety of effects on ecosystem resources that can influence the attainment of environmental, economic, and social goals. Such practices can also alter ecosystem functioning. As a result, models were developed to facilitate the assessment of potential multiple resource outcomes associated with forest management, thus providing a tool for forest planners, decisionmakers, and stakeholders.

Modeling Framework

An integrated multiple-resource assessment effort initiated by the Forest Service's North Central Forest Experiment Station was based on the successes of the Rocky Mountain Research Station's Beaver Creek program, which was designed to improve ecosystem-based, multiple-use management on National Forests. Interactive, user-friendly, multiple-resource models were developed to assist in predicting biophysical and socioeconomic impacts of alternative forest-management practices on designated landscapes (Carder et al. 1977; Larson et al. 1979; Rasmussen and Ffolliott 1981). The models were used to evaluate future impacts and then select the most appropriate practice to meet established goals within the constraints confronted.

Application in Lake States

The modular model structure adapted from the Beaver Creek program for a prototype system for the Lake States facilitated the linking of combinations of individual models to simulate impacts of a given management practice in a multiple-resource context. With a modular system, an individual component can be updated or replaced with improved models without disrupting other components.

The need for multiple-resource models for planning by the National Forests in the Lake States region was expressed by Boyce (1979) and Lundgren and Essex (1979). As a result, five models were developed as modular components in the prototype system (Ffolliott et al. 1984; Guertin et al. 1987b) known as INTErfacing Resources In Management (INTERIM). Information that formed the basis for the development of the models was obtained from the MEF and other research sites in the Lake States region. The models are

- TREES—simulates the growth and mortality of tree species and representative size classes with a modified version of GROW (Brand 1981), a subroutine of the STEMS modeling project (Belcher et al. 1982).

- WATER—simulates changes in annual water supply and snowmelt peak streamflow resulting from modifications in the type or density of forest cover prescribed by management (Bernath 1981; Bernath et al. 1982, Verry et al. 1983).

- HERBS—simulates the percentage of the ground in selected ecosystems covered by herbaceous plants (grasses, forbs, and shrubs) developed largely from work by Anderson et al. (1969), Zavitkovski (1976), Crow (1978), and Lundgren (1981).

- DEER and BIRDS—simulates relative habitat qualities for white-tailed deer and nongame birds through interpretations of ranked responses to alternative forest-management scenarios that were formulated from research by Dahlberg and Guettinger (1956), Erickson et al. (1961), and DeGraff and Evans (1979). Habitats for wildlife were assigned numerical values determined through analysis of mathematical functions that relate habitat quality to readily accessible input data.

INTERIM was used to assess potential impacts of forest-management practices on a hypothetical watershed in northern Minnesota (Ffolliott et al. 1984; Guertin et al. 1987b). These included clearcutting aspen-birch in 20 annual cuttings of 150 ha, each followed by natural regeneration (Table 15.1). Ecosystem resources considered were volume of trees, annual water yields, annual spring snowmelt flooding, percentage of herbaceous canopy cover, and habitat qualities for white-tailed deer and (collectively) nongame bird species. INTERIM was configured to generate estimates at various time scales subsequent to forest practices for initial (0) conditions and 5, 10, and 20 year intervals. Predicted end points of the respective changes are presented in Table 15.1. However, the relative changes (trends) in ecosystem resources through time likely were equally important to forest planners, managers, and decision makers.

Following the initial application of INTERIM, professionals from Federal, state, and county agencies and the private sector were surveyed to assess the utility and scope of the model packages. Respondents indicated that the multiple-resource models were useful in designing responses to land-use practices, assessing environmental impacts, and identifying research needs (Guertin et al. 1987b). They added that INTERIM would be helpful

TABLE 15.1

Effects of a Proposed Forest Management Activity on a Hypothetical Watershed, as Predicted by INTERIM

Module	Resource	Beginning	Year 5	Year 10	Year 20
TREES—total volume[a]	Aspen	73.9	65.4	49.0	2.6
	Red pine	1.7	1.7	2.0	2.6
	Lowland conifer	12.5	14.4	17.0	21.2
Area in each forest type (%)	Aspen—undisturbed	75.0	50.0	37.5	0.0
	Aspen—disturbed	0.0	25.0	37.5	75.0
	Red pine	5.0	5.0	5.0	5.0
	Lowland conifer	20.0	20.0	20.0	20.0
WATER—annual water yield	Millimeters	127.0	139.7	144.8	147.3
	Change (%)	0	10.9	15	15.9
Annual spring snowmelt flood	$m^3 s^{-1} km^{-2}$	0.109	0.095	0.81	0.075
	Change (%)	0	−3	+643	−31
HERBS—percent understory canopy		21.3	26.9	34.8	53.6
BIRDS—nongame bird species diversity index (ND)[b]	Aspen—undisturbed	9.2	8.3	7.0	na
	Aspen—disturbed	na	3.1	5.8	8.5
	Red pine	5.1	3.7	2.4	0
	Lowland conifer	0	0	0	0
DEER—white-tailed Deer habitat index (ND)[b]	Aspen—undisturbed	2.3	2.2	2.0	na
	Aspen—disturbed	na	7.6	7.6	7.6
	Red pine	3.0	3.1	3.1	3.1
	Lowland conifer	3.2	3.1	2.9	2.1

Source: Modified from Ffolliott, P.F. et al., *N. J. Appl. Forest.*, 1, 80, 1984.

[a] In 10,000's of cubic meters and including regeneration areas.

[b] Nondimensional.

in designing and implementing natural-resources inventories, developing response (capability) relationships among ecosystem resources and services, identifying informational gaps and future research needs, and assessing the environmental impacts of alternative forest-management practices.

Development and Application of Peatland Hydrologic Impact Model

Two major issues emerged in the northern Lake States during the 1970s that required more detailed hydrologic-simulation capabilities than contained in the multiple-resource models of INTERIM. First, a proposal by the Minnesota Gas Company (Minnegasco) to build a peat gasification facility in northern Minnesota would require peat extraction from 80,900 ha (Boffey 1975). This proposal triggered a comprehensive environmental impact assessment that included a study to determine the hydrologic effects of peat extraction from watersheds in northern Minnesota (Brooks and Predmore 1978). At that time, there was little information on the effects of peat mining in the region, so a field and modeling study was initiated to develop a model that could simulate the hydrologic response to peat mining. A second emerging issue was related to Minnesota's requirement that nonpoint pollution in the context of forest management be addressed. This requirement was mandated by the Water Pollution Control Act Amendments of 1972 and the Clean Water Act of 1977. Assessing nonpoint pollution across the forests of northern Minnesota would require a hydrologic model that could simulate streamflow response to forest-management options. Because most forested watersheds in northern Minnesota have both upland mineral soil and peatland components, the model was required to simulate streamflow from watersheds consisting of peatlands and upland forests.

The early stages of model development were heavily dependent on research on peatland and forest hydrology at the MEF. Scientists at the MEF also recognized the value of developing a hydrologic model of peatland–forest complexes as a means of extending research beyond MEF boundaries for broader applications in managing peatlands and forests in northern boreal forests. This research led to the development of the Peatland Hydrologic Impact Model (PHIM) (Guertin 1985) that relied heavily on MEF long-term data sets to calibrate and validate the model.

Evolution of the PHIM

Beginning with basic hydrologic concepts and drawing upon earlier research from Europe and the MEF, a conceptual model of peatland hydrology was formulated by Predmore (1978). An electric analog model that had been developed for peat bogs (Sander 1976) provided background for subsequent model development. The electric analog model was a physical representation of a peatland in which electrical processes were analogous to hydrologic processes; for example, electrical current represented the flow of water, resistance was inversely related to transmissivity of groundwater, capacitance was related to storativity, and so on. To be valid, an analog model depends on identical mathematical relationships that describe both the real system and its analog; therefore, most analog hydrologic models were converted to mathematical models (Woolhiser and Brakensiek 1982). Advances in computer technology spurred this transition. Such was the case with the PHIM, which was developed as a mathematical model.

The first step in the development of the PHIM entailed formulating a conceptual mathematical model that summarized the known hydrologic relationships of peatlands (Predmore 1978). Predmore's work helped identify the major gaps in our knowledge that constrained model development. Field experiments were initiated to address these informational gaps and develop data sets for model development and testing. More than a decade of research that ensued consisted of an iterative process by which hydrologic relationships were developed from field studies. Results were converted to mathematical algorithms, the model was tested, and studies were initiated to improve the performance of the PHIM.

The first version of the PHIM was published by Guertin (1985). Model development and applications proved valuable for developing and testing hypotheses concerning hydrologic processes in peatlands and upland–peatland watersheds (Guertin et al. 1987a). Subsequent research by Barten (1988) focused on the model subroutine that simulated upland–forest processes and responses to timber harvesting (Barten and Brooks 1988). Further programming and routing revisions by Lu (1994) and McAdams (1993) led to PHIM version 4.0 (Brooks et al. 1995). PHIM 4.0 was applied by Lu (1994) to examine the relationship between flood frequency and forest harvesting relationships, and by McAdams et al. (1993), who examined the effects of climate change on peatland water tables and flow response.

The development of the PHIM was guided by the following principles: (1) the utility of the hydrologic model for practitioners is dependent on its ability to make reasonably accurate predictions given the data environment and the purpose of the modeling effort; (2) the model is mathematical and deterministic as such models are better suited than stochastic models in predicting the hydrologic effects of a given change in vegetation, land use, or climate; (3) the model adheres to the conservation of mass principle; (4) a modular approach is the basis of the model structure so that new subroutines can be readily revised as new knowledge is gained about hydrologic processes and system response; and (5) the model would not initially be designed to simulate water-quality responses.

PHIM 4.0

PHIM version 4.0 is a deterministic, lumped-parameter, continuous-simulation model that includes five submodels: UPLAND, PEAT, MINE, CROUTE, and RROUTE (Figure 15.1). These submodels are independent of each other and are accessed by PHIMMAIN, the driver of the model. The first three submodels are used for simulating the hydrologic response of first-order watersheds consisting of forested uplands, natural peatlands, and mined peatlands. CROUTE is for channel routing, and RROUTE is for reservoir routing. As a result, five hydrological units (HU) can be simulated with PHIM: (1) forested upland, (2) peatland, (3) mined peatland, (4) channel, and (5) reservoir or lake. The PHIM is a hierarchically structured model; its subroutines and functions are used for supporting the main body of the model, plotting graphs, and the input–output interface.

The subroutine PHIMMAIN controls the flow of operations, data input, parameter adjustments, and submodel-level output. Calculations of potential evapotranspiration (PET) are also performed in this subroutine; Hamon's (1961) or Thornthwaite's (Thornthwaite and Mather 1957) equations can be used to estimate PET.

Upland Forest Subroutine (UPLAND)

The UPLAND subroutine simulates the water budget for mineral soil upland forests. Each upland area is divided into separate homogeneous slope segments, each 1 m wide at the downslope end. The soil profile within each slope segment is divided into the Shallow Subsurface Flow Layer (SSFL), the Lower Root Zone (LRZ), and the Lower Boundary Control Volume (LBCV) layer (Figure 15.2). The SSFL is separated from the LRZ on the basis of a

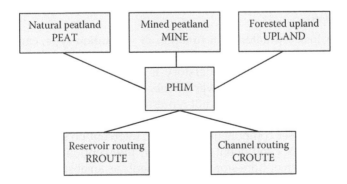

FIGURE 15.1
Submodels of the Peatland Hydrologic Impact Model (PHIM). (Modified from Guertin, D.P., Modeling streamflow response from Minnesota peatlands, PhD Dissertation, University of Minnesota, St. Paul, MN, 1985; Guertin, D.P. et al., *Nord. Hydrol.*, 18, 79, 1987a.)

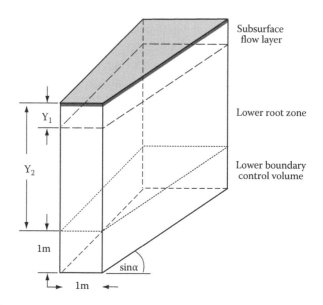

FIGURE 15.2
Soil zones represented in the UPLAND subroutine of PHIM 4.0. See text for explanation of the soil zones illustrated in the figure. (From Barten, P.K., Modeling streamflow from headwater catchments in the northern Lake States, PhD dissertation. University of Minnesota, St. Paul, MN, 1988; Brooks, K.N. et al., User Manual for the Peatland Hydrologic Impact Model (PHIM) Version 4.0. Department of Forest Resources, University of Minnesota, St. Paul, MN, 1995.)

restrictive layer (E-horizon in this case) in the soil profile that restricts vertical flow to the LRZ. The LRZ extends to the greatest soil depth in which plant roots are expected to occur. Water budget is calculated between these three layers. The following are the main algorithms in the UPLAND submodel.

Interception Loss

Interception loss estimated as a function of vegetation type, canopy cover percentage, and PET. The forest canopy is divided into the overstory (>3 m) and understory (<3 m). The understory is further divided into three segments: tall shrubs (1–3 m), lower shrubs (<1 m), and herbaceous cover. Interception of the forest litter is not represented explicitly in this model, but relationships of forest-litter interception storage, if known, can be added to the herbaceous cover component. Maximum interception storage is specified for each vegetative component by the user. Interception calculations treat the respective interception–storage components as an abstraction, conceptually as a "bucket" that must be filled before rainfall reaches the soil mantle. Rainfall is added to the interception bucket from the upper layer to the lower layer in sequence. Interception losses are the smaller of the available PET rate and the residual storage for each layer. Net precipitation equals gross rainfall minus interception storage.

Evapotranspiration Losses

Losses are from SSFL and LRZ soil. Actual evapotranspiration (AET) is determined by a three-stage algorithm that represents (1) the effect of soil-water content and retention characteristics in each zone; (2) a function that represents the transition from spring dormancy to active growth; and (3) rooting depth and distribution that characterizes soil-water depletion in the soil by forest vegetation (Barten and Brooks 1988) as follows:

$$AET_{1\,or\,2} = ROOTD_{1\,or\,2} * ETR_{1\,or\,2} * PET, \tag{15.1}$$

and

$$ETR = ETR_O + (ETR_F - ETR_O) * TRANSI, \tag{15.2}$$

where
   AET is the actual evapotranspiration (cm h^{-1} or cm d^{-1})
   TRANSI the transpiration index; an index for the spring transition period (Barten 1988)
   ROOTD the cumulative fraction of rooting in soil zones, estimated with a root distribution equation (Gale and Grigal 1987)
   ETR the evapotranspiration ratio (ETR = AET/PET) as a function of volumetric soil water content; based on Marker and Mein (1987) and Leaf and Brink (1975) (see Figure 6.6)
   Subscripts 1 and 2 represent SSFL and LRZ, respectively
   Subscripts O and F represent open area and forested area, respectively

Subsurface Flow from SSFL and Overland Flow

Simulations of flow are based on the kinematic storage model (Sloan et al. 1983). Water-balance (continuity) equations are applied to soil zones for computing discharge between the soil zones and the total discharge from the upland subwatershed. The calculations begin at SSFL and then at LRZ and end at LBCV (Figure 15.2). Within each zone, vertical components are calculated first, followed by the lateral-discharge components. At the soil surface, net rainfall (NRAIN) and snowmelt (SMELT—calculated by the temperature-index method) are the inputs to the SSFL layer. Infiltration (INF) of NRAIN and SMELT is instantaneous unless soils are saturated, frozen, or compacted. Soil saturation and soil-frost conditions are calculated by the model; compactness of the soil surface is specified by the user. When NRAIN and SMELT exceed the INF capacity, the excess water becomes overland flow.

   Lateral flow from the SSFL is calculated as a function of the hydraulic gradient and saturated hydraulic conductivity. Vertical flow between the soil zones is a function of the hydraulic gradient and the limiting hydraulic conductivity at the boundary. For estimating subsurface flow from LRZ, the

Dupuit–Forchheimer approximation is applied such that the hydraulic gradient is assumed to equal the average slope of the water table, and flow lines are horizontal. In UPLAND, the water table has a constant slope between the upper and lower hillslope, and the hydraulic gradient equals the slope of the impermeable bed. Water flow through a porous medium is quantitatively described by Darcy's law:

$$v = -\left[\left(\frac{K_x * \partial \psi}{\partial x * i}\right) + \left(\frac{K_z * \partial \psi}{\partial z * k}\right)\right], \quad (15.3)$$

where
   v is the discharge rate
   Ψ the hydraulic potential
   K the hydraulic conductivity
   x and z are the lateral and vertical flow direction, respectively
   i and k are unit vectors in the x and z directions of the Cartesian coordinate system

It is assumed that lateral hydraulic conductivity (K_x) is approximately equal to vertical hydraulic conductivity (K_z) and that its value is a function of volumetric water content that can be estimated by Campbell's (1974) method. The lateral hydraulic gradient ($\partial \psi / \partial x$) is approximately equal to the average slope of the upland subwatershed. The hydraulic potential is the sum of elevation potential (Ψ_H) and matric potential (Ψ_M). Elevation potentials are estimated by the elevation difference between the center of the soil zone and the average soil surface. Matric potential is estimated by Campbell's equation. Once Ψ_M and Ψ_H are calculated, the vertical water-flow movement can be estimated by

$$V_z = -K_z * \frac{(\psi_M + \psi_H)}{\Delta Z}, \quad (15.4)$$

where ΔZ is the elevation difference between the center of soil zones (cm).
   Subsurface flow from SSFL and LRZ combined with overland flow is the total discharge from the upland hydraulic unit.

Natural Peatland Subroutine (PEAT)

The streamflow response of a peatland area that has not been drained or mined is simulated with the PEAT submodel. Two soil zones are used to predict water storage and outflow from a peatland subwatershed (Figure 15.3). Zone 1 extends from the bottom of the outlet channel up to the soil surface (average elevation of hollows). Zone 2 extends from the bottom of the outlet channel down to the underlying mineral substratum. The datum for the

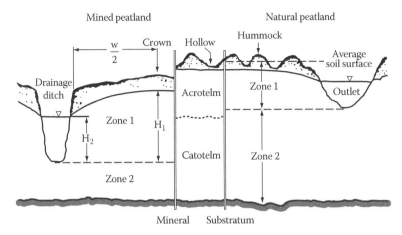

FIGURE 15.3
Soil zones represented by the PEAT subroutine for natural peatlands and soil zones represented by the MINE subroutine for mined peatlands. (Modified from Guertin, D.P. et al., *Nord. Hydrol.*, 18, 79, 1987a.)

simulated water table elevation (WTE) is defined as the bottom of the outlet channel. When water recedes below the outlet elevation, there is no outflow (Barten 1988; McAdams 1993). The following are algorithms for the PEAT submodel.

Leakage Component

Leakage to regional groundwater occurs at the interface between uplands and peatlands (the lagg). The loss of interflow from the upland to the peatland occurs as water passes through the lagg (Tracy 1997; Chapters 4 and 7). Direct vertical seepage from peat bogs is believed to be small; only a negligible amount of water is allowed to percolate directly to the regional groundwater system (McAdams 1993). Loss of interflow into the lagg is expressed as a percentage loss from the streamflow input from the upland.

Discharge Calculation from Peatlands

Discharge is a function of the peatland WTE, which is a function of water storage in the peatland. Soil-water storage in the peatland is calculated with the following continuity equation:

$$VSTOR_n = VSTOR_i + \left[\frac{NPPT + SMELT + (1+\Phi)*QIN}{100} + GWIN\text{-}PERC2\text{-}AET \right]$$

$$* \frac{AREA}{100} * \Delta t - (QOUT_i + QOUT_n) * \frac{\Delta t}{2}, \tag{15.5}$$

where
 VSTOR is the soil water storage for zone 1 and zone 2 (m^3)
 NPPT is the net precipitation (cm time^{-1})
 SMELT is the snowmelt (cm time^{-1})
 Φ is the flow from the upstream HU (upland) to the peatland that is lost
  to seepage (%)
 QIN the inflow from upstream HU (usually upland), if any (cm time^{-1})
 GWIN is the regional groundwater inflow, if any (cm time^{-1})
 PERC2 is the percolation loss at the base of zone 2 (cm time^{-1}) as approxi-
  mated by the K_{sat} of the peat
 AET is the actual ET from the peatland (cm time^{-1})
 AREA is the peatland area (m^2)
 Δt is the computation interval, 1 day or 1 h (s)
 QOUT is the discharge out of peatland ($m^3\ s^{-1}$)
 i and n are subscripts that represent beginning and end of the time step,
  respectively

The nearly flat topography, low-drainage density, and short-term detention storage of natural peatlands produce a streamflow response comparable to that of an unregulated reservoir. A dimensionless routing function represents WTE versus cumulative detention storage (CDS) and discharge versus WTE for natural peatlands (Guertin et al. 1987a). WTE is defined as the elevation above the base of zone 1 of the peatland; it is a function of the CDS of zone 1. Discharge (QOUT) from natural peatlands is a function of WTE. An iterative reservoir routing procedure is used to match relationships among $VSTOR_n$, $QOUT_i$, $QOUT_n$ versus WTE, and WTE versus CDS simultaneously. First, CDS is estimated, then WTE and $QOUT_n$, until the estimated discharge is within an allowable error. Verry et al. (1988) verified the reservoir relationship by comparing the storage–elevation–discharge of the S2 bog with that of a reservoir with the same dimensions (Figure 7.12).

Mined Peatland Submodel (MINE)

Peat mining transforms a poorly drained natural peatland into a system of well-drained fields that are devoid of vegetation. A mining operation usually includes (1) removing forest vegetation and ground cover; (2) ditching to drain the surface peat and improve trafficability for field equipment; (3) raking and grading the field surface; and (4) extracting peat. The simulation of INF and overland flow is important for mined areas.

As with the natural peat system, two soil zones (zones 1 and 2) are used to simulate storage and outflow from a mined area (Figure 15.3). Calculations of water balance and discharge for mined peatlands are based on an average field (a rectangular area with field ditches on all four sides). The area of an average field is equal to the area of HU divided by the number of fields. The standard mined-site configuration uses the approximate number of average fields draining into a main ditch. This configuration closely represents the layout of most peat-extraction operations.

If there is a frost layer in zone 1 or if the soil-water content of zone 1 is at saturation, INF is zero. Otherwise, INF is estimated with the lesser of the average INF rate (specified by the user) and the potential surface storage. Overland flow from the mined area is simulated as a function of rainfall, snowmelt, INF, and available depression storage. The latter storage can be negligible for a uniformly graded field with a high crown or can be sizeable for nearly level fields with a berm (up to 20 cm high) left at the ditch bank. Because the rate of overland flow is rapid compared to the minimum time step of 1 h, explicit routing of overland flow is necessary. Soil-frost thaw is calculated as a function of melting degree-days (MDD, maximum daily air temperature minus the base temperature). When a snowpack is represented, the available energy represented by MDD is first used to melt snow; any residual energy is used to thaw the soil frost. On the basis of field observations, it is assumed that soil frost is continuous and impermeable until the unfrozen zone reaches three-fourths of the maximum depth of frost penetration. The assumption of concrete-type frost occurrence is not made in the PEAT submodel because such soil frost is rare in natural peatlands.

Subsurface flow is assumed to be perpendicular to the drainage ditch with a flow length equal to one-half of the total field width. It is estimated with the Dupuit–Forcheimer equation for an unconfined aquifer (Fetter 1988). The Dupuit–Forchheimer approximation assumes that the hydraulic gradient ($\delta h / \delta x$) is equal to the average slope of the water table and that the flow lines are horizontal. Subsurface flow in mined areas (Gafni and Brooks 1986; Leibfried and Berglund 1986) satisfies the condition specified by Bear (1972).

Subsurface and overland flow are combined and routed through the adjacent drainage ditch for each field. Flows from field ditches are combined at the perimeter ditch and routed (CROUTE) to the outlet of the mined site. Channel or reservoir storage routing (RROUTE) can be used to simulate routing through the ditch and detention-basin systems.

Applications of the PHIM

The PHIM subroutines and overall model performance were verified with data from three watersheds in northern Minnesota: a 3758 ha undisturbed peatland (Toivola), a 155 ha mined peatland (Corona), and a 9.7 ha upland–peat bog watershed (S2) at the MEF (Guertin et al. 1987a; Figures 15.4 and 15.5). Stormflow events over a 3 year period were used to test and verify the PHIM for the Toivola and Corona peatlands. Six and five events were used to calibrate the PHIM for Toivola and Corona, respectively, and the model was verified with six and four independent storm events. For the verification events, average ratios of predicted/observed stormflow volumes and standard deviations were 0.86 (0.16) and 0.91 (0.08) for Toivola and Corona, respectively. Eleven years of streamflow records from the MEF S2 watershed

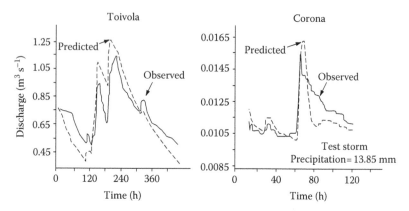

FIGURE 15.4
Predicted and observed hydrographs for stormflow simulations for Toivola (natural peatland) and Corona (mined peatland) used to verify PHIM from Guertin et al. (1987). (Modified from Guertin, D.P. et al., *Nord. Hydrol.*, 18, 79, 1987a.)

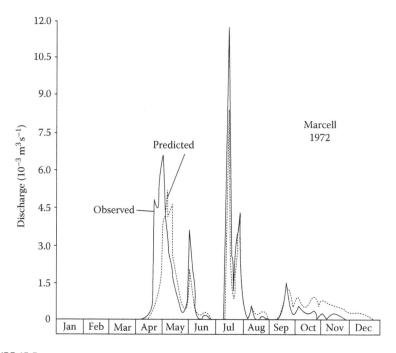

FIGURE 15.5
Predicted and observed hydrographs for annual water-yield simulations for the Marcell Experimental Forest (upland–peatland watershed). (Modified from Guertin, D.P. et al., *Nord. Hydrol.*, 18, 79, 1987a.)

were used to test and verify the PHIM. The ratios of predicted/observed annual streamflow for the six verification years averaged 1.01.

Following initial model-verification research, Barten (1988) revised the UPLAND subroutine to better represent the hydrologic relationships of forest cover. These revisions were evaluated on the basis of simulating streamflow response before and after an aspen harvest at the S4 watershed and for the unharvested S2 watershed, both at the MEF (Barten 1988; Barten and Brooks 1988). Clearcutting and regrowth of aspen on S4 were characterized in the model by changing the overstory and understory canopy, rooting depth and vertical distribution, and snowmelt rate. The PHIM provided better simulations for stormflow volumes than for peak responses, both before and after clearcutting. Results of annual streamflow simulations were better for snow-free periods than for spring snowmelt runoff (Barten and Brooks 1988). The ability to simulate the effects of forest harvesting on streamflow from upland portions of watersheds, such as those on the MEF, was considered an important attribute of the PHIM, because the model can be used broadly in the region.

Subsequent to Barten's work, Lu (1994) revised the PHIM and applied the model to simulate the effects of harvesting on annual peakflow frequency. Data collected on the S4 watershed under forested and clearcut conditions were used to calibrate and verify the model. The model was then run using 76 years of precipitation records from northern Minnesota to generate streamflow simulations under forested and clearcut conditions. Analysis of flow frequency was performed on the records of simulated maximum annual peakflow for forested and clearcut conditions. Lu (1994) found that clearcutting increased rainfall induced peak flows corresponding to recurrence intervals (RIs) of 2 years; however, the effects diminished for peak flows with RI > 10 years. Further analysis by Verry (2000) indicated that clearcutting increased annual maximum snowmelt peaks corresponding to events with RIs up to 25 years. Verry also pointed out that peaks associated with bankfull flow (RI = 1.5 years) increased by 150%; this result has important implications for stream-channel stability.

Results from the PHIM applications in addressing the effect of clearcutting on flood peaks help address questions concerning the impact of forest practices and resulting land-use changes on flooding in the northern Lake States. The debate over forest effects on floods has been and remains contentious (FAO 2005; Bradshaw et al. 2007; Eisenbies et al. 2007). Model simulations are a reasonable approach to resolve this debate as paired-watershed experiments cannot be conducted over a sufficient period to quantify effects on flooding associated with RI > 50–100 years.

Concerns about the effects of global climate change on wetlands led to a study in which the PHIM was used to simulate water table and streamflow responses of a MEF peatland to projected changes in temperature and precipitation (McAdams 1993; McAdams et al. 1993). The response of the S2 bog to projected increases in precipitation and temperature from the global climate model of the Goddard Institute of Space Studies was simulated by the PHIM.

McAdams modified the PEAT subroutine to improve the accuracy of water table responses. Results indicated that projected increases in temperature of 3°C–6°C, which were greatest from November through March, increased winter water table levels 2–4 cm and increased water tables during the snowmelt period by 8 cm. The higher temperatures resulted in earlier snowmelt and higher snowmelt runoff peaks but caused "slightly lower" water table levels in the growing season due to elevated PET in the summer. When precipitation increases of 25% were added to the temperature scenarios, there were small net increases (1 cm) in WTEs over the year and an increase in annual streamflow volumes of 17%. Apparently, the increased PET rates compensated somewhat for the increased precipitation. The magnitude of simulated water table changes due to global warming suggests that neither peatland aggradation nor carbon emissions from the peatlands would be affected significantly.

Next Generation of the PHIM

The successful applications of the PHIM in simulating hydrologic processes and responses of peatlands led Canelon et al. (2006) to consider using hydrologic algorithms in the PHIM to develop the WETland Hydrology and Water Quality model (WET-HAWQ) for applications with a broader range of wetland types. WET-HAWQ was developed under the sponsorship of the Minnesota Pollution Control Agency (MPCA) from September 2005 to June 2007. MPCA needed a model that could examine the impact of land use on the quality of wetland water. At that time, wetlands in Minnesota were being examined to determine whether they should be placed on the State's list of impaired waters. As of mid-2008, four wetlands had been placed on the list. Impairments include high nutrients and poor invertebrate Index of Biological Integrity (IBI) scores.

The WET-HAWQ model was developed by merging components from the PHIM and the SET-WET model (Lee 1999; Lee et al. 2002). The SET-WET model provided a biogeochemistry simulation capability that was not included in the PHIM, though the hydrologic-simulation capabilities of SET-WET were limited. By combining the hydrologic and hydraulic model components of the PHIM with the biogeochemistry components of SET-WET, the Microsoft Windows®-based WET-HAWQ model provided an improved wetland model (Figure 15.6) that included the following capabilities:

- Watershed schemes can comprise as many as 15 hydrologic units; any combination of wetlands and upstream inflow point-sources can be modeled. Inflows from upstream areas can be simulated or be input directly as hydrographs and pollutographs.

- Groundwater interaction is entered as a constant inflow or outflow. Future versions of the model could include the ability to compute seepage fluxes into the wetlands or INF losses from a wetland using groundwater-level data to compute head gradients between the wetland and the underlying groundwater.

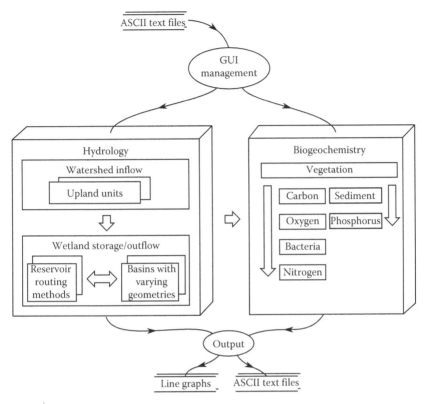

FIGURE 15.6
Content of the Windows®-based WET-HAWQ model (Canelon et al. 2006) that combined the existing DOS-based codes for water quality simulations of SET-WET (Lee et al. 2002) with the hydrologic–hydraulic codes of PHIM. (From Guertin, D.P. et al., *Nord. Hydrol.*, 18, 79, 1987a.)

- Nitrogen, phosphorus, dissolved oxygen, and sediment cycles can be simulated together or individually.
- Growth of wetland vegetation (biomass) can be simulated.
- Daily time step is used; currently, there is a limit of a 500 day period, but this could be changed in future modifications.
- Microsoft Windows graphical user interface provides enhanced input data management and facilitates analysis of output results.

Output from the WET-HAWQ model includes an outflow hydrograph and chemographs for chemical constituents from a wetland, including concentrations of dissolved organic nitrogen, nitrate, ammonium, total phosphorus, and dissolved oxygen. WET-HAWQ has been tested and validated for a wetland that consists of a series of stormwater treatment wetlands located on property of the Shakopee Mdewakantan Sioux Community south of Minneapolis. An example of model output is presented in Figure 15.7.

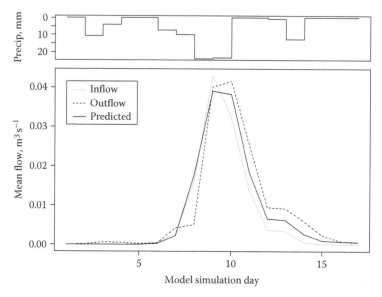

FIGURE 15.7
WET-HAWQ model simulation results of a wetland basin on property of the Mdwakatan Sioux Community south of Minneapolis.

Conclusions

Fifty years of research on the MEF has been instrumental in the development of multiple-resource models and models that can simulate the hydrologic response of forest uplands, peatlands, and other wetland types in the northern Lake States. The iterative process of field plots, laboratory, and watershed-scale experiments, and model development helped focus research that facilitated model development and testing. Models were formulated to simulate responses to land-use and climate-change scenarios that lie outside the spatial and temporal boundaries of the data sets from which they were derived. The high-quality, comprehensive, and long-term nature of the data sets obtained from the MEF have been instrumental in developing and validating mathematical relationships of models and their overall predictive capabilities. The multiple resource model INTERIM, the PHIM, and the expanded WET-HAWQ are products of the research program at the MEF. These models can be used for planning and managing forested and wetland watersheds in the northern Lake States and provide platforms and a scientific basis for future research and model development.

References

Anderson, R.C., L. Loucks, and A.M. Swain. 1969. Herbaceous response to canopy cover, light intensity, and throughfall precipitation in coniferous forests. *Ecology* 50:255–263.

Barten, P.K. 1988. Modeling streamflow from headwater catchments in the northern Lake States, PhD dissertation. St. Paul, MN: University of Minnesota.

Barten, P.K. and K.N. Brooks. 1988. Modeling streamflow from headwater areas in the northern Lake States. In: *Modeling Agricultural, Forest, and Rangeland Hydrology.* American Society of Agricultural Engineers, pp. 347–356, Chicago, IL.

Bear, J. 1972. *Dynamics of Fluids in Porous Media.* New York: Elsevier Pub.

Belcher, D.M., M.R. Holdaway, and G.J. Brand. 1982. A description of STEMS, the stand and tree evaluation modeling system. General Technical Report NC-79. St. Paul, MN: USDA Forest Service.

Bernath, S.C. 1981. TIMWAT: A planning model for the evaluation of changing land uses on water yield. Plan B Paper. St. Paul, MN: University of Minnesota.

Bernath, S.C., E.S. Verry, K.N. Brooks, and P.F. Ffolliott. 1982. Modeling water yield response to forest cover changes in northern Minnesota. In: *Hydrological Processes of Forested Areas: Proceedings of the Canadian Hydrology Symposium'82.* Fredericton, New Brunswick, Canada: Associate Committee on Hydrology, National Research Council of Canada, pp. 385–399.

Boffey, P.M. 1975. Energy: Plan to use peat as fuel stirs concern in Minnesota. *Science* 190:1066–1070.

Boyce, S.G. 1979. Management of eastern hardwood forests for multiple benefits. Research Paper SE-168. USDA Forest Service.

Bradshaw, C.J.A., S.S. Navjot, S.H. Peh, and B.W. Brooks. 2007. Global evidence that deforestation amplifies flood risk and severity in the developing world. *Global Change Biology* 13:2379–2395.

Brand, G.J. 1981. GROW—A computer subroutine that predicts the growth of trees in the Lake States' forests. Research Paper NC-207. St. Paul, MN: USDA Forest Service.

Brooks, K.N., S.-Y. Lu, and T.V.W. McAdams. 1995. User Manual for the Peatland Hydrologic Impact Model (PHIM) Version 4.0. St. Paul, MN: Department of Forest Resources, University of Minnesota.

Brooks, K.N. and S.R. Predmore. 1978. Hydrological factors of peat harvesting. Phase 2 Peat Report. St. Paul, MN: Department of Forest Resources, University of Minnesota.

Carder, D.R., F.R. Larson, J.J. Rogers, W.O. Rasumssen, and P.F. Ffolliott. 1977. Ecosystem analysis for watershed management. *Arizona Watershed Symposium* 21:22–25.

Campbell, G.S. 1974. A simple method for determining unsaturated conductivity from moisture retention data. *Soil Science* 117(6):311–314.

Canelon, D.J., J.L. Nieber, and B.N. Wilson. 2006. Development of a wetland model for TMDL assessments. Project Report. St. Paul, MN: Department of Bioproducts and Biosystems Engineering, University of Minnesota.

Crow, T.R. 1978. Biomass and production in three contiguous forests in northern Wisconsin. *Ecology* 59:262–273.

Dahlberg, B.L. and R.C. Guettinger. 1956. The white-tailed deer in Wisconsin. Technical Wildlife Bulletin 14. Wisconsin Conservation Department.

DeGraff, R.M. and K.E. Evans. 1979. Management of north central and northeastern forests for nongame birds: Workshop proceedings. General Technical Report NC-51. St. Paul, MN: USDA Forest Service.

Eisenbies, M.H., W.M. Aust, J.A. Burger, and M.B. Adams. 2007. Forest operations, extreme flooding events, and considerations for hydrologic modeling in the Appalachians—A review. *Forest Ecology and Management* 242:77–98.

Erickson, A.B., V.E. Cunvalson, M.H. Stenlund, D.W. Burcalow, and L.H. Blankenship. 1961. The white-tailed deer of Minnesota. Technical Bulletin 5. Minnesota Division of Game and Fish.

Fetter, C.W. 1988. *Applied Hydrogeology*. Columbus: Charles E. Merrill Publishing Co.

Ffolliott, P.F., K.N. Brooks, and D.P. Guertin. 1984. Multiple-resource modeling—Lake States application. *Northern Journal of Applied Forestry* 1:80–84.

Food and Agricultural Organization of the United Nations (FAO) and the Center for International Forestry Research (CIFOR). 2005. Forests and flood—Drowning in fiction or thriving on facts? RAP Publication 2005/03. Bangkok, Thailand: Regional Office for Asia and the Pacific.

Gafni, A. and K.N. Brooks. 1986. Hydrologic properties of natural versus mined peatlands. In: *Advances in Peatlands Engineering*. Ottawa, Canada: National Research Council of Canada, pp. 185–190.

Gale, M.R. and D.F. Grigal. 1987. Vertical root distributions of northern tree species in relation to successional status. *Canadian Journal of Forest Research* 17(8):829–834.

Guertin, D.P. 1985. Modeling streamflow response from Minnesota peatlands, PhD dissertation. St. Paul, MN: University of Minnesota.

Guertin, D.P., P.K. Barten, and K.N. Brooks. 1987a. The peatland hydrologic impact model: Development and testing. *Nordic Hydrology* 18:79–100.

Guertin, D.P., K.N. Brooks, and P.F. Ffolliott. 1987b. Multiple resource information and modeling: A survey of needs in the Lake States. *Northern Journal of Applied Forestry* 4:81–84.

Hamon, W.R. 1961. Estimating potential evapotranspiration. *Journal of Hydrology Division, Proceedings ASCE* 87(HY3):107–120.

Larson, F.R., P.F. Ffolliott, W.O. Rasumssen, and D.R. Carder. 1979. Estimating impacts of silvicultural management practices on forest ecosystems. In: *Best Management Practices for Agriculture and Silviculture*, ed. R.C. Loehr, D.A. Haith, M.F. Walter, and C.S. Martin. Ann Arbor, MI: Ann Arbor Science Publishers, Inc, pp. 281–294.

Leaf, C.E. and G.E. Brink. 1975. Land use simulation model of the subalpine coniferous zone. Research Paper RM-135. USDA Forest Service Rocky Mountain Forest Experiment Station.

Lee, E.R. 1999. SET-WET: A wetland simulation model to optimize NPS pollution control. MS thesis. Blacksburg, VA: Virginia Polytechnic and State University.

Lee, E.R., S. Mostaghimi, and T. Wynn. 2002. A model to enhance wetland design and optimize non-point source pollution control. *Journal of American Water Resources Association* 38(1):17–32.

Leibfried, R.T. and E.R. Berglund. 1986. Groundwater hydrology of a fuel peat mining operation. In: *Proceeding of Symposium on Advances in Peatlands Engineering*, August 25–26, 1986. Ottawa, Canada: National Research Council Canada.

Lu, S.-Y. 1994. Forest harvesting effects on streamflow and flood frequency in the northern Lake States. PhD dissertation. St. Paul, MN: University of Minnesota.

Lundgren, A.L. 1981. The effect of initial number of trees per acre and thinning densities on timber yields from red pine plantations in the Lake States. Research Paper NC-193. St. Paul, MN: USDA Forest Service.

Lundgren, A.L. and B.L. Essex. 1979. Forest resources evaluation systems—A needed tool for managing renewable resources. In: *A Generalized Forest Growth Projection System Applied to the Lake States Region*. General Technical Report NC-49. St. Paul, MN: USDA Forest Service.

Marker, M.S. and R.G. Mein. 1987. Modeling evaptranspiration from homogenous soils. *Water Resources Research* 23(10):2001–2007.

McAdams, T.V.W. 1993. Modeling water table response to climatic change in a northern Minnesota peatland. MS thesis. St. Paul, MN: University of Minnesota.

McAdams, T.V.W., K.N. Brooks, and E.S. Verry. 1993. Modeling water table response to climatic change in a northern Minnesota peatland. In: *Management of Irrigation and Drainage Systems, Integrated Perspectives. Proceedings of the National Conference on Irrigation and Drainage*. Park City, Utah, 21–23 July 1993. New York: American Society of Civil Engineers, pp. 358–365.

Predmore, S.R. 1978. Hydrologic factors of peat harvesting. Plan B Paper. St. Paul, MN: University of Minnesota.

Rasmussen, W.O. and P.F. Ffolliott. 1981. Simulation of consequences of implementing alternative natural resources policies. In: *Social and Environmental Consequences of Natural Resources Policies, With Special Emphasis on Biosphere Reserves*, tech. coord. P.F. Ffolliott and G.F. Halffter. General Technical Report RM-88. Ft. Collins, CO: USDA Forest Service, pp. 41–43.

Sander, J.E. 1976. An electric analog approach to bog hydrology. *Ground Water* 14(1):30–35.

Sloan, P.G., I.D. Moore, G.B. Coltharp, and J.D. Eigel. 1983. Modeling surface and subsurface stormflow on steeply-sloping forested watershed. Lexington, KY: Water Resources Institute, University of Kentucky.

Thornthwaite, C.W. and J.R. Mather. 1957. Instructions and tables for computing potential evapotranspiration and the water balance. *Drexel Institute Technical Publication in Climatology* 10(3).

Tracy, D.R. 1997. Hydrologic linkages between uplands and peatlands. PhD thesis. St. Paul, MN: University of Minnesota.

Verry, E.S. 2000. Water flow in soils and streams: Sustaining hydrologic function. In: *Raparian Management in Forests of the Continental Eastern United States*, ed. E.S. Verry, J.W. Hornbeck, and C.A. Dolloff. Boca Raton, FL: Lewis Publishers, pp. 99–124.

Verry, E.S., K.N. Brooks, and P.K. Barten. 1988. Streamflow response from an ombrotrophic mire. In: *Proceedings, International Symposium on the Hydrology of Wetlands in Temperate and Cold Regions*. Helsinki, Finland: International Peat Society/The Academy of Finland, pp. 52–59.

Verry, E.S., J.R. Lewis, and K.N. Brooks. 1983. Aspen clearcutting increases snowmelt and stormflow peaks in north central Minnesota. *Water Resources Bulletin* 19:59–67.

Woolhiser, D.A. and D.L. Brakensiek. 1982. Hydrologic system synthesis. In: *Hydrologic Modeling of Small Watersheds*, ed. C.R. Haan, H.P. Johnson, and D.L. Brakensiek. ASAE Monograph No. 5. St. Joseph, MI: American Society of Agricultural Engineers, pp. 3–16.

Zavitkovski, J. 1976. Ground vegetation biomass, production, and efficiency of energy utilization in some northern Minnesota forest ecosystems. *Ecology* 57:694–706.

Index